MANAGING ENERGY, NUTRIENTS, AND PESTS
IN
ORGANIC FIELD CROPS

Integrative Studies in
Water Management and Land Development

Series Editor
Robert L. France

Published Titles

MANAGING
ENERGY, NUTRIENTS,
AND PESTS
IN
ORGANIC
FIELD CROPS

Edited by
Ralph C. Martin and Rod MacRae

CRC Press
Taylor & Francis Group
Boca Raton London New York

CRC Press is an imprint of the
Taylor & Francis Group, an **informa** business

CRC Press
Taylor & Francis Group
6000 Broken Sound Parkway NW, Suite 300
Boca Raton, FL 33487-2742

First issued in paperback 2019

ISBN-13: 978-1-4665-6836-5 (hbk)
ISBN-13: 978-0-367-37891-2 (pbk)

Library of Congress Cataloging-in-Publication Data

Managing energy, nutrients, and pests in organic field crops / editors: Ralph C. Martin, Rod MacRae.
    pages cm -- (Integrative studies in water management and land development)
Includes bibliographical references and index.
ISBN 978-1-4665-6836-5 (hardcover : alk. paper)
    1. Organic farming. 2. Agriculture and energy. 3. Plant nutrients. 4. Pests--Control.
I. Martin, Ralph C. (Ralph Cameron), 1953- II. MacRae, Roderick John. III. Series: Integrative studies in water management and land development.

S605.5.M3225 2014
631.5'84--dc23   2013048165

Visit the Taylor & Francis Web site at
http://www.taylorandfrancis.com

and the CRC Press Web site at
http://www.crcpress.com

We dedicate this book to Ann Cleary—organic gardener, collector and compiler of shelves of organic information, writer, raconteur, and a decades-long member of Canadian Organic Growers—who scolded us because she loved us and because she knew we had to get it right, and that by digging deeper, we mostly could.

Contents

SECTION II   Pest Management

SECTION III   Integrating Approaches

SECTION IV   Economics, Energy, and Policy

Series Preface

Ecological issues and environmental problems have become exceedingly complex. Today, it is hubris to suppose that any single discipline can provide all the solutions for protecting and restoring ecological integrity. We have entered an age where professional humility is the only operational means for approaching environmental understanding and prediction. As a result, socially acceptable and sustainable solutions must be both imaginative and integrative in scope; in other words, garnered through combining insights gleaned from various specialized disciplines, expressed and examined together.

The purpose of the CRC Press series Integrative Studies in Water Management and Land Development is to produce a set of books that transcends the disciplines of science and engineering alone. Instead, these efforts will be truly integrative in their incorporation of additional elements from landscape architecture, land-use planning, economics, education, environmental management, history, and art. The emphasis of the series will be on the breadth of study approach coupled with depth of intellectual vigor required for the investigations undertaken.

Robert L. France
Series Editor
Professor of Watershed Management
Department of Environmental Sciences
Faculty of Agriculture
Dalhousie University
Halifax, Nova Scotia, Canada

Foreword

Traversing through the northern Canadian Prairies or American Great Plains, one is struck first by the limitless horizon and immensity of the sea of grain, and secondarily, if being literarily inclined, by remembering that some of the most highly regarded fiction—termed *prairie realism*—created on the continent in the early decades of the last century also sprung from this fertile region, as for example *My Antonia* (W. Carter), *Wild Geese* (M. Ostenso), *Giants in the Earth* (O. Rolvaag), *Grain* (R. Steed), and *Fruits of the Earth* (F. Grove). As a onetime prairie resident, I couldn't help but think of this while traversing through this book and noting that more than half the contributors have affiliations with this particular region, the oft-ascribed *breadbasket of the world*. But there is a much wider physical geography contained within these pages, applicable as it is to eastern North American and European situations as well.

However, what is most remarkable about *Managing Energy, Nutrients, and Pests in Organic Field Crops*, ably edited by Ralph C. Martin and Rod MacRae, is the catholic breadth of the conceptual geography covered within its 16 chapters. Like many, I too have stood in front of the plethora of produce and meat options available in grocery stores and deliberated on the choice as to whether or not to buy *organic*, which in my simplistic mind at the time was equated to solely meaning chemical- and GMO-free growing/rearing. But as introduced and explored in comprehensive detail in the present book, organic agriculture is of course much more, encompassing the complete and conjoined ecophysical and ecosocial systems. Indeed, by considering all of soil health, pest management, integrated growing approaches for crops, food preparation, environmental issues (including energy use and life-cycle analysis), and social issues (including economics and policy), this book represents an important contribution to the field of agroecosystem health and human well-being. As such, the book is a perfect fit for the CRC Press series Integrative Studies in Water Management and Land Development, whose specific mandate is to provide a forum in which to publish such holistic, systems-level research addressing complex issues.

And in terms of such research, readers of *Managing Energy, Nutrients, and Pests in Organic Field Crops* will learn that one of the most important justifications for compiling and publishing all the work herein is that one criticism of organic agriculture that can be leveled is that its ideological goals may sometimes be ahead of its realized pragmatics. The present book does much to correct this imbalance through its review of the history of experimentation and management and particularly through its presentation of clearly identified priorities to guide future research endeavors. By synthesizing and highlighting the practicalities of organic agriculture, Martin and MacRae, and their contributing authors, provide important and practical lessons pertaining to soil health, nutrient management, tillage reduction, organic matter decomposition, integrated pest and weed management, genetic studies, energy use, market analysis, and policy modification.

Managing Energy, Nutrients, and Pests in Organic Field Crops should do much to help move organic agriculture beyond its present niche market status in North America. Part of this comes about through the book's demonstration that as a scholarly pursuit, organic agriculture, once identified through experimental field crops and laboratory studies, is now a realized system of effective management that not only performs better than conventional approaches but that can also be more economical in the long run in terms of producing sustainable agroecosystem goods and services.

Robert L. France
Professor of Watershed Management
Department of Environmental Sciences
Faculty of Agriculture
Dalhousie University
Halifax, Nova Scotia, Canada

Acknowledgments

We would like to acknowledge the initiative of Robert L. France, Department of Environmental Sciences, Agricultural Campus, Dalhousie University, who proposed the idea for this book and offered help and guidance as we undertook this project. We are also thankful for the steady assistance of Stephanie Morket and Randy Brehm of CRC Press.

We also acknowledge the perseverance and flexibility of all chapter authors. They responded patiently to our requests, sometimes trivial, in our quest to pull so many concepts into a coherent volume.

This volume would not have had the rigor we rightly seek in our scientific community without the numerous reviewers who offered their time, analysis, and judgment to weed out the unnecessary, which prompted the editors and authors to include additional relevant material and generally sharpen the focus of the material presented in these chapters. Thank-you to all reviewers for your generous assistance.

We also acknowledge all those who funded the research discussed in this book. Directing money to research in organic agriculture has benefits for all participants in the value chain of agriculture and contributes to building capacity to meet the growing demand for organic products in Canada.

We also appreciate the contributions of numerous unnamed colleagues who helped to stimulate ideas for the chapter authors and conducted research on which much of the review material is based. Some are cited and others not, yet all warrant our appreciation.

Finally, we thank the practitioners who steadily bet their livelihoods on organic agriculture. They provide healthy products for Canadians and affirm with quiet confidence that it is possible to manage energy, nutrients, and pests in organic field crops. We hope this volume will serve you effectively.

Editors

Ralph C. Martin's formal education includes a BA and an MSc in biology from Carleton University and a PhD in plant science from McGill University. His love of teaching grew unexpectedly when he began teaching at the Nova Scotia Agricultural College, in 1990, and realized how students teach him, too. In 2001, he founded the Organic Agriculture Centre of Canada to coordinate university research and education pertaining to organic systems across Canada. In 2011, he was appointed professor and Loblaw Chair in Sustainable Food Production at the University of Guelph.

Rod MacRae, PhD, is an associate professor in the Faculty of Environmental Studies, York University, Toronto, Ontario, Canada. His research focuses on creating a national food and agriculture policy for Canada and the set of coherent and comprehensive programs required to support such a policy. He has published extensively on this topic in the academic and popular press and has also conducted numerous policy analyses for the Canadian organic food sector.

Contributors

Melissa M. Arcand
Department of Soil Science
University of Saskatchewan
Saskatoon, Saskatchewan, Canada

Yuki Audette
School of Environmental Sciences
University of Guelph
Guelph, Ontario, Canada

Keith C. Bamford
Department of Plant Science
University of Manitoba
Winnipeg, Manitoba, Canada

Dilshan Benaragama
Department of Plant Sciences
University of Saskatchewan
Saskatoon, Saskatchewan, Canada

Robert E. Blackshaw
Lethbridge Research Centre
Agriculture and Agri-Food Canada
Lethbridge, Alberta, Canada

Josée Boisclair
Platform for Innovation in Organic
    Agriculture
Research and Development Institute for
    the Agri-Environment
Saint-Bruno-de-Montarville, Québec,
    Canada

Gilles Boiteau
Potato Research Centre
Agriculture and Agri-Food Canada
Fredericton, New Brunswick, Canada

Sarah Braman
Minnesota Zoo
Apple Valley, Minnesota

Katherine Buckley
Brandon Research Centre
Agriculture and Agri-Food Canada
Brandon, Manitoba, Canada

Barbara Cade-Menun
Semiarid Prairie Agricultural Research
    Centre
Agriculture and Agri-Food Canada
Swift Current, Saskatchewan, Canada

Harun Cicek
Department of Plant Science
University of Manitoba
Winnipeg, Manitoba, Canada

Mulan Dai
Semiarid Prairie Agricultural Research
    Centre
Agriculture and Agri-Food Canada
Swift Current, Saskatchewan, Canada

Martin H. Entz
Department of Plant Science
University of Manitoba
Winnipeg, Manitoba, Canada

Rachel L. Evans
Department of Plant Science
University of Manitoba
Winnipeg, Manitoba, Canada

Ronald Ferrera-Cerrato
Area de Microbiologia
Postgrado de Edafologia
Colegio de Postgraduados
Carretera, Montecillos, Mexico

Stephen Fox
Cereal Research Centre
Agriculture and Agri-Food Canada
Winnipeg, Manitoba, Canada

Tandra Fraser
School of Environmental Sciences
University of Guelph
Guelph, Ontario, Canada

James B. Frey
Mennonite Central Committee
Akron, Pennsylvania

Caroline Halde
Department of Plant Science
University of Manitoba
Winnipeg, Manitoba, Canada

Chantal Hamel
Semiarid Prairie Agricultural Research
    Centre
Agriculture and Agri-Food Canada
Swift Current, Saskatchewan, Canada

Andrew M. Hammermeister
Faculty of Agriculture
Organic Agriculture Centre of Canada
Dalhousie University
Truro, Nova Scotia, Canada

Neil Holliday
Department of Entomology
University of Manitoba
Winnipeg, Manitoba, Canada

Pierre Hucl
Department of Plant Sciences
University of Saskatchewan
Saskatoon, Saskatchewan, Canada

Mark Juhasz
School of Environmental Design and
    Rural Development
University of Guelph
Guelph, Ontario, Canada

J. Diane Knight
Department of Soil Science
University of Saskatchewan
Saskatoon, Saskatchewan, Canada

Julia Langer
Global Threats Program
World Wildlife Fund Canada
Toronto, Ontario, Canada

Tom Lowery
Pacific Agri-Food Research Centre
Agriculture and Agri-Food Canada
Summerland, British Columbia, Canada

Derek H. Lynch
Department of Plant and Animal
    Sciences
Dalhousie University
Truro, Nova Scotia, Canada

Rod MacRae
Faculty of Environmental Studies
York University
Toronto, Ontario, Canada

Ralph C. Martin
Department of Plant Agriculture
University of Guelph
Guelph, Ontario, Canada

Shauna Mellish
PEI Department of Agriculture
Fisheries and Aquaculture
Charlottetown, Prince Edward Island,
    Canada

Karen Nelson
Faculty of Agriculture
Organic Agriculture Centre of Canada
Dalhousie University
Truro, Nova Scotia, Canada

Ron Pidskalny
Strategic Vision Consulting Ltd
Edmonton, Alberta, Canada

Kristen Podolsky
Department of Plant Science
University of Manitoba
Winnipeg, Manitoba, Canada

Hakunawadi Alexander Pswarayi
Department of Agricultural, Food and
    Nutritional Science
University of Alberta
Edmonton, Alberta, Canada

Kim Schneider
School of Environmental Sciences
University of Guelph
Guelph, Ontario, Canada

Yuying Shen
The Key Laboratory of Grassland
    Agro-Ecosystems
Lanzhou University
Lanzhou, China

Steve J. Shirtliffe
Department of Plant Sciences
University of Saskatchewan
Saskatoon, Saskatchewan, Canada

Dean Spaner
Department of Agricultural, Food and
    Nutritional Science
University of Alberta
Edmonton, Alberta, Canada

Mario Tenuta
Department of Soil Science
University of Manitoba
Winnipeg, Manitoba, Canada

Marie-Soleil Turmel
Programa de Agricultura de
    Conservación
México Centro Internacional
    de Mejoramiento de Maíz y
    Trigo - CIMMYT
Texcoco, México

Paul Voroney
School of Environmental Sciences
University of Guelph
Guelph, Ontario, Canada

Cathy Welsh
Department of Plant Science
University of Manitoba
Winnipeg, Manitoba, Canada

Alexander Woodley
School of Environmental Sciences
University of Guelph
Guelph, Ontario, Canada

Emmanuel K. Yiridoe
Department of Business and Social
    Sciences
and
Department of Economics
Dalhousie University
Truro, Nova Scotia, Canada

1 Introduction

Rod MacRae, Andrew M. Hammermeister,
and Ralph C. Martin

CONTENT

Organic farming has proven to be a viable approach to producing food, especially in field cropping. Areas planted to field crops using organic management practices, whether certified or not, continue to rise across the globe and market penetration continues to increase at rates that surpass most other food categories (Canadian Organic Trade Association, 2013; Willer et al., 2013).

But the knock against organic field crop production continues to be that strategies to manage energy, nutrients, and weeds are evolving too slowly. The tool box for organic producers is insufficiently full or too expensive to properly implement. This book tackles these criticisms in a realistic way, reviewing the state of the science to identify the strengths, weaknesses, and challenges of organic field crop production in the industrial world in a clear-headed and balanced way.

Organic farming comprises a range of approaches within the broader sustainable agriculture spectrum. In its most developed form, ecologically sustainable agriculture (including organic farming) is both a philosophy and a system of farming. It is based on a set of values that reflects an awareness of both ecological and social realities, and on a level of empowerment that is sufficient to generate responsible action. Efforts to ensure short-term viability are tested against long-term environmental sustainability, and attention to the uniqueness of every operation is considered in relation to ecological and humanistic imperatives, with an awareness of both local and global implications. It emphasizes benign designs and management procedures that work with natural processes and cycles to conserve all resources (including beneficial soil organisms and natural pest controls), and minimize waste and environmental damage, prevent problems, and promote agro-ecosystem resilience, self-regulation, evolution, and sustained production for the nourishment and optimal development of all (including rural communities both here and abroad). Special attention is paid to the relationships between soil conditions, food quality and livestock health; and livestock is cared for in the most humane way possible. In addition, organic farmers avoid the use of synthetically compounded fertilizers, pesticides, growth regulators, genetically engineered organisms (GMOs) and livestock feed additives. Instead, they rely upon crop rotations, crop residues, animal manures, legumes, green manures, off-farm organic wastes, mechanical cultivation, and mineral-bearing rocks to maintain soil fertility and productivity. Insects, weeds and other pests are managed by means of natural, cultural and biological controls. The potential of this approach, however, goes far beyond its present expression, which has largely been limited to the substitution of environmentally benign products and practices. As this new vision of what is ecologically responsible becomes established, significant development can be expected in the science and art of agro-ecosystem design and management. (Hill and MacRae, 1992)

In Canada, organic agriculture is a regulated production system driven by consumer demand and governed by standards with environmental sustainability and animal welfare at its core.

The authors are all long-time researchers in the organic field who have come to this work not because of an ideological commitment to the organic philosophy but rather because their research has revealed the strengths of the organic approach and areas requiring further work. The strengths have stimulated their interest in solving chronic farm problems. The weaknesses have tweaked their scientific curiosity. Innovations, at the heart of scientific curiosity and essential for the development of organic agriculture, must pass through an ecological filter in order to be pertinent.

Admittedly, organic farmers frequently question, even more than conventional farmers, the relevance or applicability of scientific research. This is largely due to of their recognition that organic agriculture is a product of an ecologically based management system that is highly dependent on the environment (climate and soils), crop rotation and related management practices, degree of livestock integration, and sources and forms of inputs utilized. Some also value indigenous knowledge and traditions that may not have an apparent scientific rationale (see Chapter 11). There can be tremendous variability in management practices among farmers within and between ecological regions as they respond to their own site-specific conditions, business interests, and management philosophy.

Soil, as the foundation of sound organic crop production, is very dynamic, but many of its properties take time to evolve, including the organic matter content, fertility status, biological community, and structure. Similarly, the severity of pests (i.e., insects, diseases, weeds) can be closely related to not only the ecosite but also the short- and long-term management practices that are specific to each individual farm.

While the research scene has changed considerably in Canada and the United States over the last 10 years, experience has shown that scientists must make every bit as much of a "mental transition" as the farmers do and many researchers in this book have themselves undertaken such a transition. During earlier periods of organic development, scientists engaged in this research transition made common mistakes that limited the work's utility:

- Weed management trials were carried out without the use of synthetic herbicides but still utilizing other conventional practices such as normal conventional fertilizer rates and no strategic crop rotation.
- In fertility trials, chemical fertilizers were substituted with organically approved fertilizers at the same rate of available nutrients. While input substitution in itself was not necessarily an issue, the rate of application often

was. High nutrient application rates are not often used by organic farmers due to cost and logistics but also because high levels of nutrients (especially nitrogen) can increase weed pressure and susceptibility to diseases and insects. Instead, soil building approaches are used to build fertility and release nutrients slowly.

- Land that had not been established under organic management was used in trials. Long-term management practices affect soil biology, fertility, weeds, and pathogens. Conducting research on land that had previously been under nonorganic management is less likely to produce results that are repeatable on organic farms.

- Trials comparing organic and nonorganic management practices sometimes compromised their relevance in exchange for control within the experiment. For example, the same cultivar would be used in the experiment, but in practice different cultivars may be used under organic management for ecological reasons.

- Some researchers (as well as farmers) regarded land or management practices as organic just because no synthetic fertilizers or pesticides had been used. Organic not only eliminates the use of these synthetic products but also adopts a proactive system for maintaining or improving soil quality and managing pests. Organic agriculture is not "management by neglect." In reality, organic farmers (and hence scientists) must be knowledge intensive managers with the capacity to adapt within organic standards.

Today, scientists targeting their work toward an organic management system develop an integrative ecological approach to the research, attempting to understand the ecological cause of challenges and seek ecological solutions. This awareness of the potential interactions between multiple phenomena makes the work more relevant to organic systems.

The authors of this book are primarily Canadian but mix Canadian research experiences with data and analysis from the United States and Europe. Their work in this area is framed by a number of important recent developments in the Canadian organic sector. First, the last 10 years have produced a dramatic increase in funded organic research, inspired in part by the coordinating efforts of the Organic Agriculture Centre of Canada. The most visible evidence of this shift is the creation of the Organic Science Cluster, a set of research projects costing approximately $9 million from 2009 to 2013. Concurrently, the organic sector was evolving a more strategic approach to its development that has spawned a wider range of collaborations, activities, and farmer involvement in research efforts. Particularly important is farmer participation in research priority setting with the Research and Innovation Working Group of the Organic Value Chain Roundtable (http://oacc.info/ResearchDatabase/res_ecoa.asp). Equally significant, the federal government has worked with the sector to initiate a number of marketing and regulatory projects, including the creation of the Organic Value Chain Roundtable and the implementation of the Canadian Organic Regulation. Similar types of initiatives are occurring across the industrial world and even in the global south.

This book provides the reader critical information to both assess the current state of knowledge about organic field cropping and to make these systems more viable. Each chapter summarizes the latest data and analysis from a wide range of sources, creates a comprehensive and coherent picture of the issues, and integrates agronomic, economic, and policy interpretations. In many chapters, the authors provide recent results of their own research trials.

Continuing with the themes of this introduction, Andrew M. Hammermeister, Ron Pidskalny, Karen Nelson, and Ralph C. Martin discuss the general principles guiding organic research, how organic research themes are set collaboratively in Canada, and current research priorities. The authors are very involved in this process, with the Organic Agriculture Centre of Canada, which Hammermeister directs and Martin founded, playing a central coordinating role.

Subsequent chapters are organized around many of these key principles and field crop research priorities: soil health, pest management, integrated approaches to organic field crop improvements (crop rotation design, breeding, and farm case studies), and issues of energy use, economics, and policy. Section I, "Soil Health," begins with Chapter 3 and the latest analysis of the role of arbuscular mycorrhizal fungi and phosphorus (P) nutrition. The authors, Mulan Dai, Kim Schneider, Barbara Cade-Menun, Ronald Ferrera-Cerrato, and Chantal Hamel, have extensive collective experience addressing the admittedly challenging question of P availability and uptake by organic crops. Numerous studies have identified P balances as a critical issue for long-term sustainability of organic systems (see Chapters 3 and 4), and these authors explore the limitations and potential of mycorrhiza in organic crop P nutrition.

Chapters 4 (Alexander Lake Woodley, Yuki Audette, Tandra Fraser, Melissa M. Arcand, Paul Voroney, J. Diane Knight, and Derek H. Lynch) and 5 (Derek H. Lynch) take a wider angle look at soil fertility. Woodley et al., located at universities in three different regions of Canada, examine an extensive set of strategies to improve nitrogen (N) and P fertility in organic systems, focusing particularly on green manures, residues, crop rotation design, compost, and other supplemental organic amendments. Lynch continues certain aspects of this discussion, focusing particularly on soil carbon, soil ecology, and core concepts of soil health. He introduces innovative tillage strategies that optimize soil carbon, a theme subsequently picked up by Martin Entz, Caroline Halde, Harun Cicek, Kristen Podolsky, Rachel L. Evans, Keith C. Bamford, and Robert E. Blackshaw in Chapter 6. This team, comprising researchers from Manitoba and Alberta, report on four major approaches to tillage reduction and provide some data on the effectiveness of different tactics from recent trials on the Canadian prairies.

The focus of Section II, "Pest Management," is primarily on weed and insect management, as these are the most common problems facing most field crop producers. However, many of the strategies discussed for weeds and pests are equally important for disease management (and Chapter 10 addressing cereal breeding is also pertinent). Steve J. Shirtliffe and Dilshan Benaragama in Chapter 7 review the case for, and practice of, integrated weed management in an organic farming system. Since synthetic herbicides are not permitted, the focus is on cultural methods and in-crop mechanical weed control, for which they offer critical appraisal of numerous techniques. In Chapter 8, Gilles Boiteau, Tom Lowery, and Josée Boisclair argue the importance and value of a long-term approach to insect management based on an understanding of pest biology. In their view, few farms have fully established a prevention-based farming strategy and their chapter sheds light on the dimensions of such a strategy.

Section III, "Integrating Approaches," looks at integrated approaches that address multiple field cropping challenges. In Chapter 9, Martin H. Entz, Cathy Welsh, Shauna Mellish, Yuying Shen, Sarah Braman, Mario Tenuta, Marie-Soleil Turmel, Katherine Buckley, Keith C. Bamford, and Neil Holliday report on the results of the oldest organic rotational trials in Canada—the Glenlea plots, located south of Winnipeg in Manitoba's Red River Valley—over its 20-year history. Chapter 10 (Hakunawadi Alexander Pswarayi, Stephen Fox, Pierre Hucl, and Dean Spaner) examines the challenges facing organic farmers when relying on cereals bred for conventional systems. They instead present an argument in favor of breeding strategies that complement agroecological system resilience, stimulate self-regulating internal systems, and focus on soil building and pest management with appropriate crop rotations. James B. Frey and Martin H. Entz close Section III with Chapter 11, an analysis of the experiences of 10 organic farm households from Manitoba, Saskatchewan, and Ontario. Participants represent a broad cross section of agricultural types, including grain, forages, livestock, and horticulture, or some combination of thereof. The chapter explores the broader social and ecological landscape on which these organic farming systems exist, providing a more holistic understanding of how organic farmers navigate the complexities of their operations.

Finally, Section IV, "Economics, Energy, and Policy," looks at wider-scale issues of economics, energy, and policy. In Chapter 12, Emmanuel K. Yiridoe reviews studies of the economics of energy use in organic farming systems, drawing on studies that demonstrate energy use reductions and efficiencies in ways that are economically beneficial at the farm level. Rod MacRae, Derek H. Lynch, and Ralph C. Martin then follow with two chapters reviewing the extent to which the adoption of organic farming solves current food system problems. Chapter 13 tackles environmental issues not already addressed in earlier chapters, and Chapter 14 presents consumer, economic, and rural community matters. The final chapter (Chapter 15) of Section IV is a reprint of an article by Rod MacRae, Ralph C. Martin, Mark Juhasz, and Julia Langer that was first published in 2009 in *Renewable Agriculture and Food Systems*. It presents and costs policies and programs needed to advance adoption of organic farming in Ontario.

We wrap up in Chapter 16 by discussing key conclusions and crosscutting themes for the entire book. It is important to note, however, that the value of organic research extends beyond the organic sector. Conventional farmers commonly look to organic research as they explore ways to reduce costs and to maintain the productive capacity of their soils and farms, and many of the chapters in this book provide literature and interpretation useful across the farming spectrum. It is hoped then that this book will appeal to a wide range of scientists, extension agents, and producers, all interested in enhancing the ecological performance of agriculture.

REFERENCES

Canadian Organic Trade Association. 2013. *Canada's Organic Market: National Highlights, 2013*. Ottawa, Ontario, Canada: COTA.

Hill, S.B. and R.J. MacRae. 1992. Organic farming in Canada. *Agriculture, Ecosystems & Environment* 39:71–84.

Willer, H., J. Lernoud, and L. Kilcher (eds.) 2013. *The World of Organic Agriculture: Statistics and Emerging Trends 2013*. Bonn, Germany: IFOAM and FiBL.

2 Establishing Priorities for Organic Research in Canada

*Andrew M. Hammermeister, Ron Pidskalny,
Karen Nelson, and Ralph C. Martin*

CONTENTS

2.1 INTRODUCTION

In a world of many competing interests for limited funds supporting research for agriculture and food, organic sector stakeholders and funding program managers are faced with a tremendous challenge in defining research priorities. The organic sector encompasses all aspects of agriculture including crops (fruits, vegetables, grains, pulses, oilseeds, etc.), livestock (production and welfare of dairy, poultry, beef, pork), not to mention food (storage, handling, processing, packaging, additives), and of course environmental issues (nutrient loading or depletion, sustainability, greenhouse gas emissions, energy intensity, soil conservation, biodiversity) and economic/marketing issues (consumer/market demand, cost of production, profitability). In each of these areas, the growth and development of the organic sector will depend on its ability to capture existing opportunities, create opportunities through innovation, maintain or improve competitiveness, and address barriers. So how do we set priorities for organic research? This chapter will outline the process undertaken to establish research priorities for the organic sector in Canada.

2.2   IDENTIFYING STRATEGIC RESEARCH AREAS: CONSIDERATIONS

Organic agriculture is a model of food production that is guided by principles of sustainability in terms of environment, resources, economics, and animal well-being. It is a regulated and inspected production system driven by consumer demand domestically and internationally.

In order to set research priorities, we must first understand the guiding principles of organic agriculture. There are seven general principles described in the Canadian organic standards (Canadian General Standards Board, 2006):

1. Protect the environment, minimize soil degradation and erosion, decrease pollution, optimize biological productivity, and promote a sound state of health.
2. Maintain long-term soil fertility by optimizing conditions for biological activity within the soil.
3. Maintain biological diversity within the system.
4. Recycle materials and resources to the greatest extent possible within the enterprise.
5. Provide attentive care that promotes the health and meets the behavioral needs of livestock.
6. Prepare organic products, emphasizing careful processing and handling methods in order to maintain the organic integrity and vital qualities of the products at all stages of production.
7. Rely on renewable resources in locally organized agricultural systems.

While these principles address ecological priorities for organic production, they do not specifically address economic sustainability for the farmer or the capacity of the organic sector in Canada to meet the growing domestic demand and export opportunities.

Worldwide, consumers are interested in organic food because of health issues (Yiridoe et al., 2005) and because the consumption of organic food products is perceived to be better for the environment (Certified Organic Associations of British Columbia, 2002; MacRae et al., 2002). About 7% of US organic food consumers also link personal health with environmental health and to these consumers, organic food purchases support an integrated set of values (The Hartman Group, 2000). Consumers also link food consumption to family health, and concern for their children is an important consideration among those purchasing organic food products. Affluent, well-educated, health-conscious consumers are driving demand for organic food products—and the size of this market is increasing because of demographic and income shifts (MacRae et al., 2002). Organic consumers, overall, are looking for competitively priced fresh food that looks and tastes good and is convenient to use.

The organic sector will need to improve production capacity nationally in order to meet the growing demand for organic food products.

A market research report found more than 51% of Canadian households purchased an organic product during the year prior to reporting (CBC News, 2007), and for over more than a decade, the organic sector has experienced annual double digit growth in retail sales (Canadian Organic Trade Association, 2013). While the Canadian organic sector has a history of being supported by strong consumer demand, Canadian organic producers have not been able to meet this demand (Certified Organic Associations of British Columbia, 2002). Globally, organic production is one of the fastest growing sectors of the food industry and represents a massive export opportunity for Canada's organic farmers (Reed Business Information, 2006); however, Canada's organic farmers face a well-established, competitive market for their goods. Worldwide, almost 31 million ha is certified according to organic standards on 600,000 organic farms (Willer and Yussefi, 2006). While certified organic production occurs in about 120 countries, 96% of the organic food market is in Europe and North America (Willer and Yussefi, 2006). Australia (11.8 million ha), Argentina (3.1 million ha), China (2.3 million ha), and the United States (1.6 million ha) have the largest tracts of land certified for organic production (International Federation of Organic Agriculture Movements, 2007).

In terms of economics, the number of consumers willing to pay a premium for organic food decreases as the price premium increases (Yiridoe et al., 2005). The premium the consumer is willing to pay, though, increases as the number of combinations of preferred attributes associated with the food product rises. Demand for food products also tends to depend more on the price differential between organic and conventionally grown products than on the price itself. In general, however, demand for organic food products is elastic (sensitive) in terms of prices. On the other hand, demand for organic food products is relatively inelastic (not as sensitive) in terms of consumer income. In other words, as income goes up, consumers do not purchase more organic food products and changes in consumers' income have little influence on demand.

In order for the Canadian organic sector to meet market demand for its products in a competitive manner, research priorities need to focus on the following:

1. Increasing *profitability* and *competitiveness* of Canadian organic production through *transformative innovation*
2. *Addressing barriers* limiting transition, productivity, efficiency, and market access
3. *Capturing opportunities* in new markets meeting consumer needs through a focus on social, technological, economic, environmental, and political trends
4. *Measuring indicators* of profitability and competitiveness such as changes in farm size, number, sales, costs, and net return
5. *Leading agricultural innovation in a search for* innovative, alternative solutions to changing markets, climate, pests, input costs, and resource availability
6. Considering the *feasibility* of research projects in terms of time required to put innovations into practice in commercial fields, potential utility to the end user, cost, likelihood of success, availability of scientific capacity, and *efficiencies incurred by* building on existing research programs

2.3 STRATEGIC PRIORITY PLANNING PROCESS FOR ORGANIC RESEARCH OBJECTIVES

To achieve these goals, research prioritization must be informed by social, technological, environmental, economic, and political/legal/regulatory trends, in addition to stakeholder needs, that is, a sector-wide macroenvironmental scan. Once this information is collected, a prioritization process follows. Beginning in July 2007, the Organic Agriculture Centre of Canada (OACC) initiated a national process for determining strategic areas for organic research in Canada. The process included the following steps:

1. Macroenvironmental scan (STEEP analysis [social, technological, environmental, economic, and political/legal/regulatory trends]) and strengths, weaknesses, opportunities, and threats (SWOT) analysis
2. Farmer survey
3. Prioritization process
   a. Listing of potential research questions
   b. Establishing criteria for success
   c. Rating the impact of research questions against criteria
   d. Prioritizing based on impact, likelihood of success, and cost/time
4. Evaluation of feasibility of research based on scientists available and interested in organic research in Canada, that is, the capacity to take advantage of opportunities and mitigate threats using sector research strengths while addressing options for improving upon weaknesses
5. Recommendations for strategic research directions
6. Steering committee of stakeholders–approved research program

This chapter will provide an overview of issues arising from the macroenvironmental scan as they relate to research and the research prioritization process that was followed.

2.4 MACROENVIRONMENTAL SCAN

A macroenvironmental scan was commissioned to identify the trends in organic agriculture that may influence the research prioritization process (Strategic Vision Consulting, 2009a). This "scan" included a STEEP analysis. The analysis of SWOT was also prepared by Strategic Vision Consulting (2009b) to inform the sector of how it could address the trends identified earlier. This section provides an overview of some of the trends within the Canadian organic sector placed into the context of research.

2.4.1 SOCIAL

While growth in the organic sector is driven by consumer demand, there is considerable debate between proponents and opponents of organic agriculture in a "battle for hearts and minds" for consumers. These debates center on genetic engineering, pesticides in the environment and food, animal welfare, and nutritional quality of food. Many consumers purchase organic products because they are putting a higher value on health and environmental sustainability (Certified Organic Associations of British Columbia, 2002); however, the public is largely unaware of what organic stands for, and consequently, organic food is often described in terms of what it is not, rather than what it is.

Health professionals and environmental groups tend to be supportive of organic principles and practices (Certified Organic Associations of British Columbia, 2002). However, clear claims relating to the nutritional benefits of organic food and how these benefits can be optimized require further research.

Some consumers of organic food products are moving "beyond organic." These consumers are interested in establishing a greater connection with food producers and the regions in which their food is grown. In some cases, this means that the organic consumer prefers to consume locally grown food rather than organic food that has been transported over greater distances. However, the concept of local has not been integrated into the organic standard, in part, to allow for the development of the sector without restricting the products and/or ingredients that consumers were used to purchasing.

In addition, consumer pressure for improved animal welfare standards in agriculture is growing, and large commercial retailers of food are increasing expectations for attention to animal welfare and, in some cases, setting standards (Whole Foods Market IP. L.P., 2013) for their Canadian operations (Leeder, 2012). See Chapters 13 and 14 for details on current understanding of these themes.

Stakeholders in the organic sector include consumers as well as producers of organic food products. Many consumers can be described as primarily ideologically driven (Certified Organic Associations of British Columbia, 2002; MacRae et al., 2002; Yiridoe et al., 2005), whereas producers, while ideologically driven, may also enter organic production primarily for economic reasons. The organic sector has a tradition of volunteerism, self-help, and self-determination. These characteristics of the stakeholders can provide direction to organic researchers, in-kind support in research programs, participation in surveys that inform social and natural sciences research, and direct financial support. Those motivated by ideology bring strength to the sector in their holistic view of agriculture, their willingness to invest in research that seeks alternative solutions to challenges in agriculture, and their desire to learn about how different components of the agroecosystem interact so they can manage them more effectively. This ideology, however, can also limit the options available for organic research, innovation, and management. Stakeholders motivated primarily by economics bring strength to the sector by driving a business approach to the sector that leads to greater efficiency and innovation that leads to higher profitability; however, this approach may limit the scope of interest in research, involvement in the sector as a whole, and willingness to share knowledge or innovation arising from the research.

2.4.2 TECHNOLOGICAL

The organic sector in Canada is largely knowledge driven, adopting practices and management systems that have been developed over time through improved understanding of agroecosystems. As the variety of tools and inputs available for organic producers are limited, they increasingly rely on a knowledge-based system to develop integrated solutions to problems. Pest control is a classic example where knowledge of the biology and/or physiology of the pest allows development of long-term plans to avoid, deter, resist, and manage the pest as opposed to relying on a single means of control, such as chemical, which may increase the speed of selection for resistance in the pest (Insecticide Resistance Action Committee, 2013). This knowledge-driven approach to agricultural production can lead to higher levels of innovation in breeding programs, cultural practices, and input selection. Stakeholders in the organic sector tend to be willing collaborators in on-farm research and are generally quite eager to share details of new technologies and innovations that have worked on their own farm.

Many organic producers need to lead as innovators and developers of suitable technology because of the limited scale of organic production in Canada. In 2008, certified organic operations represented about 1.5% of all Canadian farms, with organic fruit and vegetable operations accounting for about 2.3% of the Canadian total (Agriculture and Agri-Food Canada, 2008). In many cases, organic pest control products available in other parts of the world are not available in Canada because the market is too small to warrant the cost of registration.

Canadian research facilities are increasingly receptive to organic research; however, the onus is on the organic sector to take the initiative in collaborating with institutional researchers and providing the background information required to conduct organic research.

Technological support specifically for advancement of the organic value chain is limited. The number and scale of producers and volume of their output in Canada have made it difficult to entice commercial interests into establishing dedicated technologies, equipment, facilities, and systems/practices that comply with organic regulations. Examples of such would include the lack of dedicated slaughter and meat processing facilities and economic constraints to the development and/or approval of organically acceptable pesticides.

Breeding programs specifically for organic production in Canada were in their infancy in 2009, and since then, efforts have focused on wheat and oats for prairie production. There has been little or no breeding for organic in livestock or horticultural crops. Genetically engineered (GE) cultivars, particularly in canola, corn, and soybean, have become dominant in the regions where these crops are grown; the availability of non-GE cultivars has declined, coupled with increased risk of "contamination" of non-GE seed for planting and ultimately harvest and sale.

For many crops grown in Canada, yield under organic management lags behind yields under nonorganic management (Lynch, 2009; also Chapter 14), due to lower soil fertility status, lack of adapted cultivars, and/or absence of permitted pesticides. At the same time, the knowledge-based system of organic production has been suggested to produce higher yields than nonorganic production in developing nations (Bagdley et al., 2007).

With few exceptions, most new equipment is developed for large-scale commercial production that is often either too large or costly to fit into smaller-scale organic operations. Development and acceptance of technologies supporting organic agriculture such as pest control machinery and products is further advanced in Europe and the United States, placing Canadian growers at a competitive disadvantage.

2.4.3 ECONOMIC

Despite its growth, the organic sector remains relatively small, accounting for 2%–3% of the acreage and retail markets in Canada. In some respects, this small scale allows the sector to dynamically respond to changes in the market; however, it is also large enough to warrant standards development, regulation, and labeling. The small to intermediate scale of operations throughout the value chain is challenging, in that supply may be inadequate to warrant investment in dedicated production facilities or may not be produced in sufficient quantity to be of interest to larger retailers. In the prairie region and parts of Ontario and Quebec, field crop production is very important; however, close to 50% of organic producers are relatively small-scale producers selling to farmers markets. This creates challenges for research in that the size, crops, and environment vary greatly among producers, making it difficult to carry out research that will have broad relevance and impact. With limitations in funding and the number of researchers, it is impossible to address the needs of all stakeholders. Many research-funding programs offered by the federal and provincial governments require matching industry cash, which can be difficult to source in the relatively small organic sector.

The market for organic product continues to grow around the world and production is unlikely to meet the global demand. Organic markets are becoming globalized in their distribution, and as organic products become more prominent in the marketplace, the level of production and the number of competitors supplying this market will increase. Organic processing, distribution, and retail have grown to the point where there is significant consolidation of companies. This "upscaling" of processing and distribution capacity places greater pressure on product prices and management of supply. Effectively, pieces of the organic value chain such as grains are becoming commoditized (traded primarily on the basis of price and delivery capacity). The rising cost of food and conventional grain prices creates pressure on grain buyers to increase organic premiums to maintain incentives for farmers to remain in organic production. For livestock producers, higher grain prices increase the cost of feed for nonvertically integrated operations that produce their own feed. These pressures will contribute to increases in the price of organic food for consumers, unless research in the organic sector focuses, in part, on lowering the cost of production.

2.4.4 ENVIRONMENTAL

The principles of organic agriculture are grounded in sustainability. Desired outcomes of organic agriculture from an environmental perspective lie in maintaining biodiversity, avoiding contamination or degradation of the environment, and sustaining our resource base. The question is, in what ways is organic production consistent with these principles? In a review paper, Lynch (2009) determined that

> The empirical evidence presented suggests organic farming system attributes regarding cropping, floral, and habitat diversity, nutrient intensity, soil management, energy and pesticide use, etc., are sufficiently distinct as to impart potentially important environmental benefits across the indicator categories examined. More research is

needed to validate these results for the benefit of producers, consumers and policy makers as they decide the relative importance and contribution of organic farming systems to the Canadian food marketplace and agri-food sector.

Research supporting organic agriculture must quantify these benefits as well as identify practices that further achieve these principles. The diversity of organic farms, including their soils and environment as well as their management histories and production systems, makes it difficult to make specific statements about organic agriculture.

Global climate change is of great concern with potential implications for agriculture varying by region. The frequency and severity of extreme weather conditions and movement of pests into new regions is likely. Organic agriculture may prove to be a more resilient and economically viable system than conventional agricultural systems since it is less reliant on inputs produced using fossil fuels, has strong emphasis on building healthy soils with higher organic matter content, and encourages higher levels of biodiversity. These areas of research certainly require further exploration and many chapters in this book address these themes more fully.

2.4.5 POLITICAL/REGULATORY

The organic committee of the Canadian General Standards Board is continuously seeking scientific information to support the review and revision of organic standards. Regulatory requirements relating to animal welfare standards are advancing. While organic agriculture has received federal support in Canada through establishment of a Canada Organic Office, development of regulated national standards, establishment of an Organic Value Chain Roundtable, and support for a national Organic Science Cluster, ongoing support for standards maintenance and consumer education has been limited. Political support for organic production varies by region within Canada. By July 2013, Quebec, British Columbia, and Manitoba were the only provinces with intraprovincial regulation. Globally, support for organic agriculture has been highest in Europe with specific targets set in some countries. The word organic has become regulated in Canada and many other countries; equivalency agreements are being signed among countries around the world to ensure the standards are sufficiently consistent so as not to impair trade.

The national and international demand for a regulated organic production system and related products has spurred interest in organic and related sustainable production systems without the use of synthetic inputs. The principles of organic agriculture support waste utilization, nutrient recycling, and use of biodiversity and lower intensity management systems to help in avoiding pest pressure and reducing the use of off-farm inputs. This kind of production system does not suit large-scale production or support of a substantive inputs industry. As a result, there is lower investment by manufacturers and input suppliers into new products to service organic farms. Since organic production is based on crop rotation, soil building, and promoting livestock health rather than on inputs, there are few suppliers of services and inputs available to drive innovation. As such, it is imperative that public money is used to support these fundamental research programs that will be of benefit to not only the organic sector but all of agriculture.

2.5 PRIORITIZATION PROCESS

Based on information presented in the macroenvironmental scan of the Canadian organic sector, the following process was followed with the input of the Expert Committee on Organic Agriculture (ECOA). The goal was to evaluate projects based on their potential impact at the farm gate, the likelihood of success, and time required before the impact of the research would be evident at the level of the producer. The results shown in the following were used primarily as supporting documentation for identifying priority research areas in a proposal for a national strategic research initiative in organic agriculture. Other factors included; availability of researchers with interest in organic science, supporting infrastructure, focus of the government funding programs, and requirements for matching industry cash.

2.6 METHODS FOR DETERMINING SCIENCE CATEGORIES, IMPACT CRITERIA, AND WEIGHTING

The ECOA (now the Research and Innovation Working Group of the Organic Value Chain Roundtable; http://oacc.info/ResearchDatabase/res_ecoa.asp) guided and provided feedback to the prioritization process. This committee was comprised of approximately 15 individuals representing producers, retailers, processors, researchers, and government extension specialists from across Canada. A number (11–25) of research topics were identified for each of the nine subject categories based on farmer surveys (Organic Agriculture Centre of Canada, 2008) and consultations. The nine general categories (soils, plants, animals, ecological systems, sustainability, policy, markets, health and food, and general) were identified based on the scope of priority issues identified in surveys and consultations as identified earlier. The potential impact of these research topics was then ranked against up to nine criteria identified by ECOA shown in Table 2.1. Researchers, professionals, farmer groups, and other stakeholders were then asked to participate in ranking the potential impact of different research topics, along with identifying the likelihood of success and the time to carry out such the projects.

Three components of the research prioritization process are shown in Table 2.1: the research categories, evaluation criteria, and weighting values for evaluation criteria. In the first column, the list of research categories (soils, plants, sustainability, etc.) is shown. In each of these categories, we had 12–25 research questions that were derived from a combination of the survey of farmer research needs and consultations led by the OACC at meetings and conferences.

The top row of Table 2.1 shows the criteria for success that were established by ECOA. Each of the research questions was rated on a scale of 1–10 for the impact that the research question would have on each of the criteria (10 being highest). It is worth noting that the broad range of criteria is somewhat unique to organic agriculture. Other commodity groups following a similar process have been known to have only one or two criteria, typically emphasizing profitability at the farm gate (Strategic Vision Consulting Ltd., 2006a,b, 2007). This is a clear indication of the wide range of goals within organic agriculture, linked to the principles. As a result, the complexity of the decision-making model is likely to be much higher

TABLE 2.1
Weighting of Impact Criteria for Each Research Subject Area

	Farmer Gross Margin	Increase Capacity for Sales	Increase Production	Animal Welfare	Reduce Envir. Risk	Char. and Support EG&S[a]	Inform Policy Makers	Positive Social Climate	Evolution of Organic
Soils	4	4	4		3	3	1	1	1
Plants	4	4	4		3	3	1	1	1
Animals	3	3	3	5	2	2	1	1	1
Ecol. systems	2	2	2		6	6	1	1	1
Sustainability	2	2	2		6	6	1	1	1
Policy	1	1	1		3	3	4	4	4
Market	2	2	2		6	6	1	1	1
Health and food	2	2	2		6	6	1	1	1
General	1	1	1	1	1	1	1	1	1

Note: Higher numbers have higher weighting and therefore will have a larger influence on the impact factor.

a Characterization and support of environmental goods and services.

in organic agriculture. The numbers in the body of the table indicate the weighting of the research criteria, which varied depending on research category. This weighting system allowed more emphasis to be placed on selected criteria (e.g., example production, profitability, and capacity building) based on the needs perceived by ECOA. For this reason, research questions could not be compared between research categories.

The number of reviewers, which varied for each category, is indicated in the figure headings. The impact factor of a research question was calculated from all of the respondents and accounted for the weighting in Table 2.1. The results of the process are shown by category in Tables 2.2 through 2.10 and Figures 2.1 through 2.9. The first column of each table within a category shows a project identifier. The second column shows the research question that was rated. The last column in the tables shows the numerical impact factor. This is the value that was calculated from the prioritization ratings and the weighting process. The projects are listed from the largest impact to the smallest. The bubble graphs (Figures 2.1 through 2.9)

TABLE 2.2
Impact Rankings of Research Questions in the "General" Category

Identifier	Project	Impact
106	Develop integrated production systems that reduce pest pressure from weeds, insects, disease, and parasites.	6.7
102	Reduce the risk of transitioning to organic production through the development of new management systems or production aids.	6.3
105	Develop integrated production systems that increase yield and yield stability.	6.1
114	Identify key characteristics of the organic production system that are of value to the consumer.	6.1
109	Develop and/or identify livestock breeds and/or crop varieties that are adapted to organic management.	5.9
108	Identify practices that reduce energy requirements on the farm.	5.7
115	Identify key characteristics of the organic production system that are of value to the farmer.	5.6
113	Further our understanding of the complex interactions in organic production systems.	5.5
101	Attract and support new entrants to agriculture	5.1
112	Extend the length of time that the organic sector can supply high-quality Canadian produce to retail markets.	5.1
104	Develop and/or identify substances that reduce pest pressure from weeds, insects, disease, and parasites.	4.7
110	Identify ways to reduce losses or reductions in quality during the storage and handling of organic agricultural products.	4.6
111	Identify and/or develop organic products with novel traits of interest to consumers of organic products.	4.4
107	Identify practices that reduce labor requirements on the farm.	4.3
103	Develop and/or identify substances that increase yield and/or the potential for yield stability.	4.1

TABLE 2.3
Impact Rankings of Research Questions in the "Soils" Category

Identifier	Project	Impact
203	Identify integrated management practices to optimize soil quality as a substrate for crop growth.	18.0
201	Identify or develop crop rotations that sustain soil fertility and meet overall regional yield averages.	17.4
208	Identify integrated management practices for optimizing soil nitrogen in order to maximize economic crop yields.	17.3
206	Develop nutrient budgeting tools that account for whole-farm nutrient flows and/or nutrients contributed by different crops and amendments.	17.2
209	Identify the risk of soil phosphorus depletion under regionally specific organic management systems and potential solutions for organic producers.	16.6
207	Develop fertilization strategies for crops with high nutritional requirements while minimizing the environmental risk of applying excess nutrients.	16.4
204	Identify or develop organic farming techniques that can maintain soil quality and build soil organic matter with routine mechanical weed control.	16.2
202	Explore specific cover crop sequences or mixtures that interact with soil biota to stimulate plant resistance mechanisms and influence nutrient uptake.	15.3
212	Determine the amount and timing of nutrient release from different soil amendments and their efficacy in terms of improving plant nutrition.	13.5
213	Identify viable growing mediums and nutrient sources that are suitable for organic greenhouse and transplant production.	12.9
211	Determine the nature and extent of deficiencies of macronutrients as possible links to the nutritional quality of foods.	12.5
210	Determine the nature and extent of deficiencies of micronutrients and possible links to the nutritional quality of foods.	11.7
205	Conduct efficacy testing on products marketed as soil microbial stimulants or biological enhancers.	10.6

show the likelihood of success of the projects and the estimated amount of time to complete the project on the axes. The size of the bubbles represents the potential relative level of impact to the producer or end user of the technology. This is a way to visually assess the project for their potential. Large bubbles in the top left, for example, represent projects that are likely to succeed in a shorter time frame and will have larger impact. Conversely, small bubbles in the lower right have a low likelihood of success and a low impact and will take a long time to generate results for producers.

2.7   OUTCOMES

The result of the impact analysis and plotting of the impact at the farm gate against the likelihood of experimental success and time to completion was a very informative exercise. Overall, integrated and systems-level research was identified among the top selections across multiple subject categories. These results are consistent with

TABLE 2.4

Impact Rankings of Research Questions in the "Plants, Part 1" Category

Identifier	Project	Impact
326	Identify and/or develop integrated approaches to weed management.	17.7
317	Integrate approaches to insect management to reduce yield losses due to disease by >80%.	17.5
313	Integrate approaches to disease management to reduce yield losses due to disease by >80%.	16.9
311	Identify and/develop crop cultivars to reduce harvestable yield losses due to disease by >80%.	16.2
315	Identify and/or develop crop cultivars to reduce yield losses due to insects by at least 80%.	16.1
320	Identify and/or develop crop cultivars with a competitive advantage against weeds.	15.8
324	Identify and/or develop mechanical, thermal, or other weed controls to reduce losses by >80%.	15.6
325	Develop reduced- and no-tillage organic systems of weed control and design suitable machinery.	15.6
307	Develop marketable products and/or markets for cover crops used in organic management.	15.1
321	Identify and/or develop cover crops with weed suppressive abilities.	14.8
316	Identify and/or develop insect control products to reduce yield losses due to insects by >80%.	13.8
310	Identify and/or develop organically acceptable means of reducing postharvest storage losses.	13.7
312	Identify and/or develop control products to reduce yield losses due to disease by at least 80%.	13.6

the organic principles that emphasize the health of the soil, plants, and livestock through integrated and diverse systems with inherent resiliency. However, this type of research is very complex and can take a number of years before results create positive results for the producer.

Organic production covers the full spectrum of agriculture across all commodity types, agroecological regions, and farm size and along the whole value chain. Clearly, a single research program cannot cover all of the research needs of the sector. The research community needs to balance projects that can be achieved over the longer and shorter term as well as consider the probability that the results of the research will produce the required outcome once extended to the level of the producer or the end user of the research. At the same time, research projects must balance the needs of an array of crop, livestock, and food production stakeholders while maintaining a focus on a number of environmental, economic, and marketing issues. A balance of projects is needed that produce short-term impactful results, explore and capture the values (environmental, health, ethical, and social) that drive organic production, and with varying outcomes and likelihood of success are needed

TABLE 2.5

Impact Rankings of Research Questions in the "Plants, Part 2" Category

Identifier	Project	Impact
318	Develop cultural practices to activate induced systemic resistance in crops.	13.4
327	Refine soil fertility management systems to minimize weed pressure.	13.1
323	Identify or develop weed control products with at least 80% efficacy.	12.7
319	Develop insect pest thresholds for organic management systems.	12.7
314	Analyze the risk of overuse of copper or sulfur fungicides and explore potential alternatives.	12.4
308	Determine the cost of production associated with different organic vegetable farming practices.	11.0
322	Investigate weed seed bank dynamics under organic management systems.	10.5
309	Determine the effectiveness and economics of seed coatings including micronutrients, growth promoters, and/or mycorrhizae.	9.5
304	Identify long-term cropping systems and/or rotations with higher yield and economic stability under variable climatic situations.	6.7
305	Develop intercropping systems, crop sequences and cropping practices with pulse crops and other legumes to increase N fixation.	6.2
306	Conduct a detailed analysis of legumes in organic farming systems RE: strengths, weaknesses opportunities, and threats.	6.0
301	Identify or develop crop cultivars and/or crop traits for organic management with a 5% yield advantage.	5.5
302	Develop perennial grain varieties for organic management systems.	5.4
303	Develop cultivars and/or rotations that reduce phosphorus requirements by 20% but remain viable.	5.2

to most effectively advance organic agriculture. A balanced research program could include the following:

1. Short-term high-impact projects that specifically address barriers (high likelihood of creating a positive result for the end user of the research) (30%)
2. Innovative projects developing or testing new products or tools that will increase competitiveness and profitability (lower likelihood of success in creating a positive result for the end user of the research) (20%)
3. Short-term projects that characterize environmental goods and services and efficiencies in production (higher likelihood of creating a positive result for the end user of the research) (15%)
4. Integrated projects that link management with ecological interactions among soil, plants, and animals (longer term, lower likelihood of creating a positive result for the end user of the research) (35%)

Overall, the potential impact of research questions in the general category ranked relatively low. Under the "General" category, which encompassed research questions of broad interest to the development of the sector, the most impactful short-term research project would involve identifying the most key characteristics of the organic

TABLE 2.6
Impact Rankings of Research Questions in the
"Ecological Systems" Category

Identifier	Project	Impact
508	Investigate long-term effects of organic systems on crop yield, environmental impact, soil quality, nitrogen fixation, crop quality, and economic stability.	17.8
506	Determine the impact of organic production practices on greenhouse gas emissions from farms and identify ways to reduce emissions by at least 20%.	17.8
513	Identify means of reducing the energy required to produce each calorie of food by at least 5%.	17.1
511	Compare the external costs of organic and nonorganic production systems on natural capital.	16.6
510	Identify current management practices in organic systems that affect sustainability in terms of productivity, environment, and economic return.	16.6
505	Determine the impact of organic practices on soil quality, health, production capability, and/or other biological properties.	16.1
512	Determine the effect of scale and type of farm on energy use efficiency in organic systems.	15.9
503	Compare the ecological, physical, and societal impact of organic production systems on water quality to high-input nonorganic agricultural systems.	15.9
502	Develop a mechanism for increasing water use efficiency in organic crop production by at least 10%.	15.3
501	Compare the biodiversity of wildlife, flora, microfauna, and macrofauna above- and belowground in organic and high-input nonorganic systems.	15.3
504	Determine the ecological impact of manure, compost, and green manures used in organic systems compared to synthetic sources of nutrients.	14.7
507	Determine the amount of soil carbon captured by different organic production systems compared to that captured in high-input nonorganic systems.	14.6
509	Assess the role of pollinators and level of risk to pollinators in organic systems compared to pollinator roles and risk in high-input nonorganic systems.	14.4
515	Contrast long-term cumulative and interactive impacts of synthetic pesticides applied at minimal rates with few products to high rates and many products.	14.3
514	Assess the optimal use of class 1, 2, 3, and 4 land on organic farms for sustainable production of food, feed, fuel, and fiber.	14.2

production system that are of value to the consumer and the producers. This knowledge will allow research programs to be responsive to drivers in the value chain. The most impactful long-term research project relates to the development of integrated production systems that reduce pest pressure (weeds, insects, diseases, parasites, etc.). The organic principles direct producers to develop integrated systems that promote resilience to pests, and this is reinforced by restrictions in the use of synthetic pest control products imposed by the organic standards.

Under the "Soils" category, the development of integrated practices that promote soil quality may have the most impactful long-term research that could be conducted. In the short term, this could be supported by whole-farm nutrient budget research that

TABLE 2.7

Impact Rankings of Research Questions in the "Sustainability" Category

Identifier	Project	Impact
812	Investigate and identify opportunities for increasing Canadian processing of Canadian products.	14.8
814	Examine threats to the future of organic agriculture such as genetic engineering and nanotechnology.	14.2
809	Determine the feasibility of urban organic agriculture.	13.0
801	Determine the economic constraints and opportunities relevant to the viability of small- and medium-scale organic farms.	11.4
805	Develop transition and extension strategies based on a demographic analysis of Canadian farmers.	11.2
811	Identify ways to use organic agriculture to strengthen the relationship between rural and urban communities.	11.2
808	Determine the amount of vacant agricultural land by region and determine the potential for organic production of food, feed, fuel, and fiber in these areas.	11.1
815	Create linkages with researchers in developing countries to support organic agriculture as a sustainable livelihood within each of a variety of biological regions.	10.9
810	Analyze consumer behavior to determine the extent to which producer integrity has been affected in organic and high-input nonorganic agriculture.	10.6
802	Determine cost of production for both crops and animals.	10.2
803	Determine if organic farms are more profitable than high-input nonorganic farms when there is no price premium.	9.6
804	Determine local and global trends in organic price premiums by product and region and the implications for organic farmers supplying these different markets.	9.4
807	Impact of new entrants to the market and supply trends in domestic and international organic products to determine the level of risk associated with downward price pressure.	9.0
813	Determine organic agriculture's potential risk of losing government support by comparing the GDP with the Genuine Progress Index.	7.9
806	Analyze rural population demographics and the resilience in order to determine if population trends in rural communities could affect organic production capacity.	7.9

will allow producers to understand the net balance of nutrients within their farming system. The health of the soil is typically regarded as critical within organic agriculture for promoting resilient, competitive, and productive crops without the aid of synthetic pest control products. Similarly, integrated approaches to pest management rank as the top priorities under the "Plant" and "Animal" categories. These integrated approaches require systems-level research that is most often multidisciplinary in nature. In the short term, improvements in weed management tools are easier to achieve and more likely to have significant impacts in crop production and identifying the optimum land area needed for livestock productivity, health, and welfare.

TABLE 2.8
Impact Rankings of Research Questions in the "Animals" Category

Identifier	Project	Impact
404	Identify and/or develop integrated management systems that optimize health, welfare, productivity, quality, and profitability.	18.8
414	Develop pork, cattle, and/or sheep management systems that reduce the incidence of parasite productivity losses.	17.2
407	Optimize the nutritional balance of pasture and/or grain-fed livestock to improve their productivity and/or specific attributes.	17.2
415	Identify products within the management system that are effective for controlling parasites, suppressing illness/disease, and promote health.	16.6
402	Identify and/or develop livestock breeds that are suited to organic management.	16.1
406	Identify opportunities for realizing greater efficiency and profit margins in livestock.	16.1
403	Identify and/or develop organically permitted substances that are effective in managing parasites and/or disease.	15.8
412	Identify methods and/or systems to consistently reduce somatic cell counts in dairy systems to within acceptable limits.	15.2
401	Identify and/or develop livestock breeds that are resistant to parasites and/or diseases.	15.0
410	Identify the land area required per animal unit for sustainable production, optimum animal welfare, and manure management.	14.6
413	Develop an outdoor poultry management system that reduces the risk of avian flu to "negligible" levels.	14.4
409	Identify organic forage and grain feeding systems that optimize the feed conversion ratio without jeopardizing animal health or welfare.	14.3
405	Identify and/or develop emergency health care treatments for poultry and livestock that are compliant with the Canadian Organic Standards.	14.2
411	Develop a welfare index for farm animals raised in organic systems that is compatible to that used in nonorganic livestock management.	12.4
416	Develop methods of organic management to optimize the finishing of beef cattle.	12.3
408	Identify or develop feed and/or ingredients that contain organically acceptable vitamins and amino acids for livestock rations.	11.9

TABLE 2.9

Impact Rankings of Research Questions in the "Health and Food" Category

Identifier	Project	Impact
612	Determine the health risk associated with agricultural pesticide use as well as occupational, environmental, and/or food exposure to pesticides.	17.9
601	Identify key factors linking organic food quality with the quality of soils, amendments, plants, and animals in the organic production system.	16.6
603	Identify production practices that optimize the concentration of nutrients, antioxidants, and other bioconstituents in food products.	15.8
611	Analyze postharvest food storage and handling practices; develop mechanisms to increase shelf life and reduce the risk associated with postharvest storage systems.	14.8
604	Determine the effects of organically permitted pesticide use on crop physiology and product quality.	14.4
602	Determine the nutritional value, including the concentration of nutrients, antioxidants, and other bioconstituents of Canadian organic food products.	13.4
605	Determine if consumers with at least 75% of their food coming from organic sources are more resistant to illness or disease.	13.2
610	Investigate disease transmission risks from animals to humans and determine how disease transmission can be affected, controlled, and managed.	12.7
607	Determine the level of health risk reduction associated with the consumption of organic foods compared with nonorganic food products.	12.6
609	Determine the degree of risk associated with the level of microbiological activity in Canadian organic meat production when antimicrobial agents are not used.	11.7
608	Analyze food safety practices and determine whether or not the development of special safety practices would be required for organic food production systems.	11.5
606	Determine the health risk associated with the consumption of natural secondary metabolites produced by plants in response to stress from insects and diseases.	8.8

TABLE 2.10

Impact Rankings of Research Questions in the "Marketing" Category

Identifier	Project	Impact
708	Determine the impact and opportunities for alternative marketing models such as cooperatives, local marketing, and fair trade goods.	16.1
713	Identify key opportunities for import replacement and sector development in Canada.	15.3
701	Develop innovative approaches to processing or marketing organic products.	14.0
702	Determine the feasibility of developing microprocessing facilities as opposed to accessing existing infrastructure (i.e., access to value-added markets).	14.0
706	Determine the market potential for organic fibers, fuels, and pharmaceuticals and making specific adjustments to these current standards.	13.7
704	Determine the potential for organic beverage production and/or processing in Canada.	12.7
710	Identify trends and purchasing behavior for Canadian organic markets: imports and exports markets and the nature of organic food distribution channels.	12.4
711	Identify the best marketing strategies for Canadian organic products, impact of a Canada Organic Logo, and emerging "branded production or labeling systems."	12.4
709	Conduct a real-time analysis of organic market conditions that would include an analysis of products for import and export.	12.1
707	Analyze novel substances for compliance with organic standards and the potential to use these substances as productions inputs.	11.6
714	Analyze the export of organic goods from Canada in order to create strategies for the development of the organic sector in Canada.	11.5
703	Develop practices for extending the shelf life of organic produce and assess the associated trade-offs.	11.1
705	Determine the feasibility of developing a line of Canadian organic nutraceuticals and/or functional foods.	11.0
712	Determine the potential for all Canadian organic production to meet the minimum standards for a "fair trade" designation.	10.9

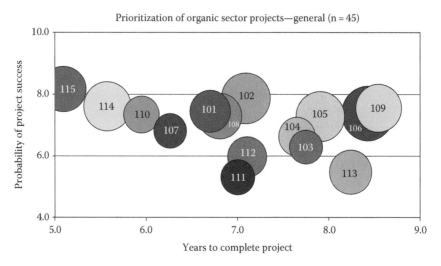

FIGURE 2.1 Research prioritization: the category of "General Research Questions."

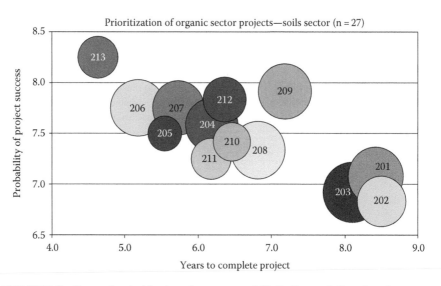

FIGURE 2.2 Research prioritization: the category of "Soils Research Questions."

The difference in the perceived level of impact among projects in the "Ecological Systems" category was relatively small. This category by its nature deals with a systems approach, but the priority here was identifying the long-term impact of organic systems on sustainability indicators.

In the "Sustainability" category, the top priority is related to the capacity for processing of Canadian organic products. This could stem from a concern that some organic products are becoming commoditized or that value adding is needed in order for Canadian products to be competitive in this market environment.

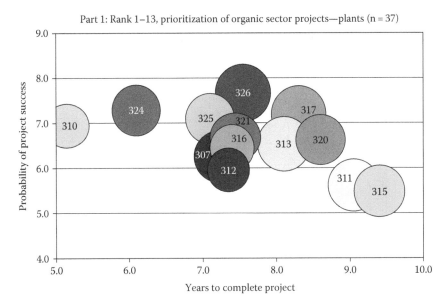

FIGURE 2.3   Research prioritization: the category of "Plants Research Questions, Part 1."

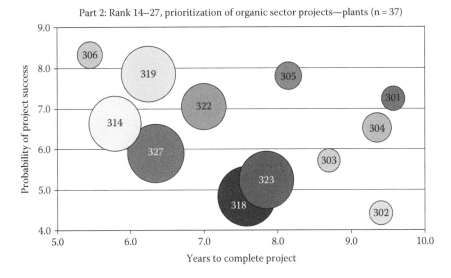

FIGURE 2.4   Research prioritization: the category of "Plants Research Questions, Part 2."

Under the "Health and Food" category, the highest ranked priority is related to studying risks associated with synthetic pesticide use. This is an issue of tremendous interest to the consumer. However, this type of research does not advance the science or development of organic agriculture. Instead, focusing on the effect of management practices on food quality and nutritional value will have great impact on consumer perception while supporting the principles of organic production.

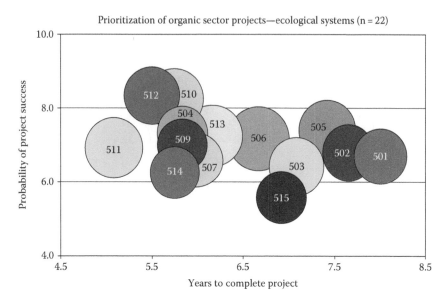

FIGURE 2.5 Research prioritization: the category of "Ecological Systems Research Questions."

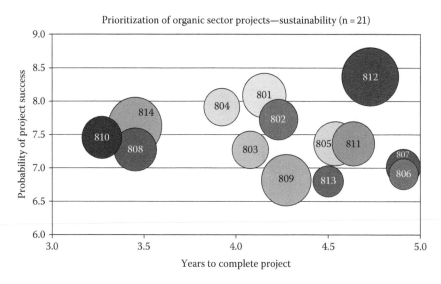

FIGURE 2.6 Research prioritization: the category of "Sustainability Research Questions."

Additional considerations in setting research priorities must include availability of suitable researchers, priorities of federal and provincial funding programs, and requirements for matching industry cash. Canada's Organic Science Cluster (http://oacc.info/OSC/osc_welcome.asp) began work in 2009 on 27 research activities in 9 subprojects with over 50 researchers receiving funding across Canada. Ultimately, the research did

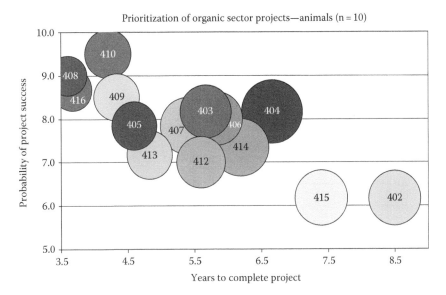

FIGURE 2.7 Research prioritization: the category of "Animals Research Questions."

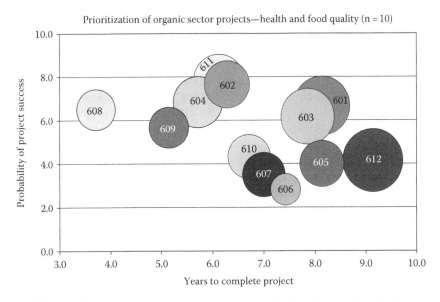

FIGURE 2.8 Research prioritization: the category of "Health and Quality Research Questions."

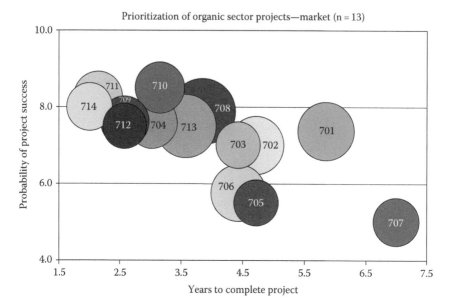

FIGURE 2.9   Research prioritization: the category of "Market Research Questions."

address many of the outcomes described earlier. Integrated research activities taking holistic approaches to management included the following: (1) studying phosphorus availability as related to the effect of management practices on mycorrhizae and phosphorus uptake by crops, (2) long-term cropping systems studies, and (3) an evaluation of dairy production systems to identify practices that optimize milk quality in conjunction with animal health, welfare, and productivity. Long-term breeding programs for wheat and oats were included, investing in cultivar development under low-input systems. Pest management is a key issue in organic agriculture. Activities relating to pest control included the following: (1) weed management under both grains and horticultural systems, (2) innovative new approaches to addressing insect and disease pressure in apples, and (3) a systematic approach to addressing parasite issues in sheep. Innovative research activities explored the feasibility of low-till organic management, development of organically acceptable preservatives for meat, and high-value greenhouse production systems designed to optimize energy use efficiency and waste reduction. Lastly, activities supporting the value-added benefits of organic production systems, specifically watershed scale modeling of the impacts of organic management and greenhouse gas emission quantification, were included. This highly diverse mix of research spanned the country and blended long-term understanding of ecological principles, with addressing short-term barriers and capturing opportunities through innovation.

2.8 CONCLUSIONS

The prioritization process was an effective means of providing useful information for strategic planning of research initiatives. Organic farmers and stakeholders clearly indicated that long-term research on integrated production systems was most likely to have the greatest impact. Integrated, systems-level research, however, is costly and can be difficult to tailor to the specific needs of a farmer. The process of ranking the many research questions against nine criteria was cumbersome and time consuming. Consolidation of these criteria into fewer categories would streamline the process but may also compromise the fair assessment of the multiple goals of organic agriculture. Lastly, carrying out this process in greater depth within subsectors on a regional basis would further focus the research into areas that will serve the needs of the stakeholders.

REFERENCES

Agriculture and Agri-Food Canada. 2008. Canada's organic industry at a glance. http://www4. agr.gc.ca/AAFC-AAC/display-afficher.do?id=1188227730017&lang=eng (accessed January 15, 2009).

Bagdley, C., J. Moghtader, E. Quintero, E. Zakem, M. Jahi Chappell, K. Avilés-Vázquez, A. Samulon, and I. Perfecto. 2007. Organic agriculture and the global food supply. *Renewable Agriculture and Food Systems* 22(2):86–108.

Canadian General Standards Board. 2006. Organic production systems general principles and management standards (last amended June 2011). Government of Canada CAN/CGSB-32.310-2006. Canadian General Standards Board, Gatineau, Quebec, Canada, 39pp.

Canadian Organic Trade Association. 2013. Canada's organic market: National highlights 2013. http://www.ota.com/pics/media_photos.171.img_filename.pdf (accessed June 3, 2013).

CBC News. 2007. Canadian consumers push up popularity of organic foods, survey finds. http://www.cbc.ca/news/story/2007/05/14/organic-food.html (accessed March 3, 2013).

Certified Organic Associations of British Columbia. 2002. British Columbia organic sector initiative strategic plan, 2002/03–2004/05. http://www.certifiedorganic.bc.ca/board/lib/exe/fetch.php?media=wiki:2001sp:2002-04_oganic_sector_inititative_strategic_plan_short.pdf (accessed March 3, 2013).

Insecticide Resistance Action Committee. 2013. Resistance definition. http://www.irac-online.org/about/resistance/ (accessed February 28, 2013).

International Federation of Organic Agriculture Movements. 2007. Nearly 31 million certified organic hectares worldwide. http://www.ifoam.org/webEdition/we_cmd.php?we_cmd%5B0%5D=show&we_cmd%5B1%5D=13925&we_cmd%5B4%5D=381 (accessed March 3, 2013).

Leeder, J. 2012. Supermarket adopts animal welfare standards for meat in Canada. *The Globe and Mail.* http://www.theglobeandmail.com/news/national/supermarket-adopts-animal-welfare-standards-for-meat-in-canada/article589686/ (accessed March 2, 2013).

Lynch, D. 2009. Environmental impacts of organic agriculture: A Canadian perspective. *Canadian Journal of Plant Science* 89:621–628.

MacRae, R., R. C. Martin, A. Macey, R. Beauchemin, and R. Christianson. 2002. A National Strategic Plan for the Canadian Organic Food and Farming Sector. http://www.organ icagcentre.ca/DOCs/reportfinal.pdf (accessed March3, 2013).

Organic Agriculture Centre of Canada. 2008. Final results of the first Canadian organic farmer survey of research needs. Nova Scotia Agricultural College, Truro, Nova Scotia, Canada. http://oacc.info/ResearchDatabase/res_strategies.asp (accessed June 1, 2013).

Reed Business Information. 2006. Australia: Organic moves for export. http://www.ferret. com.au/n/Organic-moves-for-export-n696289 (accessed January 14, 2008).

Strategic Vision Consulting Ltd. 2006a. Nitrogen strategy for Alberta. http://www.acidf.ca/index_htm_files/nitrogenan.pdf (accessed March 3, 2013).

Strategic Vision Consulting Ltd. 2006b. Nitrogen fixation—A strategy for enhancement. Alberta Pulse Growers, Saskatchewan Pulse Growers and Philom Bios/Novozymes. http://www.pulse.ab.ca/Portals/0/pdfs/N%20fixation%20-%20A%20strategy%20 for%20enhancement.pdf (accessed August 15, 2010).

Strategic Vision Consulting Ltd. 2007. Cereal competitiveness in Western Canada. A survey of cereal producers, livestock feeders and bioethanol manufacturers. http://www.acidf.ca/ index_htm_files/dsur.pdf (accessed March 3, 2013).

Strategic Vision Consulting Ltd. 2009a. An organic sector macroenvironmental scan. Prepared for the Organic Agriculture Centre of Canada, Dalhousie University, Truro, Nova Scotia, Canada, 68pp.

Strategic Vision Consulting Ltd. 2009b. Canadian organic group SWOT analysis. Prepared for the Organic Agriculture Centre of Canada, Dalhousie University, Truro, Nova Scotia, 25pp.

The Hartman Group. 2000. The organic consumer profile. Cited in MacRae, R. 2002. A national strategic plan for the Canadian organic food and farming sector. http://www. organicagcentre.ca/DOCs/reportfinal.pdf (accessed March 3, 2013).

Whole Foods Market IP. L.P. 2013. Animal welfare standards. http://www.wholefoodsmarket. com/about-our-products/quality-standards/animal-welfare-standards (accessed March 2, 2013).

Willer, H. and M. Yussefi. 2006. The world of organic agriculture—Statistics and emerging trends 2006. http://orgprints.org/5161/01/yussefi-2006-overview.pdf (accessed January 22, 2009).

Yiridoe, E. K., S. Bonti-Ankomah, and R. C. Martin. 2005. Comparison of consumer perceptions and preference toward organic versus conventionally produced foods: A review and update of the literature. *Renewable Agriculture and Food Systems* 20(4):193–205.

Section I

Soil Health

3 Arbuscular Mycorrhiza and the Phosphorus Nutrition of Organic Crops

Mulan Dai, Kim Schneider, Barbara Cade-Menun, Ronald Ferrera-Cerrato, and Chantal Hamel

CONTENTS

3.1   INTRODUCTION

Under the rules of organic certification, phosphorus (P) deficiency in crops can be an important limitation to production. Phosphorus is notoriously difficult to extract from the soil; however, many plants have evolved to form a symbiotic association with soil fungi to cope with this difficulty. This chapter considers the arbuscular mycorrhizal (AM) symbiosis as a tool to reduce the problem of P nutrition in organic production. Current knowledge of AM fungi, soil P pools, the impacts of organic production on the AM fungal resources of soils, and the contribution of AM fungi to organic crop production is reviewed. From this overview emerge avenues that may lead to improvement in the P nutrition of crops in organic production.

3.2 ARBUSCULAR MYCORRHIZAL FUNGI ARE MUTUALISTIC SOIL ORGANISMS ASSOCIATED WITH CROP PLANTS

Of the diverse microbial populations in the rhizosphere, AM fungi are especially important for plant adaptation, growth, and nutrition (Bago et al. 2000, 2001; Davies et al. 2002; Garg and Chandel 2010; Hodge et al. 2001; Johansson et al. 2004; Smith and Read 2008). The AM fungi belong to the phylum Glomeromycota, a group of obligate symbionts that form associations with plant roots (Schüβler et al. 2001). In this symbiosis, the root system provides simple carbon (C) compounds to support the metabolism of the fungi, allowing its proliferation and the completion of its life cycle (Bago et al. 2000, 2002; Lammers et al. 2001). In return, AM fungi absorb and transfer nutrients and water to the plants via their extraradical and intraradical mycelia (Bago et al. 1998; Drew et al. 2003; George et al. 1992; Smith and Read 2008). The AM fungi are abundant; their extraradical biomass may account for 5%–50% of the microbial biomass of agricultural soils (Olsson et al. 1999), where their survival and dispersion are mainly dependent on their connection with living plant root systems.

The AM fungi are important not only for their beneficial effects on plant growth and nutrition but also because of their role in plant evolution and adaptation to terrestrial ecosystems (Malloch et al. 1980; Taylor et al. 1995). According to fossil evidence, AM fungi evolved with the first land plants approximately 460 million years ago (Berbee and Taylor 1993; Brundrett 2002; Redecker et al. 2000). Today, the symbiosis is found in more than 80% of all known terrestrial plants (Smith and Read 2008).

The AM fungi improve plant uptake of macro- and micronutrients (Aguilera-Gomez et al. 1999; Higo et al. 2009). The role of AM fungi in plant uptake of P is particularly important. The extraradical hyphal networks of AM fungi take up and transport phosphate to plant roots, improving plant growth under P-limited conditions (Aguilera-Gomez et al. 1999). Synergism between AM fungi and P-solubilizing microorganisms may further increase the efficiency of soil P use by plants (Suri et al. 2011).

The AM symbiosis naturally improves the efficiency of plant production and AM fungi are seen as an important asset in organic production systems (Sousa et al. 2012); conceptually, organic production relies on efficient natural soil processes, and organically managed soils often have low fertility levels (Oberson and Frossard 2005). Therefore, AM-mediated improvements in the ability of crops to access soil P reserves may increase the productivity of organic systems.

3.3 INFLUENCE OF ORGANIC CROP PRODUCTION ON SOIL P FORMS AND POOLS

Organic farms have been identified as having low levels of plant-available P, as indicated by standard soil tests (Entz et al. 2001; Oberson and Frossard 2005; Welsh et al. 2009); this lack of P has raised concern about possible P deficiencies in organic crop production. With the use of soluble P fertilizers prohibited in organic production and the necessity for a greater focus on nutrient use efficiency, the management of soil P fertility presents a challenge in organic crop production. The role of soil microorganisms including AM fungi in promoting nutrient cycling and plant availability becomes paramount in these systems.

Understanding the forms and concentrations of P cycling in the soil, including the P in fertilizing materials, is essential for the design of sustainable organic cropping systems. Phosphorus exists in soil in various pools and forms. These P forms and pools differ in their availability to plants and microbes and in their reactivity in the environment (Condron et al. 2005; Pierzynski et al. 2005). Total soil P is the total concentration of all of the P in a soil (O'Halloran and Cade-Menun 2008). This total P can, in turn, be divided into organic P (bonded to C in some way) and inorganic P. Soil organic and inorganic P can be further divided into either specific P forms or into operationally defined pools.

Inorganic P is divided into orthophosphate, pyrophosphate, and polyphosphate (Condron et al. 2005; Pierzynski et al. 2005). Orthophosphate (HPO_4^{2-} or $H_2PO_4^{2-}$ at the pH of most soils) is the simplest P form and the only form that can be taken up by plants. All other P forms must be converted to orthophosphate for plant use. Polyphosphates are chains of orthophosphate ranging in length from two orthophosphate groups (pyrophosphate) to >100. Polyphosphates are P storage compounds for many soil microbes and can also be used as fertilizers.

Organic P is divided into orthophosphate monoesters, orthophosphate diesters, and phosphonates based on the bond of P to C (Condron et al. 2005). Orthophosphate monoesters have the general structure $ROPO_3^{2-}$ (where R is an organic moiety), with one orthophosphate per C group. This group of organic P compounds predominates in most soils and includes sugar phosphates (e.g., glucose-6-phosphate), phosphoproteins (most enzymes), mononucleotides, and inositol phosphates. Orthophosphate diesters ($R_1OROPO_2^-$, where R and R_1 are C moieties) have two C groups per orthophosphate. These include nucleic acids, phospholipids, and teichoic acid. Phosphonates differ from other organic P forms because they have a direct C–P bond (not an ester bond through O). These have the structure $[RP(O)(OH)_2]$ and include 2-aminoethyl phosphonic acid (AEP), antibiotics such as fosfomycin (produced by *Streptomyces*), and agrochemicals such as the herbicide glyphosate. Characterizing specific P forms requires either the extraction of individual forms (e.g., for phospholipids or RNA), advanced techniques such as ^{31}P nuclear magnetic resonance (NMR) spectroscopy (Cade-Menun 2005), or x-ray absorption near-edge structure (XANES) spectroscopy (e.g., Beauchemin et al. 2003; Kruse et al. 2010). There have been no published studies using these advanced techniques that investigate the effects of AM on soil P forms or P forms in organic agriculture.

Both organic and inorganic P forms exist both in the soil solution and associated with the mineral soil phase (Condron et al. 2005). These pools are shown in Figure 3.1. Inorganic P forms can be taken up by soil microbes, AM fungi, and plants and converted to organic P forms (immobilization) or stored as orthophosphate or polyphosphate. Organic P is converted back to orthophosphate by mineralization, which involves P-hydrolyzing enzymes (phosphatases). This mineralized P can either be taken up by plants and microbes from the soil solution or it can react with the soil minerals. Organic and inorganic P forms, which are predominantly anions, can be bound up ("sorbed") with cations at the soil mineral surface. Organic P forms vary in their sorption capacity; one form that has been shown to sorb strongly, thus

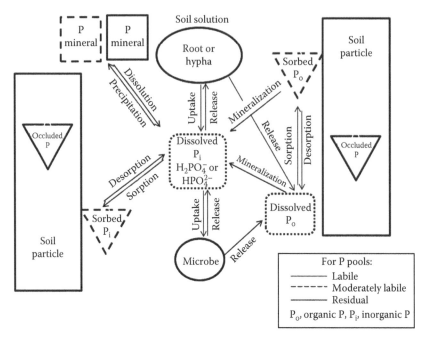

FIGURE 3.1   The cycling of pools of phosphorus (P) in the soil.

protecting it from mineralization, is phytate (*myo*-inositol hexakisphosphate), a common P storage compound in plants. Below pH 5, DNA can also sorb strongly to soil (Condron et al. 2005), which may limit its mineralization back to orthophosphate for uptake by plants and microbes. Sorbed P can be released back into the soil solution by desorption or by mineralization in the case of certain sorbed organic P forms. Orthophosphate, and, to a lesser extent, organic P forms, can also form secondary P minerals by precipitation with iron (Fe) or aluminum (Al) in low-pH soils and with calcium (Ca) in soils above pH 7 (Pierzynski et al. 2005). These minerals are released back into the soil solution by dissolution. Phosphorus that can be easily converted to orthophosphate for plant uptake is termed "labile P" or "loosely sorbed P." More tightly bound P, termed "residual P," is more difficult to convert to soluble orthophosphate. "Occluded P," found most often in highly weathered soils, is so tightly bound that it is considered completely unavailable for P cycling.

Studies using labeled P (^{32}P or ^{33}P) indicate that AM fungi take up P from the same soil sources as nonmycorrhizal plants (Richardson et al. 2011). Higher plants and soil microbes are known to produce phosphatases to mineralize organic P (Condron et al. 2005), as are ectomycorrhizal fungi (Smith and Read 2008). Although laboratory studies with artificial organic P forms have demonstrated the production of phosphatase by some AM fungi, it is unclear if this process occurs naturally in the soil (Richardson et al. 2011; Smith and Smith 2012). Higher plants and soil microbes can also release organic anions such as citrate in response to soil P deficiency (Richardson et al. 2011 and references therein). These anions can solubilize precipitated and/or sorbed orthophosphate and organic P. However, there have been no studies demonstrating the production of organic anions by AM fungi (Richardson et al. 2011). The primary benefit of AM fungi for P acquisition in organic farming systems is likely the ability of hyphae to access a larger volume of soil than roots alone for the uptake of orthophosphate (Smith and Read 2008). However, the mycorrhizal symbiosis may enhance the production of organic anions and phosphatases by the rhizosphere microbial community, in turn enhancing the availability of P for the mycorrhizal plant (Richardson et al. 2011; Smith and Smith 2012).

Operationally defining soil P involves measuring soil test P (STP, also called plant-available P) and the determination of soil P pools by fractionation (Tiessen and Moir 2008). STP is designed to predict the amount of P that the soil can provide for plant growth and thus determine the potential response of a crop to added P fertilizer (Beegle 2005). There are a number of different methods available, and their use depends on a number of factors, including soil pH, location, and preference of the soil test lab doing the analysis. Common methods are Bray P1, Mehlich 3, and Olsen P. All of these extraction procedures use colorimetric methods (e.g., Murphy and Riley 1962) to determine the concentration of reactive orthophosphate (the orthophosphate that reacts with the colored reagent in colorimetric analysis) in solution. Anion exchange resins can also be used to estimate STP.

With soil P fractionation, a soil sample is extracted with a series of reagents that increase in harshness to remove a series of soil P pools (Tiessen and Moir 2008). The original P fractionation methods, such as the Chang and Jackson method (Chang and Jackson 1957), were intended to separate inorganic P forms such as Ca phosphate and Fe phosphate. A more recently developed soil P fractionation method is the Hedley fractionation (Hedley et al. 1982). This method is designed to separate P into pools based on lability, with pools designated as "labile," "moderately labile," and "residual." The P immobilized by soil microbes is often estimated with a fumigation step in the Hedley fractionation. For all fractionation methods, total P in solution is measured after digestion. Many researchers will also measure reactive orthophosphate in solution colorimetrically, prior to digestion, and then describe the difference between digested and undigested samples as "organic P." However, caution should be taken when using this approach, both because complex inorganic P forms such as polyphosphates will be included in the "organic P" fraction and because it is more than likely that organic P will be altered with each step of a sequential fractionation (Tiessen and Moir 2008). Secondary techniques such as ^{31}P NMR spectroscopy are required to confirm the presence of organic P in any fraction, but this technique is rarely used.

Conventional farming relies on the use of chemically manufactured fertilizers, most of which are produced by converting rock phosphate to a more soluble salt form with heat and/or acid (Leikan and Achorn 2005). Rock phosphate is composed of apatite, a mineral containing orthophosphate bound to Ca; in its untreated state, this orthophosphate can dissolve only very slowly. In organic farming, external P inputs are more limited, and the use of minimally processed materials and recycled organic materials produced on farm is emphasized (Nelson and Janke 2007; Oberson and Frossard 2005; Rosen and Allan 2007). Inorganic materials used as P fertilizers in organic production include crushed but unprocessed rock phosphate and, to a lesser extent, guano and bone meal, all of which contain only orthophosphate, mainly associated with Ca. The main organic fertilizers used for organic production are manure, compost, and green manure (a crop grown and plowed back into the soil prior to seeding the main crop). Although these organic fertilizers have a high C content, analysis by ^{31}P-NMR and XANES has demonstrated that the P in these materials is predominantly orthophosphate and not organic P forms (e.g., Ajiboye et al. 2007, 2008; Cade-Menun 2011; Kruse et al. 2010; Toor et al. 2005). The exception is poultry manure, which has a very high phytate concentration because poultry cannot digest the phytate in their feed (Cade-Menun 2011 and references therein).

Although the inputs used in organic production contain orthophosphate, as do chemically manufactured fertilizers, this orthophosphate is not released as readily to the soil solution for plant uptake as is the orthophosphate in chemical fertilizers. The P in rock phosphate, for example, is most soluble in soils below pH 6 and is solubilized only if soils are sufficiently moist from rainfall or irrigation (Oberson and Frossard 2005). In organic farming, the amount of P released to the soil solution is controlled by the C:P ratio of the organic material (Alamgir et al. 2012; Leytem et al. 2005; Oberson and Frossard 2005), with more plant-available P released from organic material that is high in P relative to C. Microbial biomass and labile organic matter pools are often greater in organic than conventionally managed soils (Rosen and Allan 2007). For materials with a high C:P ratio, the microbial population can become a sink for the added P, as microbes immobilize the P for their own growth rather than releasing it to the soil solution. This immobilization may be overcome by applying an inorganic P source such as rock phosphate with an organic P source such as green manure (e.g., Arcand et al. 2010). Phytate added to soil in manure or plant residues may also be lost due to sorption.

There have been no published studies to date comparing specific P forms in soils under organic and conventional agricultural practices; studies have instead monitored changes in P pools. Most studies have shown a decrease in total P, in STP, and in labile P in organically managed systems (Lynch et al. 2012; Nelson and Janke 2007; Oberson and Frossard 2005; Welsh et al. 2009), mostly due to lower concentrations in the P sources added to organic systems and to the lower solubility of P in these materials. The pool of microbial biomass P is usually increased under organic production, as is the pool of residual P (Mäder et al. 2002; Nelson and Janke 2007; Oberson and Frossard 2005). The plant availability of this microbial P pool depends on microbial turnover, which is driven by temperature, moisture, nutrient availability, and competition with plants, food chain processes, and sorption of released P to mineral surfaces (Alamgir et al. 2012; Oberson and Frossard 2005). The changes in P pools vary with the specific management practices used by the grower in both conventional and organic systems (Lynch et al. 2012; Nelson and Janke 2007).

Other practices used in organic agriculture that would be expected to influence specific P forms include tillage and synthetic pesticide application (Nelson et al. 2010). Many conventional producers have shifted to zero till with herbicide application. In contrast, organic producers rely on tillage for weed control. Tillage has been shown to mix P forms in the soil, preventing the buildup of P at the soil surface that can occur under zero till as a result of the accumulation of decaying plant residue and P fertilizer inputs at the soil surface (Cade-Menun et al. 2010). Tilling distributes organic and inorganic P forms more uniformly through the root zone in the plow layer, which may increase uptake and sorption. In addition, many synthetic pesticides contain organophosphates, including the herbicide glyphosate and the insecticide malathion. The P forms contained in these pesticides and their breakdown products would be absent from long-term organic production systems. However, the added P from these products is expected to be very small relative to plant P needs and to fertilizer inputs.

3.4   EFFECT OF AM FUNGI ON P UPTAKE IN ORGANIC CROPS

Among the multiple roles played by AM fungi in organic systems, the improvement of plant P acquisition by AM hyphal networks is arguably the most important contribution of the AM symbiosis (Gosling et al. 2006). The contribution of AM fungi to plant P uptake in organic systems, however, has been examined in very few studies (Table 3.1). Isotopic techniques have shown that the extraradical hyphae of native AM fungi contributed 0.35%–9.5% of P uptake in winter wheat growing in a typical organic dairy farm cropping system, with lower contributions at higher soil P fertility levels (Schweiger and Jakobsen 1999). The contribution of AM fungi should be higher in mycorrhizal-dependent crops such as maize, clover, onion, strawberry, or soybean.

Some studies have shown that the AM fungi selected in organic systems may contribute better to host plant P uptake than those selected under conventional management. Inoculation with selected AM fungal strains in a conventional system resulted in 73% higher plant tissue P concentrations and substantial increases in plant productivity; no such effects of inoculation were seen in the organic system, indicating that the indigenous AM fungal community was already functioning well in the organic system, while the conventional system hosted an ineffective and insufficiently sized AM fungal community and needed remediation (Kahiluoto and Vestberg 1998). When evaluated with potted test plants, the AM fungi selected in organic systems promoted plant P uptake either better than or as well as the AM fungi selected in conventional systems, depending on the level of soil fertility and on the host plant involved (Scullion et al. 1998).

The AM fungi selected in organic systems are not always beneficial. In a recent study, a reduction in plant productivity proportional to AM fungal colonization was observed, and the negative impact on productivity was larger in organic systems than in conventional systems (Verbruggen et al. 2012b). Although findings from this experiment, which involved growing inoculated maize for 12 weeks in a very restricted soil volume, are very artificial and cannot be transposed to a field situation, they remind us that the overall impact of AM fungi on cropping systems depends on the balance between the various benefits provided by AM fungal networks to host plants and the cost of C invested by hosts to maintain these AM networks. For AM fungi to be beneficial to plant P uptake and growth, the soil P levels must be growth limiting but substantial enough to fulfill the requirements of both the plant and the fungus (Bethlenfalvay et al. 1983).

The AM fungi offer plants an important channel for P acquisition, but the provision of nutrients by AM fungi does not necessarily lead to better plant productivity (Verbruggen et al. 2012a). The AM fungi are only one component of a complex

TABLE 3.1

Influence of Organic Production System on the Contribution of AM Fungi to Crop P Uptake or Growth

Crop Species	Crop History or Rotation	Method of Study	AMF Contribution	Years in Organic Production	References
Winter wheat (*Triticum aestivum* L.)	A field in a typical dairy farm rotation of barley–white clover/grass–white clover/grass–red clover–ryegrass–winter wheat–fodder beet.	Experiment set up on a transect in one field, i.e., bricks of soil were placed on a transect; ^{32}P was used to trace plant P uptake; a fungicide was used to create controls.	AMF hyphal contribution = 0.35% to 9.5%, depending on soil extractible P level (negative correlation)	>11	Schweiger and Jakobsen (1999)
Allium ampeloprasum L.	Four grassland or grass–cereal rotations.	Spores were extracted from organically and conventionally treated soils and tested in the greenhouse.	Higher in organic (shoot P content = 1.00, 1.11, 1.53, and 4.66 times that in conventional systems)	>3	Scullion et al. (1998)
Leek (*Allium porrum* L.)	Last season was potato. Clover-rich leys were included in the rotation; Persian clover and perennial ryegrass were green manure crops.	Inoculation of soil in organic and conventional systems with *Glomus hoi* and *Glomus claroideum* concurrently with four levels of fertilization with apatite (0–640 kg/ha).	Higher in organic (no response to inoculation indicated good AMF function in organic; strong response in conventional showed impaired function)	5	Kahiluoto and Vestberg (1998)
Red clover (*Trifolium repens* L.)	Four grassland or grass–cereal rotations.	Spores were extracted from organically and conventionally treated soil and tested in the greenhouse.	Similar in organic and conventional (shoot P content = 1.10, 0.98, 1.05, and 1.07 times that in conventional)	>3	Scullion et al. (1998)
Maize (*Zea mays* L.)	Eight years of rotation including maize, wheat, grass, and grass–clover.	Soil from organic and conventional fields was used to inoculate a test plant.	Lower in organic (reduced plant growth)	>8	Verbruggen et al. (2012b)

soil system that should remain healthy. For example, the predation of AM fungi by arthropods or fungivorous nematodes can reduce the contribution of AM fungi to crop plants while increasing the C cost to the plant (Nayyar et al. 2009).

Organic systems with reduced input have lower productivity than intensive conventional systems but a higher efficiency of P use (Hildermann et al. 2010; Kahiluoto and Vestberg 1998). An important study conducted in Central Europe showed the possibility of sustaining crop production with less input in nutrient efficient organic systems (Mäder et al. 2002). After 21 years of organic production, the natural mechanisms of soil P fertility were highly effective, as shown by the very high activity of phosphatase and the high biomasses of mycorrhizal fungi and other soil biota in the organic systems. Grass–clover yields were similar under organic and mineral fertilizations, although organic P inputs were 38% lower than mineral inputs; in the organic system, potato and winter wheat yields were also temporally stable at levels 38% and 10% lower than those obtained under conventional management. These results support the theory that AM symbiosis and other natural mechanisms of soil fertility contribute to P use efficiency in organic production systems. They also show that the efficiency of organic systems can breakdown at some rotation stages; in potato, an important yield decline was attributed to potato late blight (Mäder et al. 2002).

Indirect evidence of plant reliance on AM fungi for P uptake in organic systems came from Baird (2010) who reported increased levels of AM fungal root colonization in plants grown at high plant density; the author concluded that this higher level of root colonization was the expression of a greater plant need for limited P under high competitive pressure. Equal P concentrations in plant tissues produced in soil from a conventional production system with high P fertility and in organically managed soil with low P fertility also supported the conclusion that the AM symbiosis functioned better in organic production systems (Welsh 2007).

The AM fungi are involved in plant P uptake but may also improve the efficiency of P cycling in the soil. The observation of lower P leaching losses in soil colonized by AM fungi from organic systems than in soil from conventional systems (Verbruggen et al. 2012b) highlights the role of AM fungi in agroecosystem sustainability and the positive influence of organic management on the functionality of these fungi.

3.5   HEALTH OF AM FUNGAL COMMUNITIES IN ORGANIC SYSTEMS

Studies that have characterized the health of the AM fungal communities in organic crop production indicate that organic production systems are generally more hospitable to AM fungi than conventional systems (Table 3.2). Higher levels of AM root colonization and species richness and more abundant AM spores and biomass are usually reported in organically managed systems than in conventional production systems (Gosling et al. 2010; Kahiluoto and Vestberg 1998; Oehl et al. 2003, 2004, 2010; Rasmann et al. 2009; Ryan and Ash 1999; Scullion et al. 1998; Verbruggen et al. 2010). Levels of AM fungal richness can be twice as high in organically managed fields as in conventionally managed fields (Verbruggen et al. 2010).

TABLE 3.2

Influence of Organic Crop Production on Mycorrhizal Fungi

Test Plant/Crop sp.	Crop History or Rotation	Methods	Root Colonization	AMF Spore Density	AMF Richness	Years in Organic Production	References
White clover (*Trifolium repens* L.)	Grazed pasture of white clover, perennial rye grass, and *Paspalum dilatatum* Poir.	Soil from biodynamic and conventional fields were fertilized with one of four levels of P or N and seeded with test plants in the greenhouse.	Higher in biodynamic fields; similarly influenced by fertilization in biodynamic and conventional.	—	—	17	Ryan and Ash (1999)
Perennial rye grass (*Lolium perenne* L.)	Grazed pasture of white clover, perennial rye grass, and *Paspalum dilatatum* Poir.	Soils from three pairs of fields were fertilized with one of four levels of P or N and seeded with test plants in the greenhouse.	Higher in biodynamic in 1/3 field pairs (in the low P soil); similarly influenced by fertilization in biodynamic and conventional.	—	—	17	Ryan and Ash (1999)
Clover (*Trifolium* spp.)	Various arable cereal-based, arable/horticulture, and horticulture systems.	Onions inoculated and not inoculated with farm soils were grown in the greenhouse.	Higher in organic.	Higher in organic	—	1–18	Gosling et al. (2010)

Allium ampeloprasum L.	Various organic grassland, conventional grassland/arable, and system conversion.	Spores from organic, conventional, and recently converted systems were tested in the greenhouse.	Higher in organic than conventional system, and the conversion systems were intermediate.	—	—	>9 3–9 for conversion systems	Scullion et al. (1998)
Leek (Allium porrum L.)	Last season was potato. Clover-rich leys were included in the rotation; Persian clover and perennial ryegrass were green manure crops.	Inoculated compared to noninoculated field crops grown in organic and conventional systems with four levels of apatite (0–640 kg/ha).	Higher in organic.	Higher in organic	—	5	Kahiluoto and Vestberg (1998)
18-month clover/grass meadow stand/trap plants: red clover (T. pratense), ryegrass (L. perenne L.), Plantago lanceolata L.	Biodynamic and organic rotations (7-year rotation) including perennial grass–clover meadow, potatoes, red beets, and several cereals fertilized with manure.	The AMF diversity in biodynamic and organic field management was compared to conventional rotations fertilized minerally, minerally/organically, and nonfertilized; the root colonization ability of the AMF in these systems was compared on a mixture of trap plants.	Higher in biodynamic.	Higher in organic	Higher in organic	22	Oehl et al. (2003)

(continued)

TABLE 3.2 (continued)
Influence of Organic Crop Production on Mycorrhizal Fungi

Test Plant/Crop sp.	Crop History or Rotation	Methods	Root Colonization	AMF Spore Density	AMF Richness	Years in Organic Production	References
Maize (Z. mays L.)	8 years of crop rotation with maize, wheat, grass, and grass–clover.	Inoculation with three maize-growing organic and conventional field soils.	Marginally higher in organic.	Higher in organic	Higher in organic	≥8	Verbruggen et al. (2010, 2012a,b)
Potato (Solanum tuberosum L.)	—	Pairs of organic and conventional fields growing potato were compared.	Similar in organic and conventional.	—	Similar in organic and conventional	—	Verbruggen et al. (2010)
Tomato (Lycopersicon esculentum Mill.)	Tomato-based rotation including sunn hemp and Japanese millet.	Organic field management was compared to the removal of vegetation with tillage or fumigant and with soil cover of Bahia grass or undisturbed weeds.	Lower in organic.	—	—	4	Rasmann et al. (2009)
Red clover (Trifolium pratense), ryegrass (L. perenne L.), Plantago lanceolata L.	7-year rotation including perennial grass–clover meadow, potatoes, red beets, and several cereals, fertilized with manure.	The AMF diversity in organic field management was compared to that of natural and managed grasslands, conventional rotations, and conventional maize monocrop systems; the root colonization ability of the AMF from these soils was compared on a mixture of trap plants.	Lower than grasslands, similar to monocropping.	Lower than grasslands, higher than monocultures	Lower or equal to grasslands	21	Oehl et al. (2010)

A number of reasons have been proposed to explain the greater AM activity in organic systems than in conventional systems. Land-use intensity is a major driver of AM fungal community composition (Oehl et al. 2003). Crop diversification is one of the factors leading to diverse AM fungal communities (Jansa et al. 2006). Organic cropping systems are typically highly diversified and normally include perennial crop plants such as clover or alfalfa, often mixed with grasses. The AM fungi are obligate biotrophs and benefit greatly from a perennial plant cover. Crop diversity creates a temporally and spatially heterogenous soil environment and multiple niches favoring soil microbial diversity. The AM fungal communities are different in organic and conventional systems, and, while several generalist species can be found everywhere, other species are specialized. For example, species of *Acaulospora* and *Scutellospora* were only associated with organic systems (Oehl et al. 2003 and references therein).

Low soil fertility is another component of organic systems that promotes AM fungi proliferation. The high capacity of organic systems to support AM fungi is usually associated with low soil fertility levels (Hildermann et al. 2010; Mäder et al. 2002; Ryan and Ash 1999; Scullion et al. 1998; Verbruggen et al. 2012a). With the limited choice and elevated price of the fertilizing materials allowed by organic certification and the guiding principle of reducing inputs, organically managed soils are normally minimally fertilized. This reduces the capacity of soils to supply macronutrients (Hildermann et al. 2010) and stimulates the development of AM fungi. The development of the AM symbiosis in plants is regulated by the difference between soil nutrient supply rate and plant demand for nutrients (especially for P) (Johnson 2010). The prohibition of concentrated fertilizers is another explanation for the relative abundance of AM fungi in organic systems. Chemical fertilizers directly affect extraradical AM hyphae, compounding the negative effect of high soil fertility on these fungi (Gosling et al. 2010; Gryndler et al. 2006).

The abundance of AM fungi in roots and in the soil is often but not always higher in organic systems than in conventional systems (Oehl et al. 2010). In organic systems, the supply of unexpectedly high amounts of P to crop plants may feedback negatively on AM symbiosis formation (Rasmann et al. 2009).

The dynamics of AM fungi in organic production systems remain poorly understood. The observation of both higher soil P levels and a greater abundance of AM fungal spores in organically managed soils than in corresponding conventionally managed soils (Kahiluoto and Vestberg 1998) indicates that other factors may be involved in the stimulation of AM fungal proliferation in organic systems. The presence of different AM fungal communities in different soils and production systems (Oehl et al. 2003) could explain this discrepancy.

3.6 LIMITATIONS AND POTENTIAL OF MYCORRHIZA IN ORGANIC CROP P NUTRITION

AM fungi are seen as an important component of organic agroecosystems because the AM symbiosis contributes to efficient P cycling (Verbruggen et al. 2012b). Furthermore, these fungi are regulated by soil P fertility; as such, the typically low levels of available P in soils under organic management are conducive to mycorrhizal development. However, the development and contribution of AM fungi to crop production is highly variable (Ryan and Kirkegaard 2012); while it is true that AM fungi are important channels for P acquisition by most plants, evidence shows that the provision of nutrients by AM fungi does not necessarily translate into better yields (Verbruggen et al. 2012b). We can trust that natural selection favors effective systems, but even organic agroecosystems are not natural and their AM fungal community could be ineffective. Organic systems use disrupting cropping practices such as tillage, annual cropping, and monoculture, all of which can reduce AM fungi diversity (Kabir 2005; Oehl et al. 2003). Even organic fertilization modifies soil fertility, which may influence AM fungal community composition and function.

In addition to considering the influence of cropping practices that may modify and impoverish the AM fungal resources of organic fields, we must consider how a mycorrhiza functions. The AM symbiosis is mutualistic. The AM fungi provide nutrients to host plants and improve soil health, with these services fueled by products of photosynthesis diverted from yield components. In some conditions, the C cost to the host plant may exceed the value of the services provided by AM fungi. This possibility most likely exists in systems where P availability is not limiting and the additional P access is not beneficial, where P availability is too low and the AM fungi may actually compete with the plant for scarce P reserves, or where the function of the extraradical AM fungal mycelia is impaired, for example, by fragmentation or predation (Johnson 2010). Furthermore, mycorrhiza formation may reduce plant P uptake by blocking the direct P uptake pathway of roots (Smith et al. 2011). The mycorrhizal colonization of root systems can partially or completely inactivate the pathway for direct uptake of P by roots. If the AM pathway for P uptake is less effective than the direct pathway for root P uptake, then the AM symbiosis could reduce P uptake by crop plants (Smith et al. 2011). Soil tillage, a necessary practice in organic production, disrupts the continuity of extraradical AM networks and compromises the function of the AM hyphal pathway (Kabir 2005). To define the best production strategies and maximize yields, one should be aware that mycorrhiza formation in organic fields has the potential to limit P uptake by crops in addition to being a C drain.

Organically managed fields generally possess more abundant AM fungal propagules and species richness than conventional systems (Table 3.2); consequently, AM fungi management has more potential for improving organic production than conventional production. The levels of AM root colonization of crop plants and of AM fungal diversity are typically higher in organically managed agroecosystems than conventionally managed agroecosystems (Gosling et al. 2010; Oehl et al. 2003, 2010; Rasmann et al. 2009; Scullion et al. 1998; Verbruggen et al. 2010). The restricted list of approved nutrient sources for organic production makes it difficult to maintain soil fertility on some organic farms. In general, off-farm certified nutrient sources are bulky and costly (García et al. 2012), and profitability may often be higher when less productive organic crops are produced at lower P fertility levels than conventional systems, explaining why soil P fertility tends to be low in many organic systems (Figure 3.2).

Phosphorus is not always a limiting factor in organic systems (García et al. 2012). Organic crop production systems can be highly intensive or low input to the point of being unsustainable. The intensive greenhouse production of high-value organic

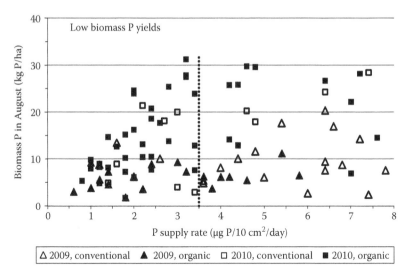

FIGURE 3.2   Relationship between P yield at maturity and P supply rate for sites with supply rates <8 µg P/10 cm²/day, as measured by PRS™ probes (Western Ag Innovations Inc., Saskatoon, Saskatchewan, Canada; Qian and Schoenau 2002). These data were obtained in a survey of 172 commercial fields seeded with wheat in the Canadian Prairie. Fifty-seven percent of the soils under organic production had a P supply rate under the threshold (3.5 µg P/10 cm²/day check soils and crops), below which P availability likely limits wheat P uptake according to segmented regression analysis. Lower biomass P accumulation occurred in the dry year of 2009 than in 2010, a wet year, indicating that the negative impact of low soil P availability is worsened in conditions of insufficient soil moisture. (Data from Dai, M. et al., Estimating the AM fungal resources of wheat fields, *Proceedings of Soils and Crops*, University of Saskatoon, Saskatoon, Saskatchewan, Canada, 2012, CD-ROM. With permission.)

fruits and vegetables close to markets and to the sites where nutrient-rich by-products are generated can be highly profitable despite a reliance on the heavy use of costly organic fertilizers (García et al. 2012). The potential contribution of AM fungi to this type of organic production might be limited by high soil P availability, although organic sources of P often have a milder impact on mycorrhiza formation and AM fungi proliferation than highly soluble mineral fertilizers (Cardarelli et al. 2010; Gryndler et al. 2006; Linderman and Davis 2004).

Animal production on mixed farms reduces the dependence of organic systems on off-farm nutrient inputs (Scialabba and Müller-Lindenlauf 2010). Soil fertility can be maintained through the use of very diversified rotations, including a high proportion of perennial cover of mixed hay or pasture, and through green manuring (Mäder et al. 2002). The roots of perennial forage cover improve soil fertility levels, especially when they contain nitrogen-fixing legumes (Havlin et al. 1999), and a large part of the nutrients exported in forage yield can be returned to the soil in the form of manure. The potential for both the contribution of AM fungi to crop production and the management of soil fertility components (including the AM fungal community) is certainly highest in cropping systems with animal husbandry.

At the other end of the spectrum, dry land farming produces a large proportion of the overall organic grains and pulses on the market. Dry land farms are often far from markets and from the source of the fertilizing material allowed by organic certification. The fertility management of soil in dry areas is often limited to green manure plow-down, most often combined with a semi-fallow,* to raise soil moisture content to levels conducive to the mineralization of the green manure (Campbell et al. 2008). Manure is largely unavailable in free-ranging animal production systems that prevail in dry regions. The productivity of these organic fields is typically low since it is limited by precipitation, or P availability following years of nutrient exports in grain yield, in the absence of fertilization. The sustainability of these organic dry land cropping systems is questionable. A survey conducted over two years on 172 organic and conventional wheat fields located throughout the Canadian Prairie provinces (Dai et al. 2012) revealed that, in good years, P availability is likely limiting in 57% of the organic fields surveyed and P limitation worsened in dry years (Figure 3.2) because water is required to mobilize P (Oberson and Frossard 2005). The high variation in wheat P uptake in fields with limiting soil P supply rates (Figure 3.2) suggests high variation in the functionality of the AM fungal communities in cultivated soils. The examination of highly P efficient organic systems shown in Figure 3.2 or elsewhere may provide avenues to improve AM efficacy in dry land farming.

3.7 IMPROVING THE CONTRIBUTION OF AM FUNGI TO ORGANIC CROP P NUTRITION AND PRODUCTIVITY

Several agronomic practices produce negative effects on AM fungi (Gosling et al. 2006; Jansa et al. 2006); although the management of production fields can be modified, practices harmful to AM fungi such as tillage or the cyclical destruction of plant cover cannot be avoided in organic crop production. The numerous reports on the impact of crop production on AM fungal communities naturally occurring in production fields (Ryan and Kirkegaard 2012) point to the inoculation of crops with selected AM fungi as a way to maintain the functionality of the AM symbiosis and to increase the contribution of AM fungi to crop P nutrition and productivity. Organic systems are often more hospitable to AM fungi than conventional production systems. Highly effective AM fungal strains introduced through inoculation can become established, positively influence root colonization levels and plant productivity, and persist in the soil until the following growing season, as was shown using molecular probes (Pellegrino et al. 2012). Highly effective AM fungal strains introduced by the inoculation of crops may be manageable in organic systems. Results of agronomic research on the development of inoculation technologies conducted in Cuba indicate that crop rotation systems involving pastures may favor the persistence of introduced superior strains of AM fungi (Ramón Rivera, personal communication).

The factors determining the receptivity of soil to AM fungi inoculants were recently reviewed (Verbruggen et al. 2012a,b). Three main factors determine the

---

* In dry areas, a green manure crop is plowed down in midsummer and the soil is left bare until the following spring. The absence of plant cover results in the accumulation in soil of the rain water received in the second part of the season.

success of the inoculation of crops with selected AM fungi: (1) the compatibility between the agroecosystem and the AM fungi introduced, (2) the capacity of the system to support AM fungi, and (3) the effect of prior occupation of roots. Evidence has shown that prior occupation of roots inhibits later infection by other AM fungi (Vierheilig et al. 2000). Thus, the establishment of highly effective AM fungi may be prevented by prior entry into roots of naturally occurring AM fungal strains. Conversely, a rapidly growing highly effective AM fungal strain triggering a strong priority effect would have good chances of becoming established despite the typically low densities of AM fungal propagules in inoculants.* In this light, the introduction of highly effective AM fungi in organic field soil through the pre-inoculation of transplants appears to be an ideal strategy (Mummey et al. 2009). In addition, the placement of inoculants in close proximity to emerging roots was shown to be associated with successful inoculation (Oliveira and Sanders 2000). A green manure plow-down or growing a nonhost crop prior to introducing a highly effective AM fungi may reduce the competition from ineffective but adapted AM fungi occupying cultivated fields, as proposed by Gosling et al. (2006). Growing host or nonhost plants prior to the AM inoculation of the AM-dependent crop strawberry in a soil with very high levels of available P did not make a difference in plant response (Stewart et al. 2005). It appears that the success of inoculation depends on several factors and is difficult to predict (Gosling et al. 2006).

Inoculating a crop species or genotype highly compatible with highly effective AM fungal strains was shown to improve the likelihood of inoculation success (Martín et al. 2009). A positive feedback loop apparently strengthens the collaboration between a host plant and a highly effective AM fungal strain. It was recently shown that plants selectively reward the AM fungi making the largest contribution to their mineral nutrition with C, thus encouraging the proliferation of these highly effective AM fungi and the persistence of highly effective AM symbioses (Kiers et al. 2011). Whether or not this mechanism could create a barrier to the introduction of highly effective AM fungi through inoculation is unknown, but plant seedlings may preferentially engage in exchanges with functional hyphal networks already present in the soil rather than investing in the construction of a functional hyphal network from the dispersed AM germlings of an introduced isolate. A positive feedback loop established in a highly effective host plant–AM fungus pair would contribute to the persistence of highly effective AM associations in perennial crops; it could also increase the abundance of highly effective AM fungal strains in an inoculated crop, with beneficial carryover effects on the crop of the following phase of the cropping system (Ramón Rivera, personal communication). Using the power of plants to manage the AM symbiosis in agroecosystems is an interesting avenue that is being explored. The selection of crop genotypes highly compatible with highly effective AM fungi is underway in certain crop breeding programs (Singh et al. 2012). However, it will be important in the development of AM inoculants and inoculation

---

* Recommended rates of AM propagule application are low. For example, Myke Pro (Premier Tech Biotechnologies, Rivière-du-Loup, Qc., Canada) contains 3200 spores/g; at the recommended rate of application, Myke Pro provides 240 spores/m^2 to the soil. ironROOTS (Novozymes Biologicals, Inc., Salem, VA) contains 67 AM fungal spores/g; at the recommended rate of application, 15 AM fungal spores are applied per m^2.

strategies to keep in mind the role of the soil environment in modifying the performance of AM fungi (Herrera-Peraza et al. 2011). In order to function effectively, AM fungi should be adapted to the soil environment where they are introduced.

Alternatively, the management of AM fungal diversity in cultivated fields could be attempted by increasing the genetic diversity of crops through the use of strategies such as intercropping* and the mixing of cultivars. Monoculture negatively affects AM fungal diversity (Oehl et al. 2003). Intercropping favors AM fungal diversity in agroecosystems (Bainard et al. 2011), but this practice may be limited by constraints imposed by incompatible agronomic usages for different crops. The use of mixtures of cultivars appears to be a simple way to maintain AM fungal diversity in agroecosystems. Different genotypes of the same crop species differentially influence root and rhizosphere inhabitants (Dunfield and Germida 2001), including AM fungi (Hannula et al. 2012). The value of cultivar mixtures has long been recognized as a simple way to reduce disease incidence through the increased diversity of disease-resistance genes (Kiær et al. 2009). By extension, mixtures of cultivars could create diversified niches for AM fungi that have host preferences (Pivato et al. 2007) and differential functionalities in association with different hosts (Burleigh et al. 2002). However, the value of cultivar mixtures in the management of AM fungi remains to be demonstrated.

Progress toward the management of AM fungi in agriculture has been slow in many countries. Interest in the AM symbiosis has been marginal and investments in agricultural sciences are influenced by market forces. Cuba and India led the development of agronomic practices enhancing the contribution of AM fungi in high- and low-input production systems (Hamel and Plenchette 2007; Herrera-Peraza et al. 2011). Recent advances in sequencing and other molecular techniques now allow the analysis of naturally occurring AM fungal communities in cultivated fields (Dai et al. 2012a, 2012b, 2013) and the monitoring of the AM fungal strains introduced in agroecosystems through inoculation (Kiers et al. 2011; Pellegrino et al. 2012). New technologies for research, increasing fertilizer prices, growing demand for biofuels and food products, and growing concern for food safety now create opportunities for the development of AM fungi–based technologies and agronomic practices relevant to organic production. Important advances in AM inoculation technology and in the management of AM fungi in agroecosystems are expected in the coming years.

3.8 CONCLUSION

Phosphorus often limits the productivity of crops under the rules of organic certification. AM fungi cannot import P into the agroecosystem but they can improve the P efficiency of crop production. Crop production affects AM fungal communities in cultivated soils; however, this effect is less pronounced in organic production systems, which are rooted in ecological principles and typically function at low levels of available P. Maximizing plant diversity and soil cover and minimizing soil disturbance and the frequency of nonhost plants in crop rotation are general recommendations for the preservation of diverse AM fungal communities in production

---

* Simultaneous production of different crop species on the same land.

fields. The introduction of highly effective AM fungal strains may also be beneficial. Organic production has attracted relatively little research attention and the development of agronomic strategies for the management of AM fungi in organic systems has been lagging. Increasing environmental concerns and the reprioritization of human societies in the face of global climate change have triggered interest in AM symbiosis and stimulated research and development relevant to organic production. Organic growers are well positioned to successfully implement agronomic strategies to increase the P use efficiency of crops through the management of AM symbioses, and exciting developments in organic production are to be expected in the coming years.

REFERENCES

Aguilera-Gomez, L., F. T. Davies Jr., V. Olalde-Portugal, S. Duray, and L. Phavaphutanon. 1999. Influence of phosphorus and endomycorrhiza (*Glomus intraradices*) on gas exchange and plant growth of chile ancho pepper (*Capsicum annuum* L. cv. San Luis). *Photosynthetica* 36: 441–449.

Ajiboye, B., O. O. Akinremi, Y. Hu, and D. N. Flaten. 2007. Phosphorus speciation of sequential extracts of organic amendments using nuclear magnetic resonance and X-ray absorption near-edge structure spectroscopies. *J. Environ. Qual.* 36: 1563–1576.

Ajiboye, B., O. O. Akinremi, Y. Hu, and A. Jürgensen. 2008. XANES speciation of phosphorus in organically amended and fertilized vertisol and mollisol. *Soil Sci. Soc. Am. J.* 72: 1256–1262.

Alamgir, M., A. McNeil, C. Tang, and P. Marschner. 2012. Changes in soil P pools during legume residue decomposition. *Soil Biol. Biochem.* 49: 70–77.

Arcand, M. M., D. H. Lynch, R. P. Voroney, and P. van Straaten. 2010. Residues from a buckwheat (*Fagopyrum esculentum*) green manure crop grown with phosphate rock influence bioavailability of soil phosphorus. *Can. J. Soil Sci.* 90: 257–266.

Bago, B., C. Azcón-Aguilar, A. Goulet, and Y. Piche. 1998. Branched absorbing structures (BAS): A feature of the extraradical mycelium of symbiotic arbuscular mycorrhizal fungi. *New Phytol.* 139: 375–388.

Bago, B., P. E. Pfeffer, and Y. Schachar-Hill. 2001. Could the urea cycle be translocating nitrogen in the arbuscular mycorrhizal symbiosis? *New Phytol.* 149: 4–8.

Bago, B., Y. Schachar-Hill, and P. E. Pfeffer. 2000. Carbon metabolism and transport in arbuscular mycorrhizas. *Plant Physiol.* 124: 949–957.

Bago, B., W. Zipfel, R. M. Williams, J. Jun, R. Arreola, P. J. Lammers, P. E. Pfeffer, and Y. Schachar-Hill. 2002. Translocation and utilization of fungal storage lipid in the arbuscular mycorrhizal symbiosis. *Plant Physiol.* 128: 108–124.

Bainard, L. D., J. N. Klironomos, and A. M. Gordon. 2011. Arbuscular mycorrhizal fungi in tree-based intercropping systems: A review of their abundance and diversity. *Pedobiologia* 54: 57–61.

Baird, J. M. 2010. Arbuscular mycorrhizal fungi colonization and phosphorus nutrition in organic field pea and lentil. *Mycorrhiza* 20: 541–549.

Beauchemin, S., D. Hesterberg, J. Chou, M. Beauchemin, R. R. Simard, and D. L. Sayers. 2003. Speciation of phosphorus in phosphorus-enriched agricultural soils using X-ray adsorption near-edge structure spectroscopy and chemical fractionation. *J. Environ. Qual.* 32: 1809–1819.

Beegle, D. 2005. Assessing soil phosphorus for crop production by soil testing. In *Phosphorus, Agriculture and the Environment*, eds. J. T. Sims and A. N. Sharpley, pp. 123–143. Madison, WI: Soil Science Society of America.

Berbee, M. L. and J. W. Taylor. 1993. Dating the evolutionary radiations of the true fungi. *Can. J. Bot.* 71: 1114–1127.

Bethlenfalvay, G. J., H. G. Bayne, and R. S. Pacovsky. 1983. Parasitic and mutualistic associations between a mycorrhizal fungus and soybean: The effect of phosphorus on host plant-endophyte interactions. *Physiol. Planta.* 57: 543–548.

Brundrett, M. C. 2002. Coevolution of roots and mycorrhizas of land plants. *New Phytol.* 154: 275–304.

Burleigh, S. H., T. Cavagnaro, and I. Jakobsen. 2002. Functional diversity of arbuscular mycorrhizas extends to the expression of plant genes involved in P nutrition. *J. Exp. Bot.* 53: 1593–1601.

Cade-Menun, B. J. 2005. Characterizing phosphorus in environmental and agricultural samples by ^{31}P nuclear magnetic resonance spectroscopy. *Talanta* 66: 359–371.

Cade-Menun, B. J. 2011. Characterizing phosphorus in animal waste with solution ^{31}P NMR spectroscopy. In *Environmental Chemistry of Animal Manure*, ed. Z. He, pp. 275–299. New York: Nova Science Publishers, Inc.

Cade-Menun, B. J., M. R. Carter, D. C. James, and C. W. Liu. 2010. Phosphorus forms and chemistry in the soil profile under long-term conservation tillage: A phosphorus-31 nuclear magnetic resonance study. *J. Environ. Qual.* 39: 1647–1656.

Campbell, C. A., R. P. Zentner, P. Basnyat, R. DeJong, R. Lemke, R. Desjardins, and M. Reiter. 2008. Nitrogen mineralization under summer fallow and continuous wheat in the semi-arid Canadian prairie. *Can. J. Soil Sci.* 88: 681–696.

Cardarelli, M., Y. Rouphael, E. Rea, and G. Colla. 2010. Mitigation of alkaline stress by arbuscular mycorrhiza in zucchini plants grown under mineral and organic fertilization. *J. Plant Nutr. Soil Sci.* 173: 778–787.

Chang, S. C. and M. L. Jackson. 1957. Fractionation of soil phosphorus. *Soil Sci.* 84: 133–144.

Condron, L. M., B. L. Turner, and B. J. Cade-Menun. 2005. Chemistry and dynamics of soil organic phosphorus. In *Phosphorus, Agriculture and the Environment*, eds. J. T. Sims and A. N. Sharpley, pp. 87–121. Madison, WI: Soil Science Society of America.

Dai, M., M. Sheng, E. Bremer, Y. He, H. Wang, and C. Hamel. 2012a. Estimating the AM fungal resources of wheat fields. In *Proceedings of Soils and Crops*, University of Saskatoon, Saskatoon, Saskatchewan, Canada, CD-ROM.

Dai, M., Hamel, C., St. Arnaud, M., He, Y., Grant, C.A., Lupwayi, N. Z., Janzen, H. H., Malhi, S. S., Yang, X., and Zhou, Z. 2012b. Arbuscular mycorrhizal fungi assemblages in Chernozem great groups revealed by massively parallel pyrosequencing. *Can. J. Microbiol.* 58: 81–92.

Dai, M., Bainard, L. D., Hamel, C., Gan, Y., Lynch, D. 2013. Impact of land use on arbuscular mycorrhizal fungal communities in rural Canada. *Appl. Environ. Microbiol.* doi: 10.1128/AEM.01333-13.

Davies, F. T. Jr., V. Olalde-Portugal, L. Aguilera-Gomez, M. J. Alvarado, R. Ferrera-Cerrato, and T. W. Boutton. 2002. Alleviation of drought stress of chile ancho pepper (*Capsicum annuum* L. cv. San Luis) with arbuscular mycorrhiza indigenous to Mexico. *Sci. Hortic.* 92: 347–359.

Drew, E. A., R. S. Murray, S. E. Smith, and I. Jakobsen. 2003. Beyond the rhizosphere: Growth and function of arbuscular mycorrhizal external hyphae in sands of varying pore sizes. *Plant Soil* 251: 105–114.

Dunfield, K. E. and J. J. Germida. 2001. Diversity of bacterial communities in the rhizosphere and root interior of field-grown genetically modified *Brassica napus*. *FEMS Microbiol. Ecol.* 38: 1–9.

Entz, M. H., R. Guilford, and R. Gulden. 2001. Crop yield and soil nutrient status on 14 organic farms in the eastern portion of the northern Great Plains. *Can. J. Plant Sci.* 81: 351–354.

García, M. C., A. Belmonte, F. Pascual, T. García, A. Simón, M. L. Segura, G. Martín, D. Janssen, and I. M. Cuadrado. 2012. Economic evaluation of cucumber and french bean production: Comparing integrated and organic crop production management. *Acta Hortic.* 930: 115–120.

Garg, N. and S. Chandel. 2010. Arbuscular mycorrhizal networks: Process and functions. A review. *Agron. Sustain. Dev.* 30: 581–599.

George, E., K.-U. Häussler, D. Vetterlein, E. Gorgus, and H. Marschner. 1992. Water and nutrient translocation by hyphae of *Glomus mosseae*. *Can. J. Bot.* 70: 2130–2137.

Gosling, P., A. Hodge, G. Goodlass, and G. D. Bending. 2006. Arbuscular mycorrhizal fungi and organic farming. *Agric. Ecosyst. Environ.* 113: 17–35.

Gosling, P., A. Ozaki, J. Jones, M. Turner, F. Rayns, and G. D. Bending. 2010. Organic management of tilled agricultural soils results in a rapid increase in colonisation potential and spore populations of arbuscular mycorrhizal fungi. *Agric. Ecosyst. Environ.* 139: 273–279.

Gryndler, M., J. Larsen, H. Hrselová, V. Rezácová, H. Gryndlerová, and J. Kubát. 2006. Organic and mineral fertilization, respectively, increase and decrease the development of external mycelium of arbuscular mycorrhizal fungi in a long-term field experiment. *Mycorrhiza* 16: 159–166.

Hamel, C. and C. Plenchette, eds. 2007. *Mycorrhizae in Crop Production*. Binghampton, NY: Haworth Press.

Hannula, S. E., H. T. S. Boschker, W. de Boer, and J. A. van Veen. 2012. ^{13}C pulse-labeling assessment of the community structure of active fungi in the rhizosphere of a genetically starch-modified potato (*Solanum tuberosum*) cultivar and its parental isoline. *New Phytol.* 194: 784–799.

Havlin, J. L., J. D. Beaton, S. L. Tisdale, and W. L. Nelson. 1999. *Soil Fertility and Fertilizers: An Introduction to Nutrient Management*, 6th edn., 499pp. Upper Saddle River, NJ: Prentice Hall.

Hedley, M. J., J. W. B. Stewart, and B. S. Chauhan. 1982. Changes in inorganic and organic soil phosphorus fractions induced by cultivation practices and by laboratory incubations. *Soil Sci. Soc. Am. J.* 46: 970–976.

Herrera-Peraza, R. A., C. Hamel, F. Fernández, R. L. Ferrer, and E. Furrazola. 2011. Soil-strain compatibility: The key to effective use of arbuscular mycorrhizal inoculants? *Mycorrhiza* 21: 183–193.

Higo, M., K. Isobe, D.-J. Kang, K. Ujiie, R. A. Drijber, and R. Ishii. 2009. Inoculation with arbuscular mycorrhizal fungi or crop rotation with mycorrhizal plants improves the growth of maize in limed acid sulfate soil. *Plant Prod. Sci.* 13: 74–79.

Hildermann, I., M. Messmer, D. Dubois, T. Boller, A. Wiemken, and P. Mäder. 2010. Nutrient use efficiency and arbuscular mycorrhizal root colonisation of winter wheat cultivars in different farming systems of the DOK long-term trial. *J. Sci. Food Agric.* 90: 2027–2038.

Hodge, A., C. D. Campbell, and A. H. Fitter. 2001. An arbuscular mycorrhizal fungus accelerates decomposition and acquires nitrogen directly from organic material. *Nature* 413: 297–299.

Jansa, J., A. Wiemken, and E. Frossard. 2006. The effects of agricultural practices on arbuscular mycorrhizal fungi. In *Function of Soils for Human Societies and the Environment*, eds. E. Frossard, W. E. H. Blum, and B. P. Warkentin, pp. 89–115. London, U.K.: Geological Society.

Johansson, J. F., L. R. Paul, and R. D. Finlay. 2004. Microbial interactions in the rhizosphere and their significance for sustainable agriculture. *FEMS Microbiol. Ecol.* 48: 1–13.

Johnson, N. C. 2010. Resource stoichiometry elucidates the structure and function of arbuscular mycorrhizas across scales. *New Phytol.* 185: 631–647.

Kabir, Z. 2005. Tillage or no-tillage: Impact on mycorrhizae. *Can. J. Plant Sci.* 85: 23–29.

Kahiluoto, H. and M. Vestberg. 1998. The effect of arbuscular mycorrhiza on biomass production and phosphorus uptake from sparingly soluble sources by leek (*Allium porrum* L.) in Finnish field soils. *Biol. Agric. Hortic.* 16: 65–85.

Kiær, L. P., I. M. Skovgaard, and H. Østergård. 2009. Grain yield increase in cereal variety mixtures: A meta-analysis of field trials. *Field Crops Res.* 114: 361–373.

Kiers, E. T., M. Duhamel, Y. Beesetty, J. A. Mensah, O. Franken, E. Verbruggen, C. R. Fellbaum et al. 2011. Reciprocal rewards stabilize cooperation in the mycorrhizal symbiosis. *Science* 333: 880–882.

Kruse, J., W. Negassa, N. Appathurai, L. Zuin, and P. Leinweber. 2010. Phosphorus speciation in sequentially extracted agro-industrial by-products: Evidence from X-ray absorption near edge structure spectroscopy. *J. Environ. Qual.* 39: 2179–2184.

Lammers, P. J., J. Jun, J. Abubaker, R. Arreola, A. Gopalan, B. Bago, C. Hernandez-Sebastia et al. 2001. The glyoxylate cycle in the arbuscular mycorrhizal fungus. Carbon flux and gene expression. *Plant Physiol.* 127: 1287–1298.

Leikan, D. F. and F. P. Achorn. 2005. Phosphate fertilizers: Production, characteristics and technologies. In *Phosphorus, Agriculture and the Environment*, eds. J. T. Sims and A. N. Sharpley, pp. 23–50. Madison, WI: Soil Science Society of America.

Leytem, A. B., B. L. Turner, V. Raboy, and K. L. Peterson. 2005. Linking manure properties to phosphorus solubility in calcareous soils: Importance of the manure carbon to phosphorus ratio. *Soil Sci. Soc. Am. J.* 69: 1516–1524.

Linderman, R. G. and E. A. Davis. 2004. Evaluation of commercial inorganic and organic fertilizer effects on arbuscular mycorrhizae formed by *Glomus intraradices*. *HortTechnology* 14: 196–202.

Lynch, D. H., N. Halberg, and G. D. Bhatta. 2012. Environmental impacts of organic agriculture in temperate regions. *CAB Rev.* 7: 1–17. doi 10.1079/PAV SNNR20129010.

Mäder, P., A. Fließbach, D. Dubois, L. Gunst, P. Fried, and U. Niggli. 2002. Soil fertility and biodiversity in organic farming. *Science* 296: 1694–1697.

Malloch, D. W., K. A. Pirozynski, and P. H. Raven. 1980. Ecological and evolutionary significance of mycorrhizal symbioses in vascular plants (a review). *Proc. Natl. Acad. Sci. USA* 77: 2113–2118.

Martín, G. M., R. Rivera, L. Arias, and M. Rentería. 2009. Effect of *Canavalia ensiformis* and arbuscular mycorrhizae on corn crops. *Cuban J. Agric. Sci.* 43: 185–192.

Mummey, D. L., P. M. Antunes, and M. C. Rillig. 2009. Arbuscular mycorrhizal fungi preinoculant identity determines community composition in roots. *Soil Biol. Biochem.* 41: 1173–1179.

Murphy, J. and J. P. Riley. 1962. A modified single solution method for determination of phosphate in natural waters. *Anal. Chem. Acta* 27: 31–36.

Nayyar, A., C. Hamel, G. Lafond, B. D. Gossen, K. Hanson, and J. Germida. 2009. Soil microbial quality associated with yield reduction in continuous-pea. *Appl. Soil Ecol.* 43: 115–121.

Nelson, A. G., J. C. Froese, and M. H. Entz. 2010. Organic and conventional field crop soil and land management practices in Canada. *Can. J. Plant Sci.* 90: 339–434.

Nelson, N. O. and R. R. Janke. 2007. Phosphorus sources and management in organic production systems. *HortTechnology* 17: 442–453.

Oberson, A. and E. Frossard. 2005. Phosphorus management for organic agriculture. In: *Phosphorus, Agriculture and the Environment*, eds. J. T. Sims and A. N. Sharpley, pp. 761–779. Madison, WI: Soil Science Society of America.

Oehl, F., E. Laczko, A. Bogenrieder, K. Stahr, R. Bösch, M. van der Heijden, and E. Sieverding. 2010. Soil type and land use intensity determine the composition of arbuscular mycorrhizal fungal communities. *Soil Biol. Biochem.* 42: 724–738.

Oehl, F., E. Sieverding, K. Ineichen, P. Mäder, T. Boller, and A. Wiemken. 2003. Impact of land use intensity on the species diversity of arbuscular mycorrhizal fungi in agroecosystems of Central Europe. *Appl. Environ. Microbiol.* 69: 2816–2824.

Oehl, F., E. Sieverding, P. Mäder, D. Dubois, K. Ineichen, T. Boller, and A. Wiemken. 2004. Impact of long-term conventional and organic farming on the diversity of arbuscular mycorrhizal fungi. *Oecologia* 138: 574–583.

O'Halloran, I. P. and B. J. Cade-Menun. 2008. Total and organic phosphorus. In *Soil Sampling and Methods of Analysis*, 2nd edn., eds. M. R. Carter and E. G. Gregorich, pp. 265–291. Boca Raton, FL: Canadian Society of Soil Science/CRC Press.

Oliveira, A. A. R. and F. E. Sanders. 2000. Effect of inoculum placement of indigenous and introduced arbuscular mycorrhizal fungi on mycorrhizal infection, growth, and dry matter in *Phaseolus vulgaris*. *Trop. Agric.* 77: 220–225.

Olsson, P. A., I. Thingstrup, I. Jakobsen, and E. Baath. 1999. Estimation of the biomass of arbuscular mycorrhizal fungi in a linseed field. *Soil Biol. Biochem.* 31: 1879–1887.

Pellegrino, E., A. Turrini, H. A. Gamper, G. Cafà, E. Bonari, J. P. W. Young, and M. Giovannetti. 2012. Establishment, persistence and effectiveness of arbuscular mycorrhizal fungal inoculants in the field revealed using molecular genetic tracing and measurement of yield components. *New Phytol.* 194: 810–822.

Pierzynski, G. M., R. W. McDowell, and J. T. Sims. 2005. Chemistry, cycling and potential movement of inorganic phosphorus in soils. In *Phosphorus, Agriculture and the Environment*, eds. J. T. Sims and A. N. Sharpley, pp. 53–86. Madison, WI: Soil Science Society of America.

Pivato, B., S. Mazurier, P. Lemanceau, S. Siblot, G. Berta, C. Mougel, and D. Van Tuinen. 2007. *Medicago* species affect the community composition of arbuscular mycorrhizal fungi associated with roots. *New Phytol.* 176: 197–210.

Qian, P. and J. J. Schoenau. 2002. Practical applications of ion exchange resins in agricultural and environmental soil research. *Can. J. Soil Sci.* 82: 9–21.

Rasmann, C., J. H. Graham, D. O. Chellemi, L. E. Datnoff, and J. Larsen. 2009. Resilient populations of root fungi occur within five tomato production systems in southeast Florida. *Appl. Soil Ecol.* 43: 22–31.

Redecker, D., R. Kodner, and L. E. Graham. 2000. Glomalean fungi from the Ordovician. *Science* 289: 1920–1921.

Richardson, A. E., J. P. Lynch, P. R. Ryan, E. Delhaize, F. A. Smith, S. E. Smith, P. R. Harvey et al. 2011. Plant and microbial strategies to improve the phosphorus efficiency of agriculture. *Plant Soil* 349: 121–151.

Rosen, C. J. and D. L. Allan. 2007. Exploring the benefits of organic nutrient sources for crop production and soil quality. *HortTechnology* 17: 422–430.

Ryan, M. and J. Ash. 1999. Effects of phosphorus and nitrogen on growth of pasture plants and VAM fungi in SE australian soils with contrasting fertiliser histories (conventional and biodynamic). *Agric. Ecosyst. Environ.* 73: 51–62.

Ryan, M. H. and J. A. Kirkegaard. 2012. The agronomic relevance of arbuscular mycorrhizas in the fertility of Australian extensive cropping systems. *Agric. Ecosyst. Environ.* 163: 37–53.

Schüßler, A., D. Schwarzott, and C. Walker. 2001. A new fungal phylum, the Glomeromycota phylogeny and evolution. *Mycol. Res.* 105: 1413–1421.

Schweiger, P. F. and I. Jakobsen. 1999. Direct measurement of arbuscular mycorrhizal phosphorus uptake into field-grown winter wheat. *Agron. J.* 91: 998–1002.

Scialabba, N. E.-H and M. Müller-Lindenlauf. 2010. Organic agriculture and climate change. *Renew. Agric. Food Syst.* 25: 158–169.

Scullion, J., W. R. Eason, and E. P. Scott. 1998. The effectivity of arbuscular mycorrhizal fungi from high input conventional and organic grassland and grass-arable rotations. *Plant Soil* 204: 243–254.

Singh, A. K., C. Hamel, R. M. DePauw, and R. E. Knox. 2012. Genetic variability in arbuscular mycorrhizal fungi compatibility supports the selection of durum wheat genotypes for enhancing soil ecological services and cropping systems in Canada. *Can. J. Microbiol.* 58: 293–302.

Smith, S. E., I. Jakobsen, M. Grønlund, and F. A. Smith. 2011. Roles of arbuscular mycorrhizas in plant phosphorus nutrition: Interactions between pathways of phosphorus uptake in arbuscular mycorrhizal roots have important implications for understanding and manipulating plant phosphorus acquisition. *Plant Physiol.* 156: 1050–1057.

Smith, S. E. and D. J. Read. 2008. *Mycorrhizal Symbiosis*, 3rd edn. San Diego, CA: Academic Press.

Smith, S. E. and F. A. Smith. 2012. Fresh perspectives on the roles of arbuscular mycorrhizal fungi in plant nutrition and growth. *Mycologia* 104: 1–13.

Sousa, C. S., R. S. C. Menezes, E. V. S. B. Sampaio, F. Oehl, L. C. Maia, M. S. Garrido, and F. S. Lima. 2012. Occurrence of arbuscular mycorrhizal fungi after organic fertilization in maize, cowpea and cotton intercropping systems. *Acta Sci. Agron.* 34: 149–156.

Stewart, L. I., C. Hamel, R. Hogue, and P. Moutoglis. 2005. Response of strawberry to inoculation with arbuscular mycorrhizal fungi under very high soil phosphorus conditions. *Mycorrhiza* 15: 612–619.

Suri, V. K., A. K. Choudhary, G. Chander, T. S., Verma, M. K., Gupta, and N. Dutt. 2011. Improving phosphorus use through co-inoculation of vesicular arbuscular mycorrhizal fungi and phosphate-solubilizing bacteria in maize in an acidic Alfisol. *Commun. Soil Sci. Plant Anal.* 42: 2265–2273.

Taylor, T. N., W. Remy, H. Hass, and H. Kerp. 1995. Fossil arbuscular mycorrhizae from the early Devonian. *Mycologia* 87: 560–573.

Tiessen, H. and J. O. Moir. 2008. Characterization of available P by sequential extraction. In *Soil Sampling and Methods of Analysis*, 2nd edn., eds. M. R. Carter and E. G. Gregorich, pp. 293–306. Boca Raton, FL: Canadian Society of Soil Science/CRC Press.

Toor, G. S., J. D. Peak, and J. T. Sims. 2005. Phosphorus speciation in broiler litter and turkey manure produced from modified diets. *J. Environ. Qual.* 34: 687–697.

Verbruggen, E., E. T. Kiers, P. N. C. Bakelaar, W. F. M. Röling, and M. G. A. van der Heijden. 2012b. Provision of contrasting ecosystem services by soil communities from different agricultural fields. *Plant Soil* 350: 43–55.

Verbruggen, E., W. F. M. Röling, H. A. Gamper, G. A. Kowalchuk, H. A. Verhoef, and M. G. A. van der Heijden. 2010. Positive effects of organic farming on below-ground mutualists: Large-scale comparison of mycorrhizal fungal communities in agricultural soils. *New Phytol.* 186: 968–979.

Verbruggen, E., M. G. A. Van der Heijden, M. C. Rillig, and E. T. Kiers. 2012a. Mycorrhizal fungal establishment in agricultural soils: Which factors determine inoculation success? *New Phytol.* 197: 1104–1109.

Vierheilig, H., J. M. Garcia-Garrido, U. Wyss, and Y. Piché. 2000. Systemic suppression of mycorrhizal colonization of barley roots already colonized by AM fungi. *Soil Biol. Biochem.* 32: 589–595.

Welsh, C. M. 2007. Organic crop management can decrease labile soil P and promote mycorrhizal association of crops. Master thesis, Department of Soil Science, University of Manitoba, Winnipeg, Manitoba, Canada.

Welsh, C. M., M. Tenuta, D. Flaten, J. Thiessen Martens, and M. Entz. 2009. High yielding organic crop management decreases plant available but not total soil phosphorus. *Agron. J.* 101: 1027–1035.

4 Nitrogen and Phosphorus Fertility Management in Organic Field Crop Production

*Alexander Woodley, Yuki Audette,
Tandra Fraser, Melissa M. Arcand, Paul Voroney,
J. Diane Knight, and Derek H. Lynch*

CONTENTS

4.1   INTRODUCTION

We know that nature's stores, rich as they once were on the Great Plains, will not always sustain us without our own re-investments. The nutrient reserves will not persist without us returning what we extract; the diversity of biota in and on the soil will not flourish without our conscious effort to uphold it; the life-giving humus will not re-build without us restoring what is decayed or eroded away; the land's quiet services of filtering water and air will not continue without us paying heed to them. (H.H. Janzen, 2010)

Organic field crop production in Canada has expanded over the past few decades in all regions. Ongoing productivity challenges are closely linked to needs to develop more systematic, regionally appropriate, and validated methods for nutrient and weed management. The following chapter reviews our understanding to date of approaches to manage nitrogen (N) and phosphorus (P) for organic field crop production in Canada.

Any discussion of fertility management and productivity in specific organic farming systems must include consideration of that system's sustainability or ability to minimize environmental impacts while maintaining an economically viable production level. Agriculture is a key driver of environmental pressures and ecosystem degradation globally, through its impact on water use, loss of habitat, climate change, and pollution from excess N and P. The increase in agricultural intensification over the past 40 years, and widespread and intensive use of synthetically produced N and P fertilizers, is oversaturating many of the world's agricultural areas, affecting many nontarget ecosystems. This includes greatly increased risk and incidence of contamination to surface and groundwaters by N, P, and pesticides. In eastern Canada, an increasing acreage of agricultural soils has been classified as "at high risk of being a source of nitrate-N losses to water" (de Jong 2007; see also CCME 2002; Erisman et al. 2008; Lynch et al. 2012a).

On cropped organic farms, environmental benefits with respect to energy use, soil quality, biodiversity, and reduced off-farm nutrient impacts are closely linked to reliance on increased spatial and temporal diversity associated with more complex rotations, legume biological nitrogen fixation (BNF) and organic matter inputs, and reduced overall nutrient intensity (Lynch et al. 2012a). In some jurisdictions, most notably Europe, government and consumer support for organic farming partially reflect support for the environmental benefits that can be derived from organic management.

Whole-farm nutrient (NPK) budgeting (imports–exports) is a useful tool to gauge farming system sustainability, nutrient loading, and farm efficiency of nutrient use as affected by management approach (Figure 4.1a and b). A study of 15 organic dairy farms in Ontario found farm nutrient loading was greatly reduced under commercial

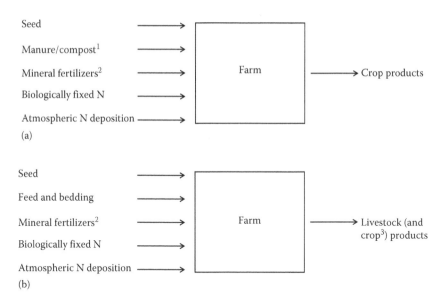

FIGURE 4.1 (a) Farm-scale nutrient inputs and outputs for an organic field crop farm without livestock. (b) Farm-scale nutrient inputs and outputs for an organic livestock farm. [1]Imported manures and composts used by some organic crop producers in Canada in regions where available. [2]Permitted mineral nutrient inputs such as PR or Sul-Po-Mag used by some producers. [3]If also a cash crop producer.

organic dairy production compared with more intensive confinement-based livestock systems but varied significantly with management strategy (Roberts et al. 2008). Annual farm nutrient surpluses averaged 75 (N), 1 (P), and 11 (K) kg ha^{-1} year^{-1} across all farms but increased with farm livestock density and decreased with greater farm feed self-sufficiency (Roberts et al. 2008). To date, similar studies have not been conducted on stockless organic farms producing field crops.

The challenge of managing N in organic systems to sustain adequate productivity and nitrogen-use efficiency (NUE), while minimizing N losses, should not be understated. Synchronizing N availability with crop demand in organic systems is difficult as the supply of N to the crop from organic amendments, green manures (GMrs), and crop residues varies with climatic conditions and among years (Askegaard et al. 2011; Lynch et al. 2012b). As discussed in the following, options with respect to GMrs and amendments for organic field crop producers also vary substantially with cropping region in Canada.

In organic field crop production systems, there are limited options for addition of mineral forms of nutrients, such as P, and the specialization of many organic grain producers in Canada has created potentially unique challenges with respect to management of P. For this reason, we initially review the issue of soil P status of organic field crop farms in the following section. This is followed by broader treatment of N and P management for organic field crop production within sections addressing the following: (1) annual legume GMrs and pulse crop residues for supplying N and P in the Canadian Prairies; (2) crop N supply in humid regions as affected by rotation, GMr type, and management; (3) composts, manures, and anaerobic digestates for N and P supply; (4) efficacy of microbial inoculants and phosphate rock (PR) for enhancing phosphorus supply; and lastly (5) future directions and research needs.

4.1.1   SOIL P STATUS OF CANADIAN ORGANIC FARMS PRODUCING FIELD CROPS

Phosphorus is a component of genetic material used for energy transfer in organisms (White and Hammond 2008), making it an essential macronutrient for the development and vigor of all plants. It is essential for plant growth and development including its role in phosphatidyl linkages between RNA and DNA, in carbon fixation (ribulose bisphosphate carboxylase/oxygenase), and in phospholipids essential to the structure and functioning of cell membranes. Even small deficiencies affect root growth, biomass and seed production, plant disease resistance, and N-fixing capabilities of legumes (White and Hammond 2008). Soil P is slowly released into soil solution from three main forms: organic, calcium-bound inorganic, and iron- or aluminum-bound inorganic (Brady and Weil 2002). The balance between the two inorganic reserves of P is largely controlled by soil pH where calcium-bound inorganic P prevails in alkaline soils and iron- or aluminum-bound inorganic P prevails in acidic soil.

In many field crop systems, P is exported with the grain or hay or indirectly via farm livestock products, resulting in a depletion of soil test phosphorus (STP) levels over time. In organic systems, where use of orthophosphate fertilizer is prohibited, there are limited options for P addition. Although animal manure can be a valuable source of P, farmers in North America often specialize in either grains or livestock, not both (Russelle et al. 2007), and logistics may prohibit regular manure applications. Although PR is allowed for field application under Canadian organic standards (Government of Canada 2006b), the release of available P is extremely slow in many Canadian soils, greatly limiting its effectiveness. This is discussed in more detail in Section 4.5.3.

Few extensive surveys of STP levels on organic farms in Canada have been published and are limited to Saskatchewan, Manitoba, and Ontario (Table 4.1). Low plant-available phosphorus levels have been reported in organically managed soils across Canada (Entz et al. 2001; Martin et al. 2007; Roberts et al. 2008; Knight et al. 2010; Main et al. 2013). In 1991, Entz et al. (2001) initiated a survey of crop yield and soil nutrient levels on 14 farms in the eastern Great Plains. From 42 fields, STP levels were combined based on similar soil characteristics (texture, pH, and soil organic matter [SOM]) with all but one mean value below 18 kg PO_4-P ha^{-1}. The fine sand to fine sandy loam class with pH 7.6 and SOM 3.9% (n = 7) had a mean value of 34, ranging from 24 to 54 kg PO_4-P ha^{-1}. Knight et al. (2010) found similar levels of STP in surface soil (0–15 cm) collected from 60 organic fields across Saskatchewan (5.7–30.7 kg P ha^{-1}), with perennial systems being lowest. Although a direct comparison cannot be made between the studies, they do highlight the risk of continuously exporting grain and hay off-farm without returning nutrients to the systems.

TABLE 4.1

Summary of Studies Reporting Estimated Available Soil Phosphorus on Organic Farms and Research Plots in Canada and Northern United States

Location	Description	Years Organic at Initiation	Sampling Period (Years)	STP Values (kg P ha⁻¹)	Soil Test	Depth (cm)	Reference
Manitoba Saskatchewan North Dakota, United States	9 farms, 42 fields	>5	1	**15.6** (4–54)[a,b]	n/a	0–15	Entz et al. (2001)
Saskatchewan	60 fields	>5	1	**18.0** (5.7–30.7)	Modified	0–15	Knight et al.
				13.1 (4.0–28.6)	Kelowna	15–30	(2010)
				12.2 (4.0–28.3)		30–45	
Ontario	15 fields	>10	1	**12** (<10 to >20)[c]	Olsen P	0–15	Roberts et al. (2008)
Ontario	18 fields	>10	2	**17.4** (7.4–31.4)[d]	Olsen P	0–15	Main et al. (2013)
Saskatchewan	Field trial	6	6	**8.7** (7.8–9.4)	Mehlich	0–15	Malhi et al.
				7.2 (5.6–8.1)		15–30	(2009)
				2.1 (1.3–2.7)		30–60	
				0.9 (0.7–1.3)		60–90	
Montana US	Field trial	4	1	**18**[c]	Olsen P	0–15	Miller et al. (2008)
Manitoba	Field trial	13	1	**21.7**[e], **8.0**[f], **11.3**[g]	Olsen P	0–15	Welsh et al. (2009)
Manitoba (Glenlea)	Field trial	18	1	**23.4**[e], **3.6**[f]	Olsen P	0–15	Bell et al.
				4.5, 1.7		15–30	(2012)
				1.5, 1.1		30–60	
Manitoba (Carmen)	Field trial	9	1	**8.4**[e], **8.8**[f]	Olsen P	0–15	Bell et al.
				2.4, 2.7		15–30	(2012)
				1.3, 1.9		30–60	

[a] Values in bold are mean values with the range given in parenthesis when available.

[b] Value is labeled as PO_4^- in paper by Entz et al. (2001).

[c] Value reported as mg kg⁻¹.

[d] Originally reported as mg P kg⁻¹.

[e] Grain-only rotation (spring wheat–pea–spring wheat–flax).

[f] Forage–grain rotation (spring wheat–alfalfa–alfalfa–flax).

[g] Forage–grain with composted beef manure applied in 2002 only.

At a long-term research trial at Scott, SK, Malhi et al. (2009) reported STP to typically be deficient in the organic system. The experiment compared 6-year rotations with three diversity levels managed organically or conventionally with high inputs or conventionally with low inputs. The experiment was established in 1995. Soil samples were extracted to 90 cm from 2001 to 2006. They reported no effect of cropping sequence or input level on M3 extractable P in any of the years even though the three organic systems had P balances of −40, −41, and −22 kg P ha^{-1}, respectively, over 12 years. It is likely that P is being converted from less-soluble forms to plant-available forms (Zhang et al. 2004), but the effect on total P was not reported in this study.

The Glenlea Long-Term Crop and Management Study at Glenlea, MB, demonstrated similar trends, especially in the forage–grain rotation where the forage is exported and no amendments are added. Welsh et al. (2009) revealed STP levels below the agronomic threshold in the grain-forage rotation after 12 years (Table 4.1). A one-time application of composted beef manure did not result in adequate replenishment, with a net P balance of −94 kg P ha^{-1} from 1992 to 2004. A significant effect of both management and rotation on total P was present. However, using a modified Hedley sequential extraction (Hedley et al. 1982), they concluded that the more recalcitrant P pools were not affected by management.

Although low concentrations of STP are widely reported on organic farms, farmers may not always observe a relationship between STP and yield. Conventional soil tests may be adequate indicators for management practices including fertilizer and manure applications but do not consider important microbial and plant-mediated processes that make less-soluble P available for plant uptake in low-input organic systems. In addition, even within conventional farming systems, the relationship between STP and crop yield under field conditions is often very poor, due to the sensitivity of P supply and demand, to environmental conditions. For example, a recent follow-on study to that of Roberts et al. (2008) on forage fields on organic dairy farms in Ontario found no consistent relationship between STP level and perennial legume (alfalfa and clover) productivity and BNF (Main et al. 2013). It has been suggested that organic management encourages greater enzymatic activity and increased P turnover and mineralization (Oberson and Frossard 2005), and this may be an important consideration when interpreting STP results. Since STP may not provide sufficient information for predicting P availability, other useful measurements for understanding P dynamics in organic systems include P fluxes, whole-farm P budgets, total P, organic P relative to total P, crop vigor, and mycorrhizal colonization. In addition, more research into rates and frequency of manure application for P replenishment may assist in balancing P budgets and improving the long-term sustainability of organic farming systems.

Roberts et al. (2008) reported STP levels from 15 long-term organic dairy farms in Ontario. The overall weighted farm average STP level was 12 mg kg^{-1}; seven of the farms had average STP levels in the low to very low range (<10 mg kg^{-1}) and five farms in the moderate range (10–20 mg kg^{-1}). Whole-farm nutrient budgets were close to zero for P for many of these dairy farms and in some cases negative.

Manure may provide adequate amounts of nutrients required for plant growth. In an organic–conventional comparison study in Nebraska, Wortman et al. (2011) found STP levels in the organic-animal manure (OAM)-based system increased from 70 mg P kg^{-1} in 1996 to 165 mg P kg^{-1} in 2008, while levels in the organic forage (OFG)-based system where alfalfa was included and the hay removed remained relatively stable. A nutrient budget demonstrated a total P balance of 779 kg P ha^{-1} over 12 years in the OAM system, where manure was applied based on N requirements for 2 out of the 4 years of the rotation, compared to 327 and −212 kg P ha^{-1} for the OFG and conventional systems, respectively. While integrated crop–livestock systems should be considered as an option to reduce the negative environmental issues linked to intensive feedlot operations while improving soil fertility and profitability and contributing to long-term sustainability of the farm (Russelle et al. 2007), as discussed further in Section 4.4.6, over-reliance on manures for nutrient supply can lead to excessive P application rates in organic systems. This is particularly important given that soils overloaded with manure may cause environmental problems on entering waterways through runoff and leaching.

4.2 ANNUAL LEGUME GREEN MANURES AND PULSE CROP RESIDUES FOR SUPPLYING N AND P IN THE CANADIAN PRAIRIES

Nutrient management is a major challenge for organic producers in the Canadian Prairies, specifically the replacement of nutrients exported through the harvest and sale of grain, hay, and fiber. It is well established that animal manures are excellent sources of nutrients (Schoenau et al. 2010), but the predominance of grain farms in the region and the typically large farm size means that livestock numbers are not high enough to supply manure to the entire farm at levels necessary to meet nutrient requirements. A survey of 60 organically managed fields farmed by 39 producers across Saskatchewan reported extremely low frequency of manure application (Knight et al. 2010). Over the six management years that the questionnaire addressed (360 field years), only six fields had animal manure applied, despite 39% of the farms reporting raising cattle and/or other livestock. Although not addressed directly in the survey, it was assumed that farms with livestock were unable to produce enough manure to maintain fertility on the whole farm. Instead of amending soils with animal manure, producers in the Prairie Provinces rely on plant sources for nutrients such as legume GMrs, which supply N through BNF.

4.2.1 Green Manuring

The benefit of GMrs depends on the GMr species, environmental conditions, and the management practices used. Desirable characteristics of GMrs include rapid accumulation of biomass, ample nutrient uptake, adequate BNF, and minimal maintenance required during the growth period (Fageria and Baligar 2005). For instance, N and P concentrations of the crop and the decomposition rate will affect nutrient release from the GMr (Lupwayi et al. 2006, 2007; Olson-Rutz et al. 2010). Generally, higher plant tissue N and P concentrations and, hence, narrower C/N or C/P ratios result in net mineralization and nutrients that are readily available for uptake (Lupwayi et al. 2006, 2007). Lupwayi et al. (2006) compared N release patterns (N mineralization) from decomposing residues of red clover (*Trifolium pratense* L.) GMr and field pea (*Pisum sativum* L.), canola (*Brassica napus* L.), and monoculture wheat (*Triticum aestivum* L.) in Beaverlodge, AB. Nitrogen release ranged from 22% from the wheat to 71% from the red clover. While the study used conventional management, it demonstrated the tremendous difference in decomposition and mineralization rates from different residue sources. However, rapid and extensive mineralization from legume residues also means a rapid turnover of SOM and not necessarily a net accumulation or buildup of SOM (see Chapter 5 for further discussion of *net* SOM changes under organic field cropping).

The ultimate agronomic success of a GMr crop is the growth and yield of the subsequent crop in rotation. In general, GMrs increase yields compared to continuous cereal rotations (Rice et al. 1993; Brandt 1996; Bullied et al. 2002, Shirtliffe and Knight 2006). Similarly, wheat grain N content increased 21%–35% following GMrs compared to following continuous wheat. However, when compared to fallow, GMrs are frequently not as productive in yield and/or N content (Zentner et al. 1996; Pikul et al. 1997; Brandt 1999). A Manitoba study examined green mulch applications to cereal crops for their effect on moisture conservation, N contribution, and weed suppression (Wiens et al. 2006). Application of 3.9–5.2 Mg ha^{-1} alfalfa mulch to wheat produced grain yields that were equivalent to those produced with the application of 20 and 60 kg N ha^{-1}. Furthermore, oat (*Avena sativa* L.) grown in the second year produced higher grain yield and had higher N uptake than the ammonium nitrate treatment, indicative of mineralization occurring into the second year. Although not widely used, perennial legume mulches have the potential to play a significant role in organic systems.

Legume GMrs (plough-down crops) are one of the most common nutrient management tools used by organic producers in the Prairie Provinces. Crops grown as GMrs are terminated while still green and incorporated into the soil or left on the soil surface as a mulch to decompose. Because water is frequently the most limiting factor for crop production on the prairies, crops for GMrs are usually terminated early in the growing season, to conserve water and enable late season replenishment of the soil water supply (Figure 4.2). Innovative approaches to termination of GMrs that reduce reliance on tillage are being explored in organic field crop production systems (see special issue of the journal Renewable Agriculture and Food Systems; volume 27). A "biological no-till" cover crop roller system was developed first in Brazil and an early model further tested by USDA-ARS in Alabama (Ashford and Reeves 2003; Mirsky et al. 2012). In Manitoba, Vaisman et al. (2011) found the added soil cover (50%–90% cover) benefits provided by rolled GMr mulches of pea, or pea and oat, were offset by the negative impact of reduced spring soil temperatures and reduced available N, ultimately resulting in reduced yields of the subsequent wheat crop. In a study at two sites in Saskatchewan on a Black Chernozemic clay loam soil, Shirtliffe and Johnson (2012) found wheat yields to be just as good when a field pea GMr was terminated by roller crimping as when it was tilled or mowed. However, in contrast to Vaisman et al. (2011), tillage was applied in the spring prior to planting of the following wheat crop.

Early incorporation of GMrs means that the young, green, minimally lignified tissues that are susceptible to rapid decomposition are unlikely to contribute significantly to long-term development of stable SOM. However, a GMr that decomposes rapidly is an excellent source of immediately available nutrients. In contrast, a slowly decomposing GMr will have little effect on immediately available nutrients but a larger long-term effect on SOM content, thus contributing to residual nutrient release.

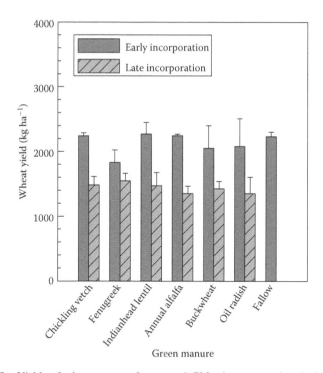

FIGURE 4.2 Yields of wheat grown after annual GMrs incorporated early in the season (early flowering) or incorporated late in the season (late budding) at a site in east central Saskatchewan (Dark Brown soil zone).

On the Canadian Prairies, organic producers use perennial, biennial, or annual legume GMrs to manage soil fertility. However, the use of perennial legumes risks depleting soil moisture, reducing the yield of subsequent crops in dry regions (Meyer 1987; Hesterman et al. 1992), and tends to be restricted to the moist regions of the prairies. Biennial yellow sweet clover (*Melilotus officinalis* L.) is the most widely grown GMr crop on prairie organic farms. Early incorporation is recommended for water conservation and also to maximize soil nitrate levels (Foster 1990). Late incorporation of sweet clover can lead to low concentrations of spring available N (Foster 1990), thought to result from the slow decomposition of lignified residues.

Annual GMrs are becoming more popular on prairie farms because these crops provide more flexibility than biennial or perennial crops. However, the use of annual GMrs requires a year without cash crop production, and therefore the GMr must be managed properly to maximize the benefit. The shallow root systems of annual legumes extract water from shallower depths than perennial and biennial legumes (Biederbeck and Bouman 1994). As an example, soil water after harvest of a crop following an early-incorporated lentil GMr was only slightly lower than after summer fallow. Most importantly, returning N to the soil is of little benefit if the amount of water necessary to grow the GMr limits production of the succeeding cash crop.

TABLE 4.2

Amounts of N Fixed by Annual GMr Crops in Different Soil Zones in the Canadian Prairies

GMr	Soil Zone	N Fixed (kg N ha⁻¹)	Study
Black lentil	Brown	18; 7[a]	Biederbeck et al. (1996); Alberta Agriculture (1993)
Chickling vetch	Brown	49	Biederbeck et al. (1996)
Field pea	Brown	40; 12[a]	Biederbeck et al. (1996); Alberta Agriculture (1993)
Flat pea	Brown	10[a]	Alberta Agriculture (1993)
Black lentil	Dark brown	15; 8	Biederbeck et al. (1996); Alberta Agriculture (1993)
Field pea	Dark brown	40; 48	Biederbeck et al. (1996); Alberta Agriculture (1993)
Flat pea	Dark brown	9	Alberta Agriculture (1993)
Faba bean	Dark brown	41	Biederbeck et al. (1996)
Black lentil	Black	10; 23–39; 20–40	Townley-Smith et al. (1993); Thiessen Martens et al. (2005); Alberta Agriculture (1993)
Field pea	Black	53–80	Alberta Agriculture (1993)
Flat pea	Black	16; 17–46	Townley-Smith et al. (1993); Alberta Agriculture (1993)
Annual alfalfa	Black	1	Townley-Smith et al. (1993); Alberta Agriculture (1993)
Chickling vetch	Black	29–43	Thiessen Martens et al. (2005)

[a] Rainfall was 25% of normal.

A number of annual legumes have been evaluated as GMr for semiarid regions, especially in the Dry Brown soil zone, where perennial legumes are not suitable. In general, the amount of N_2 fixed and the overall success of the various crops depend in part on the prairie soil zone in which the crop is grown (Table 4.2). Major GMrs evaluated include feed pea (Biederbeck and Bouman 1994; Biederbeck et al. 1998; Lawley 2004), Tangier flat pea (*Lathyrus tingitanus* L.) (Biederbeck and Bouman 1994; Biederbeck et al. 1998), black lentil (*Lens culinaris*) (Biederbeck and Bouman 1994; Biederbeck et al. 1998; Brandt 1999; Lawley 2004; Thiessen Martens et al. 2005), chickling vetch (*Lathyrus sativus* L.) (Biederbeck and Bouman 1994; Biederbeck et al. 1998; Lawley 2004; Thiessen Martens et al. 2005), annual alfalfa (*Medicago sativa* Leyss) (Townley-Smith et al. 1993), and faba bean (*Vicia faba* L.). Nitrogen fixation rates for faba bean are reported to be higher than field pea and can be up to 350 kg N ha⁻¹ (Rochester et al. 1998). Limited research has been conducted on faba bean in organic systems, but its poor competitive ability with weeds and intolerance to drought (De Costa et al. 1997; Hollinger 2010) appears to limit its usefulness. A further obstacle to its adoption in organic systems has been the high cost of its large seed, although one cultivar developed for Saskatchewan, CDC SSNS1, was selected for small seed size (Oomah et al. 2011).

Of the major annual legumes evaluated as GMr crops, field pea (Biederbeck et al. 1996), chickling vetch (Biederbeck et al. 1996; Lawley 2004), and Indianhead lentil (Lawley 2004) are the superior crops. Of the three, lentil is the most economical due to its low seed cost and small seed size, but pea accumulated the highest amounts of total N and therefore had the largest impact on soil fertility (Lawley 2004). Overall field pea was the best choice for organic production in Saskatchewan (Lawley 2004). Chickling vetch produced comparable amounts of biomass and accumulated comparable amounts of N as pea and lentil but, because of its large seed size and high seed cost, was the least economical of the three pulse crops (Lawley 2004). Unless seed is from an on-farm source, it may not be an affordable GMr option.

Limited research has targeted growing nonlegume GMr crops for improving availability of P to the subsequent crop. Nonleguminous GMr crops can influence soil fertility by extracting nutrients from deep in the soil and redistributing them to the plough-down layer. Organic acids released from the roots of some plants, and the stimulation of rhizosphere activity (Vance et al. 2003), contribute to enhanced solubilization and uptake of P. Buckwheat and oilseed radish (*Raphanus sativus*) are reported to have excellent P uptake mechanisms (Aman and Amberger 1989; Wang et al. 2008) and in the past were encouraged as GMr crops targeting P availability. However, in a side-by-side comparison of buckwheat, oilseed radish, fall rye (*Secale cereale* L.), and a variety of annual legumes (Knight and Shirtliffe 2003), the legumes were as good or better at supplying P to a subsequent wheat crop (Figure 4.3). Hence, it appears unnecessary to grow crops such as buckwheat or oilseed radish specifically for P uptake, as legumes are also able to perform this function. Also, organically grown legumes such as field pea and lentil had arbuscular mycorrhizal fungi (AMF) colonization levels between 70% and 85% to contribute to the extraction of soil P when grown in organically managed soils (Baird 2007).

4.2.2 Managing Residues

Organic producers in the prairies are encouraged to have 40% of their land seeded to a legume in any year (Smith and Groenen 2000). This includes GMrs and cereals underseeded with legumes as well as forage legumes and pulse crops grown as cash crops. Careful management of legume crop residues is encouraged to capitalize on BNF. The quantity and quality of crop residues influence the rate and pattern of decomposition by soil microorganisms, and environmental conditions such as soil moisture and temperature affect decomposition, mineralization, and immobilization rates by affecting the activity of the soil microbial populations. A number of studies have reported pulse crops supplying modest amounts of N to subsequent crops through decomposition of residues (Welty et al. 1988; Wright 1990a,b; Badaruddin and Meyer 1994; Beckie and Brandt 1997; Miller et al. 2002).

The C/N ratios of residues are important. Typically, C/N ratio is negatively correlated with net N mineralization. When the C/N ratio of a substrate (residue) is higher than that of the microorganisms decomposing it, the fraction of organic N that is mineralized is less than the fraction of organic C that is dissimilated

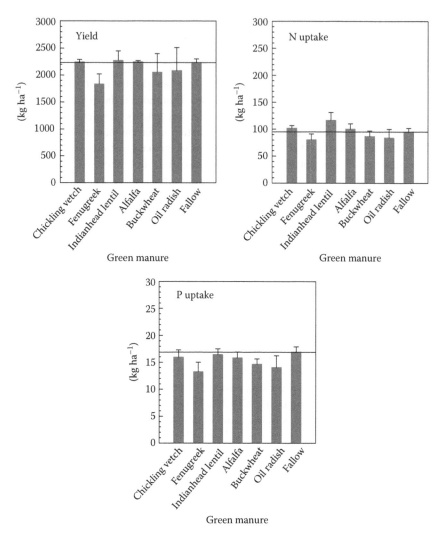

FIGURE 4.3 Productivity and nutrient uptake by wheat grown the year after the annual GMr crops at a site in east central Saskatchewan (Dark Brown soil zone). The horizontal line on the graphs indicates wheat grown after a fallow year.

(Janssen 1996). In practical terms, this means that when residue C/N ratios are wide, net immobilization of N occurs, and when C/N ratios are narrow, net mineralization occurs. A C/N ratio of approximately 20 is often considered to be where immobilization and mineralization are balanced (Loeck et al. 2012). The heavy reliance on legumes in organic rotations for managing N fertility means that overall, good-quality residues with narrow C/N are incorporated.

Averaged over eight site years at two locations in southern Saskatchewan, lentil, dry bean (*Phaseolus vulgaris* L.), and dry pea stubbles contributed 5, 6, and 9 kg ha^{-1} more soil N in the 0–120 cm soil depth than wheat stubble. The average yield of wheat grown on pulse crop stubble was 21% greater than wheat grown on wheat stubble. As one example, an 18-year study at Swift Current, SK, comparing lentil–wheat and wheat–wheat rotations, the lentil–wheat rotation yielded higher grain protein in 11 of 18 years (Zentner et al. 2002). Of four annual grain legumes tested in Manitoba field experiments, field pea provided the largest and most consistent N benefit to a succeeding wheat crop (Przednowek et al. 2004). Soybean provided relatively little N benefit, and chickpea and dry bean were inconsistent although chickpea performed well under drought conditions (Przednowek et al. 2004). Walley et al. (2007) examined previously published data to assess the contribution of various pulse crop species to soil N. Those pulse crops that can achieve relatively high levels of BNF, such as field pea and lentil, are more likely to contribute positively to the overall N economy of the system; pulse crops that achieve only modest levels of BNF such as chickpea and dry bean are likely to be either N neutral or contribute to an N deficit. Reported values for BNF were extremely variable reflecting differences in the inherent productivity of the soil as well as variations in local climate and weather. It should be stressed that while it is important to manage pulse crop residues carefully, it is unlikely that these residues will be able to fully replace N removed through harvest of seed.

4.3 CROP N SUPPLY IN HUMID REGIONS AS AFFECTED BY ROTATION AND GREEN MANURE TYPE AND MANAGEMENT

Organic agriculture is characterized by longer and more complex rotations that use perennial forages, cover crops, and pulse crops to supply N that would be otherwise provided by inorganic fertilizers in conventional systems (Doltra et al. 2011). Enhanced cropping diversity tends to increase both the amount and variation in quality of residues, including root residues, and such "substrate diversity" can enhance soil mineralizable C and N pools (Sanchez et al. 2004; Chapter 5). Regardless, supplying field crops with adequate plant-available N (PAN) at the correct growth stage can be one of the main challenges to achieving crop productivity in organic systems (Berry et al. 2002).

Perennial forages in rotation provide numerous benefits to SOM and overall soil quality, reduce compaction, improve weed suppression, and supply significant quantities of N to a succeeding crop (Thorup-Kristensen et al. 2012). Rotations are typically designed so that the crop with the greatest N demand, such as corn (*Zea mays* L.), which can require 150 kg N ha^{-1}, follows a leguminous crop such as alfalfa. Alfalfa is the highest yielding perennial forage crop grown commonly in the US corn belt and much of Canada (Baute et al. 2002). In the US corn belt, alfalfa with stand densities of 43–53 plants m^{-3} is assigned a N credit of 168 kg N ha^{-1}, which can satisfy corn N demand (Yost et al. 2012). In Ontario, alfalfa is commonly grown with timothy grass (*Phleum pratense* L.) and bird's-foot trefoil (*Lotus corniculatus* L.) and perennial forages. Those fields that contain over 50% legumes (typically alfalfa or red clover) are credited 100 kg N ha^{-1} for the succeeding season crop (Baute et al. 2002).

GMrs and cover crops used in rotations can either conserve soil N through plant uptake or contribute to soil N by N$_2$ fixation. Leguminous winter cover crops are often relayed or interseeded with small grains to provide additional N to the system.

Red clover is the most common interseeded cover crop used in eastern Canada; some other alternative crops grown are hairy vetch (*Vicia villosa* Roth), field pea, and sweet clover. Liebman et al. (2012) conducted a study in Iowa evaluating the effect of intercropped alfalfa and red clover into oats on a succeeding corn crop. The two cover crops provided a yield improvement of 25%–63% as compared to the nonfertilized oat control. The fertilizer replacement values were estimated at 87–184 kg N ha^{-1} for red clover and 70–121 kg N ha^{-1} for alfalfa. Hairy vetch can have N biomass accumulation as high as 150–172 kg N ha^{-1} (Clark et al. 1997; Teasdale et al. 2012; Table 4.3).

The amount of N supplied by GMrs will vary widely based on species, region and planting date, soil type, and timing and type of termination (Liebman et al. 2012). For overwintering and perennial GMrs, significant biomass accumulation can occur in the spring following the fall growth; this can be desirable as the crop will take up rapidly mineralizing soil N, which commonly occurs in early spring. This must be balanced by timely plough down prior to planting to avoid a potential period of soil N immobilization during the initial decomposition of the GMr. There is also a trend of increasing C/N ratio as the crop matures (Crandall et al. 2005), furthering the likelihood for N immobilization. Because leguminous and nonleguminous broad-leaf GMrs have lower C/N ratios than cereals, time of plough down can be delayed prior to planting as initial mineralization rates will be higher than cereals (Clark et al. 1997). Management adjustments are required when using rye or oats as a cover crop due to their slow N mineralization (Thorup-Kristensen and Dresboll 2010). Vyn et al. (2000) observed a yield decline in corn following wheat interseeded with oat and rye cover crops, which was attributed to immobilization of soil mineral N.

Lynch et al. (2012b) recently provided an extensive review on the influence of rotations, GMrs, and amendments on supplying N for organic potato (*Solanum tuberosum* L.) productivity and quality in eastern Canada. The effect of N supply

---

TABLE 4.3
Dry Matter Yield and N Uptake by Cover Crops in Humid Regions

GMr	Dry Wt (ton ha^{-1})	N Content (kg ha^{-1})	Site	Termination Date	Study
Hairy vetch	5.10	172	Maryland, United States	Average (late April–June)	Teasdale et al. (2012)
	3.15	117	Maryland, United States	Average (late March–April)	Clark et al. (1994)
Cereal rye	2.13	31	Maryland, United States	Average (late March–April)	Clark et al. (1994)
Buckwheat	2.1–3.7	52–65	Quebec, Canada	May	N'Dayegamiye and Tran (2001)
Red clover	1.9	71.5	Ontario, Canada	May	Vyn et al. (2000)
	0.93	146	Ontario, Canada	Late October	Bruulsema and Christie (1987)
Oilseed radish	0.95	19	Ontario, Canada	Late fall	Vyn et al. (2000)

and GMr type on pests (wireworm, Colorado potato beetle) and population dynamics is also discussed (Boiteau et al. 2008). On commercial farms in Canada, organic potatoes are typically included in extended (4–5 year or more) rotations that include small grains and leguminous forages or GMrs. Adequate N supply is critical for potato canopy development, tuber initiation, and yield. Inadequate N results in low yields and tuber size, while excessive N supply results in excessive haulm (vegetative) growth and delayed tuber initiation and maturation leading to reduced yields. Delayed maturation also exposes the crop to greater risk of losses due to late blight. In humid regions such as eastern Canada, after crop harvest and during the fall and winter months, any residual soil mineral N (RSMN) can be susceptible to denitrification and is usually lost by leaching as nitrate (de Jong et al. 2007; Lynch et al. 2008a; Sharifi et al. 2009).

A high-yielding potato crop is estimated to require between 2.5 and 5.9 kg of N per Mg of yield. Finckh et al. (2006) suggests NUE may be improved under low-input and organic systems due to potentially improved efficiency in use of sunlight. In general, while the N supply to the crop varies with climatic conditions and among years, organic potato yields in the range of 25–30 Mg ha^{-1} can be obtained without added N supplementation and where the rotation or GMr supplies 100–130 kg N ha^{-1} (Sullivan et al. 2007; Liu et al. 2008; Lynch et al. 2008a,b, 2012b). Lynch et al. (2012b) report on 6 years of research trials in Atlantic Canada on organic potatoes, with yields in the range of 30–35 Mg ha^{-1} and plant N uptake 100–125 kg N ha^{-1} following either red clover or hairy vetch, with the added benefit of RSMN remaining low. Delaying ploughing of a red clover GMr until spring increased crop N supply by approximately 20–30 kg N ha^{-1} (Lynch et al. 2008b). In contrast, in a 13-year University of Maine trial, a 1-year alfalfa/timothy GMr supplied only 99 kg N ha^{-1} (Sharifi et al. 2008).

A growing range of commercially produced organic amendments, such as composts and dehydrated pelletized manures, are changing options for providing supplemental N for organic producers. In organic potato production systems, supplying supplemental N through either composts or dehydrated manures improved both total and marketable tuber yield, although RSMN and risk of leached nitrates also increased as rates of amendment increased (Lynch et al. 2008a, 2012b). A "catch" cover crop, if planted early enough, can reduce nitrate leaching losses by uptake of RSMN into the plant biomass. A cover crop can also prevent erosion, which can be a source of sediment N loss. A cover crop can increase the infiltration rate of water into soil and protect the structure of the soil from the impact of rainfall (Delgado 1999). In the following season, the crop is ploughed under, and otherwise potentially lost N can be mineralized to provide PAN to the succeeding crop (Thorup-Kristensen and Dresboll 2010). Cereals such as oat and rye have an expansive rooting system excellent for N scavenging and reducing soil erosion, with rye achieving biomass accumulation between 0.8 and 2.9 Mg ha^{-1} in the northeastern United States (Snapp et al. 2005). Shipley et al. (1992) examined the residual soil N uptake by several winter cover crops planted following corn harvest on a field site in Maryland during the course of the fall and winter. The following April, the aboveground biomass N of the cover crops indicated that the cereal rye was the most efficient N scavenger with a 45% recovery of the fall residual soil N, followed by annual ryegrass (*Lolium multiflorum* Lam.) (27%), hairy vetch (10%), and crimson clover (*Trifolium incarnatum* L.) (8%). Cereal catch crops tend to be the most cold tolerant and will mineralize N the most slowly due to a relatively high C/N ratio. Broadleaf nonleguminous catch crops such as oilseed radish and forage radish (*R. sativus* var. *longipinnatus*) have deep tap roots capable of sequestering N rapidly in the fall, in the plough layer (Thorup-Kristensen 2001). A 3-year lysimeter study in Maryland comparing forage radish, oilseed radish, rape (*B. napus* L.), and rye as potential crops for N retention found that the forage radish accumulated the greatest quantity of soil N in the fall. Alternatively, the radish crops were freeze-killed in the winter, and on sandy soil, the decomposition of the radishes in the early spring caused accumulation of NO_3-N in the soil to depth. The authors suggest that in coarse textured soils, cover crops such as rape and rye that survive the winter and continue to accumulate N from the soil in the spring may be a more environmentally wise choice (Dean and Weil 2009).

Mirsky et al.'s (2012) report on 21 site years of trials across the mid-Atlantic region of the United States where corn and soybean were planted in no-till rolled mulch following fall-seeded hairy vetch and fall rye, respectively. GMr biomass achieved >8 Mg ha^{-1} in 50% of the trials, an amount considered optimum for weed suppression (ranging from 4 to 8 Mg ha^{-1} otherwise). In North Dakota, in contrast, biomass production of these two GMrs rarely exceeded 5 Mg DM ha^{-1}. Where moisture is not limiting, manipulating time of planting, seeding rate, and fertilization regime are recommended to improve GMr biomass production. When combined with subsurface-banded poultry litter applied as a supplement to corn, and the weed suppression benefits provided through optimizing a vetch GMr mulch, a multitactical strategy combining GMrs and fertility supplements for weed and fertility management for some organic grain production systems is emerging. The implications, with respect to soil quality, of adoption of no-till GMr termination strategies are discussed in Chapter 5.

4.4   COMPOSTS, MANURES, AND ANAEROBIC DIGESTATES FOR N AND P SUPPLY

4.4.1   PERMITTED SOIL AMENDMENTS

The use of manures and composts, regardless of source, as amendments within organic systems must follow the regulated organic production protocols as described in *Organic Production Systems; General Principles and Management Standards* (Government of Canada 2006a). Producers with livestock on farm must utilize all available manure before applying manure acquired off-site. The application timing, rate, and method must strive to minimize runoff in surface waters and have minimal contribution to groundwater contamination. Fresh manure must be incorporated into the soil no later than 90 days before the harvesting of a crop that does not contact the soil and that is used for human consumption (e.g., corn, small grain) and 120 days for crops having edible parts that contact the soil or soil particles (potatoes, carrots, leafy vegetables, etc). To be considered compost, the composting process must maintain a temperature of 55°C for at least 4 consecutive days, with proper mixing to ensure the entirety of the material has undergone the heating process. Composts must also be derived from permitted feedstocks such as animal manures, animal by-products, plants, and plant by-products (e.g., sawdust, grass clippings, card-board). Processed animal manures such as those dehydrated by heat treatment and/ or mechanical processing are also permitted as long as the source material is an acknowledged permitted substance.

4.4.2 Nitrogen Mineralization and Plant Availability of N from Soil Amendments

Conversion of organic forms of N to plant-available inorganic forms is a key component of the N-cycle in any terrestrial ecosystem. Of agronomic importance is whether there is net mineralization or net immobilization of N occurring in the soil. As is the case with GMr quality (see discussion in Section 4.2.1), the C/N ratio of an organic soil amendment has traditionally been an indicator of whether immobilization or mineralization will occur (Qian and Schoenau 2002). A commonly used threshold suggests that amendments with a C/N ratio >20 will cause net immobilization and ratios less than 20 will cause net mineralization (Loecke et al. 2012). This is a generalized rule, and in many cases, especially in field trials, net immobilization/ mineralization may occur at lower or higher ratios.

A comprehensive incubation and field study performed by Gale et al. (2006) compared PAN between several fresh and composted materials (broiler litter, dairy solids, and yard trimmings) (Table 4.4). C/N ratios between 4:1 and 15:1 were determined to be approximate indicators of PAN in both the field and lab trials. As C/N increased, PAN decreased. Soil laboratory incubations are often an initial step in determining the kinetics of N release from a particular organic source. However, while incubations offer more control of external variables than field experiments, they limit the applicability of using the data generated at the field scale, as environmental factors can have a drastic influence on mineralization rates (Paul 2007). Approaches that consider not only the C/N ratio but also the quality of the C (i.e., whether comprised of more recalcitrant material or not) to more closely determine mineralization rates and PAN have also been considered (Wichuk and McCartney 2010). An extensive incubation study by Lashermes et al. (2010) examined 273 organic sources including composted manure, animal by-products and yard wastes, and their mineralization of N over a 28-day period. Organic N content, the soluble organic C, cellulose, and lignin fractions were identified as key parameters for predicting potential N availability and not the more traditional C/N ratio. In practice, however, it is important for producers to know the feedstock used to create the compost and not rely solely on the chemical composition from compost quality analyses.

4.4.3 Manures

An inevitable consequence of livestock production is the generation of manure. Manure can be both a source of nutrients for crop production and a serious environmental concern. Statistics Canada (2008) estimated over 180 million tonnes of manure was produced in 2006. The majority of the manure was produced by cattle (72%) with beef contributing the most, followed by swine (9%), and poultry (3%).

TABLE 4.4

Amendment N Analysis and PAN on Selected Field, Incubation, and Simulation Models of Various Organic Amendment Sources

Amendment	Total N g kg^{-1}	NH$_4$-N g kg^{-1}	PAN % of Total N	C/N Ratio	Study	Study
Fresh poultry manure	26	14.73	57		Field	Munoz et al. (2008)
Pelleted poultry manure	35	8.6	53		Field	Munoz et al. (2008)
Composted poultry manure	10	2.5	14		Field	Munoz et al. (2008)
Broiler litter	35–41	4–8	27–54	9–10	Field	Gale et al. (2006)
Broiler compost	37–42	4.6–8.7	28–40	8–9	Field	Gale et al. (2006)
Composted dairy manure	5.3	0.1	4		Field	Munoz et al. (2008)
Dairy solids	13–21	0.6–2.5	5–17	20–32	Field	Gale et al. (2006)
Dairy solid compost	19–20	0.4–0.7	–2 to 16	19–20	Field	(Gale et al., 2006)
Yard trimmings	16–24	4–2.3	13–18	11–15	Field	Gale et al. (2006)
Yard trimmings compost	13–20	1.6–0.3	–10 to 19	12–22	Field	Gale et al. (2006)
Composted beef manure	8.25	0.9	20	9.5	Field	Eghball and Power (1999a)
Fresh beef manure	11.67	7.5	38	13	Field	Eghball and Power 1(999a)
Composted turkey litter	32.5	4.2	33	10.15	Estimation[a]	Woodley (2012)

Source	Total N (%)	NH$_4$-N (g L^{-1})	PAN (kg per 1000 L)	C/N ratio	Study	Reference
Liquid swine manure	4	267	2.7		Estimation	Brown (2008)
Liquid dairy manure	3.5	152	2.2		Estimation	Brown (2008)
Liquid beef manure	3.1	133	1.6		Estimation	Brown (2008)

Source	Total N (%)	NH$_4$-N (%)	PAN (kg tonne^{-1})	C/N ratio	Study	Reference
Fresh turkey litter	2.74	0.8	13		Estimation	Brown (2008)
Horse manure	0.50	0.07	1.3		Estimation	Brown (2008)
Biosolids (dewatered)	3.76	0.34	13.3		Estimation	Brown (2008)
Sheep manure	0.80	0.23	2.9		Estimation	Brown (2008)

[a] PAN estimation generated by NMAN 2.1 software modeling based on manure composition.

The intensity of manure production and proportion of livestock producing the manure vary widely geographically. The main areas of manure production are southern and central Alberta, southwestern Ontario, and southeastern Quebec. In Alberta, the manure is predominantly from cattle production, whereas in other areas of Canada, it is more dispersed among livestock (Statistics Canada 2008). A 2001 survey of Canadian farms by Statistics Canada (2004) indicated that 35.4% of producers applied manure in the fall, followed by 33.2% applying manure in the spring, 25.9% in the summer, and 5.5% in the winter. The type, rate, and timing of application vary widely based on manure source, farm size, and soil type.

Poultry manure is commonly applied as a litter that often contains feathers and bedding material. The N is approximately 80% uric acid and urea. The uric acid and urea are rapidly hydrolyzed to NH_4^+ (Mahimairaja et al. 1994). If pH is high, the NH_4^+ quickly converts into NH_3. The NH_3 can easily be lost from the system through volatilization and, in some cases, within 7–10 days, this loss can account for 48% of the total N excreted (Mahimairaja et al. 1994). Poultry manure generally contains the highest proportion of total N of any animal manure source (Table 4.4). Due to the high availability of inorganic N, there is potential for salt injury "burning" to a crop if poultry manure is not applied appropriately (Nahm 2003). Immediate incorporation of poultry manure is recommended.

The vast majority of swine manure N (97%) is land applied as a liquid slurry (Environment Canada 2007). NH_4-N can account for 27%–45% of the total N in liquid form (Brown 2008). Dairy manure at excretion can have between 10% and 20% of the total N in the form of urea. Due to the high NH_4-N, volatilization loss is a major concern at application time. Injection or immediate incorporation can significantly reduce losses due to volatilization. Applying the manure during the cool period in the morning and at low winds is another suggested practice. Using a knife injection method for application of liquid manure has the lowest atmospheric N losses to the environment and is the most valuable to crop growth. In contrast, composted solid beef and dairy manure have most of the N assimilated into the organic fraction (Table 4.4).

The chemical composition of manures varies from farm to farm and even during different growth stages of the livestock making generalized recommendations a challenge. This is highlighted by Davis et al. (2002) who looked at the inherent variability of manure stockpiles. In this study, researchers sampled 6–10 livestock operations for each manure type, subsampling each farm ten times, and then compared the results to each farm and published data. The numbers of samples required for a 95% confidence interval for total N were 55, 19, and 17 for chicken, beef, and dairy manures, respectively. Dou et al. (2001) sampled dairy, swine, and broiler manure from several farms with varied handling systems and tested the manures for total N and ammoniacal N immediately prior to field application. Farms that agitated their manure prior to loading needed only three to five samples of manure to achieve a ±10% variance from the mean, for total N. With no agitation, approximately 40 samples were required before a ±10% variance from the mean was attained. While regional databases are sometimes available depending on province (Brown 2008), for improved confidence in the nutritive value of individual manures, farm-specific, repeated sampling and historical record keeping are recommended to create a baseline for use in best manure-management practices.

4.4.4 COMPOST

Composting is the enhancement of the natural process of decomposition, converting organic material such as C and N into more stable forms mediated by microorganisms (Paul 2007). The traditional method of composting is a turned system. Regular turning of the compost allows improved aeration and more homogenous decomposition of the organic matter. There are three typical stages in the composting process: mesophilic, thermophilic, and the maturing stage (Paul 2007). In the first stage of the composting process that is dominated by mesophilic bacteria and fungi, large quantities of labile organic C (sugars, proteins, etc.) undergo rapid decomposition, causing elevated temperatures (20°C–45°C). Temperatures continue to rise, with the ideal temperature reaching between 52°C and 60°C, indicating the thermophilic stage. During this stage, the maximum breakdown of organic material occurs and at temperatures exceeding 55°C, human pathogens are destroyed (Bernal et al. 2009). Such high temperatures will also sterilize weed seeds that are often a concern when applying fresh manures (Larney et al. 2006a). As the availability of easily decomposable organic matter decreases, the metabolic heat generated by the microbial population begins to decline. The mesophilic bacteria re-establish, actinomycetes flourish, and in combination with fungi, decomposition of more resistant materials (lignin, cellulose, etc.) occurs during the maturing stage (Bernal et al. 2009).

During the composting process, typically 50% of the C is lost to the atmosphere as CO_2. Ideally, minimal N is volatilized and the majority is incorporated in the microbial biomass as organic N. Nitrogen lost through volatilization can be significant during the composting process (Larney et al. 2006b, 2008). Larney et al. (2006b) observed up to 54.5% N mass balance loss during the composting of beef manure as compared to a 22.5% loss when manure was stockpiled. When composting animal manures with high inorganic N contents, it is common to co-compost with a feedstock that has a high C/N ratio. General guidelines recommend a starting C/N ratio of 25:1–30:1. The high C/N ratio feedstocks such as straw or woodchips will help immobilize the inorganic N that would otherwise be susceptible to loss through NH_3 volatilization and to some extent leaching (Bernal et al. 2009).

In a study examining the effectiveness of different bulking agents in preventing NH_3-N loss when composting poultry manure, wheat straw was the most effective, reducing loss by 33.5% as compared to woodchips (20.6%) (Larney et al. 2008). The high C/N ratio and lignin content prevented significant N immobilization, whereas the comparatively more decomposable wheat straw allowed a microbial population to increase to a level where NH_4-N was immobilized from the system (Mahimairaja et al. 1994). The feedstock or "bulking agent" can also improve aeration within the compost pile and change overall particle size; both factors can have a significant effect on N dynamics (Larney et al. 2006a). Traditionally, the feedstock choice for cattle manure was cereal straw, which was used as a bedding material. By-products from the forestry industry such as sawdust and woodchips are now being used as viable bulking agents for compost (Larney et al. 2006a). There are suggestions that feedstock with high soluble C content decreases N losses during the composting process (Lynch et al. 2006). During the composting processes, it is expected that the

C/N ratio will decline and NO_3-N levels will increase. This suggests that while C is being lost, N is changing from NH_4-N through microbial assimilation (Paul 2007).

4.4.5 ANAEROBIC DIGESTATES

While more established in Europe and Asia (Moller and Muller 2012), there is increasing interest in Canada in the use of anaerobic digestion for biogas production and ultimately energy generation (Frigon et al. 2012) on-farm. As the name implies, feedstock added to these systems undergoes decomposition in oxygen-limited systems. This process produces mainly CO_2 and methane (CH_4), which are often referred to as "biogas." While the production of renewable energy is often the main goal of installing these facilities, the process also produces digested materials referred to as digestates. Digestates are permissible in organic systems and fall under the category of a raw manure. Such digestates can be used as an amendment similar to compost (Alburquerque et al. 2012). The digestates can be applied as a slurry and PAN estimates are often based on NH_4-N concentrations (Moller and Muller 2012). However, digestates produced from municipal waste feedstock are prohibited for use in Canada (Government of Canada 2006a). Digestates created with high degradability such as poultry and swine manures tend to have high NH_4-N/N ratios.

4.4.6 MANURE AND COMPOST APPLICATION TO FIELD CROPS

There are arguments to be made either way for choosing manure or compost as a nutrient source on organic farms. If producers are transporting an amendment from off-site for application, then compost is arguably a more efficient method, as they contain more concentrated nutrients due to significantly less C and typically less water than fresh manures such as manure slurries (Miller et al. 2009). In addition, there are fewer odors associated with the application of compost. Also, composts have reduced potential for N loss through volatilization at the time of application, although substantial volatilization of NH_3, often at rates much greater than from stockpiled manure, can occur during the composting process (Larney et al. 2006b). In evaluating amendments for N supplying ability to a crop, manure and composts can have markedly different N release characteristics. For example, a 9-year study by Miller et al. (2010) compared the N availability of fresh beef cattle manure (FM), composted beef manure with straw bedding (CS), and composted beef manure with woodchip bedding (CW). Overall, the FM increased the soil N level the most but varied from year to year. The significant differences found from year to year were likely due to soil moisture and temperature variations. These two parameters can ultimately be the major drivers affecting N mineralization (Paul 2007).

Composts have been touted as offering improved N synchrony with crop N demands as compared to fresh manures. Studies have examined this hypothesis with varied success. Loecke et al. (2012) compared fresh versus hoop house composted solid swine manure on corn N uptake, with the bedding derived from corn stalks. In the first growing season, the fresh swine manure immobilized soil inorganic N for 9 weeks following application, while the composted swine manure caused immobilization for only 2–4 weeks. Nitrogen uptake in the aboveground biomass was 25%

higher in the composted treatments contrasted with the fresh manure-amended soil, with no differences in yield. In contrast, in the second season, the freshly manured soil mineralized significantly more N in the first 10–12 weeks than that of the compost.

Consideration of timing of application and total amount applied over time is important when using composts. Gale et al. (2006) observed immobilization of N for 30–60 days following incorporation of composted dairy solids, highlighting the importance of applying amendments far enough ahead of planting to avoid N-limiting soil conditions, although over time, composts generally contribute to the slow-release pool of N and other nutrients (Sanchez et al. 2004).

Caution should be taken if compost is applied at rates meant to provide the entire N requirement of the crop. Composts derived from animal manures tend to be high in P and K in relation to PAN as compared to fresh manures. Repeated application of composts for N goals can cause P loading in the soil and potential for P contamination to the surrounding environment (Reider et al. 2000). The use of leguminous crops in rotation and basing compost rates on P or K requirements is recommended. For example, Wortman et al. (2011) in Nebraska used composted beef manure ($\sim$30 Mg ha^{-1} applied once in 11 years) to offset potential P deficiencies in their organic perennial-grains cropping system. Lynch et al. (2004) demonstrated how the legume (red clover) component of a mixed forage (red clover–timothy) stand acted as an effective "N-buffer," maintaining forage yields and accommodating, through BNF, varying N supply from diverse composts or lower compost (and thus P and K) application rates. Where imported manures and composts are used, combining adequate legume composition in the cropping system thus reduces reliance on these materials and avoids excess P supply. In some provinces, such as Quebec, P-based nutrient management regulations prohibit excessive buildup of P in soil from continuous manure or compost use, beyond established soil test thresholds.

4.4.7 RESIDUAL N EFFECTS OF AMENDMENTS AND ESTIMATING DERIVED ECONOMIC BENEFIT

It is important to consider the residual contributions of N from amendments to the soil in the seasons after the initial application. Composts release a small fraction of the total N in the first season but can show sustained release in the following years. Paul and Beauchamp (1993) showed that composted cattle manure caused the lowest yield response in the first growing season as compared to urea or liquid dairy manure, but caused the highest response to yield in the second residual year. Reider et al. (2000) reported first-year corn yield declines with broiler and dairy composts as compared to inorganic fertilizer, while in the second and third year with repeated application, the composts achieved comparable yields. Eghball and Power (1999b) estimated N availability for fresh and composted beef manure at 40% and 15%, respectively, of the total N (Table 4.4). In the second year, 18% and 8% of total N was estimated available for the fresh and composted beef manure, respectively. The carryover effect of amendment application can last for decades. Reeve et al. (2012) showed a 1.6-fold higher total organic C, higher microbial biomass, and higher plant-available P and K in soils that received composted dairy manure (50 Mg ha^{-1}) for 16 years prior, compared to the control on dryland wheat. Lynch et al. (2006), in Atlantic Canada, reported composts applied to forage crops slowly mineralized at rates of approximately 5% per year.

Understanding the continued contribution amendments have on crop yields will affect economic considerations. Composts are an expensive input with commercial prices ranging up to \$137–\$160 Mg DM^{-1} (Endelman et al. 2010; Woodley 2012). The economically optimal rate (EOR) used commonly in conventional agriculture is the point at which any additional application of amendment would no longer return a yield increase large enough to balance the cost of the input. The challenge in organic farming is incorporating the carryover effects of the composts. If there is a yield impact from residual composts on the succeeding crop (Paul and Beauchamp 1993; Eghball and Power 1999b; Eghball et al. 2004; Reeve et al. 2012), then the EOR should be modified to incorporate this effect, as the input costs remain the same but the yield response now represents a second crop (Endelman et al. 2010).

Grain protein content can garner price premiums or penalties depending on the crop, grain purchaser, and end use. There is a correlation between increasing N rates and increasing grain N content (Olesen et al. 2009; Woodley 2012). Attempts should be made to include these correlations into the EOR equations. However, the complex nature of N management in organic systems can never be fully represented in these equations, although there is merit in attempting to establish an economic evaluation. The merit is to make confident and informed purchasing decisions that benefit the farmer and organic sector as a whole, especially in the ever-expanding market of amendment options available to organic producers.

4.4.8   Soil P Dynamics in Soils Receiving Manures and Composts

There is limited potential for altering manure and compost characteristics based on the targeted soil or plant needs. Application of manures can cause a significant accumulation of P in soils since the amount of manure applied to agricultural fields is generally based on the N requirement of the crop. In the year of application, about 40% of the P in manure is available and another 40% is available in subsequent years (Brown 2008). Table 4.5 gives an indication of the typical quantity of available nutrients from various livestock types (Brown 2008). The data show that the average N/P ratios for liquid and solid manures and compost are 2.3, 1.0, and 0.5, respectively; however, a typical N/P uptake ratio for corn is 7.5 (Sato et al. 2005). Therefore, if manure or compost is applied to meet crop N requirements, a surplus of P is applied. If manure or compost is applied intermittently on this basis, this surplus P is an excellent source for several crops (Main et al. 2013). However, if manure or compost is applied annually to meet crop requirements, a substantial surplus of P will accumulate.

Excess P from manure application can move deeper in soils by association with colloidal clay or OM and may leach from the topsoil to groundwater (Eghball et al. 1996; Stockdale et al. 2002; Ashjaei et al. 2010). Although the presence of P in groundwater is not, on its own, a major threat to human health or environmental quality, groundwater P can eventually drain into surface water, where it causes eutrophication. Therefore, as noted previously, to avoid applying excess P and

TABLE 4.5

Typical Amounts of Available Nutrients from Different Types of Organic Nutrient Sources[a]

Manure Types	Dry Matter (%)	Available N[b] (kg 1000 L⁻¹ or kg ton⁻¹)	Available P₂O₅ (kg 1000 L⁻¹ or kg ton⁻¹)	Available K₂O (kg 1000 L⁻¹ or kg ton⁻¹)	# Samples (kg 1000 L⁻¹ or kg ton⁻¹)
Liquid manures					
Cattle	7.9	1.6	0.7	2.3	85
Dairy	8.4	1.8	0.8	2.6	948
Hog	3.7	2.7	1.2	1.9	1160
Poultry	10.5	5.8	2.8	3.2	137
Milk-fed veal	1.5	0.55	0.2	1.9	3
Solid manures					
Cattle	28.4	1.9	2.2	6.1	184
Dairy	25.0	1.8	1.5	5.3	174
Hog	30.2	4.0	4.3	6.0	61
Poultry	55.3	10.6	11.0	13.4	809
Sheep	33.8	2.9	2.6	8.4	57
Horse	37.4	1.3	1.4	4.6	41
Grain-fed veal	28.8	2.2	1.7	5.1	18
Composts					
All types (average)	38.9	2.0	4.1	8.9	63
Cattle	38.3	1.7	2.6	11.9	29

[a] Data from organic nutrients sources analysis provided from Brown (2008).

[b] Nitrogen based on spring application, incorporated within 24 h.

prevent P-rich leachate entering water systems, the additional supply of N by utilization of rotations containing legumes as cash crops, GMrs or forages, is necessary (Lynch et al. 2004; Schroder et al. 2011).

Phosphorus is the most immobile, inaccessible, and unavailable nutrient among all macronutrients in soils (Holford 1997). Phosphorus exists as both organic and inorganic forms in the soil and almost always exists as phosphate minerals in nature. Phosphorus minerals vary in their solubility, which is highly dependent on pH of the soil solution, the surface area of soil particles, and the concentration of minerals such as Ca, Al, and Fe in soils that can interact with P (Section 4.1.1). Figure 4.4 shows the interaction between various soil P pools and their transformations in organically amended soils receiving animal manures and composts. Understanding the distribution of P forms in manures and composts, as well as how these organic amendments impact soil chemistry, is crucial to managing organically amended soils effectively.

Animal manures and composts contain significant amounts of both inorganic and organic P. The majority of P found in these amendments is generally inorganic P (Tables 4.6 and 4.7), and the distribution of both organic and inorganic P forms

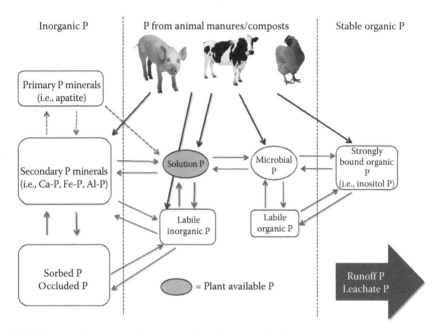

FIGURE 4.4 Suggested soil P dynamics under organically managed soils. Dashed arrows express slow transformations. (Modified from Sharpley, A., in *Handbook of Soil Sciences, Resource Management and Environmental Impacts*, 2nd edn., CRC Press, Boca Raton, FL, pp. 11–15, 2012.)

varies widely depending on its source, the animal's physiology and diet, bedding materials, and the method of storage and preparation (Leytem and Westermann 2005; Ashjaei et al. 2010). In a study of seven different composts, 73%–96% of total P was inorganic P (Gagnon et al. 2012). Gagnon et al. (2012) also reported that the proportion of total P as inorganic P in soils amended with these composts was not significantly different from those receiving synthetic fertilizers, even though the proportion of labile P in compost-applied soils (51%–58%) was significantly less than that in fertilizer-applied soils (82%). The labile P derived from a range of amendments generally followed the pattern: liquid manure–amended soils > solid manure– or compost-amended soils (Leytem and Westermann 2005). He et al. (2009) reported that there were significant amounts of moderately soluble Ca- and Mg-P species in organic dairy manure, which were gradually released as plant-available P in soils. This result is consistent with various manure-amended soils (Sharpley et al. 2004; Sato et al. 2005), where high contents of P and Ca in poultry and swine manure–applied soils were found as compared to untreated soils. The large amounts of Ca, which exist in manures and composts, often increase soil pH, and the added bicarbonates and organic acids with carboxyl and phenolic hydroxyl groups increase buffering effects (Sharpley and Moyer 2000). In addition, organic acids may delay the crystallization and transformation of stable Ca–P forms such as hydroxyapatite by chemical bonding or adsorption onto the

TABLE 4.6

Distribution of P Forms in Various Manures/Composts Detected by Solution ^{31}P NMR Spectroscopy in NaOH-EDTA Extractant

P Forms	Manure Types	Proportion of Total P (%)	References
Inorganic P			
Phosphate	Swine feces, slurry	47–94	Leytem et al. (2004), Leytem and Thacker (2008), Leytem and Thacker (2010), Sharpley and Moyer (2000), Turner (2004)
	Cattle manure	62–70	He et al. (2007), Turner (2004)
	Dairy manure	79–89	Hill and Cade-Munum (2009), Sharpley and Moyer (2000)
	Dairy compost	92	Sharpley and Moyer (2000)
	Poultry litter	34–91	He et al. (2007), Hill and Cade-Munum (2009), Sharpley and Moyer (2000), Turner (2004)
	Poultry compost	87	Sharpley and Moyer (2000)
Pyrophosphate	Swine feces	<3	Leytem et al. (2004), Leytem and Thacker (2008), Leytem and Thacker (2010), Turner (2004)
	Cattle feces	1–7	He et al. (2007), Turner (2004)
	Poultry litter	< 1	He et al. (2007), Hill and Cade-Munum (2009), Turner (2004)
Organic P			
Phosphate monoester (inositol phosphate)	Swine feces	0–46	Leytem et al. (2004), Leytem and Thacker (2008), Leytem and Thacker (2010), Turner (2004)
	Cattle manure	5–6	He et al. (2007), Turner (2004)
	Poultry litter	3–51	He et al. (2007), Hill and Cade-Munum (2009), Turner (2004)
Phosphate diesters	Swine feces	0–2	Leytem et al. (2004), Leytem and Thacker (2008), Turner (2004)
	Cattle manure	4–11	He et al. (2007), Turner (2004)
	Poultry litter	0–7	He et al. (2007), Hill and Cade-Munum (2009), Turner (2004)

crystalline surfaces (Sato et al. 2005). Therefore, animal manures and composts can affect the chemistry of soil P altering both the amount and the distribution of the various P fractions (Ohno et al. 2005; Ashjaei et al. 2010) resulting in the P chemistry in organically amended soils differing from soils amended by synthetic fertilizers.

The organic matter in manures and composts increases the soil microbiological population and activity, which can play an important role in solubilizing inorganic P

TABLE 4.7

Distribution of P Forms in Various Manures/Compost-Amended Soils Detected by Solution ^{31}P NMR Spectroscopy in NaOH-EDTA Extractant

P Forms	Manure Types	Proportion of Total P (%)	References
Inorganic P			
Phosphate	Swine slurry	80.9	Dou et al. (2009)
	Dairy manure	78–89.5	Dou et al. (2009), Hansen et al. (2004)
	Poultry manure	81–91	Dou et al. (2009)
	Duck manure	93	Dou et al. (2009)
	Spent mushroom compost	45	Dou et al. (2009)
Pyrophosphate	Swine slurry	1.33	Dou et al. (2009)
	Dairy manure	0.43–1.8	Dou et al. (2009), Hansen et al. (2004)
	Poultry manure	0.08–0.69	Dou et al. (2009)
	Duck manure	0.35	Dou et al. (2009)
	Spent mushroom compost	4.75	Dou et al. (2009)
Polyphosphate	Swine slurry	0	Dou et al. (2009)
	Dairy manure	0–2.15	Dou et al. (2009)
	Poultry manure	0.71–1.25	Dou et al. (2009)
	Duck manure	1.51	Dou et al. (2009)
	Spent mushroom compost	0	Dou et al. (2009)
Organic P			
Phosphate monoester	Swine slurry	8.41	Dou et al. (2009)
(inositol phosphate)	Dairy manure	0–16.2	Dou et al. (2009), Hansen et al. (2004)
	Poultry manure	3.24–7.3	Dou et al. (2009)
	Duck manure	1.81	Dou et al. (2009)
	Spent mushroom compost	n/a	Dou et al. (2009)
Others	Swine slurry	9.40	Dou et al. (2009)
	Dairy manure	4.73–19.40	Dou et al. (2009)
	Poultry manure	4.60–9.66	Dou et al. (2009)
	Duck manure	5.1	Dou et al. (2009)
	Spent mushroom compost	47.24	Dou et al. (2009)
Phosphate diesters	Swine slurry	0	Dou et al. (2009)
	Dairy manure	0–1.53	Dou et al. (2009), Hansen et al. (2004)
	Poultry manure	0–0.36	Dou et al. (2009)
	Duck manure	0	Dou et al. (2009)
	Spent mushroom compost	2.88	Dou et al. (2009)

(Stockdale et al. 2002) and mineralizing organic P. Certain organic P forms are transformed into inorganic P forms by hydrolysis, catalyzed by phosphatase enzymes. Oberson et al. (1996) reported that the activity of acid phosphatases is higher in organically managed soils. Organic P speciation in manures, composts, and organically amended soils has been studied intensively by several researchers using solution ^{31}P nuclear magnetic resonance (NMR) spectroscopy technique (Tables 4.6 and 4.7). Organic P is classified into phosphate esters, phosphonates, and phosphoric acid anhydrides. Phosphate esters are the dominant organic P species in soils and can be subclassified into phosphate monoesters and phosphate diesters. Phosphate monoesters mainly occur as inositol phosphate, with small amounts of sugar phosphates, phosphoproteins, and mononucleotides (Turner et al. 2005). Phosphate diesters typically comprise less than 10% of total organic P in soils and include nucleic acids (DNA and RNA), phospholipids, and teichoic acids. Table 4.7 shows that substantial proportions of the P are inositol P, especially in swine and poultry manures. Inositol P, which has high charge density, adsorbs strongly to soils to form stable phytate complexes (Dou et al. 2009). While P accumulation is assumed to occur in organically amended soils as phytate, accumulations of phytate have not been observed in manure-applied soils (He et al. 2007; Hill and Cade-Munum 2009). In P-saturated soils, such as long-term manure-applied soils, immobilization of P may be limited, and leaching or runoff losses of phytate may occur (He and Dou 2010).

4.4.9 Weed Proliferation

A discussion on N and P supply from manures, composts, and other organic amendments would not be complete without noting also the potential impact these materials can have on weed proliferation in organic field crops. There have been several studies comparing compost and manure on weed biomass and proliferation. A rotation study examining zero-input rotations and manure input rotations (50 NH_4-N kg ha^{-1} swine and dairy slurry) with and without catch crops showed weed biomass increased in all three of the field sites. In every rotation treatment, a mean grain-yield decrease of 0.2 Mg DM ha^{-1} in winter rye and winter wheat was attributed to weed pressure (Olesen et al. 2009). A study comparing composts with two different C/N ratios and cattle slurry found that total seed bank density was lowest in plots receiving the low C/N ratio compost and highest in the cattle slurry. It was suggested that the increase in microbial biomass found in the compost as compared to the slurry could be attributed to microbial seed deterioration during the composting process (DeCauwer et al. 2011). A variable rate study examining yield response between fresh broiler manure and composted turkey litter on organic white winter wheat by Woodley (2012) found significant biomass differences for Canada fleabane (*Conyza canadensis*) based on nutrient source and rate. High rates of fresh broiler litter (10–20 t ha^{-1}) caused significant weed growth that caused a yield decline in the wheat as compared to lower rates of application. The plots receiving composted turkey litter at similar application rates exhibited far less weed presence, suggesting the Canada fleabane growth is responsive to the inorganic N available in the fresh manure. It is important for producers to have an indication of weed biomass potential based on amendment, for timely weed management practices.

4.5   EFFICACY OF MICROBIAL INOCULANTS AND PHOSPHATE ROCK FOR ENHANCING PHOSPHORUS SUPPLY

Currently, there are no readily soluble P amendments identified in the Canadian organic standards that can be used in organic farming systems in Canada (Government of Canada 2006a,b). External inputs of P are limited to the application of compost, animal manure, and mineral amendments, such as PR. Because plants take up P from the soil solution as inorganic phosphate ions (HPO_4^{2-} and $H_2PO_4^{2-}$) and probably do not take up organic P forms directly (Harvey et al. 2009), organic P must be mineralized and inorganic P solubilized prior to plant uptake. Inoculants containing phosphate-solubilizing microorganisms (PSMs) are used in an attempt to improve the solubility of sparingly soluble forms of soil P or PR, and inoculants with AMF are used to increase the P uptake ability of crops. Implementing combinations of strategies, such as combining PR application with inoculation of beneficial microorganisms, probably has the best chance of improving P supply to organically managed crops compared to using one strategy alone. It is important to note that any management strategy aimed at solubilizing P is only effective in the short term and will accelerate long-term depletion of P from the system.

4.5.1   Phosphate-Solubilizing Microbial Inoculants

PSMs are ubiquitous in soils, comprising up to 50% of total bacterial populations and 0.5% of total fungal populations (Khan et al. 2009). The most important P-solubilizing bacteria include those of the *Bacillus* and *Pseudomonas* genera, while important P-solubilizing fungi include *Aspergillus* and *Penicillium* genera (Khan et al. 2009). Despite being less numerous than phosphate-solubilizing bacteria, phosphate-solubilizing fungi are more effective at solubilizing mineral P (Khan et al. 2009). The mechanism for P solubilization by microorganisms lies in their ability to produce organic acids, which act to solubilize P through chelation, in which organic acids form stable complexes with cations that are bound to insoluble P forms (Jones 1998) or through ligand exchange, where the organic anion competes with phosphate ions that are adsorbed to iron and aluminum oxides (Trolove et al. 2003). Nitrogen source also can affect P dissolution (Asea et al. 1988; Whitelaw et al. 1999) and is dependent on the PSM used (Asea et al. 1988).

Other mechanisms that may account for enhanced plant P uptake following inoculation with PSM include changes in root morphology (Gulden and Vessey 2000; Vessey and Heisinger 2001), increased nodulation (Gleddie 1993; Rice et al. 2000; Abd-Alla et al. 2001), and positive interactions with AMF (Kucey 1987; Asea et al. 1988). However, it should be noted that dual inoculation of pea with *R. leguminosarum* and *Penicillium bilaiae*, although increasing dry matter production, was also reported to decrease BNF compared to inoculation with *R. leguminosarum* alone (Downey and van Kessel 1990).

Inoculant formulations of PSM have been manufactured for agricultural production in many countries worldwide. In particular, *Penicillium* spp., which are free-living fungi isolated from a wide range of plant rhizospheres and soils (Harvey et al. 2009), have been used in commercial products. In Canada, *P. bilaiae* is the active ingredient in JumpStart® manufactured and marketed by Novozymes BioAg Group (Saskatoon, SK). It has been sold in Canada commercially since 1990 in various formulations and under different trade names including PB50® and Provide® (Leggett et al. 2007). The Novozymes' *P. bilaiae* products were not developed specifically for use in organic agriculture; therefore, components of the product may prohibit their use on organic farms depending on the standard of the local certification agency (Novozymes 2012).

In controlled-environment experiments, inoculation with *P. bilaiae* has improved growth and/or P uptake by many crop species including wheat (Kucey 1987, 1988; Gleddie et al. 1991; Anstis 2004), canola (Kucey and Leggett 1989), field bean (Kucey 1987; Gleddie 1993), pea (Downey and van Kessel 1990; Gleddie 1993) lentil (Gleddie 1993), and pasture legumes (Beckie et al. 1998; Rice et al. 2000). Reports from field studies investigating *P. bilaiae* applications yielded variable results. Hnatowich et al. (1990) and Gleddie et al. (1991) reported positive responses to *P. bilaiae* application in the field, whereas Chambers and Yeomans (1990, 1991) observed no statistically significant response. Takeda and Knight (2006) demonstrated that *P. bilaiae* grew better and solubilized more P from PR under neutral conditions than acidic conditions because of enhanced production of organic acids under neutral conditions. While currently untested, this may mean that *P. bilaiae* inoculants perform better in the moderately alkaline soils common to organic farms across Canada (Martin et al. 2007).

Responsiveness of a soil to fertilizer P application appears to be a partial determinant of responsiveness to *P. bilaiae* application. For instance, Bullock et al. (1990) found that of 46 sites used to grow wheat, seven sites were responsive to fertilizer P, and three of these seven sites responded positively to *P. bilaiae* application. Also, Androsoff et al. (1991) reported no effect of *P. bilaiae* inoculation of wheat at 13 sites that were also unresponsive to fertilizer P. More recently, Karamanos et al. (2010) undertook an analysis of 47 experiments carried out between 1989 and 1995 in which hard red spring wheat was grown with four rates of P fertilizer, with or without *P. bilaiae*. Wheat grown at 33 sites responded to P fertilization, wheat at 14 sites responded to *P. bilaiae* inoculation, but at only five of these sites was the response of wheat positive (Karamanos et al. 2010).

None of the aforementioned studies were performed specifically on organically managed land but are included to demonstrate the tremendous variability and unpredictability of these inoculants. It should be noted that inoculation of legumes with rhizobial inoculants similarly produced positive yield responses only one-third to one-half of the time (Vessey 2004) with reported success rates varying from 0% for dry bean grown in southern Alberta (McKenzie et al. 2001) to 100% for chickpea grown in Saskatchewan (Kyei-Boahen et al. 2002). Expecting positive yield and/or nutrient uptake responses to any biological amendment 100% of the time is unrealistic considering the effect that local soil and climate conditions can have on the growth of organisms.

4.5.2 ARBUSCULAR MYCORRHIZAL INOCULANTS

The use of commercial AMF inoculants is an emerging technology in field crop production in Canada (Jin et al. 2012). For a more detailed discussion of AMF and their role in organic field crop production, please refer to Chapter 3.

The production of commercial AMF inoculants historically has been limited by the technology used to produce the inoculants. Traditionally, "nurse plants" are grown with the desired AMF species and spores, hyphae, and infected roots harvested for use as inoculants. Recent advances using an in vitro culture protocol result in production of higher numbers of spores (Gianinazzi and Vosatka 2004) that can be commercially produced. However, not many species of AMF are culturable with this system, resulting in commercial inoculants being limited mainly to a single AMF species, namely, *Glomus intraradices* (Schwartz et al. 2006; Antunes et al. 2009). Considering that mycorrhizal effects depend on plant and AMF genotype interaction (Hamel and Strullu 2006), it seems unlikely that the limited AMF strains in commercial inoculants will be effective on a wide range of crop species.

The low P levels occurring on organic farms should encourage AMF colonization of susceptible crops and indeed in a field study in Saskatchewan, roots of organically grown field pea and lentil were highly colonized, with levels averaging 80% in field pea and 83% in lentil (Baird et al. 2010). These results suggest that organically grown grain legumes would not benefit from inoculation with a commercial AMF inoculant. A 3-year field study conducted in Alabama reported improved grain yields and enhanced uptake of N, P, and K into the grain in corn inoculated with AMF compared to controls (Adesemoye et al. 2008). The study was not conducted in an organically managed system but is one of only a few studies conducted in North America examining the efficacy of commercial inoculants under field conditions. Corkidi et al. (2004) compared ten commercial mycorrhizal inoculants on corn grown under nursery conditions in a soil-based medium and two soil-less substrates. Mycorrhizal colonization ranged from 0% to 50% and was influenced by growing media. However, colonization did not enhance plant growth. Only plants inoculated with products that did not promote mycorrhizal colonization increased their growth compared to noninoculated controls suggesting some other growth promotion mechanism was functioning independent of improved nutrient acquisition through the hyphal network. Jin et al. (2012) reported that inoculation of field pea with *Glomus irregulare* was ineffective at enhancing biomass and N and P uptake. In contrast, mixed species inoculation with *G. irregulare*, *Glomus mosseae*, and *Glomus clarum* promoted biomass production, N and P uptake, and N_2 fixation in field pea.

As it stands, compelling evidence is lacking to justify widespread use of AMF inoculants currently on the market for organic production systems. As the technology develops to produce inoculants with different AMF species than are currently available, and mixtures of species, the value of these inoculants should become clearer. Although the relationship between AMF species and plant species does not appear to be as specific as the relationship between *Rhizobium* species (and indeed strain) and plant species, it is clear that genotype interactions occur (Hamel and Strullu 2006), which could be exploited. However, exploiting AMF–plant symbioses will not improve P fertility in the long term and indeed will deplete the system of P more rapidly. Instead, implementing an integrated strategy, which combines occasional P input application as required along with inoculation of beneficial microorganisms, has the best chance of improving P supply to organically managed crops.

4.5.3 Phosphate Rock

PR with low heavy-metal content is permitted for use in organic production (Government of Canada 2006b). Despite its approval, organic farmers have been slow to adopt its use (Knight and Shirtliffe 2003) presumably due to a lack of evidence of its effectiveness in supplying P to crops. Moreover, PR is a nonrenewable resource—it is mined primarily for production of synthetic P fertilizers, and at current rates of extraction, depletion of commercial reserves is expected in 50–100 years (Cordell et al. 2009). Therefore, application of PR may run counter to the principle of nonexhaustion of natural resources held by proponents of organic agriculture (Lammerts van Bueren et al. 1999). Nevertheless, it is approved for use in organic farming systems as an external source of P and its efficacy must be evaluated.

PR, as the name suggests, is a ground form of rock with high phosphorus content. However, there are a wide range of ores that contain a variety of phosphate minerals with diverse mineralogy and chemistry and, hence, chemical reactivity (van Straaten 2002). Chemical reactivity is a measure of the PR's ability to release P for plant uptake. The properties of the rock that influence P supply to crops such as total P, reactivity, and particle size can vary greatly from deposit to deposit and even within deposits (Armiger and Fried 1957). The degree of carbonate substitution in the apatite mineral is an indicator of PR reactivity and can be determined by x-ray diffraction (van Straaten 2002). For practical purposes, the percentage of P in the rock that is soluble in 2% citric acid provides an indication of the relative solubility of PR since it correlates well with agronomic effectiveness (Chien et al. 2011). Whereas soluble P fertilizers such as triple superphosphate are nearly 100% soluble in 2% citric acid, the solubility of rocks considered to be medium to highly reactive can range between 3.9% and 5.7% in 2% citric acid (Chien et al. 2011). Phosphorus release from relatively high-reactivity PR is best in acid soils with low P and Ca concentrations, conditions common in tropical soils, but not typical of most agricultural soils in Canada.

Phosphorus from PR is supplied to plants through dissolution of the apatite mineral and the subsequent release of orthophosphate ions into the soil solution, which may then be taken up by plants. In soils that have pH higher than 6 and high concentrations of Ca^{2+} and orthophosphate, dissolution of PR will be significantly slowed (Bolan et al. 1997) regardless of geological or geographical origins of the rock (Fardeau et al. 1988). Soil properties, particularly pH, Ca^{2+}, buffering capacity, SOM content, and moisture holding capacity influence PR dissolution (Anderson et al. 1985; Bolland et al. 2001). In Canada, the direct application of PR without any other complementary amendment is unlikely to be effective because most soils under organic management are alkaline (Martin et al. 2007). Therefore, the effective use of PR must rely on enhancing the mechanisms that promote PR dissolution either prior to application or in situ.

Suggestions have been made that PR may be used as an external source of P to augment soil P levels on organic farms (Haynes 1992), especially those farms in regions like the Canadian Prairies, where livestock manures are not readily applied (Knight et al. 2010). In a survey of nine organic farms in the United Kingdom, P surpluses on five of the farms were attributed to additions of PR (Berry et al. 2003). However, it is unclear if applications of the PR contributed to P nutrition of the crops.

Field studies evaluating the effectiveness of PR to supply P to crops on organic farms in temperate regions have been limited. In a 2-year field study on an organic farm in Ontario, Canada, yields of a buckwheat GMr crop were not improved with the application of locally available igneous PR sources or with a sedimentary PR from Florida. However, the tissue P content of the crop increased with the application of a Florida sedimentary PR (2.8% soluble in 2% citric acid) relative to the nonamended control but only at very high application rates (400–800 kg P ha^{-1}) (Arcand et al. 2010). These high rates are cost prohibitive especially over the short term. Similarly, field application of a very low solubility sedimentary PR from Idaho applied at rates of 3.1 and 7.7 kg available P ha^{-1} (corresponding to 19 and 48 kg total P ha^{-1}) did not increase buckwheat yield, P uptake, or tissue P content of buckwheat (Rick et al. 2011). However, averaged over buckwheat, yellow mustard, and spring pea crops, P uptake (kg P ha^{-1}) increased with PR application (Rick et al. 2011), suggesting the effective supply of P from this PR of relatively low solubility was crop specific.

4.5.4 STRATEGIES TO ENHANCE P RELEASE FROM PHOSPHATE ROCK

PR dissolution is more efficient in the presence of plants than in the absence of plants (Bolan et al. 1997), and crop species may vary considerably in their ability to take up P from PR (Flach et al. 1987). Careful crop selection can therefore have an overwhelming effect on the extent of P uptake from the dissolution of PR. The primary mechanisms for plant-mediated dissolution of PR involve chelation and ligand exchange involving organic acid anions that are exuded from plant roots or to rhizosphere acidification due to root exudation of organic anions or H$^+$ (Hinsinger 2001). Root exudation of organic anions resulting in dissolution of PR has been well documented particularly for nonmycorrhizal plants such as white lupin (*Lupinus albus* L.) that possess proteoid roots (Hinsinger and Gilkes 1995) and *Brassica* spp., such as rape (Hoffland 1992). Some plants can induce acidification of the rhizosphere as a result of an alkaline uptake pattern—this has been documented for species such as buckwheat (van Ray and van Diest 1979; Bekele et al. 1983; Flach et al. 1987; Haynes 1992; Zhu et al. 2002) and actively N$_2$-fixing legumes (Aguilar and van Diest 1981). In addition, while there is little evidence for the direct solubilization of PR by AMF (Antunes et al. 2007; Ngwene et al. 2010), AMF used in combination with PSM may enhance yield and the plant uptake of P, particularly in soils with high OM (Barea et al. 2002).

As noted previously, GMr crops are an integral component of crop rotations on organic farms. However, selection of GMr crops that possess a mechanism for solubilizing PR may have the potential for improving P supply to subsequent crops. On an organically managed field study in Montana, Rick et al. (2011) found that P uptake was increased in mustard (*Brassica rapa* L.) and spring pea with application of an Idaho PR at 3.1 and 7.7 kg available P ha^{-1}. In contrast, P uptake of buckwheat did not appear to improve with addition of PR. The strong P uptake response for mustard was likely due to root exudation of organic anions (Rick et al. 2011), which enhance PR solubility through chelation and/or ligand exchange (Richardson et al. 2009), while the contribution of organic anions to rhizosphere acidification may be negligible (Hinsinger 2001). In contrast, PR dissolution by buckwheat and pea occurs due to root exudation of H$^+$ to counterbalance the net excess of cations taken up by the root

(Hinsinger 2001). Therefore, the authors suggest that dissolution of PR by organic anions was more effective than rhizosphere acidification in the calcareous soil due to its strong pH buffering capacity (Rick et al. 2011). Indeed, rapeseed—another *Brassica* species that has been shown to exude organic anions that can act to increase PR solubility (Hoffland 1992)—was found to access P from a low-reactive PR even in an alkaline soil (Chien et al. 2003). Therefore, the interaction between crop, soil, and PR will affect whether crops in a GMr–PR system can improve P supply to other crops. However, at least in the short term, application of PR even to GMrs that can access P from the PR has not yet proven to provide adequate P to subsequent crops. More long-term studies employing a variety of crop species in different soils and using the most reactive PR available may provide insight into whether the P that is taken up by the crop in the GMr phase is recycled into soil inorganic or organic P pools that are less recalcitrant than P held in the original apatite mineral. Overall, it is unlikely that the majority of plant species, beyond white lupin, red clover, and rape, exude enough organic acids to effectively solubilize PR (Hinsinger 2001).

Much of the work examining PSM has also been tested in conjunction with PR application either in laboratory incubations or in field trials (Kucey 1988; Kucey and Leggett 1989; Takeda and Knight 2006). The mechanisms for PR solubilization are the same as when inoculants of PSM are applied to soil alone—production of organic acids and release of H[+] by PSM promote PR dissolution. However, PSM that can solubilize PR in laboratory incubations may not be effective in the field (Richardson 2001).

Pretreatment of PR with PSM prior to field application may hold the most promise for the use of PR for delivering P to plants, regardless of plant species. A strain of *Aspergillus niger* (ATCC 9142) selected for its superior ability to produce citric acid showed consistently promising PR solubilization for three PR sources (Schneider and van Straaten 2007). Pretreatment of PR with organic wastes as a C source and inoculation with effective PSM prior to field application also has been proposed as an alternative to direct inoculation and PR application in the field and has been shown to increase crop yields in pot studies (Vassilev et al. 1996, 1998).

In organic farming systems, nonbiological amendments for improving PR dissolution are limited. However, dual application with elemental sulfur, which is allowed under organic certification standards, may improve the efficacy of PR (Chien et al. 2011). Because PR dissolution is enhanced under acidic conditions, biological oxidation of elemental S to H_2SO_4 can provide the conditions for increased solubilization of PR (Chien et al. 2011); however, this will depend on soil type and is effective in increasing available P in relatively acidic soils ($pH_{Ca} < 5.3$) (Evans and Condon 2009).

4.6 FUTURE DIRECTIONS AND RESEARCH NEEDS

The previous review of advances in strategies for N and P fertility management across organic field crop production systems in Canada suggests effective combinations of GMrs, and supplementary organic amendments can be developed and adapted to regional conditions and production systems, which enhance organic field-crop productivity, without compromising on maintaining a low environmental footprint of these systems.

However, there is a need for more research to examine the net benefits to soil labile N pools of GMrs of varying type and duration combined with advances in practical tools to gauge soil N supply within these production systems. These latter tools may include more frequent use of, for example, simple tools such as whole plant bioassays of potato N uptake (Lynch et al. 2012b) perhaps extended to other crops. Most organic producers have not attempted to gauge the N supply from soil and GMr residues as affected by their management system, and this simple approach could be more widely promoted. Other techniques showing some promise include use of ion-exchange resins and membranes as a means to assess PAN in spring and aid with decisions regarding further supplementation as shown for timothy (Ziadi et al. 1999), canola (Qian and Schoenau 2005), and organic potato systems (Sharifi et al. 2009). As discussed, innovative approaches to termination of GMrs that reduce reliance on tillage (see Chapter 6 as well) are increasingly being explored in organic field crop production systems, with promising multitactical strategies for weed and fertility management for some organic grain production systems emerging, which combine rolled GMrs and targeted, sometimes banded, fertility supplements (Mirsky et al. 2012).

On the basis of STP data collated to date, the P status of organic field crop production systems is potentially of increasing concern with respect to the potential for deficits and P surpluses alike. This undoubtedly varies by region due to differences in access to manure and other supplemental P sources. Whole-farm nutrient budgeting conducted on these farms, which captures the diversity of management strategies for field cropping and their influence on farm P status, would provide valuable information to help direct improvements in farm P sustainability. Fundamental ongoing work on soil organic and inorganic P composition and their dynamics under long-term organic field cropping is also providing important insights as to the degree to which biologically mediated organic P turnover, and mycorrhizal–plant relationships, can be relied upon to compensate for the often low farm P flows and low soluble P status of organically managed soils. Reliance on PR alone is unlikely to be a practical solution to this potential deficiency, and advances in inoculant technologies for mycorrhizae and phosphate-solubilizing microbial inoculants will no doubt also contribute to a combination of future strategies to improve P fertility and productivity of both GMrs and cash crops in organic field crop production systems.

REFERENCES

Abd-Alla, M.H., S.A. Omar, and S.A. Omar. 2001. Survival of rhizobia/bradyrhizobia and a rock-phosphate-solubilizing fungi *Aspergillus niger* on various carriers from some agro-industrial wastes and their effects on nodulation and growth of faba bean and soybean. *J. Plant Nutr.* 24:261–272.

Adesemoye, A.O., H.A. Torbert, and J.W. Kloepper. 2008. Enhanced plant nutrient use efficiency with PGPR and AMF in an integrated nutrient management system. *Can. J. Microbiol.* 54:876–886.

Aguilar, S. and A. van Diest. 1981. Rock-phosphate mobilization induced by the alkaline uptake pattern of legumes utilizing symbiotically fixed nitrogen. *Plant Soil* 61:27–42.

Alberta Agriculture. 1993. Legume green manuring. http://www1.agric.gov.ab.ca/$department/deptdocs.nsf/all/agdex133#benefits (accessed September 12, 2012). Alberta Agriculture and Rural Development.

Alburquerque, J.A., C. de la Fuente, and M.P. Bernal. 2012. Chemical properties of anaerobic digestates affecting C and N dynamics in amended soils. *Agric. Ecosyst. Environ.* 160:15–22.

Aman, C. and A. Amberger. 1989. Phosphorus efficiency of buckwheat (*Fagopyrum esculentum*). *Zeitschrift Pflanzenernahrung Bodenkunde* 152 (2):181–189.

Anderson, D.L., W.R. Kussow, and R.B. Corey. 1985. Phosphate rock dissolution in soil: Indications from plant growth studies. *Soil Sci. Soc. Am. J.* 49:918–925.

Androsoff, G., C. van Kessel, and R. Karamanos. 1991. Yield of PB50 inoculated and phosphorus fertilized wheat. In *Proceedings of the Soils and Crops Workshop*, pp. 186–193. University of Saskatchewan, Saskatoon, Saskatchewan, Canada.

Anstis, S.T. 2004. *Penicillium radicum*: Studies on the mechanisms of plant growth promotion in wheat. PhD dissertation, The University of Adelaide, Adelaide, South Australia, Australia.

Antunes, P.M., A.M. Koch, K.E. Dunfield, M.M. Hart, A. Downing, M.C. Rillig, and J.N. Klironomos. 2009. Influence of commercial inoculation with *Glomus intraradices* on the structure and functioning of an AM fungal community from an agricultural site. *Plant Soil* 317:257–266.

Antunes, P.M., K. Schneider, D. Hillis, and J.N. Klironomos. 2007. Can the arbuscular mycorrhizal fungus *Glomus intraradices* actively mobilize P from rock phosphates? *Pedobiologia* 51:281–286.

Arcand, M.M., D.H. Lynch, R.P. Voroney, and P. van Straaten. 2010. Residues from a buckwheat (*Fagopyrum esculentum*) green manure crop grown with phosphate rock influence bioavailability of soil phosphorus. *Can. J. Soil Sci.* 90:257–266.

Armiger, W.H. and M. Fried. 1957. The plant availability of various sources of phosphate rock. *Soil Sci. Soc. Am. J.* 21:183–188.

Asea, P.E.A., R.M.N. Kucey, and J.W.B. Stewart. 1988. Inorganic phosphate solubilization by two *Penicillium* species in solution culture and soil. *Soil Biol. Biochem.* 20:459–464.

Ashford, D.L. and D.W. Reeves. 2003. Use of a mechanical roller-crimper alternative kill method for cover crops. *Am. J. Altern. Agric.* 18:37–45.

Ashjaei, S., H. Tiessen, and J.J. Schoenau. 2010. Correlation between phosphorus fractions and total leachate phosphorus from cattle manure– and swine manure– amended soil. *Commun. Soil. Sci. Plant. Anal.* 41:1338–1349.

Askegaard, M., J.E. Olesen, I.A. Rasmussen, and K. Kristensen. 2011. Nitrate leaching from organic arable crop rotations is mostly determined by autumn field management. *Agric. Ecosyst. Environ.* 142:149–60.

Badaruddin, M. and D.W. Meyer. 1994. Grain legume effects on soil nitrogen, grain yield, and nitrogen nutrition of wheat. *Crop Sci.* 34:1304–1309.

Baird, J.M. 2007. Optimal seeding rates for organic production of field pea and lentil. MSc thesis, University of Saskatchewan, Saskatoon, Saskatchewan, Canada.

Baird, J.M., F.L. Walley, and S.J. Shirtliffe. 2010. Arbuscular mycorrhizal fungi colonization and phosphorus nutrition in organic field pea and lentil. *Mycorrhiza* 20:541–549.

Barea, J.M., M. Toro, M.O. Orozco, E. Campos and R. Azcón. 2002. The application of isotopic (^{32}P and ^{15}N) dilution techniques to evaluate the interactive effect of phosphate-solubilizing rhizobacteria, mycorrhizal fungi and *Rhizobium* to improve the agronomic efficiency of rock phosphate for legume crops. *Nutr. Cycl. Agroecosyst.* 63:35–42.

Baute, T., A. Hayes,, I. McDonald, and K. Reid. 2002. Agronomy guide for field crops. In *"Publication 811"*[2]. F. a. R. A. Ministry of Agriculture, Queen's Printer, ed. Queen's Printer for Ontario, Toronto, Ontario, Canada, pp. 50–82.

Beckie, H.J. and S.A. Brandt. 1997. Nitrogen contribution of field pea in annual cropping systems. 1. Nitrogen residual effect. *Can. J. Plant Sci.* 77:311–322.

Beckie, H.J., D. Schlechte, A.P. Moulin, S.C. Gleddie, and D.A. Pulkinen. 1998. Response of alfalfa to inoculation with *Penicillium bilaii* (Provide). *Can. J. Plant Sci.* 78:91–102.

Bekele, T., B.J. Clino, P.A.E. Ehlert, A.A. van der Mass, and A. van Diest. 1983. An evaluation of plant borne factors promoting the solubilization of alkaline rock phosphates. *Plant Soil* 75:361–378.

Bernal, M.P., J.A. Alburquerque, and R. Moral. 2009. Composting of animal manures and chemical criteria for compost maturity assessment. A review. *Bioresour. Technol.* 100:5444–5453.

Berry, P.M., E.A. Stockdale, R. Sylvester-Bradley, L. Philipps, K.A. Smith, E.I. Lord, C.A. Watson, and S. Fortune. 2003. N, P and K budgets for crop rotations on nine organic farms in the UK. *Soil Use Manag.* 19:112–118.

Berry, P.M., R. Sylvester-Bradley, L. Philipps, D.J. Hatch, S.P. Cuttle, F.W. Rayns, and P. Gosling. 2002. Is the productivity of organic farms restricted by the supply of available nitrogen? *Soil Use Manag.* 18:248–255.

Biederbeck, V.O. and O.T. Bouman. 1994. Water use by annual green manure legumes in dryland cropping systems. *Agron. J.* 86:543–549.

Biederbeck, V.O., O.T. Bouman, J. Looman, A.E. Slinkard, L.D. Bailey, W.A. Rice, and H.H. Janzen. 1998. Productivity of four annual legumes as green manure in dryland cropping systems. *Agron. J.* 85:1035–1043.

Biederbeck, V.O., W.A. Rice, C. van Kessel, L.D. Bailey, and E.C. Huffman. 1996. Present and potential future nitrogen gains from legumes in major soil zones in the Prairies. In *Proceedings of Saskatchewan Soils and Crops 1996.* pp. 441–445. University of Saskatchewan, Saskatoon, Saskatchewan, Canada.

Boiteau, G., D.H. Lynch, and R.C. Martin. 2008. Influence of fertilization on the Colorado potato beetle, *Leptinotarsa decemlineata* (Say), in organic potato production. *Econ. Ent.* 37: 575–585.

Bolan, N.S., J. Elliott, P.E.H. Gregg, and S. Weil. 1997. Enhanced dissolution of phosphate rocks in the rhizosphere. *Biol. Fertil. Soils* 24:169–174.

Bolland, M.D.A., R.J. Gilkes, and R.F. Brennan. 2001. The influence of soil properties on the effectiveness of phosphate rock fertilisers. *Soil Res.* 39:773–798.

Brady, N.C. and R.R. Weil. 2002. *The Nature and Properties of Soils*, 13th edn. Prentice Hall, Upper Saddle River, NJ.

Brandt, S.A. 1996. Alternatives to summerfallow and subsequent wheat and barley yield on a dark brown soil. *Can. J. Plant Sci.* 76(2):223–228.

Brandt, S.A. 1999. Management practices for black lentil green manure for the semi-arid Canadian prairies. *Can. J. Plant Sci.* 79, 11–17.

Brown, C. 2008. Available nutrients and value for manure from various livestock types. F. a. R. A. Ontario Ministry of Agriculture, ed. Queen's Printer for Ontario, Toronto, Ontario, Canada, http://www.omafra.gov.on.ca/english/crops/facts/08-041.htm

Bruulsema, T.W. and B.R. Christie. 1987. Nitrogen contribution to succeeding corn from alfalfa and red clover. *Agron. J.* 79:96–100.

Bullied, W.J., M.H. Entz, S.R. Smith Jr., and K.C. Bamford. 2002. Grain yield and N benefits to sequential wheat and barley crops from single-year alfalfa, berseem and red clover, chickling vetch and lentil. *Can. J. Plant Sci.* 82:53–65.

Bullock, P., L. Cowell, and C. van Kessel. 1990. *Penicillium bilaji* (PB50) and phosphorus fertilizer responses of yield of wheat and barley grown on stubble and summerfallow. In *Proceedings of the Soils and Crops Workshop.* pp. 81–88. University of Saskatchewan, Saskatoon, Saskatchewan, Canada.

CCME. 2002. Linking water science to policy: Effects of agricultural activities on water quality. In *Proceedings of a CCME Sponsored Workshop.* Canadian Council of Ministers of the Environment, Quebec City, Quebec, Canada.

Chambers, J.W. and J.C. Yeomans. 1990. The influence of PB-50 (*Penicillium bilaji* inoculant) on yield and phosphorus uptake by wheat. In *Proceedings of the Annual Manitoba Soil Science Meeting.* pp. 283–293. University of Manitoba, Winnipeg, Manitoba, Canada.

Chambers, J.W. and J.C. Yeomans. 1991. The influence of PB-50 on crop availability of phosphorus from soils and fertilizer as determined by ^{32}P dilution. In *Proceedings of the Annual Manitoba Soil Science Meeting*. pp. 375–387. University of Manitoba, Winnipeg, Manitoba, Canada.

Chien, S., L. Prochnow, S. Tu, and C. Snyder. 2011. Agronomic and environmental aspects of phosphate fertilizers varying in source and solubility: An update review. *Nutr. Cycl. Agroecosyst*. 89:229–255.

Chien, S.H., G. Carmona, J. Henao, and L.I. Prochnow. 2003. Evaluation of rape response to different sources of phosphate rock in an alkaline soil. *Commun. Soil Sci. Plant Anal*. 34:1825–1835.

Clark, A.J., A.M. Decker, and J.J. Meisinger. 1994. Seeding rate and kill date effects on hairy vetch-cereal rye cover crop mixtures for corn production. *Agron. J*. 86:1065–1070.

Clark, A.J., A.M. Decker, J.J. Meisinger, and M.S. McIntosh. 1997. Kill date of vetch, rye, and a vetch-rye mixture: I. Cover crop and corn nitrogen. *Agron. J*. 89:427–434.

Cordell, D., J.O. Drangert, and S. White. 2009. The story of phosphorus: Global food security and food for thought. *Global Environ. Change*: 19:292–305

Corkidi, L., E.B. Allen, D. Merhaut, M.F. Allen, J. Downer, J. Bohn, and M. Evans. 2004. Assessing the infectivity of commercial mycorrhizal inoculants in plant nursery conditions. *J. Environ. Hortic*. 22:149–154.

Crandall, S.M., M.L. Ruffo, and G.A. Bollero. 2005. Cropping system and nitrogen dynamics under a cereal winter cover crop preceding corn. *Plant Soil* 268:209–219.

Davis, J.G., K.V. Iversen, and M.F. Vigil. 2002. Nutrient variability in manures: Implications for sampling and regional database creation. *J. Soil Water Conserv*. 57:473–478.

De Cauwer, B., T. D'Hose, M. Cougnon, B. Leroy, R. Bulcke, and D. Reheul. 2011. Impact of the quality of organic amendments on size and composition of the weed seed bank. *Weed Res*. 51:250–260.

De Costa, W.A.J.M., M.D. Dennett, U. Ratnaweera, and K. Nyalemegbe. 1997. Effects of different water regimes on field-grown determinate and indeterminate faba bean (*Vicia faba* L.). I. Canopy growth and biomass production. *Field Crops Res*. 49:83–93.

de Jong R, J.Y. Yang, C.F. Drury, E.C. Huffman, V. Kirkwood, and X.M. Yang. 2007. The indicator of risk of water contamination by nitrate-nitrogen. *Can. J. Soil. Sci*. 87:179–188.

Dean, J.E. and R.R. Weil. 2009. Brassica cover crops for nitrogen retention in the mid-Atlantic Coastal Plain. *J. Environ. Qual*. 38:520–528.

Delgado, J.A. 1999. *Soil Quality and Soil Erosion*. Soil and Water Conservation Society, CRC Press, Ankeny, IA.

Doltra, J., M. Laegdsmandand, and J.E. Olesen. 2011. Cereal yield and quality as affected by nitrogen availability in organic and conventional arable crop rotations: A combined modeling and experimental approach. *Eur. J. Agron*. 34:83–95.

Dou, Z., D.T. Galligan, R.D. Allshouse, J.D. Toth, C.F. Ramberg, and J.D. Ferguson. 2001. Manure sampling for nutrient analysis. *J. Environ. Qual*. 30:1432–1437.

Dou, Z., C.F. Ramberg, J.D. Toth, Y. Wang, A.N. Sharpley, S.E. Boyd, C.R. Chen, D. Williams, and Z.H. Xu. 2009. Phosphorus speciation and sorption-desorption characteristics in heavily manured soils. *Soil Sci. Soc. Am. J*. 73:93–101

Downey, J. and C. van Kessel. 1990. Dual inoculation of *Pisum sativum* with *Rhizobium leguminosarum* and *Penicillium bilaji*. *Biol. Fertil. Soils* 10:194–196.

Eghball, B., G.D. Binford, and D.D. Baltensperger. 1996. Phosphorus movement and adsorption in a soil receiving long-term manure and fertilizer application. *J. Environ. Qual*. 25:339–1343.

Eghball, B., D. Ginting, and J.E. Gilley. 2004. Residual effects of manure and compost applications on corn production and soil properties. *Agron. J*. 96:442–447.

Eghball, B. and J.F. Power. 1999a. Composted and noncomposted manure application to conventional and no-tillage systems: Corn yield and nitrogen uptake. *Agron. J*. 91:819–825.

Eghball, B. and J.F. Power. 1999b. Phosphorus- and nitrogen-based manure and compost applications: Corn production and soil phosphorus. *Soil Sci. Soc. Am. J.* 63:895–901.

Endelman, J.B., J.R. Reeve and D.J. Hole. 2010. Economically optimal compost rates for organic crop production. *Agron. J.* 102:1283–1289.

Entz, M.H., R. Guilford, and R. Gulden. 2001. Crop yield and soil nutrient status on 14 organic farms in the eastern portion of the northern great plains. *Can. J. Plant Sci.* 81:351–354.

Environment Canada. 2007. National inventory report 1990–2005: Greenhouse gas sources and sinks in Canada. Greenhouse Gas Division, Gatineau, Quebec, Canada.

Erisman, J.W., M.A. Sutton, J. Galloway, Z. Klimont, and W. Winiwarter. 2008. How a century of ammonia synthesis changed the world. *Nat. Geosci.* 1:636–639.

Evans, J. and J. Condon. 2009. New fertiliser options for managing phosphorus for organic and low-input farming systems. *Crop Pasture Sci.* 60:152–162.

Fageria, N.K. and V.C. Baligar. 2005. Enhancing nitrogen use efficiency in crop plants. In *Advances in Agronomy*. A.G. Norman, ed. Vol. 88, pp. 97–185. Elsevier Academic Press, San Diego, CA.

Fardeau, J.-C., C. Morel, and M. Jahiel. 1988. Does long contact with the soil improve the efficiency of rock phosphate? Results of isotopic studies. *Nutr. Cycl. Agroecosyst.* 17:3–19.

Finckh, M.R., E. Schulte-Geldermann, and C. Bruns. 2006. Challenges to organic potato farming: Disease and nutrient management. *Potato Res.* 49: 27–42.

Flach, E.N., W. Quak, and A. van Diest. 1987. A comparison of the rock phosphate-mobilising capacities of various crop species. *Trop. Agric.* 64:347–352.

Foster, R.K. 1990. Effect of tillage implement and date of sweet clover incorporation on available soil N and succeeding spring wheat yields. *Can. J. Plant Sci.* 70: 269–277.

Frigon, J.-C., C. Roy, and S. Guiot. 2012. Anaerobic co-digestion of dairy manure with mulched switchgrass for improvement of the methane yield. *Bioprocess Biosys. Eng.* 35: 341–349.

Gagnon, B., I. Demers, N. Ziadi, M.H. Chantigny, L.E. Parent, T.A. Forge, F.J. Larney, and K.E. Buckley. 2012. Forms of phosphorus in composts and in compost-amended soils following incubation. *Can. J. Soil. Sci.* 92:711–721.

Gale, E.S., D.M. Sullivan, C.G. Cogger, A.I. Bary, D.D. Hemphill, and E.A. Myhre. 2006. Estimating plant-available nitrogen release from manures, composts, and specialty products. *J. Environ. Qual.* 35:2321–2332.

Gianinazzi, S. and M. Vosatka. 2004. Inoculum of arbuscular mycorrhizal fungi for production systems: Science meets business. *Can. J. Bot.* 82:1264–1271.

Gleddie, S.C. 1993. Response of pea and lentil to inoculation with the phosphate-solubilizing fungus *Penicillium bilaii* (PROVIDE™). In *Proceedings of the Soils and Crops Workshop*. pp. 47–52. University of Saskatchewan, Saskatoon, Saskatchewan, Canada.

Gleddie, S.C., G.L. Hnatowich, and D.R. Polonenko. 1991. A summary of wheat response to PROVIDE (*Penicillium bilaji*) in western Canada. In *Proceedings of the Alberta Soil Science Workshop*. pp. 306–313. Lethbridge, Alberta, Canada.

Government of Canada. 2006a. *Organic Production Systems General Principles and Management Standards*. National Standard of Canada CAN/CGSB-32.310-2006.

Government of Canada. 2006b. *Organic Production Systems Permitted Substances Lists*. National Standard of Canada CAN/CGSB-32.311-2006.

Gulden, R.H. and J.K. Vessey. 2000. *Penicillium bilaii* inoculation increases root-hair production in field pea. *Can. J. Plant Sci.* 80: 801–804.

Hamel, C. and D.G. Strullu. 2006. Arbuscular mycorrhizal fungi in field crop production: Potential and new direction. *Can. J. Plant Sci.* 86: 941–950.

Hansen J.C., B.J. Cade-Menun, and D.G. Strawn. 2004. Phosphorus speciation in manure-amended alkaline soils. *J. Environ Qual.* 33:1521–1527.

Harvey, P.R., R.A. Warren, and S. Wakelin. 2009. Potential to improve root access to phosphorus: The role of non-symbiotic microbial inoculants in the rhizosphere. *Crop Pasture Sci.* 60:144–151.

Haynes, R.J. 1992. Relative ability of a range of crop species to use phosphate rock and monocalcium phosphate as P sources when grown in soil. *J. Sci. Food Agric.* 60:205–211.

He, A. and Z. Dou. 2010. *Phosphorus Forms in Animal Manure and the Impact on Soil P Status.* Nova Science Publishers, Inc, New York.

He, Z., B.J. Cade-Menum, G.S. Toor, A. Fortuna, C.W. Honeycutt, and J.T. Sims. 2007. Comparison of phosphorus forms in wet and dried animal manures by solution phosphorus-31 nuclear magnetic resonance spectroscopy and enzymatic hydrolysis. *J. Environ. Qual.* 36:1086–1095.

He, Z., C.W. Honeycutt, T.S. Griffin, B.J. Cade-Menun, P.J. Pellechia, and Z. Dou. 2009. Phosphorus forms in conventional and organic dairy manure identified by solution and solid state P-31 NMR spectroscopy. *J. Environ. Qual.* 38:1909–1918.

Hedley, M.J., J.W.B. Stewart, and B.S. Chauhan. 1982. Changes in inorganic and organic soil phosphorus fractions induced by cultivation practices and by laboratory incubations. *Soil Sci. Soc. Am. J.* 46:970–976.

Hesterman, O.B., T.S. Griffin, P.T. Williams, G.H. Harris, and D.R. Christenson. 1992. Forage legume-small grain intercrops: Nitrogen production and response of subsequent corn. *J. Prod. Agric.* 5:340–348.

Hill, J.E. and B.J. Cade-Munum. 2009. Phosphorus-31 nuclear magnetic resonance spectroscopy transect study of poultry operations on the Delmarva Peninsula. *J. Environ. Qual.* 38:130–138.

Hinsinger, P. 2001. Bioavailability of soil inorganic P in the rhizosphere as affected by root-induced chemical changes: A review. *Plant Soil* 237:173–195.

Hinsinger, P. and R. Gilkes. 1995. Root-induced dissolution of phosphate rock in the rhizosphere of lupins grown in alkaline soil. *Soil Res.* 33:477–489.

Hnatowich, G.L., S.C. Gleddie, and D.R. Polonenko. 1990. Wheat responses to PB-50 (*Penicillium bilaji*), a phosphate-inoculant. In *Proceedings for the 1990 Great Plains Soil Fertility Conference,* Denver, CO.

Hoffland, E. 1992. Quantitative evaluation of the role of organic acid exudation in the mobilization of rock phosphate by rape. *Plant Soil* 140:279–289.

Holford, I.C.R. 1997. Soil phosphorus: Its measurement, and its uptake by plants. *Aust. J. Soil. Res.* 35:227–239.

Hollinger, J. 2010. Growing legume crops for nitrogen. Manitoba Agriculture, Food and Rural Initiatives, Winnipeg, Manitoba, Canada. http://www.gov.mb.ca/agriculture/organic/org03s03.html (accessed December 6, 2011).

Janssen, B.H. 1996. Nitrogen mineralization in relation to C:N ratio and decomposability of organic materials. *Plant Soil* 181:39–45.

Janzen, H.H. 2010. Foreword. In *Recent trends in Soil Science and Agronomy in the Northern Great Plains of North America.* S.S. Malhi et al., eds. Research Signpost, Trivandrum, Kerala, India.

Jin, H., J.J. Germida, and F.L. Walley. 2012. Impact of arbuscular mycorrhizal fungal inoculants on subsequent arbuscular mycorrhizal fungi colonization in pot-cultured field pea (*Pisum sativum* L.). *Mycorrhiza*, 23:45–59.

Jones, D.L. 1998. Organic acids in the rhizosphere—A critical review. *Plant Soil* 205:25–44.

Karamanos, R.E., N.A. Flore, and J.T. Harapiak. 2010. Re-visiting use of *Penicillium bilaii* with phosphorus fertilization of hard red spring wheat. *Can. J. Plant Sci.* 90: 265–277.

Khan, M.S., A. Zaidi, and P.A. Wani. 2009. Role of phosphate solubilizing microorganisms in sustainable agriculture—A review. In *Sustainable Agriculture.* E. Lichtfouse et al. (eds.). pp. 551–570. Springer, Amsterdam, the Netherlands.

Knight, J.D., Buhler, R., J.Y. Leeson et al. 2010. Classification and fertility status of organically managed fields across Saskatchewan, Canada. *Can. J. Soil Sci.* 90: 667–678.

Knight, J.D. and S. Shirtliffe. 2003. Saskatchewan organic on-farm research: Part I: Farm survey and establishment of on-farm research infrastructure. Saskatchewan Agriculture Development Fund Final Report 2001-0016, http://www.agr.gov.sk.ca/apps/adf/adf_admin/reports/20010016.pdf (accessed September 1, 2012). Saskatchewan Ministry of Agriculture, Regina, Saskatchewan, Canada.

Kucey, R.M.N. 1987. Increased phosphorus uptake by wheat and field beans inoculated with a phosphorus-solubilizing *Penicillium bilaji* strain and with vesicular-arbuscular mycorrhizal fungi. *Appl. Environ. Microbiol.* 53:2699–2703.

Kucey, R.M.N. 1988. Effect of *Penicillium bilaji* on the solubility and uptake of P and micronutrients from soil by wheat. *Can. J. Soil Sci.* 68:261–270.

Kucey, R.M.N. and M.E. Leggett. 1989. Increased yields and phosphorus uptake by Westar Canola (*Brassica napus* L.) inoculated with a phosphate-solubilizing isolate of *Penicillium bilaji. Can. J. Soil Sci.* 69:425–432.

Kyei–Boahen, S., F.L. Walley, and A.E. Slinkard. 2002. Evaluation of rhizobial inoculation methods for chickpea. *Agron. J.* 94: 851–859.

Lammerts van Bueren, E.T., M. Hulscher, J. Jongerden, M. Haring, J.D. van Mansvelt, and A.M.P. Ruivenkamp, 1999. *Sustainable Organic Plant Breeding—A Vision, Choices, Consequences and Steps.* Louis Bolk Institute, Driebergen, the Netherlands. 60pp.

Larney, F.J., K.E. Buckley, X. Hao, and P.W. McCaughey. 2006b. Fresh, stockpiled, and composted beef cattle manure: Nutrient levels and mass balance estimates in Alberta and Manitoba. *J. Environ. Qual.* 35:1844–1854.

Larney, F.J., A.F. Olsen, J.J.Mi0ller, P.R. DeMaere, F. Zvomuya, and T.A. McAllister. 2008. Physical and chemical changes during composting of wood chip-bedded and straw-bedded beef cattle feedlot manure. *J. Environ. Qual.* 37:725–735.

Larney, F.J., D.M. Sullivan, K.E. Buckley, and B. Eghball. 2006a. The role of composting in recycling manure nutrients. *Can. J. Soil Sci.* 86:597–611.

Lashermes, G., B. Nicolardot, V.Parnaudeau, L. Thuries, R. Chaussod, M.L. Guillotin, M. Lineres et al. 2010. Typology of exogenous organic matters based on chemical and biochemical composition to predict potential nitrogen mineralization. *Bioresour. Technol.* 101:157–164.

Lawley, Y. 2004. Determining optimum plant population densities for three annual green manure crops under weedy and weed-free conditions. MSc thesis, University of Saskatchewan, Saskatoon, Saskatchewan, Canada.

Leggett, M., J. Cross, G. Hnatowich, and G. Holloway. 2007. Challenges in commercializing a phosphate-solubilizing microorganism: *Penicillium bilaiae*, a case history. In *First International Meeting on Microbial Phosphate Solubilization.* E. Velazquez and C. Rodriquez-Barrueco (eds.). pp. 215–222. Springer, New York.

Leytem, A.B. and P.A. Thacker. 2008. Fecal phosphorus excretion and characterization from swine fed diets containing a variety of cereal grains. *J. Anim. Vet. Adv.* 7:113–120.

Leytem, A.B. and P.A. Thacker. 2010. Phosphorus utilization and characterization of excreta from swine fed diets containing a variety of cereal grains balanced for total phosphorus. *J. Anim. Sci.* 88:1860–1867.

Leytem, A.B., B.L. Turner, and P.A. Thacker. 2004. Phosphorus composition of manure from swine fed low-phytate grains: Evidence for hydrolysis in the animal. *J. Environ. Qual.* 33:2380–2383.

Leytem, A.B. and D.T. Westermann. 2005. Phosphorus availability to barley from manures and fertilizers on a calcareous soil. *Soil Sci.* 170:401–412.

Liebman, M., R.L. Graef, D. Nettleton, and C.A. Cambardella. 2012. Use of legume green manures as nitrogen sources for corn production. *Renew. Agr. Food Syst.* 27:180–191.

Liu, K., A.M. Hammermeister, D.G. Patriquin, and R.C. Martin. 2008. Assessing organic potato cropping systems at the end of the first cycle of four-year rotations using principal component analysis. *Can. J. Soil Sci.* 88:543–552.

Loecke, T.D., C.A. Cambardella, and M. Liebman. 2012. Synchrony of net nitrogen mineralization and maize nitrogen uptake following applications of composted and fresh swine manure in the Midwest US. *Nutr. Cycl. Agroecosyst.* 93:65–74.

Lupwayi, N.Z., G.W. Clayton, J.T. O'Donovan, K.N. Harker, T.K. Turkington, and Y.K. Soon. 2006. Nitrogen release during decomposition of crop residues under conventional and zero tillage. *Can. J. Soil Sci.* 86:11–19.

Lupwayi, N.Z., G.W. Clayton, J.T. Donovan, K.N. Harker, T.K. Turkington, and Y.K. Soon. 2007. Phosphorus release during decomposition of crop residues under conventional and zero tillage. *Soil Till. Res.* 95:231–239.

Lynch, D.H., E. Clegg, J. Owen, and D. Burton. 2008b. Greenhouse gas emissions from organic crop management in humid region eastern Canada. In *Proceedings of 'Organic Agriculture and Climate Change'*, April 17–18, Enita Clermont, Lempdes, France.

Lynch, D.H., N. Halberg, and G.D. Bhatta. 2012a. Environmental impacts of organic agriculture in temperate regions. *CAB Rev.* 7:1–17

Lynch, D.H., M. Sharifi, A. Hammermeister et al. 2012b. Nitrogen management in organic potato production. In *Sustainable Potato Production: Global Case Studies*. Zhongi H., R.P. Larkin and C.W. Honeycutt (eds). pp. 209–231. Springer Science+Business Media, New York.

Lynch, D.H., R.P. Voroney, and P.R. Warman. 2004. Nitrogen availability from composts for humid region perennial grass and legume-grass forage production. *J. Environ. Qual.* 33:1509–1520.

Lynch, D.H., R.P. Voroney, and P.R. Warman. 2006. Use of ^{13}C and ^{15}N natural abundance techniques to characterize carbon and nitrogen dynamics in composting and in compost-amended soils. *Soil Biol. Biochem.* 38:103–114.

Lynch, D.H., Z.M. Zheng, B.J. Zebarth et al. 2008a. Organic amendment effects on tuber yield, plant N uptake and soil mineral N under organic potato production. *Renew. Agric. Food Syst.* 23:250–259.

Mahimairaja, S., N.S. Bolan, M.J. Hedley, and A.N. Macgregor. 1994. Losses and transformation of nitrogen during composting of poultry manure with different amendments: An incubation experiment. *Bioresour. Technol.* 47:265–273.

Main, M., D.H. Lynch, R.P. Voroney, and S. Juurlink. 2013. Soil phosphorus effects on forage harvested and nitrogen fixation on Canadian organic dairy farms. *Agron. J.* 105:1–9.

Malhi, S.S., S.A. Brandt, R. Lemke et al. 2009. Effects of input level and crop diversity on soil nitrate-N, extractable P, aggregation, organic C and N, and nutrient balance in the Canadian Prairie. *Nutr. Cycl. Agroecosyst.* 84:1–22.

Martin, R.C., D.H. Lynch, B. Frick, and P. van Straaten. 2007. Phosphorus status on Canadian organic farms. *J. Sci. Food Agric.* 87:2737–2740.

McKenzie, R.H., A.B. Middleton, K.W. Seward, R. Gaudiel, C. Wildschut, and E. Bremer. 2001. Fertilizer responses of dry bean in southern Alberta. *Can. J. Plant Sci.* 81:343–350.

Meyer, D.W. 1987. Sweet clover: An alternative to fallow for set-aside acreage in eastern North Dakota. *N.D. Farm Res.* 44:3–8.

Miller, J.J., B. Beasley, C.F. Drury, and B.J. Zebarth. 2009. Barley yield and nutrient uptake for soil amended with fresh and composted cattle manure. *Agron. J.* 101:1047–1059.

Miller, J.J., B.W. Beasley, C.F. Drury, and B.J. Zebarth. 2010. Available nitrogen and phosphorus in soil amended with fresh or composted cattle manure containing straw or wood-chip bedding. *Can. J. Soil Sci.* 90:341–354.

Miller, P.R., J. Waddington, C.L. McDonald, and D.A. Derksen. 2002. Cropping sequence affects wheat productivity on the semiarid northern Great Plains. *Can. J. Plant Sci.* 82:307–318.

Mirsky, S.B., M.R. Ryan, W.S. Curran et al. 2012. Conservation tillage issues: Cover-crop based organic rotational no-till grain production in the mid Atlantic region, USA. *Renew. Agric. Food Syst.* 27:31–40.

Moller, K. and T. Muller. 2012. Effects of anaerobic digestion on digestate nutrient availability and crop growth: A review. *Eng. Life Sci.* 12:242–257.

Munoz, G.R., K.A. Kelling, K.E. Rylant, and J. Zhu. 2008. Field evaluation of nitrogen availability from fresh and composted manure. *J. Environ. Qual.* 37:944–955.

Nahm, K.H. 2003. Evaluation of the nitrogen content in poultry manure. *Worlds Poult. Sci. J.* 59:77–88.

N'Dayegamiye, A. and T.S. Tran. 2001. Effects of green manures on soil organic matter and wheat yields and N nutrition. *Can. J. Soil Sci.* 81, 371–382.

Ngwene, B., E. George, W. Claussen, and E. Neumann. 2010. Phosphorus uptake by cowpea plants from sparingly available or soluble sources as affected by nitrogen form and arbuscular-mycorrhiza-fungal inoculation. *J. Plant Nutr. Soil Sci.* 173:353–359.

Novozymes. 2012. Information on inoculants for organic production. *Novozymes BioAg.* Saskatoon, Saskatchewan, Canada.

Oberson, A., J.M. Besson, N. Maire, and H. Sticher. 1996. Microbiological transformation in soil organic phosphorus transformations in conventional and biological cropping systems. *Biol. Fertil. Soils.* 21:138–148.

Oberson, A. and E. Frossard. 2005. Phosphorus management for organic agriculture. In *Phosphorus: Agriculture and the Environment.* Sims, J.T. and A.N. Sharpley (eds.). pp. 761–779. ASA, CSSA and SSSA, Madison, WI.

Ohno, T., T.S. Griffin, M. Liebman, and G.A. Porter. 2005. Chemical characterization of soil phosphorus and organic matter in different cropping systems in Maine, U.S.A. *Agric. Ecosyst. Environ.* 105:625–634.

Olesen, J. E., M. Askegaard, and I.A. Rasmussen. 2009. Winter cereal yields as affected by animal manure and green manure in organic arable farming. *Eur. J. Agron.* 30:119–128.

Olson-Rutz, K., C. Jones, and P. Miller. 2010. Soil nutrient management on organic grain farms in Montana. Available at http://landresources.montana.edu/soilfertility/ PDFbyformat/publication%20pdfs/Org%20Soil%20Fert%20Mgt%20EB0200. pdf (accessed December 6, 2011). Department of Land Resources & Environmental Sciences. Montana State University Extension, Bozeman, MT.

Oomah, B.D., G. Luc, C. Leprelle, J.C.G. Drover, J.E. Harrison, and M. Olson. 2011. Phenolics, phytic acid, and phytase in Canadian-grown low-tannin faba bean (*Vicia faba* L.) genotypes. *J. Agric. Food Chem.* 59:3763–3771.

Paul, E.A. 2007. *Soil Microbiology, Ecology and Biochemistry*, 3rd edn. Academic Press, New York.

Paul, J.W. and E.G. Beauchamp. 1993. Nitrogen availability for corn in soils amended with urea, cattle slurry, and solid and composted manures. *Can. J. Soil Sci.* 73:253–266.

Pikul, J.L., J.K. Aase, and V.L. Cochran. 1997. Lentil green manure as fallow replacement in the semiarid northern Great Plains. *Agron. J.* 89(6):867–874.

Przednowek, D.W.A., M.H. Entz, B. Irvine, D.N. Flaten, and J.R. Thiessen Martens. 2004. Rotational yield and apparent N benefits of grain legumes in southern Manitoba. *Can. J. Plant Sci.* 84:1093–1096.

Qian, P. and J.J. Schoenau. 2002. Availability of nitrogen in solid manure amendments with different C:N ratios. *Can. J. Soil Sci.* 82:219–225.

Qian, P. and J.J. Schoenau. 2005. Use of ion-exchange membrane to assess nitrogen-supply power of soils. *J. Plant Nutr.* 28:2193–2200.

Reeve, J.R., J.B. Endelman, B.E. Miller, and D.J. Hole. 2012. Residual effects of compost on soil quality and dryland wheat yield sixteen years after compost application. *Soil Sci. Soc. Am. J.* 76:278–285.

Reider, C.R., W.R. Herdman, L.E. Drinkwater, and R. Janke. 2000. Yields and nutrient budgets under composts, raw dairy manure and mineral fertilizer. *Compost Sci. Util.* 8:328.

Rice, W.A., N.Z. Lupwayi, P.E. Olsen, D. Schlechte, and S.C. Gleddie. 2000. Field evaluation of dual inoculation of alfalfa with *Sinorhizobium meliloti* and *Penicillium bilaii*. *Can. J. Plant Sci.* 80:303–308.

Rice, W.A., P.E. Olsen, L.D. Bailey, V.O. Biederbeck, and A.E. Slinkard. 1993. The use of annual legume green–manure crops as a substitute for summerfallow in the Peace River region. *Can. J. Soil Sci.* 73:243–252.

Richardson, A., J.-M. Barea, A. McNeill, and C. Prigent-Combaret. 2009. Acquisition of phosphorus and nitrogen in the rhizosphere and plant growth promotion by microorganisms. *Plant Soil* 321:305–339.

Richardson, A.E. 2001. Prospects for using soil microorganisms to improve the acquisition of phosphorus by plants. *Funct. Plant Biol.* 28:897–906.

Rick, T., C. Jones, R. Engel, and P. Miller. 2011. Green manure and phosphate rock effects on phosphorus availability in a northern Great Plains dryland organic cropping system. *Org. Agric.* 1:81–90.

Roberts, C.J., D.H. Lynch, R.P. Voroney, R.C. Martin, and S.D. Juurlink. 2008. Nutrient budgets of Ontario organic dairy farms. *Can. J. Soil Sci.* 88:107–114.

Rochester, I.J., M.B. Peoples, G.A. Constable, and R.R. Gault. 1998. Faba beans and other legumes add nitrogen to irrigated cotton cropping systems. *Aust. J. Exp. Agric.* 38:253–260.

Russelle, M.P., M.H. Entz, and A.J. Franzluebbers. 2007. Reconsidering integrated crop-livestock systems in North America. *Agron. J.* 99:325–334.

Sanchez, J.E., R.R. Harwood, T.C. Willson et al. 2004. Managing soil carbon and nitrogen for productivity and environmental quality. *Agron. J.* 96: 769–775.

Sato, S., D. Solomon, C. Hyland, Q.M. Ketterings, and J. Lehmann. 2005. Phosphorus speciation in manure and manure-amended soils using XANES spectroscopy. *Environ. Sci. Technol.* 39:7485–7491.

Schneider, K.D. and P. van Straaten. 2007. Solubilization of phosphate rock using *Aspergillus niger*: Potential for an alternative P-fertilizer. htpps://www.organicagcentre.ca/ResearchDatabase/res_phosphate_rock.asp/ (accessed September 25, 2012).

Schoenau, J.J., C.C. Carley, C. Stumborg, and S.S. Malhi. 2010. Strategies for maximizing crop recovery of applied manure nitrogen in the Northern Great Plains of North America. In *Recent Trends in Soil Science and Agronomy Research in the Northern Great Plains of North America*. S.S. Malhi (ed.). pp. 95–108. Research Signpost, Trivandrum, India.

Schroder, J.J., A.L. Smith, D. Cordell, and A. Rosemarin. 2011. Improved phosphorus use efficiency in agriculture: A key requirement for its sustainable use. *Chemosphere* 84:822–831.

Schwartz, M.W., J.D. Hoeksema, C.A. Gehring, N.C. Johnson, J.N. Klironomos, L.K. Abbott, and A. Pringle. 2006. The promise and the potential consequences of the global transport of mycorrhizal fungal inoculum. *Ecol. Lett.* 9:501–515.

Sharifi, M., D.H. Lynch, B.J. Zebarth, Z. Zheng, and R.C. Martin. 2009. Evaluation of nitrogen supply rate measured by in situ placement of plant root simulator probes as a predictor of nitrogen supply from soil and organic amendments in potato crop. *Am. J. Potato Res.* 86:356–366.

Sharifi, M., B.J. Zebarth, D.L. Burton, C.A. Grant, and G.A. Porter. 2008. Organic amendment history and crop rotation effects on soil nitrogen mineralization potential and soil nitrogen supply in a potato cropping system. *Agron. J.* 100:1562–1572.

Sharpley, A. 2012. Phosphorus availability. In *Handbook of Soil Sciences, Resource Management and Environmental Impacts*, 2nd edn. pp. 11-14–11-36. CRC Press, Boca Raton, FL.

Sharpley, A. and B. Moyer. 2000. Phosphorus forms in manure and compost and their release during simulated rainfall. *J. Environ. Qual.* 29:1462–1469.

Sharpley, A., W. Richard, P.J. McDowell, and A. Kleinman. 2004. Amounts, forms, and solubility of phosphorus in soils receiving manure. *Soil. Sci. Soc. Am. J.* 68:2048–2057.

Shipley, P.R., J.J. Messinger, and A.M. Decker. 1992. Conserving residual corn fertilizer nitrogen with winter cover crops. *Agron. J.* 84:869–876.

Shirtliffe, S.J. and E.N. Johnson. 2012. Progress towards no-till organic weed control in western Canada. *Renew. Agric. Food Syst.* 27:60–67.

Shirtliffe, S.J. and J.D. Knight. 2006. Saskatchewan organic on-Farm research: Part II: Soil fertility and weed management. Saskatchewan Agriculture Development Fund Final Report 2002-0198 (accessed September 1, 2012) http://www.agr.gov.sk.ca/apps/adf/adf_admin/reports/20020198.pdf. Saskatchewan Ministry of Agriculture, Regina, Saskatchewan, Canada.

Smith, G. and Groenen, W. (eds). 2000. *Organic Farming on the Prairies.* Saskatchewan Organic Directorate, Grand Valley Press, Moose Jaw, Saskatchewan, Canada.

Snapp, S.S., S.M. Swinton, R. Labarta, D. Mutch, J.R. Black, R. Leep, J. Nyiraneza, and K. O'Neil. 2005. Evaluating cover crops for benefits, costs and performance within cropping system niches. *Agron. J.* 97:322–332.

Statistics Canada. 2004. Manure Management in Canada. 1(2) [Online]. Available: http://publications.gc.ca/collections/Collection/Statcan/21-021-M/21-021-MIE2004001.pdf? (January 27, 2013).

Statistics Canada. 2008. A geographical profile of livestock manure production in Canada, 2006. [Online] Available: http://www.statcan.gc.ca/pub/16-002-x/2008004/article/10751-eng.htm#a3 (January 27, 2013).

Stockdale, E.A., M.A. Shepherd, S. Fortune, and S.P. Cuttle. 2002. Soil fertility in organic farming systems—Fundamentally different? *Soil Use Manage.* 18:301–308.

Sullivan, D.M., J.P.G. McQueen, and D.A. Horneck. 2007. Estimating nitrogen mineralization in organic potato production. Factsheet produced by Oregon State University Extension Service. 8pp.

Takeda, M. and J.D. Knight. 2006. Enhanced solubilization of rock phosphate by *Penicillium bilaiae* in pH-buffered solution culture. *Can. J. Microbiol.* 52:1121–1129.

Teasdale, J.R., S.B. Mirsky, J.T. Spargo, M.A. Cavigelli, and J.E. Maul. 2012. Reduced-tillage organic corn production in a hairy vetch cover crop. *Agron. J.* 104:621–628.

Thiessen Martens, J.R., M.H. Entz, and J.W. Hoeppner. 2005. Legume cover crops with winter cereals in southern Manitoba: Fertilizer replacement values for oat. *Can. J. Plant Sci.* 85:645–648.

Thorup-Kristensen, K. 2001. Are differences in root growth of nitrogen catch crops important for their ability to reduce soil nitrate-N content, and how can this be measured? *Plant Soil* 230: 185–195.

Thorup-Kristensen, K. and D.B. Dresboll. 2010. Incorporation time of nitrogen catch crops influences the N effect for the succeeding crop. *Soil Use Manage.* 26:27–35.

Thorup-Kristensen, K., D.B. Dresboll, and H.L. Kristensen. 2012. Crop yield, root growth, and nutrient dynamics in a conventional and three organic cropping systems with different levels of external inputs and N re-cycling through fertility building crops. *Eur. J. Agron.* 37:66–82.

Townley-Smith, L., A.E. Slinkard, L.D. Bailey, V.O. Biederbeck, and W.A. Rice. 1993. Productivity, water-use and nitrogen-fixation of annual-legume green-manure crops in the Dark Brown soil zone of Saskatchewan. *Can. J. Plant Sci.* 73:139–148.

Trolove, S.N., M.J. Hedley, G.J.D. Kirk, N.S. Bolan, and P. Loganathan. 2003. Progress in selected areas of rhizosphere research on P acquisition. *Soil Res.* 41:471–499.

Turner, B.L., B.J. Cade-Menun, L.M. Condron, and S. Newman. 2005. Extraction of soil organic phosphorus. *Talanta* 66:294–306.

Vaisman, I., M.H. Entz, D.N. Flaten et al. 2011. Blade roller–green manure interactions on nitrogen dynamics, weeds, and organic wheat. *Agron. J.* 103:879–89.

van Ray, B. and A. van Diest. 1979. Utilization of phosphate from different sources by six plant species. *Plant Soil* 51:577–589.

van Straaten, P. 2002. *Rocks for Crops: Agrominerals of Sub-Saharan Africa.* ICRAF, Nairobi, Kenya, 338pp.

Vance, C.P., C. Uhdo-Stone, and D.L. Allan. 2003. Phosphorus acquisition and use: Critical adaptations by plants for securing a nonrenewable resource. *New Phytol.* 157:423–447.

Vassileva, M., N. Vassilev, and R. Azcon. 1998. Rock phosphate solubilization by *Aspergillus niger* on olive cake-based medium and its further application in a soil–plant system. *World J. Microbiol. Biotechnol.* 14:281–284.

Vassilev, N., I. Franco, M. Vassileva, and R. Azcon. 1996. Improved plant growth with rock phosphate solubilized by *Aspergillus niger* grown on sugar-beet waste. *Bioresour. Technol.* 55:237–241.

Vessey, J.K. 2004. Benefits of inoculating legume crops with rhizobia in the northern great plains. Online. *Crop Manage* doi: 10.1094/CM-2004-0301-04-RV.

Vessey, J.K. and K.G. Heisinger. 2001. Effect of *Penicillium bilaii* inoculation and phosphorus fertilisation on root and shoot parameters of field-grown pea. *Can. J. Plant Sci.* 81: 361–366.

Vyn, T.J., J.G. Faber, K.J. Janovicek, and E.G. Beauchamp. 2000. Cover crop effects on nitrogen availability to corn following wheat. *Agron. J.* 92:915–924.

Walley, F.L., G.W. Clayton, P.R. Miller, P.M. Carr, and G.P. Lafond. 2007. Nitrogen economy of pulse crop production in the Northern Great Plains. *Agron. J.* 99:1710–1718.

Wang, G., M. Ngouajio, and D.D. Warncke. 2008. Nutrient cycling, weed suppression, and onion yield following brassica and sorghum sudangrass cover crops. *Horttechnology* 18(1):68–74.

Welsh, C., M. Tenuta, D.N. Flaten, J.R. Thiessen-Martens, and M.H. Entz. 2009. High yielding organic crop management decreases plant-available but not recalcitrant soil phosphorus. *Agron. J.* 101(5):1027–1035.

Welty, L.E., L.S. Prestbye, R.E. Engel, R.A. Larson, R.H. Lokerman, R.S. Speilman, J.R. Sims, L.I. Hart, G.D. Kushnak, and A.L. Dubbs. 1988. Nitrogen contribution of annual legumes to subsequent barley production. *Appl. Agric. Res.* 3:98–104.

White, P.J. and J.P. Hammond. 2008. Phosphorus nutrition of terrestrial plants. In *The Ecophysiology of Plant-Phosphorus Interactions.* White, P.J. and J.P. Hammond (eds.). pp. 51–81. Springer Science+Business Media, New York.

Whitelaw, M.A., T.J. Harden, and K.R. Helyar. 1999. Phosphate solubilisation in solution culture by the soil fungus *Penicillium radicum. Soil Biol. Biochem.* 31:655–665.

Wichuk, K.M. and D. McCartney. 2010. Compost stability and maturity evaluation—A literature review. *Can. J. Civil Eng.* 37:1505–1523.

Wiens, M.J., M.H. Entz, R.C. Martin, and A.M. Hammermeister. 2006. Agronomic benefits of alfalfa mulch applied to organically managed spring wheat. *Can. J. Plant Sci.* 86:121–131.

Woodley, A. 2012. Management of soil N fertility for organic cereal production. Unpublished doctoral dissertation. School of Environmental Science, University of Guelph, Guelph, Ontario.

Wortman, S.E., T.D. Galusha, S.C. Mason et al. 2011. Soil fertility and crop yields in long-term organic and conventional cropping systems in Eastern Nebraska. *Renew. Agric. Food Syst.* 27:200–216.

Wright, A.T. 1990a. Quality effects of pulses on subsequent cereal crops in the northern prairies. *Can. J. Plant Sci.* 70:1013–1021.

Wright, A.T. 1990b. Yield effects of pulses on subsequent cereal crops in the northern prairies. *Can. J. Plant Sci.* 70:1013–1032.

Yost, M.A., J.A. Coulter, M.P. Russelle, C.C. Sheaffer, and D.E. Kaiser. 2012. Alfalfa nitrogen credit to first-year corn: Potassium, regrowth, and tillage timing effects. *Agron. J.* 104:953–962.

Zentner, R.P., C.A. Campbell, V.O. Biederbeck, and F. Selles. 1996. Indianhead black lentil as green manure for wheat rotations in the Brown soil zone. *Can. J. Plant. Sci.* 76(3):417–422.

Zentner, R.P., D.D. Wall, C.N. Nagy et al. 2002. Economics of crop diversification and soil tillage opportunities in the Canadian prairies. *Agron. J.* 94: 216–230.

Zhang, T.Q., A.F. MacKenzie, J. Lafond et al. 2004. Soil test phosphorus and phosphorus fractions with long-term phosphorus addition and depletion. *Soil Soc. Am. J.* 68:519–528.

Zhu, Y.G., Y.Q. He, S.E. Smith, and F.A. Smith. 2002. Buckwheat (*Fagopyrum esculentum* Moench) has high capacity to take up phosphorus (P) from calcium (Ca)-bound source. *Plant Soil* 239:1–8.

Ziadi, N., R.R. Simard, G. Allard, and J. Lafond. 1999. Field evaluation of anion exchange membranes as a N soil testing method for grasslands. *Can. J. Soil Sci.* 79:281–294.

5 Sustaining Soil Organic Carbon, Soil Quality, and Soil Health in Organic Field Crop Management Systems

Derek H. Lynch

CONTENTS

5.1 INTRODUCTION

Improved soil quality and minimizing soil degradation have long been considered two central elements of a suite of positive broader environmental and ecological attributes from organic farming found within organic standards (CGSB, 2006). Promoting the health and biological characteristics of the soil, which play a key role in the linked processes of decomposition and nutrient cycling, while providing other ecosystem services, is similarly recognized as a key goal of organic cropping systems (Stockdale and Watson, 2009; Lynch et al., 2012a). Surveys of organic farmers (see Chapter 2) also reflect these central interests of organic farm management. Within the topic of soil management, research focused on the optimum use of green manures (GMrs) and management approaches "to improve existing soil life" were the top-ranked priorities among Canadian organic producers' research interests (OACC, 2008). Recent reviews such as Gomiero et al. (2011) and Lynch et al. (2012a) have examined soil quality within organic farming in temperate regions as a component of a broader summary of environmental and ecological system impacts including biodiversity, energy use and global warming potential, and nutrient loading. In the following, we focus specifically on the impacts of organic field crop management on soil abiotic and biotic quality, drawing primarily, but not exclusively, on studies conducted in Canada and the United States. Soil fertility and nutrient (nitrogen and phosphorus) dynamics under organic field cropping is covered in Chapter 4. This topic is approached with subsections on (1) soil organic matter, (2) labile fractions of soil organic matter, (3) aggregate dynamics, (4) soil ecology and soil health, (5) water infiltration and water holding capacity, and (6) reduced and zero-tillage approaches and integrating livestock—implications for soil quality—followed by a concluding commentary on (7) future directions and research needs.

5.2   SOIL ORGANIC MATTER

Soil organic matter is a key driver of soil quality and soil health and along with soil type and texture, acidity and salinity, the amount, distribution, and quality of soil organic matter has the greatest impact on soil properties. Surveys of organic farmers show a remarkable degree of appreciation of the dynamic nature of soils, the complex interactions at play that are influenced by management, and the central role of soil organic matter in maintaining the physical and biological aspects of soil quality and soil health (OACC, 2008; Powlson et al., 2011). Indeed, soil organic matter has a range of constituents or attributes including total soil C and N, labile fractions of C and N, microbial biomass, carbohydrates, and soil enzyme activity (Gregorich et al., 1994).

Within any given management system, the soil total organic C (SOC) level is a function of the influence of the net effect of the processes of C deposition from crop residues and organic amendments versus C losses from soil respiration and SOC decay (Janzen, 2006), with tillage and soil disturbances being key management factors promoting C losses. Parton et al. (1996) suggest a minimum average of 2.00 Mg C ha^{-1} input per year requirement to maintain SOC levels in cropping systems. Potato residues, for example, provide much less than this, at approximately 1100 kg C ha^{-1} (Angers et al., 1999). In Canadian and US agricultural systems generally, projected gains in SOC over the past few decades have previously been attributed to a reduction in the use of summer fallow and in particular to the adoption of no-till and minimum-tillage practices (Smith et al., 1997). More recent analyses, however, have indicated that no-till does not always contribute to gains in SOC, especially under cooler and/or wetter environmental conditions (Yang et al., 2008; Ogle et al., 2012).

Many authors have proposed organic management systems as a means of promoting SOC gains (Smith, 2004; Mondelaers et al., 2009; Niggli et al., 2009; Gomiero et al., 2011). At the same time, organic field crop production systems continue to be criticized for their continued reliance on mechanical tillage for incorporation of GMrs and weed control, as it is expected these practices contribute to the depletion of SOC (Trewevas, 2004). Can organic field crop systems be promoted as viable options for storage of SOC due to a greater return of crop residue C as GMrs and perennial crops in rotation, regardless of the added tillage employed? The use of GMrs is much more prevalent among Canadian Prairie region organic grain producers when compared with conventional grain growers (84% vs. 6%), while organic farmers continue to use tilled summer fallow more than conventional producers (52% vs. 6%) (Nelson et al., 2010). However, the *net* effect on SOC of such added C return, combined with added tillage, within organic management systems has rarely been systematically examined. Alternatively, are organic systems possibly neutral with respect to SOC storage, where the added C return (and C turnover) in these systems instead plays a key role in stimulating nutrient dynamics and soil biological life? Organic systems have been suggested (Mader et al., 2002; Flieβbach et al., 2007) to promote a shift in soil microbial composition and the efficiency of the decomposer population resulting in proportionately more of incoming residue C being converted to SOC instead of lost to the atmosphere as CO_2. Finally, are new advances in reduced or no-tillage systems for organic field crop production influencing these SOC dynamics? This section, and Sections 5.3 and 5.7, will tackle each of these aspects of SOC dynamics within organic field crop systems in turn.

Prior to evaluating our current understanding of the net effect of organic farming systems on total SOC, a number of points regarding the complexity of the issue are worth mentioning briefly (Lynch et al., 2012a,b). With respect to methodology, studies vary with respect to whether SOC levels are being reported for surface soils only or more ideally SOC levels throughout the entire soil profile. For example, the recent study of Bell et al. (2012) from southern Manitoba is one of few to examine SOC stocks throughout the soil profile (0–120 cm) and demonstrated that C deposition (and N and P maintenance) differences between farming systems can be just as important below the annual crop rooting zone (0–30 cm). Accurate comparisons of farming system influence on SOC should also monitor changes in soil bulk density to compare SOC changes on an equivalent soil mass basis. Where manure or compost is applied in the farm system under study, appropriate recognition of this return, rather than actual gain, of C must be acknowledged (i.e., chosen manure or compost input rates must be credible and correspond to the production systems' capacity for assimilating C [Leifeld et al., 2009]). Where the organic plots receive large amounts of imported organic matter (e.g., manure from a disproportionate livestock production that could not have been sustained by the crop production in the tested crop rotation), SOC results should be interpreted with caution. Conversely, some long-term studies have reported for organic field crop systems where no manure or compost is returned, which, although it is reflective of organic production practices in the region, contributes to reduced system productivity and relative SOC depletion overtime. Finally, all soils have a limit to SOC storage, a capacity strongly influenced by soil texture (Hassink, 1997), which may limit any gains due to cropping system management.

A summary of medium- to long-term studies comparing field crop production systems' influence on SOC is presented in Table 5.1. These include some excellent recent studies from Canada and the United States. Teasdale et al. (2007) in the United States reported on a 9-year comparison of selected minimum-tillage strategies for production of corn, soybean, and wheat conducted at USDA-ARS Beltsville, MD. The study included four management systems, including (1) an organic system using cover crops and manure for nutrients and reliance on chisel plowing for tillage and postplanting mechanical cultivation for weed control, (2) a standard no-tillage system

TABLE 5.1

Comparison of Soil Organic Carbon Storage of Organic and Conventional Field Cropping Systems from Long-Term Studies

Authors	Region	Type of Study	Study Period (Years)	Org<Conv	Org=Conv	Org>Conv
Mahli et al.	Canada	Comparative field trial	12		√	
Bell et al.	Canada	Comparative field trial	18	13%–15%[a]		
Pimentel et al.	United States	Comparative field trial	22			20%–25%[b]
Teasdale et al.	United States	Comparative field trial	9			19%[c]
Wortman et al.	United States	Comparative field trial	11		√	
Robertson et al.	United States	Comparative field trial	8	12%[d]		
Leifeld et al.	Switzerland	Comparative field trial	27		√	
Kirchmann et al.	Sweden	Comparative field trial	18			16%
Hathaway-Jenkins et al.	England	Paired farm study	1–58[d]		√	
Chirinda et al.	Denmark	Comparative field trials	11		√	

[a] SOC stocks reported to 120 cm depth. Differences between farming systems were smaller for the alfalfa/crop (13%) compared to annual crop (15%) rotations.

[b] Higher gains (25%) were recorded for the "organic animal" then "organic legume" (20%) system.

[c] Compared to a no-till treatment.

[d] Variation in length of time in which organic farms (n=16) were managed organically.

with recommended N inputs and herbicides, (3) a no-till cover crop (hairy vetch and rye) system with reduced herbicide and N inputs, and (4) a no-tillage crown vetch living mulch system. In the organic system, cow manure was applied to maintain soil nutrient levels while allowing for estimated N credits for the crimson clover cover crop. Despite the use of tillage in the organic systems, at the end of the study, SOC and N concentrations were greatest at all depth intervals (to 30 cm) for the organic cropping regime and 19% and 23% greater, respectively, than found for the no-tillage system. This also resulted in improved soil productivity under the organic plots. More recently, Wortman et al. (2011) reported on a long-term (11 years) cropping systems study in Nebraska. The conventional system consisted of fertilized soybean, winter wheat, corn, and sorghum, while the organic regime utilized composted beef manure (applied once) in place of fertilizer or an alfalfa forage, prior to corn and sorghum. At the end of the study (SOC levels were reported for 0–15 cm depth only), each cropping regime had largely equivalent SOC concentrations. Drinkwater et al. (1998) considered qualitative differences in crop residues in the organic system, which included a vetch GMr, to be critical to SOC gains in a US corn–soybean cropping systems study. Pimentel et al. (2005) reported SOC levels after a 22-year study in Pennsylvania, United States, which employed an "organic legume"-based (reliant on leguminous GMrs) and an "organic animal"-based system. Gains in surface soil (0–30 cm) SOC under organic management ranged from 20% to 25% compared to conventional management (Table 5.1).

In Canada, Mahli et al. (2009) reported on surface soil amount (Mg ha^{-1}) of SOC and TN following 11 years of a study on the dark brown loam soil at Scott, Saskatchewan. The study compared three levels of pesticide and fertility inputs: organic, reduced, and high. The reduced system utilized no-till and integrated pest management. Lentil (*Lens culinaris* Medikus) was the GMr used in the organic system but no organic fertility inputs were applied. These input regimes were applied across cropping systems alternately (i) low in diversity, (ii) diversified annual grains (which included oilseed and pulse crops), and (iii) a diversified annual perennial system (which included forage crops along with oilseed rape, wheat, and barley). Treatments failed to differ in SOC and TN when measured in two surface soil depth increments (0–7.5 and 7.5–15 cm). Bell et al. (2012) recently reported on soil profile (0–120 cm) SOC and nutrient levels following two long-term (9 and 18 years) comparative farming systems studies in southern Manitoba, Canada. Treatments included annual crop and alfalfa/crop rotations managed both organically and conventionally. No manure, compost, or other fertility inputs were applied to the organic cropping treatments. Subsoil layers (30–120 cm) contributed 45%–60% of the SOC stocks with, interestingly, a higher proportion of profile SOC in the surface (0–30 cm) soils under organic management. Across both sites, profile SOC amounts were lower by 25–30 Mg C ha^{-1} under organic management. This was attributable to lower net primary productivity and thus lower return of shoot and root residue C to soil with organic management (1.5 vs. 2.3 Mg C ha^{-1} year^{-1}) over the experimental period, regardless of rotation sequence. Using short phases (1.5 years) of alfalfa, cut and removed as hay, in the grain cropping rotation failed to maintain SOC throughout the soil profile (estimated C return to soil was equal to that for annual grain rotations within both management systems) and also depleted soil available P levels more than the annual grain rotations. Notably, compost recently applied to half of the organic alfalfa plots at the longer-term study site (the Glenlea site farming systems study) has resulted in a substantial increase in alfalfa forage productivity (M. Entz, pers. comm.).

In Europe, Leifeld et al. (2009) conducted a detailed study (including soil fractionation, radiocarbon dating, and modeling with the carbon model RothC) of SOC levels following 27 years of the DOK trial in Switzerland (Mader et al., 2002). That study includes three organically fertilized treatments under conventional, organic, and biodynamic management and two systems with or without mineral fertilizer. After 27 years, topsoil (0–20 cm) SOC levels were found to have declined equally for all treatments. Kirchmann et al. (2007) also found in Sweden that SOC concentrations decreased in both organic and conventional systems after 18 years but less so in the organic due to higher C inputs and lower soil pH values. The authors concluded, however, that organic farming appeared not to be an option for sequestering SOC.

Organic potato farms in Atlantic Canada utilize extended (5 years) rotations including legume cover crops compared with much more frequent cropping of potatoes (and associated tillage) in conventional systems (Angers et al., 1999; Lynch et al., 2008). On four organic farm sites, Nelson et al. (2009) found that SOC levels were sustained by these extended rotations, ranging from 30 to 38 Mg C ha for surface (0–15 cm) soil, across all farm sites and rotation phases comparable to that found in adjacent permanent pastures. A recent 5-year trial compared two GMrs' (red clover [*Trifolium pratense* L.] vs. a mixture of oats/peas and vetch [*Avena sativa* L./*Pisum sativum* L./*Vicia villosa* L.] on potato and carrot [*Daucus carota* ssp. sativus] and bean [*Phaseolus lunatus* L.]) productivity, partially reported on in Lynch et al. (2012b). The supplemental use of composts at moderate rates (5–10 Mg ha^{-1} DWt basis) applied just during the potato phase (i.e., 1 year in 5) provided some nutrient benefits but also provided the additional benefit of changing the C balance (respired CO_2–C input) to positive (data not shown). The total amount of organic C added to the soil through imported composts was 1.1 and 2.7 Mg C ha^{-1} over the 5 years.

Few studies have explored in detail root exploration of the soil profile by crop and GMr, an area of tremendous scope for further research. Thorup-Kristensen et al. (2012) in Denmark examined over 4 years vegetable and small grain crop yields, nutrient dynamics, and root exploration (using glass minirhizotrons) as affected by management system (one conventional and three organic systems). Crop rotations were identical in all systems but varied in GMrs and catch crop utilization and degree of external inputs (slurry vs. fertilizer) and pest management. Average annual N imports were 149, 85, 25, and 25 for the conventional (C), organic import-based (O1), organic fall soil cover (O2), and fall cover plus intercrop system (O3), respectively. Organic crop yields averaged 82% of conventional. The addition of GMrs in the rotation significantly increased soil profile (0–2.4 m) exploration by roots from 21% in C to 38% in O2. It was notable also that some crops (white cabbage) produced greater root density than oats and fall rye. Fodder radish had higher root density to 2.5 m depth, greater than all other rotation crops.

Evidence for enhanced soil microbial activity, but not necessarily SOC gain, under organic field crop management was provided by a recent study from Denmark by Chirinda et al. (2010). Over a period of 11 years, four different organic 4-year rotations cropping winter wheat, spring barley, and potatoes were compared with an inorganic fertilizer-based rotation. The organic rotations differed in the type and application of manure, presence of catch crops (perennial ryegrass or mixtures of ryegrass and chicory and clover species), and whether whole-year grass–clover GMrs ("leys") were included. While organic management returned C to soil at rates 18%–91% greater than the conventional rotation, after 11 years, SOC levels (0–30 cm depth) were similar across all systems. This was attributable to increased measured microbial biomass and activity (soil respiration of CO_2), which increased with C input rates. This enhanced microbial activity was attributed significantly to improved soil quality (reduced bulk density and larger soil pore volume) with the organically managed soils, which reduced the onset of anaerobic conditions in wet soils. Mader et al. (2002) had also earlier reported on enhanced microbial activity under organic management and further suggested that long-term management influenced soil decomposer community C resource utilization efficiency (specifically metabolic quotient or CO_2 respired per unit microbial biomass C) due to shifts in soil microbial functional diversity.

Leifeld (2012) recently examined data from four previously published studies to further examine this aspect. Fifteen of nineteen soils showed higher specific respiration rates (i.e., CO_2 respired per unit SOC) suggesting a larger labile pool fraction under organic management. Improved microbial efficiency (or reduced maintenance requirement) of C resource use with organic management, however, was found in only eleven out of the nineteen soils. Leifeld (2012) suggests that nutrient depletion in organic systems, through a mechanism known as nutrient mining, may partially explain the decline in C use efficiency and increased respiration of soil heterotrophs in some organically managed soils. Thus, targeted maintenance of nutrient supply can be considered critical not only for aboveground biomass (crop and GMr) productivity and thus residue C return to soil, but potentially for enhancing decomposer efficiency. As noted earlier, studies vary widely in the degree of organic fertilizer returned within organically managed regimes.

Leifeld and Fuhrer (2010), after analyzing 68 datasets from 32 peer-reviewed publications primarily from Europe and the United States, found that after conversion, SOC in organic systems increased annually by 2.2% on average, whereas in conventional systems, SOC did not change significantly. Importantly, where the amount of organic fertilizer within organic cropping regimes exceeded that applied under conventional, the annual increase in SOC was 2.91% (± 1.29, $n = 37$). In 20 of these cases, where the amount of organic fertilizer applied in the organic system was considered to exceed the cropping systems' productivity, SOC increased significantly by an even greater 4.85% (± 2.29; $n = 20$) under organic management. In contrast, SOC changes under organic management were not significant for trials where the amount of organic fertilizer was similar to conventional (0.40 ± 0.24, $n = 13$) or when organic fertilizer was applied according to the systems' productivity where SOC changed by only 0.23% (± 0.29, $n = 14$). Studies included both experimental plot- and farm-scale comparisons with a mean experimental duration of 17.6 years. Most studies only reported SOC concentrations rather than amounts thus neglecting possible changes in soil bulk density, and results were usually only for the plow layer also.

Gattinger et al. (2012) similarly found that when datasets from 74 long-term studies (avg. 14.4-year duration) internationally were restricted to less than 50% of these studies also reporting soil bulk densities and external C and N inputs, differences in topsoil (0–20 cm) C sequestration rates (Mg C ha^{-1} year^{-1}) between organic and nonorganic management were insignificant.

Incorporating an optimum level of fertility input can thus be considered critical to medium- and long-term goals of maintaining soil organic matter levels through their impact on system productivity and possibly also soil microbial efficiency. In regions where organic amendments are not routinely available to organic producers, even occasional applications of slow nutrient release inputs such as composts can be beneficial in this regard as shown by Wortman et al. (2011) in Nebraska. Composts and manures can be inherently variable in nutrient supplying ability and perennial legumes, though biological nitrogen fixation can also provide an "N-buffer" capacity to accommodate this variability as demonstrated by Lynch et al. (2004). In turn, the compost contributes to a slow-release particulate pool of organic C and N and other nutrients (Sanchez et al., 2004; Lynch et al., 2006).

Leguminous GMrs can also play a critical role in supplying N and C to organic cropping systems. However, agroecosystem influences on GMr primary productivity, and in particular, constraints influenced by moisture limitations must also be considered. For example, under typical growing conditions in the Canadian Prairie region black soil zone, annual legume GMrs can generally produce 5000 kg DM ha^{-1}. Under drier regions as found in the brown soil zone, biomass production of the same annual GMr could generally be expected to reach only 2500 kg DM ha^{-1} (Bell et al., 2012). Weeds can also be an important contributor to noncrop biomass and SOC gains under organic systems compared to conventional management regimes (Sanchez et al., 2004; Teasdale et al., 2007).

In summary, bearing in mind the discussed challenge of variability in trial design and location and composition of organic farming system employed (i.e., impact of GMr productivity, return of manure/compost, fertility regime) and methodology employed for gauging SOC differences, the consensus of the data from medium- to long-term studies comparing field crop management systems (Table 5.1) suggests that organic management at least sustains SOC when compared with conventional field crop production. These conclusions are broadly consistent with those of similar reviews of this topic such as Leifeld and Fuhrer (2010). In addition, organic cropping systems fairly consistently appear to enhance microbial activity. While Leifeld (2012) argues that this is evidence of "C losses" from organically managed soils, it is reflective of the more routine return, and enhanced turnover, of fresh or labile organic matter fractions, discussed in the following section, and their benefits to soil biophysical and biological properties. As noted by Janzen (2006), "organic matter has the most benefit, biologically, when it decays."

5.3   LABILE FRACTIONS OF SOIL ORGANIC MATTER

Qualitative differences in the composition of organic matter in terms of the residue itself and resultant soil organic matter fractions or "pools" may be as important as discussions of the total SOC in understanding the impact of field crop systems on soil quality. Increasing the diversity of cropping tends to increase both the amount and variation in quality of residues, and the period in which roots are active and returned to soil, and such "substrate diversity" can enhance soil-mineralizable C and N pools (Sanchez et al., 2004). These labile and biologically active soil organic matter fractions are important for the management of cropping systems both for enhancing nutrient cycling and maintaining soil quality, including the support of microbial activity and aggregation (Marriott and Wander, 2006a). In addition, labile fractions (including particulate organic matter (POM) and density-derived [light fraction organic matter—LFOM] organic matter fractions) can be more sensitive indicators of management-induced changes in soil C and N, due to the high background of total SOC and total N (TN) in soil and its natural variability (Gregorich et al., 1997; Angers et al., 1999; Lynch et al., 2005a,b; Marriot and Wander, 2006a). Willson et al. (2001) found that POM was closely related to soil N mineralization and is a simple enough procedure to be included in routine testing.

Marriot and Wander (2006a) examined soil organic matter quality in surface (0–25 cm) soils from nine farming systems trials across a range of soil and climatic conditions in the United States, including manure-plus-legume-based organic, legume-based organic, and conventional farming systems. Organic rotations were at all sites longer and more diverse than the conventional management regimes. POM was fractionated at >53 µm after whole soil dispersion. While both organic input systems increased total SOC and TN compared to conventional in these selected trials, these differences were much more pronounced when POM-C and POM-N concentration (g kg^{-1} soil) was compared, with both organic management regimes enriching these fractions by 30%–40%. Subsequent analyses (Marriott and Wander, 2006b) of subdivisions of the POM as free-light POM (FPOM < 1.6 g cm^{-3}) not occluded within soil aggregates, and intra-aggregate occluded POM (OPOM < 2.0 g cm^{-3}), found that the OPOM but not the FPOM, the former fraction considered to have slower turnover in soil due to its physical protection within aggregates, was more abundant in the organic systems. This was attributed partly to greater aggregation under organic management as a result of enhanced residue return. Thus, organic management simultaneously appeared to have enhanced both the labile N in soil and also the amount of POM protected within aggregates. Nelson et al. (2009) tracked POM-C changes across each phase of a 5-year potato/grain/leguminous forage rotations on four organic farm sites, compared with an undisturbed perennial pasture at each site. In contrast to soil biological properties (discussed in the following texts), rotation phase was found to have had no significant effect on soil (0–15 cm) physical and chemical properties, including LFOM and mineralizable C. After total dispersion of soil, LFOM as a percentage of SOC varied from 6.2% to 9% of total SOC.

Miller et al. (2008) in Montana found that the inclusion of a GMr of winter pea in a 4-year rotation with cash crops of winter wheat/lentil and barley resulted in an increase of 23% in soil-mineralizable N, but lower soil nitrate levels, when compared with a fertilized but diversified (including spring pea and winter wheat) no-till conventional grain (corn and sunflower) crops regime (less diversified no-till systems were also included). The organic system received no organic fertilization, however, and soil test P dropped by 14% under organic management. The organic winter wheat yields matched that found under conventional no-till and economic returns were equal after 4 years for both systems. Interestingly, in the conventional no-till systems, summer drought appeared to limit yield more strongly than in the organic system as determined by lower grain harvest indices and grain density. Spring soil moisture depth after the pea GMr was equal to or greater than that under stubble in all other rotations, and combined with the lower soil nitrate levels, the slower rate of winter wheat biomass accumulation under organic management appeared to have utilized soil water more efficiently.

The dynamics of compost C in soil often differs from crop residues and compost C can persist for many years as a slowly decomposing pool in the free POM as shown in a study of organic fertilization of forage in Atlantic Canada by Lynch et al. (2005b). Two years after compost application to perennial clover–grass forages at rates of less than 13 Mg DM ha^{-1} year^{-1}, three different composts promoted gains of up to 9.7 Mg C ha^{-1}. The POM-C in composted plots accounted for 67% of total SOC compared to 39% of SOC for the unfertilized control. In addition, the litter fraction (i.e., >2000 μm), a fraction often discarded in initial soil processing, accounted for 14% of SOC in composted plots (and 41% of the gains in SOC for some treatments) compared to 4% of SOC for the control. Thus, coarsely sieving the soil, and quantifying the litter fraction, is critical in avoiding underestimations of SOC and TN in organic management regimes where compost is applied. The composts were determined to be decomposing very slowly in soil with turnover rates (k) ranging from 0.06 to 0.09 year^{-1} (Lynch et al., 2005b, 2006).

5.4   AGGREGATE DYNAMICS

Both abiotic factors (soil physical properties including texture and inorganic compounds) and biotic factors (microorganisms, soil fauna, and plant roots) influence aggregate formation and their dynamics in soil (Six et al., 2004). Collins et al. (2011) found that soil aggregation (>250 μm) was best predicted by soil texture, particularly clay content, rather than soil biological properties, in 81 samples taken across 25 ha of an organic vegetable farm in Washington state. Manure additions and GMr residues are also known to increase soil aggregation (Angers et al., 1999; Marriott and Wander, 2006b) although this is less the case for more recalcitrant organic amendments such as composts (Lynch et al., 2005b). Aggregate size distribution may also influence the functioning and stability of soil food webs (Briar et al., 2011).

How do soil organic matter quality, distribution, and soil physical properties differ between organic and conventional arable farming systems when compared with perennial pastures or grassland? In the Netherlands, Pulleman et al. (2005) examined soil (0–20 cm) organic matter distribution and microaggregate dynamics in an organic and conventional arable system (both comprising cereals, potatoes, and sugar beets, in rotation with 2–3 years of grass, with the organic system receiving animal manure instead of fertilizer) compared with a permanent pasture (grazed and receiving fertilizer and manure). Equal amounts of microaggregates (53–250 μm), which accounted for the majority of total SOC, were found to be present for all systems, but under pasture, the greater fraction of SOC and greater stability of microaggregates suggested greater organic stabilization under this regime, attributed to greater observed earthworm activity (52% of volume of soil appeared worm-worked in thin sections) and reduced mechanical disturbance. Between arable systems, total SOC and earthworm activity was greater (28% vs. 8% volume of soil) under organic management as was the amount of slake-resistant macroaggregates (1.7×) under organic management. Organic farming did not significantly affect microaggregate stabilization of SOC compared to conventional management, however, and differences were small compared to permanent pasture effects.

Few studies have directly measured soil erodibility as affected by organic and conventional arable cropping. Mahli et al. (2009), in the 11-year study at Scott, Saskatchewan, described in Section 5.2 earlier, found that the proportion of wind-erodible fine dry aggregates (<1.3 mm) was most affected by cropping system diversity and not inputs/management regime (i.e., organic or conventional) and lowest for the diversified grain system that included forages. In contrast, wet aggregate stability (of the 1.3–2.00 mm fraction) was greatest under the reduced input/tillage management regime attributed to greater measured SOC in these aggregates. Soil structure generally (as measured by mean weight diameter of aggregates) was greatest under more diversified cropping systems together with reduced tillage.

In Washington, United States, the topsoil of a winter wheat field organically managed for more than 50 years was found to be 16 cm deeper than that on the adjacent conventionally managed field. Soil (a Naff silt loam) and subsoil conditions were otherwise identical. The conventional 2-year rotation of fertilized winter wheat followed by spring pea differed on the organic farm in that a GMr crop of Austrian winter pea (*P. sativum* ssp. *arvense* L. Poir) was also included in place of fertilizer inputs. Tillage operations, including fertilizer application, were less in the organic system overall. Winter wheat yields averaged 8% less on the organic farm. In addition to higher SOC and polysaccharide content, lower modulus of rupture (related to surface crusting hardness) was found under organic production. Polysaccharides are binding agents in soil aggregate formation. Soil water erosion losses (determined by a method of measuring cross sections of rills in soil) at 34.2 t ha^{-1} greatly exceeded (>3×) tolerable rates (11.2 t ha^{-1} year^{-1}) to sustain long-term productivity and environmental quality under conventional management, while at only 8.3 t ha^{-1} were within tolerable levels on organic farms (Reganold et al., 1987).

In the long-term DOK trial in Switzerland, the aggregate stability, which is inversely related to erosion susceptibility, when determined by means of percolation, was significantly better in the organic plots. In the field, soil detachability tests using vertical splash boards only partially highlighted these differences, however. The improved aggregate stability under organic management was related to enhanced microbial biomass and earthworm activity (Siegrist et al., 1998; Mader et al., 2002).

5.5   SOIL ECOLOGY AND SOIL HEALTH

The unique characteristics of organic production regimes, and specifically the routine return of C of varying quality to soil in organic systems through extended and more diversified rotations, perennial crops, and GMrs, are potentially most important in its ecological effect in maintaining soil health and biodiversity and mitigating the disruptive effects of more intensive cropping systems on micro- and mesofaunal communities (Birkhofer et al., 2008; Postma-Blaauw et al., 2010; Gomiero et al., 2011; Lynch et al., 2012a). However, it must be also noted that differences between management practices within a given farming system can be just as influential, and sometimes more influential, than the effects of the farming system per se (Stockdale and Watson, 2009; Elmholt and Labouriau, 2005). Collins et al. (2011), in a study designed to look at the relationship between N mineralization potential, aggregate stability, and soil physical, chemical, and biological properties on an organic vegetable farm in Washington state, found soils on the farm with a greater microbial biomass, SOC, and N mineralization potential, formed a homogenous group that was particularly high under undisturbed pastures. Soil pH, for example, has been suggested to have a predominant influence on soil microbial community diversity and functioning (Nelson and Spaner, 2010). In Holland, Postma-Blaauw et al. (2010) found that agricultural intensification affected the abundance of taxonomic groups with larger body size (earthworms, microarthropods, enchytraeids, and nematodes) more than protozoa, bacteria, and fungi. Finally, for some organisms, factors such as year-to-year variation (Osler et al., 2008 [mites]; Schneider et al., 2010 [fungal community]) or surrounding habitat composition and landscape type (Lynch et al., 2012a [arthropods]) can sometimes equally dominate their observed dynamics.

In Atlantic Canada, earthworm abundance and biomass, and soil microbial quotient, were shown to particularly benefit from extended (5 years) rotations on organic potato farms, compared with much more frequent cropping of potatoes (and associated tillage) in conventional systems (Nelson et al., 2009). Earthworm and microbial populations recovered from marked reductions during potato cropping to levels found in adjacent permanent pastures after 3–4 years of the rotation on the organic farms. In the long-term DOK trial in Switzerland, Siegrist et al. (1998) similarly found that earthworm biomass and density, as well as the population diversity, were significantly greater on the organic plots than on the conventional plots. Irmler (2010) found that changes in earthworm populations during conversion to organic farming were not related to SOC levels or soil pH alone, while as noted in Section 5.2 earlier, a recent long-term study in Denmark by Chirinda et al. (2010) found enhanced microbial biomass and activity (respiration) linked to increased C input rates under organic management that were not reflected in SOC gains.

A number of other studies have found greater microbial biomass and activity, and enhanced earthworm populations, in organically managed field cropping systems. Bolton et al. (1985) reported higher levels of urease, phosphatase, and dehydrogenase and significantly higher soil microbial biomass under long-term organic grain cropping in Washington state (see also Reganold et al., 1987 earlier). In Switzerland, Birkhofer et al. (2008) found that long-term organic farming and also the application of farmyard manure promoted soil quality, microbial biomass, and earthworms. The increased prey, combined with greater weed coverage, in turn, promoted more aboveground generalist predators and subsequent pest (aphid) control. In Germany, Flohre et al. (2011) found that soil microbial biomass and earthworm species richness increased under organic farming as landscape structural complexity simplified, and predation pressure was apparently reduced, while the opposite was true under conventional farming.

Is soil microbial community composition and diversity strongly influenced by farming system? In a broad look at the soil fungal community, Elmholt and Labouriau (2005) examined soil fungal abundance (using selective plate media and data adjusted for soil clay content) across 29 sites in Denmark. These included organic, "conventional with manure" and "conventional with fertilizers" field cropping sites. The organic system differed primarily in the diversification of the rotation and inclusion of legumes. Farming systems could not be separated on the basis of fungal abundance (CFU mg^{-1} dry soil) due to large variation within each farming system. Similarly, Widmer et al. (2006), using molecular and nonmolecular techniques, found that bacterial community structure was only weakly influenced by farming systems within the DOK trials in Switzerland, when compared to the effects of manure and crop rotations. Gauging fungal community structure within the same trial, Schneider et al. (2010) found that temporal effects (the variation in fungal communities between 2000 and 2007 samplings) and individual crops had greater influence than farming systems, although community richness remained largely stable. An earlier study by Wander et al. (1995) found, using phospholipid fatty acid analysis (PLFA) of soils after 14 years of the Rodale Farming Systems Trial, that the viable microbial communities were very similar in all three farming systems. However, all soils had been collected in November following corn harvest. Soil respiration and ^{13}C isotope fate studies suggested a larger and more heterogeneous population under organic management. In western Canada, Nelson et al. (2011a) attributed greater microbial richness and diversity (as measured by PFLA analyses) under organic wheat management to be due to soil management and history of compost use, plus greater weed populations in agreement with Birkhofer et al. (2008). Thus, cropping systems that promote vegetative diversity, including polycultures or weeds, may enhance aboveground and belowground diversity. Plant roots exude carbon (at up to 3 Mg C ha^{-1}) and nutrients into the rhizosphere zone, which can influence soil microbial population density and diversity. In turn, this microbial community facilitates turnover and flow of these rhizodeposits within the soil. Kong and Six (2012) examined how 14 years of different farming systems (conventional, low-input, and organic maize–tomato production) influenced soil community structure (using PLFA techniques) as influenced by addition of ^{13}C-labeled hairy vetch root tissue. The microbial community composition processing the incoming root C was found to be similar across all farming systems, as was the relative subsequent distribution of root-derived C in the rhizosphere and bulk soil.

Thus, while the soil microbial community composition appears equally responsive to crop sequences and temporal shifts as to effects of farming system per se and is likely generally resilient to farming system impacts in agreement with Postma-Blaauw et al. (2010), selected key constituents of the soil microbial community may be more sensitive to field crop management regimes. Arbuscular mycorrhizal fungi (AMF) play such a wide range of roles, from enhancing plant uptake of a range of nutrients, stabilization of soil aggregates, contributions of carbon to soil, and influences on the soil microbial community that they are often considered central to sustainable production systems (Hamel and Strullu, 2006; Nelson and Spaner,

2010). A number of studies have found that organic crop management promotes mycorrhizal colonization (Galvez et al., 2001; Mader et al., 2002; Welsh et al., 2006; Gosling et al., 2010).

Within the long-term Glenlea cropping systems study in Manitoba, Welsh et al. (2006) found enhanced VAM colonization and increased mycorrhizal spore populations under organic systems as also noted by Galvez et al. (2001) at the Rodale Institute in Pennsylvania. Gosling et al. (2010) found that AM spores and root colonization potential increased rapidly, that is, within 2 years of conversion to tilled organic arable (cereal based) or horticultural production when compared with paired conventional fields in 11 sites in the United Kingdom.

Recent research, utilizing molecular techniques, found significant differences in mycorrhizal species composition between paired long-term organic and conventional dairy farms in Ontario, although this may reflect both aboveground (differences in forage species composition including legume content) and edaphic (especially soil test P levels) differences between systems (Schneider et al., 2011). Weeds may play an important role also in hosting mycorrhiza during rotation phases with nonmycorrhizal crops or during the overwintering period (Nelson and Spaner, 2010). Differences in crop cultivars appear to also play an important role. Nelson et al. (2011a) found that the wheat cultivar, AC Superb, resulted in higher levels of mycorrhizal fungi in soil (1.97%) than four other older cultivars (1.32%–1.43%).

In an interesting avenue of research, Grossman et al. (2011), using molecular techniques, found that organic management was a strong driver of rhizobial richness and diversity within soybean production systems in New York state. Linking the relationship of this diversity to effective biological nitrogen fixation remains a research goal.

Studies of farming system influence on nematode community structure to date have primarily focused on organic vegetable cropping rather than field crop production systems (Briar et al., 2011; Nair et al., 2012). An exception is the study of Birkhofer et al. (2008) from the DOK trial in Switzerland. That study found that bacterivorous nematodes were most abundant, and nematode communities tended to be more diverse, in systems receiving farmyard manure. Fungivorous nematodes were far less abundant than bacterivorous and plant-feeding nematodes in all systems. As noted earlier, Widmer et al. (2006) had earlier similarly found, within this trial, the bacterial community structure to be only weakly influenced by farming system, when compared to the effects of manure and crop rotations.

The DOK long-term trial also recently provided an opportunity to conduct one of the few studies to date on farming system influence on protists (Heger et al., 2012). While results were based on a spring sampling alone, testate amoebae (Arcellinida and Euglyphida) and diatoms (Bacillariophyta) appeared to respond to farming system treatments. Diatom community structure differed between organic and the two conventional systems, with the greatest diversity found in the biodynamic system. While testate amoeba abundance was about five times higher in biodynamic than in conventional systems, richness and diversity did not differ between treatments.

With respect to aboveground soil fauna, Boutin et al. (2009), in Ontario, found that while overall arthropod abundance was influenced by organic versus conventional farming system, richness was not. Local factors (plant composition and management regime) strongly influenced arthropod composition as did habitats in the surrounding landscape. Beneficial arthropods were more abundant in woody hedge-rows while phytophagous arthropods were more abundant in crop fields. Among taxa, vegetative diversity appears most consistently influenced by farming system (Lynch et al., 2012a), and in Europe, Ponce et al. (2011) observed a strong link between this enhanced plant diversity (especially of insect-pollinated weeds) and arthropod diversity, with arthropod abundance and richness 43% and 6% greater under organic management.

Others concurred that landscape complexity (quantified as percent grassland cover) was more important than farmland management to Carabid beetle diversity or found no difference in ground beetle abundance and richness between farming systems (Purtauf et al., 2005; Winqvist et al., 2011). In one of the few studies of mites assemblages as influenced by farming system, Osler et al. (2008), in Alberta, exam-ined, over 2 years, the effect of crop diversity within organic and conventional crop-ping systems. Established 4–5 years previously, organic rotations comprised either two crops (sweet clover [*Melilotus officinalis* (L.) Pall., Norgold] rotated with wheat) or four crop species [wheat–peas–flax–sweet clover], while conventional sequences consisted of wheat or summer fallow every other year or a 4-year rotation of wheat–peas–flax–summer fallow. In general, differences were found in mite assemblages both within and between cropping systems, but these results were not consistent across years, suggesting an over-riding sensitivity to environmental conditions.

With the advent and increasing use of no-till GMr termination and field crop seed-ing techniques, how will this shift in amount and distribution of crop residues and reduction in tillage influence micro- and mesofaunal communities in organic sys-tems? As discussed by Mirsky et al. (2012), grain stand losses from seed and seedling damage by herbivorous insects such as wireworms (larvae of Coleoptera: Elateridae), seed-corn maggot (*Delia platura* and related species, Diptera: Anthomyiidae), and cutworms (Lepidoptera: Noctuidae) have been observed to increase in some mid-Atlantic region cover crop and no-till grain systems. Mulch residue quality differ-ences will likely also affect the microbial biomass and functional diversity of the soil microbial community (Bending et al., 2002). Different tillage regimes in organic field cropping will also differentially affect earthworm populations as shown in a 3-year study at three field crop sites across France by Peigné et al. (2009). Reduced tillage in organic systems is also expected to benefit Carabids, assuming that adequate shelter zones such as grass field margins are also provided (Legrand et al., 2011).

In summary, while research over the past decade in particular has greatly improved our understanding of soil ecology as affected by farming system, much more needs to be done to improve our understanding of the mechanisms underlying shifts in spe-cific components of above- and belowground micro- and mesofaunal communities at the field and farm scale, including those affected by type and intensity of organic production, and in turn the link of these changes to specific agroecosystems services.

5.6   WATER INFILTRATION AND WATER HOLDING CAPACITY

Agriculture uses over 70% of global freshwater supplies (Lynch et al., 2012a) and the efficient use of declining water resources, along with development of produc-tion systems resilient to climatic adversity, is critical components of agricultural sustainability and food security (Jägerskog and Jønch Clausen, 2012). Reganold et al. (1987), in the paired study of grain farms in Washington state described earlier, attributed greater moisture content (15.9% vs. 9%) in the organically farmed soil to improved soil organic matter status under organic management. Enhanced soil organic matter storage can contribute also to yield stability and production system resilience (Lynch, 2009). Pimentel et al. (2005) reported that conservation of soil moisture and water resources via enhanced soil organic matter under organic management promoted higher corn yields in 5 dry years in a 10-year Rodale Institute Farming Systems Trial in Pennsylvania. In Maine, Mallory and Porter (2007) found, over a 13-year study, that management systems that improved soil quality, through the application of compost and GMrs, enhanced potato yield stability (i.e., year-to-year variation in yields). Yields of nonirrigated potatoes grown in 2-year rotations in the amended system were much less negatively influenced by adverse conditions, particularly poor rainfall.

In the long-term study at Scott, Saskatchewan, reported earlier by Mahli et al. (2009), growing season precipitation was very low (averaging just 181 mm) with severe droughts in 4 of these years, and the relative performance recorded for the organic rotations may be attributed to some resilience under these conditions. In the long-term DOK trial in Switzerland, water holding capacity was found to be enhanced by 20%–40% under organic management (Siegrist et al., 1998; Mader et al., 2002).

Gomiero et al. (2011) conclude, on the basis of a review of a number of comparative studies, that under drought conditions, organic crops can outyield conventional systems by 70%–90% primarily due to both improved water capture and water holding capacity of soils under organic management. By some estimates, every 1% of soil organic matter retains 10,000–11,000 L of water ha^{-1}. Mirsky et al. (2012), in a discussion of no-till termination of GMrs and organic grain production, noted that systems that may delay GMr termination such as with roller crimpers may result in subsequent spring moisture depletion for the grain crop. This can be offset by the GMr residue mulch benefits in conserving summer soil moisture, with this trade-off influenced by cover crop species and local soil and weather conditions.

Oquist et al. (2006) in Minnesota examined how alternative field crop production (without fertilizer or pesticides and including certified organic management), established since 1989, affected soil physical properties including SOC, bulk density, moisture retention, and saturated hydraulic conductivity. Alternative cropping practices included the use of diverse longer rotations compared to conventional systems comprised of short (2 years) corn–soybean rotations plus use of manure in place of fertilizer. Primary tillage regimes were similar for both systems. Bulk density was improved moderately (3%) in all soil horizons by alternative management, while saturated hydraulic conductivity (45.5 vs. 18.1 cm day^{-1}) and water retention were significantly improved under the alternative cropping regimes. It was concluded that these changes in soil physical properties, in addition to crop benefits, translate into differences in runoff and soil profile water movement that in turn will provide environmental benefits with respect to water quality.

5.7 REDUCED TILLAGE AND ZERO-TILLAGE APPROACHES TO ORGANIC FIELD CROP PRODUCTION AND INTEGRATING LIVESTOCK: IMPLICATIONS FOR SOIL QUALITY

Innovative approaches to termination of GMrs that reduce reliance on tillage are increasingly being explored in organic field crop and vegetable production systems (Lynch et al., 2012a). A recent special issue of the journal *Renewable Agriculture and Food Systems* (Volume 27), with papers from Europe (Gadermaier et al., 2012), the United States (Carr et al., 2012; Delate et al., 2012; Luna et al., 2012; Mirsky et al., 2012; Reberg-Horton et al., 2012), and Canada (Shirtliffe and Johnson, 2012), was devoted to this topic. Some studies on this novel topic of reduced and no-till systems for organic field crop production have also examined impacts on soil quality parameters.

A "biological no-till" system that combined cover crops and a crop roller was developed first in Brazil and an early model was developed by USDA-ARS in Alabama (Ashford and Reeves, 2003; Mirsky et al., 2012). Hepperly (2008) reported on substantial SOC gains from such a "blade roller" or "roller crimper" for GMr termination at the Rodale Institute when compared to conventional no-till and standard organic management. In Manitoba, Vaisman et al. (2011) studied the impact of rolling or tilling, or their combination at flowering, of pea, or pea and oat GMr, on weeds, soil N dynamics, and yields of the subsequent wheat crop. The added soil cover (50%–90% cover) provided by GMr mulches with rolling compared to less than 5% cover with tillage may provide important soil and water conservation benefits.

The observed negative impact of reduced spring soil temperatures and reduced available N in this study, however, points to the important trade-offs that would have to be resolved for adoption of this practice in regions with cool or wet spring conditions. In conventional no-till systems also, Ogle et al. (2012) reported that reduced overall productivity in cool and/or wetter regions reduces C return to soil especially, although this is sometimes offset by reduced rates of SOC decomposition. Shirtliffe and Johnson (2012), in a study at two sites in Saskatchewan on black chernozemic clay loam soil, found wheat yields to be just as good when a field pea GMr was terminated by roller crimping as when it was tilled or mowed. However, in this study, in contrast to that of Vaisman et al. (2011), tillage was used in the spring prior to planting of the following wheat crop.

In some regions, cover crop biomass often exceeds the >8.0 Mg DM ha^{-1} considered optimum for annual weed suppression. Mirsky et al. (2012) report on 21 site years of trials across the mid-Atlantic region of the United States where corn and soybean were planted in GMr mulch following fall-seeded hairy vetch and fall rye, respectively. GMr biomass achieved >8 Mg DM ha^{-1} in 50% of the trials (range from 4 to 8 Mg ha^{-1} otherwise). In North Dakota, in contrast, biomass production of these two GMrs rarely exceeded 5 Mg DM ha^{-1} (Carr et al., 2012). This is consistent with the general range of biomass productivity for annual legume GMrs of 2.5 Mg DM ha^{-1} in the Canadian Prairie region brown soil zone and 5.0 Mg DM ha^{-1} in the black soil zone suggested by Bell et al. (2012). Where moisture is not limiting, manipulating time of planting and seeding rate and fertilization regime is recommended to improve GMr biomass production. When combined with diversified rotations in high GMr biomass-producing regions, these systems may enhance SOC sequestration (Mirsky et al., 2012).

Luna et al.'s (2012) report on the benefits of a zone or strip tillage system for sweet corn production in the Western United States (California, Oregon, and Washington states). In this system, only 35%–40% of the soil surface is tilled, while surface residues are retained between tilled zones to reduce soil erosion. Yields in a 4-year trial were identical for both zone-tilled and conventional tillage systems. Previous trials in the region with the no-till crimper roller system had produced very inconsistent results for a range of field crops. The requirement for specialized tillage equipment remains an important barrier to adoption of zone tillage, however (Luna et al., 2012).

As has been well documented in conventional no-till systems, reducing tillage in organic systems will shift the distribution or stratification in soil of SOC, nutrients, and soil biological activity. In Europe, Gadermaier et al. (2012) found after 6 years of reduced tillage (chisel plowing and superficial [0–5 cm] incorporation of grass–clover leys, compared to moldboard plowing) that SOC, microbial biomass and activity, and soluble P and K were all increased in the surface (0–10 cm) soil.

A number of studies have called for greater integration of livestock in organic cropping systems in the Great Plains of the United States and Canadian Prairie region (Russelle et al., 2007; Miller et al., 2008; Gomiero et al., 2011; Thiessen Martens and Entz, 2011) to improve farm nutrient balances and provide a range of additional ecosystem services. The use of livestock to graze GMr phases is also an active area of on-farm and station research in various regions in Canada (R. Jannasch pers comm.; M. Entz pers. comm.). Some studies suggest that improved grazing management, including the use of legumes, can be not only a cost-effective option but can promote substantial SOC gains on the extensive acreage of often degraded, permanent grasslands in North America or when compared with haying of perennial forages (Lynch et al., 2005a; Franzluebbers and Stuedemann, 2009; Wiltshire et al., 2010).

5.8   FUTURE DIRECTIONS AND RESEARCH NEEDS

This chapter provides a review of the biophysical and biological components of soil as influenced by organic field crop management. It examines various attributes of soil quality and soil health, including total and labile soil organic matter, aggregate dynamics, water infiltration and water holding capacity, and soil biological activity and diversity. While management practices play a key role in influencing improvements in many of these attributes, it is important to note the degree to which inherent edaphic properties such as texture can be equally influential, a distinction that Collins et al. (2011) refer to as "dynamic soil quality" versus "inherent soil quality." As this review suggests also, in many cases, specific beneficial management practices are clearly at least as influential on soil quality attributes than farming systems per se, although it can be argued that organic farming, through its reliance on "feeding the soil" approaches, inherently promotes and requires many of these practices (such as extended and diverse rotations). It is also important to note that in any discussion of soil quality goals, there are inherently trade-offs that must be considered, for example, increases in decomposers in soil and soil TN (and SOC) may enhance N_2O emissions (Griffiths et al., 2010; Powlson et al., 2011).

With respect to other future directions and research needs, the fundamental importance of an integrated approach to sustaining soil fertility in organic field crop production to maintain system productivity, carbon and nutrient cycling, and soil quality is clearly evident from the overview provided here. The advent of reduced and zero-tillage regimes for organic field cropping and increasing interest in reintegrating the coupling of livestock systems to GMr phases (see Chapter 9) are exciting and relatively new developments that will impact in novel ways on the flows, distribution, and dynamics of carbon and nutrients within these production systems. Thorup-Kristensen et al. (2012) have demonstrated the important linkage between crop and GMr root exploration of the soil profile and benefits with respect to nitrate sponging capacity of different crop/GMr combinations and ultimately N use efficiency, which should be further explored.

Documenting a minimum dataset (Gregorich et al., 1994) of parameters to assess changes in soil quality or soil health in response to management practices in situ can be labor intensive and costly. The response of a suite of soil parameters, or of a sole bioindicator organism, especially if developed as a lab method, may have promise as an alternative and more viable means to gauge the soil quality and health of agricultural soils. POM, as noted earlier, for example, may act as a simple procedure with which to help gauge N mineralization potential (Willson et al., 2001). The Cornell Soil Health Test (CSHT) is a farmer-oriented soil quality assessment tool consisting of 15 soil indicators recently developed that includes physical, chemical, and biological soil properties (Idowu et al., 2009). Nelson et al. (2011b) modified an established ISO standard laboratory soil ecotoxicology test using the Collembola bioindicator *Folsomia candida* and found, in initial trials, a more positive bioindicator response (with respect to changes in body growth and reproduction) when exposed to organically managed compared to conventionally managed soils.

Substantial gaps exist in our understanding of how shifts in soil micro- and mesofaunal communities at the field and farm scale, as affected by specific management practice or type of organic regime, are linked to specific agroecosystems services. An improved understanding of how, in the context of climate change, organic field crop management influences soil biotic and abiotic components and their resilience will undoubtedly grow as an important and exciting new research direction.

Finally, organic producers are fundamentally managers of the processes of decomposition. Thus, perhaps the most fundamental future challenge is whether organic cropping system management can become refined enough in its understanding and practices to develop a suite of SOC fractions in soil that both retain stored SOC and at the same time "feed the soil," that is, provide inputs of energy and nutrients for soil microbial populations synchronized to provide both crop and broader ecosystem benefits. As noted by Janzen (2006), such a balanced approach is an urgent and laudable goal.

REFERENCES

Angers, D.A., L.M. Edwards, J.B. Sanderson et al. 1999. Soil organic matter quality and aggregate stability under eight potato cropping sequences in fine sandy loam of Prince Edward Island Canadian. *J. Soil Sci.* 79:411–417.

Ashford, D.L. and D.W. Reeves. 2003. Use of a mechanical roller-crimper alternative kill method for cover crops. *Am. J. Altern. Agric.* 18:37–45.

Bell, L.W., B. Sparling, M. Tenuta et al. 2012. Soil profile carbon and nutrient stocks under long-term conventional and alfalfa-crop rotations and re-established grassland. *Agric. Ecosyst. Environ.* 158:156–163.

Bending, G.D., M.K. Turner, and J.E. Jones. 2002. Interactions between crop residue and soil organic matter quality and the functional diversity of soil microbial communities. *Soil Biol. Biochem.* 34:1073–1082.

Birkhofer, K., T.M. Bezemer, J. Bloem et al. 2008. Long-term organic farming fosters below and aboveground biota: Implications for soil quality, biological control and productivity. *Soil Biol. Biochem.* 40:2297–2308.

Bolton Jr., H., L.F. Elliott, R.I. Papendick et al. 1985. Soil microbial biomass and selected soil enzyme activities: Effect of fertilization and cropping practices. *Soil Biol. Biochem.* 17:297–302.

Boutin, C., P.A. Martin, and A. Baril. 2009. Arthropod diversity as affected by agricultural management (organic and conventional farming), plant species, and landscape context. *Ecoscience* 16:492–501.

Briar, S., S. Fonte, I. Park et al. 2011. The distribution of nematodes and soil microbial communities across soil aggregate fractions and farm management systems. *Soil Biol. Biochem.* 43:904–914.

Carr, P., R. Anderson, Y. Lawley et al. 2012. Organic zero till in the northern US Great Plains region: Opportunities and obstacles. *Renew. Agric. Food Syst.* 27:12–20.

CGSB. 2006. Organic production systems general principles and management standards. Canada General Standards Board/SCC (Standards Council of Canada) [Amended October 2008] National Standard of Canada. CAN/CGSB-32.310-2006. Canadian General Standards Board, Gatineau, Quebec, Canada, p. 36.

Chirinda, N., J.E. Olesen, J.R. Porter et al. 2010. Soil properties, crop production and greenhouse gas emissions from organic and inorganic fertilizer-based arable cropping systems. *Agric. Ecosyst. Environ.* 139:584–594.

Collins, D.P., C.G. Cogger, A.C. Kennedy et al. 2011. Farm-scale variation of soil quality indices and association with edaphic properties. *Soil. Sci. Soc. Am. J.* 75:580–590.

Delate, K., D. Cwach, and C. Chase. 2012. Organic no tillage system effects on soybean, corn and irrigated tomato production and economic performance in Iowa, USA. *Renew. Agric. Food Syst.* 27:49–59.

Drinkwater, L.E., P. Wagoner, and M. Sarrantonio. 1998. Legume-based cropping systems have reduced carbon and nitrogen losses. *Nature* 396:262–265.

Elmholt, S. and R. Labouriau. 2005. Fungi in Danish soils under organic and conventional farming. *Agric. Ecosyst. Environ.* 107:65–73.

Fließbach, A., H. Oberholzer, L. Gunst et al. 2007. Soil organic matter and biological soil quality indicators after 21 years of organic and conventional farming. *Agric. Ecosyst. Environ.* 118:273–284.

Flohre, A., M. Rudnick, G. Traser et al. 2011. Does soil biota benefit from organic farming in complex vs. simple landscapes? *Agric. Ecosyst. Environ.* 141:210–214.

Franzluebbers, A.J. and J.A. Stuedemann. 2009. Soil-profile organic carbon and total nitrogen during 12 years of pasture management in the Southern Piedmont USA. *Agric. Ecosyst. Environ.* 129:28–36.

Gadermaier, F., A. Berner, A. Fließbach et al. 2012. Impact of reduced tillage on soil organic carbon and nutrient budgets under organic farming. *Renew. Agric. Food Syst.* 27:68–80.

Galvez, L., D.D. Douds Jr., L.E. Drinkwater et al. 2001. Effect of tillage and farming system upon VAM fungus populations and mycorrhizas and nutrient uptake of maize. *Plant Soil* 228:299–308.

Gattinger, A., A. Muller, H. Matthias et al. 2012. Enhanced top soil carbon stocks under organic farming. *PNAS* 109:18226–18231.

Gomiero, T., D. Pimentel, and M.G. Paoletti. 2011. Environmental impact of differential agricultural management practices: Conventional vs. organic agriculture. *Crit. Rev. Plant Sci.* 30:95–124.

Gosling, P., A. Ozakib, J. Jones et al. 2010. Organic management of tilled agricultural soils results in a rapid increase in colonisation potential and spore populations of arbuscular mycorrhizal fungi. *Agric. Ecosyst. Environ.* 139:273–279.

Gregorich, E.G., M.R. Carter, D.A. Angers, C.M. Monreal, and B.H. Ellert. 1994. Towards a minimum data set to assess soil organic matter quality in agricultural soils. *Can. J. Soil Sci.* 74: 367–385.

Gregorich, E.G., M.R. Carter, J.W. Doran et al. 1997. Biological attributes of soil quality. In *Developments in Soil Science; Soil Quality for Crop Production and Ecosystem Health.* eds. E.G. Gregorich and M.R. Carter, pp. 81–113. Elsevier, New York.

Griffiths, B.S., B.C. Ball, T.J. Daniell et al. 2010. Integrating soil quality changes to arable agricultural systems following organic matter addition, or adoption of a ley-arable rotation. *Appl. Soil Ecol.* 46:43–53.

Grossman, J.M., M.E. Schipanski, T. Sooksanguan et al. 2011. Diversity of rhizobia in soybean [*Glycine max* (Vinton)] nodules varies under organic and conventional management. *Appl. Soil Ecol.* 50:14–20.

Hamel, C. and D.G. Strullu. 2006. Arbuscular mycorrhizal fungi in field crop production: Potential and new direction. *Can. J. Plant Sci.* 86:941–950.

Hassink, J. 1997. The capacity of soils to preserve organic C and N by their association with clay and silt particles. *Plant Soil* 191: 77–87.

Hathaway-Jenkins, L.J., R. Sakrabani, B. Pearce et al. 2011. A comparison of soil and water properties in organic and conventional farming systems in England. *Soil Use Manage.* 27:133–142.

Heger, T.J., F. Straube, and E.A.D. Mitchell. 2012. Impact of farming practices on soil diatoms and testate amoebae: A pilot study in the DOK-trial at Therwil, Switzerland. *Eur. J. Soil Biol.* 49:31–36.

Hepperly, P. 2008. Food and agriculture offer world of opportunity to combat global greenhouse gases. Oral presentation at the *International Conference on Organic Agriculture and Climate Change*, Enita Clermont, France, April 17–18, 2008.

Idowu, O.J., H.M. van Es, G.S. Abawi et al. 2009. Use of an integrative soil health test for evaluation of soil management impacts. *Renew. Agric. Food Syst.* 24:214–224.

Irmler, U. 2010. Changes in earthworm populations during conversion from conventional to organic farming. *Agric. Ecosyst. Environ.* 135:194–198.

Jägerskog, A. and T. Jønch Clausen. 2012. Feeding a thirsty world challenges and opportunities for a water and food secure future. Report Nr. 31. SIWI, Stockholm, Sweden.

Janzen, H.H. 2006. The soil carbon dilemma: Shall we hoard it or use it? *Soil Biol. Biochem.* 38:419–424.

Kirchmann, H., L. Bergstrom, T. Katterer et al. 2007. Comparison of long-term organic and conventional crop-livestock systems on a previously nutrient-depleted soil in Sweden. *Agron. J.* 99:960–972.

Kong, A.Y.Y. and J. Six. 2012. Microbial community assimilation of cover crop rhizodeposition within soil microenvironments in alternative and conventional cropping systems. *Plant Soil* 356:315–330.

Legrand, A., C. Gaucherel, J. Baudry et al. 2011. Long-term effects of organic, conventional, and integrated crop systems on Carabids. *Agron. Sustain. Dev.* 31:515–524.

Leifeld, J. 2012. How sustainable is organic farming? *Agric. Ecosyst. Environ.* 150:121–122.

Leifeld, J. and J. Fuhrer. 2010. Organic farming and soil carbon sequestration: What do we really know about the benefits? *Ambio* 39:585–599.

Leifeld, J., R. Reiser, and H.R. Oberholzer. 2009. Consequences of conventional versus organic farming on soil carbon: Results from a 27-year field experiment. *Agron. J.* 101:1204–1218.

Luna, J.M., J.P. Mitchell, and A. Shrestha. 2012. Conservation tillage for organic agriculture: Evolution toward hybrid systems in the western USA. *Renew. Agric. Food Syst.* 27:21–30.

Lynch, D.H. 2009. Environmental impacts of organic agriculture: A Canadian perspective. *Can. J. Plant Sci.* 89:621–628.

Lynch, D.H., M. Sharifi, A. Hammermeister et al. 2012b. Nitrogen management in organic potato production. In *Sustainable Potato Production: Global Case Studies*. eds. Zhongi, H., R.P. Larkin, and C.W. Honeycutt, pp. 209–231. Springer Science+Business Media, New York.

Lynch, D.H., N. Halberg, and G.D. Bhatta. 2012a. Environmental impacts of organic agriculture in temperate regions. *CAB Rev.* 7(No. 010):1–17.

Lynch, D.H., R.D.H. Cohen, A. Fredeen et al. 2005a. Management of Canadian prairie region grazed grasslands: Soil C sequestration, livestock productivity and profitability. *Can. J. Soil Sci.* 85:183–192.

Lynch, D.H., R.P. Voroney, and P.R. Warman. 2004. Nitrogen availability from composts for humid region perennial grass and legume-grass forage production. *J. Environ. Qual.* 33:1509–1520.

Lynch, D.H., R.P. Voroney, and P.R. Warman. 2005b. Soil physical properties and organic matter fractions under forages receiving composts, manure or fertilizer. *Compost Sci. Util.* 13:252–261.

Lynch, D.H., R.P. Voroney, and P.R. Warman. 2006. Use of ^{13}C and ^{15}N natural abundance techniques to characterize carbon and nitrogen dynamics in composting and in compost-amended soils. *Soil Biol. Biochem.* 38:103–114.

Lynch, D.H., Z.M. Zheng, B.J. Zebarth et al. 2008. Organic amendment effects on tuber yield, plant N uptake and soil mineral N under organic potato production. *Renew. Agric. Food Syst.* 23:250–259.

Mader, P., A. Fliessbach, D. Dubois et al. 2002. Soil fertility and biodiversity in organic farming. *Science* 296:1694–1697.

Mahli, S.S., S.A. Brandt, R. Lemke et al. 2009. Effects of input level and crop diversity on soil nitrate-N, extractable P, aggregation, organic C and N, and nutrient balance in the Canadian Prairie. *Nutr. Cycl. Agroecosyst.* 84:1–22.

Mallory, E.B. and G.A. Porter. 2007. Potato yield stability under contrasting soil management strategies. *Agron. J.* 99:501–510.

Marriott, E.E. and M.M. Wander. 2006a. Total and labile soil organic matter in organic and conventional farming systems. *Soil Sci. Soc. Am. J.* 70:950–959.

Marriott, E.E. and M. Wander. 2006b. Qualitative and quantitative differences in particulate organic matter fractions in organic and conventional farming systems. *Soil Biol. Biochem.* 38:1527–1536.

Miller, P.R., D.E. Buschena, C.A. Jones et al. 2008. Transition from intensive tillage to no-tillage and organic diversified annual cropping systems. *Agron. J.* 3:591–599.

Mirsky, S.B., M.R. Ryan, W.S. Curran et al. 2012. Conservation tillage issues: Cover-crop based organic rotational no-till grain production in the mid-Atlantic region, USA. *Renew. Agric. Food Syst.* 27:31–40.

Mondelaers, K., J. Aertsens, and G. Van Huylenbroeck. 2009. A meta-analysis of the differences in environmental impacts between organic and conventional farming. *Br. Food J.* 111:1098–1119.

Nair, A. and M. Ngouajio. 2012. Soil microbial biomass, functional microbial diversity, and nematode community structure as affected by cover crops and compost in an organic vegetable production system. *Appl. Soil Ecol.* 58:45–55.

Nelson, A. and D. Spaner, 2010. Cropping systems management, soil microbial communities, and soil biological fertility: A review. In *Genetic Engineering, Biofertilization, Soil Quality and Organic Farming.* ed. Lichtfouse, E., pp. 217–242. Springer Science+Business Media, New York.

Nelson, A.G., J.C. Froese, and M.H. Entz. 2010. Organic and conventional field crop soil and land management practices in Canada. *Can. J. Plant Sci.* 90:339–343.

Nelson, A.G., S. Quideau, B. Frick et al. 2011a. Spring wheat genotypes differentially alter soil microbial communities and wheat breadmaking quality in organic and conventional systems. *Can. J. Plant Sci.* 91:485–495.

Nelson, K.L., G. Boiteau, D.H. Lynch et al. 2011b. Influence of agricultural soils on the growth and reproduction of the bio-indicator *Folsomia candida. Pedobiology* 54:79–86.

Nelson, K.L., D.H. Lynch, and G. Boiteau. 2009. Assessment of changes in soil health throughout organic potato rotation sequences. *Agric. Ecosyst. Environ.* 131:220–228.

Niggli, U., A. Fliessbach, P. Hepperly et al. 2009. Low greenhouse gas agriculture: Mitigation and adaptation potential of sustainable farming systems (Rev. 2-2009). FAO, Rome, Italy.

OACC. 2008. *Final Results of the First Canadian Organic Farmer Survey of Research Needs.* Organic Agriculture Centre of Canada, Nova Scotia Agricultural College, Truro, Nova Scotia, Canada.

Ogle, S.M., A. Swan, and K. Paustian. 2012. No-till management impacts on crop productivity, carbon input and soil carbon sequestration. *Agric. Ecosyst. Environ.* 149:37–49.

Oquist, K.A., J.S. Strock, and D.J. Mulla. 2006. Influence of alternative and conventional management practices on soil physical and hydraulic properties. *Vadose Zone J.* 5:356–364.

Osler, G.H.R., L. Harrison, D.K. Kanashiro et al. 2008. Soil microarthropod assemblages under different arable crop rotations in Alberta, Canada. *Appl. Soil Ecol.* 38:71–78.

Parton, W.J., D.S. Ojima, and D.S. Schimel. 1996. Models to evaluate soil organic matter storage and dynamics. In *Structure and Organic Matter Storage in Agricultural Soils.* eds. M.R. Carter and B.A. Stewart, pp. 421–448. Lewis Publishers/CRC Press, Boca Raton, FL.

Peigné, J., M. Cannavaciuolo, Y. Gautronneau et al. 2009. Earthworm populations under different tillage systems in organic farming. *Soil Tillage Res.* 104:207–214.

Pimentel, D., P. Hepperly, J. Hanson et al. 2005. Environmental, energetic, and economic comparisons of organic and conventional farming systems. *Bioscience* 55:573–582.

Ponce, C., C. Bravo, D.G. Leon et al. 2011. Effects of organic farming on plant and arthropod communities: A case study in Mediterranean dryland cereal. *Agric. Ecosyst. Environ.* 141:193–201.

Postma-Blaauw, M., R.G.M. deGoede, J. Bloem et al. 2010. Soil biota community structure and abundance under agricultural intensification and extensification. *Ecology* 91:460–473.

Powlson, D.S., P.J. Gregory, W.R. Whalley et al. 2011. Soil management in relation to sustainable agriculture and ecosystem services. *Food Policy* 36:453–467.

Pulleman, M.M., J. Six, N. Van Breemen et al. 2005. Soil organic matter distribution and microaggregate characteristics as affected by agricultural management and earthworm activity. *Eur. J. Soil Sci.* 56:453–467.

Purtauf, T., I. Roschewitz, J. Dauber et al. 2005. Landscape context of organic and conventional farms: Influences on carabid beetle diversity. *Agric. Ecosyst. Environ.* 108:165–74.

Reberg-Horton, S.C., J.M. Grossman, T.S. Kornecki, A.D. Meijer, A.J. Price, G.T. Place, and T.M. Webster. 2012. Utilizing cover crop mulches to reduce tillage in organic systems in the southeastern USA. *Renew. Agric. Food Syst.* 27:41–48.

Reganold, J., L. Elliott, and Y. Unger. 1987. Long term effects of organic and conventional farming on soil erosion. *Nature* 330: 370–372.

Robertson, G.P., E.A. Paul, and R.R. Harwood. 2000. Greenhouse gases in intensive agriculture: Contributions of individual gases to the radiative forcing of the atmosphere. *Science* 289:1922–1925.

Russelle, M.P., M.H. Entz and A.J. Franzluebbers. 2007. Reconsidering integrated crop—Livestock systems in North America. *Agron. J.* 99:325–334.

Sanchez, J.E., R.R. Harwood, T.C. Willson et al. 2004. Managing soil carbon and nitrogen for productivity and environmental quality. *Agron. J.* 96: 769–775.

Schneider, K., P. Voroney, and D.H. Lynch. 2011. Phosphorus availability in organic dairy farm soils: A closer look at the role of soil biology. Paper presented at *Plant Canada 2011*, Saint Mary's University Halifax, Halifax, Nova Scotia, Canada, July 17–21, 2011.

Schneider, S., M. Hartmann, J. Enkerli et al. 2010. Fungal community structure in soils of conventional and organic farming systems. *Fungal Ecol.* 3:215–224.

Shirtliffe, S.J. and E.N. Johnson. 2012. Progress towards no-till organic weed control in western Canada. *Renew. Agric. Food Syst.* 27:60–67.

Siegrist, S., D. Staub, L. Pfiffner et al. 1998. Does organic agriculture reduce soil erodibility? The results of a long-term field study on loess in Switzerland. *Agric. Ecosyst. Environ.* 69:253–264.

Six, J., H. Bossuyt, S. Degryze et al. 2004. A history of research on the link between (micro) aggregates, soil biota, and soil organic matter dynamics. *Soil Tillage Res.* 79:7–31.

Smith, P. 2004. Carbon sequestration in croplands: The potential in Europe and the global context. *Eur. J. Agron.* 20:229–236.

Smith, W.N., P. Rochette, C. Monreal et al. 1997. The rate of carbon change in agricultural soils in Canada at the landscape level. *Can. J. Soil Sci.* 77:219–229.

Stockdale, E.A. and C.A. Watson. 2009. Biological indicators of soil quality in organic farming systems. *Renew. Agric. Food Syst.* 24:308–318.

Teasdale, J.R., C.B. Coffman, and R.W. Mangum. 2007. Potential long-term benefits of no-tillage and organic cropping systems for grain production and soil improvement. *Agron. J.* 99:1297–1305.

Thiessen Martens, J.R. and M.H. Entz. 2011. Integrating green manure and grazing systems: A review. *Can. J. Plant Sci.* 91:811–824.

Thorup-Kristensen, K., D.B. Dresbøll, and H.L. Kristensen. 2012. Crop yield, root growth, and nutrient dynamics in a conventional and three organic cropping systems with different levels of external inputs and N re-cycling through fertility building crops. *Eur. J. Agron.* 37:66–82.

Trewevas, A. 2004. A critical assessment of organic farming-and-food assertions with particular respect to the UK and the potential benefits of no-till agriculture. *Crop Prot.* 23:757–781.

Vaisman, I., M.H. Entz, D.N. Flaten et al. 2011. Blade roller–green manure interactions on nitrogen dynamics, weeds, and organic wheat. *Agron. J.* 103:879–889.

Wander, M.M., D.S. Hedrick, D. Kaufman et al. 1995. The functional significance of the microbial biomass in organic and conventionally managed soils. *Plant Soil* 170:87–97.

Welsh, C., T. McGonigle, M. Entz et al. 2006. Organic crop management decreases labile P, promotes mycorrhizal colonization, and increases spore populations. *Can. J. Plant Sci.* 86:1414–1415.

Widmer, F., F. Rasche, M. Hartmann et al. 2006. Community structures and substrate utilization of bacteria in soils from organic and conventional farming systems of the DOK long-term field experiment. *Appl. Soil Ecol.* 33:294–307.

Willson, T.C., E.A. Paul, and R.R. Harwood. 2001. Biologically active soil organic matter fractions in sustainable cropping systems. *Appl. Soil Ecol.* 16:63–76.

Wiltshire, K., K. Delate, M. Wiedenhoeft et al. 2010. Incorporating native plants into multifunctional prairie pastures for organic cow-calf operations. *Renew. Agric. Food Syst.* 26:114–126.

Winqvist, C., J. Bengtsson, T. Aavik et al. 2011. Mixed effects of organic farming and land-scape complexity on farmland biodiversity and biological control potential across Europe. *J. Appl. Ecol.* 48:570–579.

Wortman, S., T.D. Galusha, S.C. Mason et al. 2011. Soil fertility and crop yields in long-term organic and conventional cropping systems in Eastern Nebraska. *Renew. Agric. Food Syst.* 25:281–295.

Yang, X.M., C.F. Drury, M.M. Wander et al. 2008. Evaluating the effect of tillage on carbon sequestration using the minimum detectable difference concept. *Pedosphere* 18:421–430.

6 Reduced Tillage in Organic Cropping Systems

Martin H. Entz, Caroline Halde,
Harun Cicek, Kristen Podolsky, Rachel L. Evans,
Keith C. Bamford, and Robert E. Blackshaw

CONTENTS

6.1 INTRODUCTION

Farmers and scientists have long searched for cropping systems that require less tillage. Economic benefits of less tillage include less machinery and fuel use, while biological benefits are associated with soil and water conservation. The soil cover in no-till systems is important for water conservation (Jalota and Prihar, 1998) and weed suppression (Teasdale et al., 2012). However, the lack of tillage also increases the potential for weeds, especially in organic systems.

Canadian scientists began testing no-till organic systems more than 100 years ago (Janzen, 2001). The dust bowl years of the 1930s spurred further innovation. Implements such as the wide-blade cultivator (also called the Noble blade after its inventor, Charles Noble) (Figure 6.1) and the rod weeder were developed to facilitate weed control with minimal soil disturbance. Other approaches during this preherbicide period included the Lister plow and crop rotation with perennial and biennial crops (R. Brust, University of Manitoba, pers. comm).

FIGURE 6.1 A wide-blade (Noble blade) cultivator implement. The blade travels below the soil cutting off roots of weeds and green manure plants. (Courtesy of University of Manitoba, Ian N. Morrison Research Farm, Carman, Manitoba, Canada.)

There is one other historical fact that we must get right before proceeding. The latest innovation—the blade roller—has created a great deal of excitement and optimism for moving organic no-till forward. It is important that we acknowledge the people who invented this tool—the South American small-holder farmer.

6.2 FOUR APPROACHES TO LESS TILLAGE

There are currently four approaches for reducing tillage in organic agriculture (Figure 6.2). The first approach, arguably the simplest, is to substitute conventional herbicides with organically acceptable herbicides. This approach has been largely unsuccessful to date at least as far as grain production in Canada is concerned (E. Johnson, Agriculture and Agri-Food Canada, pers. comm.). The second

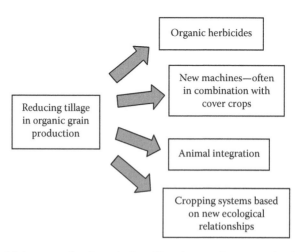

FIGURE 6.2 Main approaches for reducing tillage in organic crop production in Canada.

approach is the development of new machines for weed control in organic no-till and reduced tillage systems. In some cases, these new machines provide the necessary weed control but leave the soil less disturbed, for example, the wide-blade cultivator (Figure 6.1). In other cases, the new tool is used in combination with cover crops (e.g., blade roller). This second approach is the focus on much ongoing research in Canada. A third approach to reduce tillage in organic grain production involves livestock integration, where livestock are used to manage weeds and decompose crop residues. The final approach is to change the crop ecosystem so that tillage for weed control and seedbed preparation is no longer needed. Examples include perennial grains, intercropping, and agroforestry. In these cases, cropping systems are "redesigned to function on the basis of a whole new set of ecological relationships" (Gliessman, 2010). The third and fourth approaches have received almost no research attention in Canada and require further conceptualization and field-based experimentation.

6.3 OBSERVATIONS OF ORGANIC NO-TILL INITIATIVES

Reduced tillage organic systems were recently reviewed by Carr et al. (2012) and some recent work is summarized in Table 6.1. In Europe, the focus has been to shift from inversion (plow) to less intensive tillage (Mader and Berner, 2012). European researchers are wrestling with an important obstacle—that complete elimination of tillage seriously reduced N mineralization, thereby reducing crop yields. In North America, much of the research focus has been on reducing tillage in the green manure cover crop phase. At Rodale, for example, systems have been developed that allow direct seeding of both corn and soybeans into mechanically killed, fall-planted cover crops (Jeff Moyer, pers. comm.). Researchers in Alberta tested no-till suppression of sweet clover, arguably the most important legume cover crop in dry-land organic farming. Results showed that terminating clover with mowing instead of tillage provided superior weed and erosion control (Blackshaw et al., 2010). Other recent Canadian work has involved the deliberate use of biological mulches in potato and small grain production (Boyd et al., 2001). Ridge tillage, promoted in conventional agriculture before glyphosate-tolerant crops were available, is being used on organic farms in Quebec (Table 6.1). Ridge tillage allows for significant tillage reduction in row crop years plus it facilitates more rapid soil warming and hence more rapid N mineralization.

6.4 BLADE ROLLER

With the development of the blade roller, we now have a new tool for reducing till-age in Canadian organic cropping systems. Recent research (Vaisman et al., 2011; Shirtliffe and Johnson, 2012) shows that this machine is effective at killing common "Canadian" green manure crops such as pea, fababean, and oat. However, it has been observed that replacing tillage with the roller decreases wheat yields due to limited N availability (Vaisman et al., 2011). Therefore, some Canadian experiences mirror those in Europe—that complete elimination of tillage under organic cropping conditions reduces N mineralization. An earlier Canadian study showed that terminating green manure crops with the roller reduced N availability and grain yield in

TABLE 6.1
Some Initiatives in Reduced Tillage Systems for Organic Agriculture

Crop or Cropping System	Process	Reference
Eliminate plowing	Use sweep cultivation instead of inversion plow.	Berner et al. (2008)
Fall rye before soybean	Rye blade rolled at full flower, then soybean no-till seeded into mechanically killed mulch.	Wilson (2005)
Red clover underseeded to spring grain	Red clover mowed in autumn for Canada thistle control. Combination of competition from clover plants and mowing reduces thistle growth.	Lukashyk et al. (2008)
Green manure cover crop planted for a full season	Blade rolling at full flower. Creates mulch that reduces weed growth.	Vaisman et al. (2011), Carr et al. (2012), Shirtliffe and Johnson (2012)
	Use mowers instead of blade roller to suppress green manure.	Shirtliffe and Johnson (2012)
Underseed wheat with biennial sweet clover	Killing sweet clover at 80% flower stage by mowing at 30 cm height. Weed suppression equal to full tillage.	Blackshaw et al. (2010)
Plant-based mulches	Chop forage and spread on grain or potato land for dual purpose of nutrient supply and weed suppression.	Boyd et al. (2001), Wiens et al. (2006)
Ridge tillage	Ridges are created in autumn and row crop (e.g., corn) planted on top of ridge following spring. Ridges facilitate more rapid soil warming that helps warm season crop production.	La Coop Agrobio du Québec (pers. comm.)
Sheep grazing for weed control	Various approaches including sheep grazing weeds in wheat stubble and sheep selectively grazing weeds in organic lupins.	Chris Penfold (1998). Waite Institute; Roseworthy campus organic program (pers. comm.)
Pigs to control weeds	Research in Sweden has shown that 7 pigs/ha for 30 days controlled quack grass.	Sigurd Håkansson, Swedish University of Agricultural Sciences, Uppsala (pers. comm.)

a manner similar to glyphosate (Table 6.2). These observations suggest that some intervention may be required to increase the N available to grain crops following legume green manures in organic no-till systems. Of course, the simplest intervention is tillage, which is proven to speed up N mineralization. But are there other strategies that would work equally well to make N available without inverting soil? Perhaps light grazing? Perhaps a different green manure legume species?

A second study compared different green manure species on the basis of how well they were killed with the blade roller and the N supply to a following spring wheat

TABLE 6.2

Effect of Green Manure Termination Method on Dry Matter Yield, Nitrogen Content, and Grain Yield of Two Consecutive Crops Following the Green Manure Year

Termination Method (TM)	Wheat Dry Matter			Plant Height cm	Wheat Grain			Oat Grain Yield
	Nitrogen %	Yield kg/ha	Nitrogen kg N/ha		Nitrogen %	Yield kg/ha	Nitrogen	
Tillage	1.39 a	6951	97.0 a	94	3.40 a	2625 a	89.5 a	3681 a
Rolling	1.24 b	6467	80.5 b	97	3.12 b	2225 b	69.5 b	3378 b
Glyphosate	1.33 a	6304	84.3 b	99	3.29 a	2086 b	68.5 b	3542 a
ANOVA				Pr>F				
	0.0009	n.s.	0.0010	n.s.	<0.0001	<0.0001	<0.0001	<0.0001

Notes: Pea green manure at 4138 kg/ha dry matter. Numbers followed by different letters are significantly different (P < 0.05).

crop. The results showed N uptake in wheat following roller termination was reduced for most species (pea, lentil, chickling vetch, fababean, and pea/oat). The exception was hairy vetch, where N uptake in wheat was similar for the rolled and tilled systems (Entz and Bamford, unpublished). It was interesting to observe that sometimes the roller did not effectively kill shorter crops. For example, lentil green manure crops sometimes were not tall enough to be crimped effectively so that plants were killed. Roller designs for shorter vegetation are required.

6.5 COMPARING TRADITIONAL AND NEW MACHINES FOR GREEN MANURE MANAGEMENT

Podolsky et al. (2012) compared the wide-blade cultivator with two other reduced tillage treatments (flail mower and blade roller) and standard disc tillage. The termination treatments were applied to a spring-planted pea/barley green manure crop and spring wheat was planted in the following year. The blade roller resulted in lower wheat yields than the tillage treatment in 3 of 4 site years (Table 6.3). Tilling the rolled green manure in late autumn (blade roll plus tillage treatment) improved wheat yields at some sites, presumably because of greater N availability. The wide-blade treatment resulted in fewer weeds, greater soil cover in late autumn (data not shown), and wheat yields that were equal to or greater than the blade roll plus tillage treatment (Table 6.3). The wide-blade cultivator also provided more N to the following wheat crop than the blade roller (data not shown), suggesting this tool may provide just enough soil disturbance to enhance N mineralization. One of the challenges with the wide-blade cultivator is soil penetration into high biomass cover crops.

The flail mower was included in this study since it was hypothesized that shredding cover crop biomass would increase N release and subsequent availability.

TABLE 6.3

Effect of Green Manure Termination Method on Main Crop Grain Yield at Carman (2011, 2012) and Lethbridge (2011, 2012)

	Carman		Lethbridge	
	2011	2012	2011	2012
	Buckwheat	Wheat	Wheat	Wheat
Treatment	(kg/ha)	(kg/ha)	(kg/ha)	(kg/ha)
Tillage	792 a[a]	2599 a	4489	4065 b
Blade rolling	331 c	1039 cd	4413	3001 d
Blade roll + tillage	635 abc	2250 ab	3884	3594 c
Wide-blade cultivation	566 abc	1582 bc	4275	4255 ab
Flail mower	400 bc	558 d	4551	4035 b
P > F	0.0026	<0.0001	0.2589	<0.0001

[a] Means within a column followed by different letters are significantly different at P < 0.05 according to Tukey's HSD; means within a column not followed by letters are not significantly different.

However, Podolsky et al. (2012) observed more weeds in the flail mowed systems, and this heavy weed burden often reduced wheat yields the following year (Table 6.3).

6.6 LONG-TERM ORGANIC NO-TILLAGE: A CANADIAN CASE STUDY

Both no-till and organic farmers often pose the question, "Can we develop an organic no-till system?" Such an experiment was started in Carman, Manitoba, in 2008. The rotation includes a barley/hairy vetch green manure every third year in a 6-year rotation. The barley/hairy vetch green manure was selected because of its superior weed suppressing abilities (Halde et al., 2012).

Total biomass production in the barley/hairy vetch green manure was approximately 7000 kg/ha in both 2008 and 2011 (Table 6.4). The barley/hairy vetch mixture was blade-rolled in mid-July (Figure 6.3a). Rolling killed the barley, but the hairy vetch continued to grow until freeze-up, allowing more biomass accumulation. Mulch biomass at planting, the following spring, averaged about 4000 kg/ha in both years (2009 and 2012) (Table 6.4). Six years of previous research demonstrated that spring-planted hairy vetch is effectively winter-killed in Manitoba; this also occurred in the present study (Figure 6.3b). In southern Alberta, on the other hand, spring-planted hairy vetch overwintered in the two years when it was tested (Molnar and Blackshaw, unpublished).

In springtime, the winter-killed mulch consisted of a layer of barley residue, covered by a layer of hairy vetch residue (Figure 6.3c). This amount of biomass

TABLE 6.4

Hairy Vetch/Barley Biomass Production and Yield of the Following Flax (2009) and Spring Wheat (2012) Crop for the Long-Term No-Till Organic Experiment during the First 5 Years

	2008	2009	2009	2010	2011	2012	2012
	Total mulch aboveground biomass in October (kg/ha)	Total mulch aboveground biomass in May (kg/ha)	Flax grain yield (kg/ha)	Oat grain yield (kg/ha)	Total mulch aboveground biomass in October (kg/ha)	Total mulch aboveground biomass in May (kg/ha)	Spring wheat grain yield (kg/ha)
Tillage systems (T)							
NT				2608	8027	4233	2245
CT				2923	2238	686	3327
Management systems (M)							
ORG				2814	4753	2304	2155
CONV				2718	5511	2615	3416
T×M							
CONV CT				2850	2922	813	3431.47a
CONV NT				2585	8100	4416	3401.34a
ORG CT	7593	4476	2265a	2996	1553	558	3221.83a
ORG NT	7593	4476	1984b	2632	7954	4050	1088.34b

Note: In 2010, the experiment was split, and conventional no-till and conventional tilled systems were included for comparison purposes.

Abbreviations: NT, no-till; CT, conventional tillage; CONV, herbicides and fertilized used; ORG, organic.

appeared sufficient for weed suppression in the second and third organic no-till crops as suggested by relatively high organic no-till flax yields in 2009 and oats yields in 2010 (Table 6.4). Grain crops are seeded into the dead mulch using an offset no-till disc seeder (Figure 6.3d).

The fifth organic no-till crop was spring wheat seeded into the barley/hairy vetch mulch in 2012. The no-till organic treatment produced only 1000 kg/ha of grain yield (one-third of the tilled organic treatment) owing to competition from perennial weeds, mostly dandelion and some Canada thistle (Figure 6.3e). This observation suggests that long-term organic no-till has its limitations. Fall rye was seeded into the wheat stubble in September 2012, and then sheep were used for preemergence weed control (Figure 6.3f). It is hoped that the competitive nature of fall rye will allow for a successful sixth organic no-till crop. The objective is to maintain this long-term study and test alternative weed management tools if the weeds become too serious. Options for weed control include grazing with pigs or use of a wide-blade cultivator.

6.7 FALL COVER CROPS

Fall-seeded cover crops are gaining popularity in organic cropping systems. Much previous research has focused on growing soybeans after blade-rolled rye

(a)   (b)

(c)

FIGURE 6.3 Long-term organic no-till system at Carman, Manitoba. (a) Initial blade rolling of hairy vetch barley (2008); (b) flax, spring 2009 (right side no-till, left side tilled); (c) mulch present in spring of 2009.

(d)   (e)

(f)

FIGURE 6.3 (continued)   Long-term organic no-till system at Carman, Manitoba. (d) flax emerging through mulch (2009); (e) plots in 2012 (left side organic no-till; right side conventional no-till); (f) sheep grazing weeds after wheat harvest in no-till organic plots (2012).

(Wilson, 2005) and corn after blade-rolled hairy vetch (Teasdale et al., 2012). There is often a yield penalty from using the fall-planted cover crop in such a system (Wilson, 2005; Mischler et al., 2010) because of competition for water and seeding delays (grain crop seeding delayed to wait for ideal time for blade roll cover crop). However, sometimes (Teasdale et al., 2012) organic corn production in a reduced tillage blade-rolled cover crop system can provide similar yields to those in a traditional tillage-based system.

Evans et al. (2012) tested the effect of fall-planted cover crops on organic dry bean (*Phaseolus*) productivity in Manitoba. Both annual and biennial plants were tested under tilled and no-till conditions. The winter annual cover crops (fall rye and winter wheat) reduced bean yield (Table 6.5), presumably due to competition for water and nutrient resources. Bernstein et al. (2011) reported 24% lower yields

TABLE 6.5

Effect of Fall-Planted Cover Crops on Dry Bean Yield in Two Seasons at Carman, Manitoba

		2011			2012	
Cover Crop	Cover Crop Biomass (Fall 2010) (kg/ha)	Cover Crop Biomass (at Bean Seeding) (kg/ha)	Dry Bean Yield[a] (kg/ha)	Cover Crop Biomass (Fall 2011) (kg/ha)	Cover Crop Biomass (at Bean Seeding) (kg/ha)	Dry Bean Yield[a] (kg/ha)
Fall rye	404	6921	515	1457	4288	551
Winter wheat	332	3987	569	1524	3499	457
Barley	325		941	1687		1133
Oats	395		914	1488		1431
Oilseed radish	233		999	1398		1195
Control	155		673	480		1292

[a] Bean yields are averaged across tillage treatments since there was not a significant cover crop×tillage interaction.

in no-till vs. tilled systems when fall rye was grown as a cover crop before organic soybeans. They also observed few differences between mowed and blade-rolled rye.

Some fall-seeded annual cover crop species improved bean yield relative to the no cover crop control (Table 6.5), demonstrating the value of fall cover crops in organic bean production. A further advantage was soil erosion protection provided by the cover crop biomass in late autumn (Table 6.5).

This research should provide an incentive to consider fall-planted cover crops for production of other organic pulse crops, such as peas, lentil, and fababean. Matus et al. (1997) reported that no-till pea and lentil crops had greater biological N fixation than crops grown in tilled systems. Greater fixation was partly attributed to improved soil health, an attribute also associated with cover crops.

6.8 ANIMAL INTEGRATION

Grazing livestock can facilitate tillage reduction in a number of ways. The first is weed management; weed control with grazing is already used by some organic farmers (e.g., Kopke, 2007, University of Bonn, pers. comm). There is limited literature on grazing in arable organic systems as most previous studies have focused on weed control in perennial pastures (Popay and Field, 1996). There is also merit in multispecies systems (e.g., cattle and sheep) for vegetation management. Even Canada thistle, one of the organic agriculture's worst weeds, can be suppressed with grazing (pers. field observations).

A second benefit of grazing in organic systems is nutrient cycling. Nutrient availability of manure and urine is typically greater than the original plant material. Therefore, grazing may offer a solution to low N mineralization sometimes associated with no-till organic systems. Further, what is the potential of light grazing in cover crop systems (rye and soybean) where the physical and chemical attributes of the rye reduce organic no-till soybean yields? This is fodder for future research.

In water-limited areas, grazing may be one way to limit transpirational water use by cover crops (Gardner and Faulkner, 1991). Cover crop termination timing is an important consideration on the Canadian prairies though current approaches tend to offer only an all or nothing solution; the cover crop is completely killed with tillage or it remains completely alive (Biederbeck et al., 1993). Grazing may be one way to "regulate" this water use better.

Cicek et al. (2012) compared grazing and tillage in an organic green manure–wheat system in southern Manitoba. Two green manure species were used (oat and pea/oat mixture) and green manures were terminated either with sheep grazing or tillage. Grazing did not affect the following wheat yield (Table 6.6) though grazing did increase autumn soil nitrate levels compared to ungrazed plots (data not shown). Similar organic wheat yields in grazed vs. ungrazed plots indicate that animal gains can be achieved (through green manure grazing) without sacrificing wheat productivity. This leads to improved economic performance of the system (Thiessen Martens and Entz, 2012). Gardner and Faulkner (1991) recommend that catch crops be used when high N content green manure cover crops are grazed. Late-season catch crops have the added benefit of providing soil erosion protection, something that is reduced with grazing.

TABLE 6.6

Grain Yield of Wheat as Influenced by Green Manure Cover Crop Type and Termination Management

Management	Crop	Wheat Yield 2010 (kg/ha)	Wheat Yield 2012 (kg/ha)
Grazed	Oat	3136	2830
	Pea/oat	3603	2976
Incorporated	Oat	2950	2567
	Pea/oat	3605	2830
Crop (C)		0.0354[a]	0.2466
Management (M)		0.5629	0.1152
C × M		0.5578	0.6203

Source: Cicek, H. et al., Investigating soil NO_3–N and plant N uptake when green manures are grazed, in *Proceedings of the Canadian Organic Science Conference*, Winnipeg, Canada, February 2012, p. 70.

[a] LSD for Crop 2010 is 488.

It is common on mixed farms to graze grain stubble or corn stover with ruminant animals. In one experiment, rye stubble was grazed with sheep (mainly for weed control), and N uptake by the following spring wheat crop was increased by 14% due to this grazing (Thiessen Martens and Entz, unpublished). This observation strengthens the incentive to exploit grazing for nutrient management in organic farming systems.

6.9 CONCLUSIONS

This chapter has considered traditional and new approaches for reducing tillage in organic grain-based cropping systems. We have learned that we are not the first generation of scientists, agronomists, or farmers trying to wean organic cropping system from tillage. Indeed, there are lessons and tools from the past that are proving useful today. There have also been exciting new developments such as the blade roller. Significant progress has been made with the roller in only a few short years, and future research to optimize the plant–roller combination will undoubtedly yield important results.

Organic farming systems are site specific, relying on the ecology of the region, the farm, and the field. As such, reduced tillage systems need to be flexible, in order to adjust to the changing conditions within these farm ecosystems. The good news is that a set of complementary tools are available for farmers to use. These include cover crops, wide-blade cultivators, mowers, blade rollers, and grazing animals. In the future, we can add perennial grain crops and trees to this mix. These tools will need to be optimized for specific conditions, and additional research is needed to fully exploit their utility and effectiveness.

We have also provided glimpses of cropping systems that reduce reliance on tillage in new ways. For example, using grazing animals for strategic vegetation management has potential for reducing soil tillage in organic grain production. Most field agronomists have been reluctant to include livestock directly in the field agronomy experiments. Future research, and indeed future student training, should emphasize integrated systems that include a broader range of creatures in the production system.

REFERENCES

Berner, A., I. Hilderman, A. Fliessbach, L. Pfiffner, U. Niggli, and P. Mader. 2008. Crop yield and soil quality response to reduced tillage under organic management. *Soil Tillage Res.* 101:89–96.

Bernstein, E.R., J.L. Posner, D.E. Stoltenberg, and J.L. Hedtcke. 2011. Organically managed no-tillage rye–soybean systems: Agronomic, economic, and environmental assessment. *Agron. J.* 103: 1169–1179.

Biederbeck, V.O., O.T. Bouman, J. Looman, A.E. Slinkard, L.D. Bailey, W.A. Rice, and H.H. Janzen. 1993. Productivity of four annual legumes as green manure in dryland cropping systems. *Agron. J.* 85:1035–1043.

Blackshaw, R.E., L.J. Molnar, and J.R. Moyer. 2010. Sweet clover termination effects on weeds, soil water, soil nitrogen, and succeeding wheat yield. *Agron. J.* 102:634–641.

Boyd, N.S., R. Gordon, S.K. Asiedu, and R.C. Martin. 2001. The effects of living mulches on tuber yields of potato (*Solanum tuberosum L.*). *Biol. Agric. Hortic.* 18: 203–220.

Carr, P.M., R.L. Anderson, Y.E. Lawley, and P.R. Miller. 2012. Organic zero-till in the northern Great Plains Region: Opportunities and obstacles. *Renew. Agric. Food Syst.* 27:12–20.

Cicek, H., M.H. Entz, J.R. Thiessen Martens, and K. Bamford. 2012. Investigating soil NO_3–N and plant N uptake when green manures are grazed. p. 70. In *Proceedings of the Canadian Organic Science Conference*, Winnipeg, Canada, February 2012.

Evans, R., M.H. Entz, and Y.E. Lawley. 2012. Cereal cover crops for early season weed control in organic field beans. p. 108. In *Proceedings of the Canadian Organic Science Conference*, Winnipeg, Canada, February 2012.

Gardner, J.C. and D.B. Faulkner. 1991. Use of cover crops with integrated crop–livestock production systems. pp. 185–191. In W.L. Hargrove (ed.), *Cover Crops for Clean Water*. Soil and Water Conservation Society, Ankeny, IA.

Gliessman, S.R. 2010. The framework for conversion. pp. 3–14. In S.R. Gliessman and M. Rosemeyer (eds.), *The Conversion to Sustainable Agriculture*. CRC Press, Baton Rouge, FL.

Halde, C., M.H. Entz, R.H. Gulden, A.M. Hammermeister, K.H. Ominski, and M. Tenuta. 2012. Using mulches to reduce tillage in organic grain production in western Canada. pp. 104. In *Proceedings of Canadian Organic Science Conference*, Winnipeg, Canada, February 2012.

Jalota, S.K. and S.S. Prihar. 1998. *Reducing Soil Water Evaporation with Tillage and Straw Mulching*. Iowa State University Press, South State Avenue, Ames, IA, 50014.

Janzen, H.H. 2001. Soil science on the Canadian prairies—Peering into the future from a century ago. *Can. J. Soil Sci.* 81: 489–503.

Lukashyk, P., M. Berg, and U. Köpke. 2008. Strategies to control Canada thistle (*Cirsium arvense*) under organic farming conditions. *Renew. Agric. Food Syst.* 23:13–18.

Mader, P. and A. Berner. 2012. Development of reduced tillage systems in organic farming in Europe. *Renew. Agric. Food Syst.* 27:7–11.

Matus, A., D.A. Derksen, F.L. Walley, H.A. Loeppky, and van Kessel, C. 1997. The influence of tillage and crop rotation on nitrogen fixation in lentil and pea. *Can. J. Plant Sci.* 77: 197–200.

Mischler, R., S.W. Duiker, W.S. Curran, and D. Wilson. 2010. Hairy vetch management for no-till organic corn production. *Agron. J.* 102: 355–362.

Podolsky, K., M.H. Entz, and R.E. Blackshaw. 2012. Comparing reduced tillage implements for termination of cover crops. p. 109. In *Proceedings of the Canadian Organic Science Conference*, Winnipeg, Canada, February 2012.

Popay, I. and R. Field. 1996. Grazing animals as weed control agents. *Weed Technol.* 10:217–231.

Shirtliffe, S.J. and E.N. Johnson. 2012. Progress towards no-till organic weed control in western Canada. *Renew. Agric. Food Syst.* 27:60–67.

Teasdale, J.R., S.B. Mirsky, J.T. Spargo, M.A. Cavigelli, and J.E. Maul. 2012. Reduced-tillage organic corn production in a hairy vetch cover crop. *Agron. J.* 104:621–628.

Thiessen Martens, J.R. and M.H. Entz. 2011. Integrating green manure and grazing systems: A review. *Can. J. Plant Sci.* 91:811–824.

Vaisman, I., M.H. Entz, D.N. Flaten, and R.H. Gulden. 2011. Blade roller–green manure interactions on nitrogen dynamics, weeds, and organic wheat. *Agron. J.* 103(3):879–889.

Wiens, M.J., M.H. Entz, R.C. Martin, and A.M. Hammermeister. 2006. Agronomic benefits of alfalfa mulch applied to organically managed spring wheat. *Can. J Plant Sci.* 2006, 86:121–131.

Wilson, D. 2005. Choosing cover crops for no-till organic beans. http://www.rodaleinstitute. org/choosing_cover_crops (accessed October 1, 2013).

Section II

Pest Management

7 Sometimes You Need a Big Hammer: Evaluating and Appraising Selected Nonherbicidal Weed Control Methods in an Integrated Weed Management System

Steve J. Shirtliffe and Dilshan Benaragama

CONTENTS

7.1   INTRODUCTION

Integrated weed management (IWM) is an approach to weed control that includes a variety of agronomic practices. It has long been recommended as a means to reduce reliance on herbicides, reduce crop losses from weeds, and protect the environment (Swanton and Murphy, 1996). Liebman and Gallandt's seminal work *Many Little Hammers: Ecological Management of Crop-Weed Interactions* (1997) was for many, including us, an agroecological call to arms. In it, they used a metaphor to explain that in order to control weeds effectively, we cannot rely solely on one big hammer (herbicides) but instead must use many little hammers (several different nonherbicidal methods in addition to herbicides). They, and many researchers after them, provided extensive lists of cultural, physical, and biological control methods that farmers should adopt when practicing IWM (e.g., Buhler, 2002). Nevertheless, relatively few farmers seem to have implemented IWM (Llewellyn, 2007). We believe that farmers reject IWM because they perceive it to be an overly complex approach. What is needed is an assessment of different nonherbicidal weed control methods to determine their relative effects and their potential as part of an IWM approach. To use Liebman and Gallandt's analogy, we need to determine which hammers are big and likely to work predictably, which hammers are small and won't always work, which hammers hit the mark when used together, and which hammers will only hit your thumb and are best left in the toolbox.

This chapter assesses nonherbicidal weed control methods for use in extensive annual cropping systems. Discussions are limited to methods that can be used immediately before or during annual crop production; crop rotations and other long-term methods of weed control are not discussed. We hypothesize that using multiple nonherbicidal weed control methods will result in a higher crop yield than can be obtained using only a single weed control method. To support this hypothesis, the chapter does the following:

- Appraises cultural weed control methods that are applicable for annual crop production in grain-based cropping systems
- Explores the biological or ecological functions that promote weed control in each nonherbicidal method
- Reviews relevant literature to determine the reported efficacy of each method
- Assesses the benefits and drawbacks of using multiple nonherbicidal weed control methods
- Discusses the applicability of these methods beyond organic agriculture

Most discussions and examples in this chapter are drawn from research into organic weed control methods in western Canada. Organic crop production systems often have lower crop yields compared to conventional systems (Seufert et al., 2012), partly due to greater weed abundance and the prohibition of synthetic herbicides

(Leeson et al., 2000; Bond and Grundy, 2001; Entz et al., 2001). The first principle of weed management in organic systems is to establish a vigorous crop stand that is able to preempt resources from weeds by occupying above- and belowground space (Kolb and Gallandt, 2012). Therefore, the main weed control strategies used in organic cropping systems are those methods that enhance crop competitive ability combined with long-term crop rotations.

Nonherbicidal weed control methods can be subdivided into the following three classes: (1) cultural methods, which modify the culture or agronomy of crop production; (2) mechanical methods, which use physical means to remove or damage weeds; and (3) biological methods, which use living organisms to control weeds. Despite extensive research into inundative biological weed control (Boyetchko and Peng, 2004; Boyetchko and Rosskopf, 2006), no methods have as yet been successfully commercialized for use for a major weed in annual cropping systems in western Canada. Therefore, this chapter discusses cultural and mechanical weed control methods, but not biological methods.

7.2 CULTURAL WEED CONTROL METHODS

Cultural weed control methods are used during the year of crop production in an attempt to modify the competitive ability of the crop relative to the weed. Crop competitive ability refers to the crop's capacity to compete against weeds. It has two mechanisms. The first mechanism is known as crop interference (Harper, 1977), or net competitive effect (Goldberg, 1990). This refers to the ability of the crop to suppress weeds. A crop that can suppress weeds is able to reduce the weed's emergence, biomass, and seed production (Jordan, 1993). The second mechanism is the crop competitive response, or crop tolerance to weed competition. This occurs when the crop is able to tolerate the weed effect on crop emergence, biomass, and yield (Jordan, 1993). Therefore, plants can be superior competitors either by rapidly depleting a resource or resources or by being able to continue growing under limited resource conditions (Goldberg, 1990).

Weed competition greatly affects crop yield. Hence, attempts to alter the competitive balance in favor of the crop could help to maximize yield. Crop competitive ability can be enhanced by both genetic (Lemerle et al., 1995; Huel and Hucl, 1996) and agronomic (Koscelny et al., 1990; Mohler, 2001b) measures, which together provide a greater advantage in a given environment. This section will discuss four cultural weed control methods for increasing a crop's competitive ability: (1) selecting crops with greater competitive ability, (2) adjusting the crop seeding rate, (3) choosing a large crop seed size, and (4) adjusting the spatial arrangement of the crop.

7.2.1 SELECTING CROPS WITH GREATER COMPETITIVE ABILITY

7.2.1.1 Premise

The premise of this cultural weed control method is that crops differ in their ability to compete with weeds. Farmers should be able to increase weed suppression and crop yield by choosing crop types or varieties with greater competitive ability.

In general, this principle is valid, as numerous studies have found that crop species as well as crop varieties differ in competitive ability.

Differences in relative crop competitive ability have been widely demonstrated in many crop species. Pavlychenko and Harrington (1934) assessed crops in western Canada for their competitive ability against wild oat (*Avena fatua* L.) and ranked them, from most competitive to least competitive, as follows: barley (*Hordeum vulgare* L.) > spring rye (*Secale cereale* L.) > wheat (*Triticum aestivum* L.) > flax (*Linum usitatissimum* L.). In Australia, Lemerle et al. (1995) ranked crop competitive ability against ryegrass (*Lolium rigidum* L.) and found, in order of decreasing competitive ability, that oat > cereal rye > triticale (x *Triticosecale*) > oilseed rape (*Brassica napus* L.) > spring wheat > spring barley > field pea (*Pisum sativum* L.). Therefore, in Australian conditions, barley appears to have less competitive ability than wheat depending on the cultivar and seasonal conditions. In Denmark, Melander (1993) ranked peas and oilseed rape as less competitive against weeds than winter rye, wheat, and barley. In the United Kingdom, oat was found to be most competitive with cleavers (*Galium aparine* L.), followed by barley and then wheat (Seavers and Wright, 1999). These disparities in the ranking of competitive ability could be mainly due to environmental conditions. Disparities in ranking among studies are likely due to environmental conditions, crop morphology, life cycle, and canopy development (Van Hemmst, 1985). Crop competitive ability has been mostly studied in conventional systems and not in organic systems. Mason et al. (2007a) studied barley and spring wheat cultivars differing in height, tillering, and maturity under organic conditions and found that barley cultivars were superior in weed suppression and crop tolerance to wheat cultivars.

Within a crop species, varieties and cultivars often differ in their competitive ability against weeds. For example, differences for competitive ability among crop cultivars grown in the Canadian prairies have been identified for wheat (Blackshaw, 1994; Huel and Hucl, 1996), barley (O'Donovan et al., 2000), oat (*Avena sativa* L.) (Wildeman, 2004; Benaragama et al., in press), pea (Spies et al., 2011), and canola (Zand and Beckie, 2002; Harker et al., 2003). Among three spring wheat cultivars—Neepawa (hard red spring wheat), HY320, and HY355 (Canada prairie spring wheat)—the genotype Neepawa was found to be the most weed suppressive (Kirkland and Hunter, 1991). O'Donovan et al. (2000) documented that among six barley cultivars, Seebe, AC Lacombe, and Harrington were more competitive and had approximately 25% less weed biomass than the less competitive cultivars. Using competitive wheat varieties under weedy conditions increased yield by 7%–9% over noncompetitive varieties (Hucl, 1998). In organic conditions, use of a competitive oat cultivar had a 22% yield advantage over the noncompetitive cultivar (Benaragama and Shirtliffe, 2013).

7.2.1.2 Plant Traits Important for Competitive Ability

Among cereals crops grown in western Canada, plant height has been found to be the major trait attributed to greater competitive ability against numerous weeds. In spring wheat cultivars, crop height and competitive ability were positively correlated (Huel and Hucl, 1996). Among 29 barley cultivars, a semidwarf hull-less cultivar was generally less competitive than full height and hulled cultivars (Watson et al., 2006).

Among the pea cultivars evaluated, leafy and long-vined cultivars were superior to semi-leafless grain-type cultivars (Figure 7.1; Spies et al., 2011). Among the oat cultivars tested, tall leafy-type forage cultivars were superior in competitive ability against wild oat (Wildeman, 2004; Benaragama et al., in press). However, other traits such as early seedling growth have been found to be associated with competition in some short crop cultivars (Benaragama et al., in press). Similarly, in wheat and barley varieties in western Canada, differences in seedling establishment among the varieties tended to influence competitive ability more than plant height or interception of photosynthetically active radiation (O'Donovan et al., 2005). For more details on this topic, see Chapter 10.

7.2.1.3 Critical Appraisal of Selecting Crops with Greater Competitive Ability

Several issues prevent farmers from applying the principles of selecting crops with greater competitive ability. Farmers usually select crop species with high financial returns and strong market demand. For example, fall rye is the most competitive annual crop grown in western Canada (Blackshaw et al., 2002). However, because of low grain prices and subsequent poor financial returns, it is planted on only 0.24% of conventional farmland and 3% of organic farmland in Canada (Macey, 2010; Statistics Canada, 2012).

Farmers are often unable to choose competitive crop varieties because they do not know which varieties are the most competitive among existing cultivars. The process of testing crop competitive ability is expensive because special crop trials are required. Because of the expense associated with running these trials, they are conducted on an ad hoc basis. Furthermore, because these trials take a long time, their results are often obtained after the crop variety is released. Because crop varieties tend to have a limited lifespan in the market, the investment required to assess competition may not be worthwhile, as the evaluated varieties may have been replaced with superior varieties by the time competition is evaluated.

The differences in competitive ability between crop species and genotypes are difficult to predict and vary between environments. Harker et al. (2011) found that the relative competitive ability of annual grain crops and canola differed depending upon environment. Furthermore, Wildeman (2004) and Benaragama et al. (in press) found no consistent difference in competitive ability of most oat genotypes between different environments. Variation in competitive ability between environments probably occurs because of a genotype by environment (GxE) interaction. This well-known phenomenon occurs when certain plant genotypes perform better in some environments than others. The GxE interaction is further complicated when considering that weeds potentially have their own GxE interactions. Thus, the extent of a crop's competitive ability with one weed species is potentially affected by the environment, crop genotype, GxE interaction, weed species by environment interaction, and interaction of the crop and weed with each other. Clearly, all these potential interactions will add greater variability and less predictability to the competitive effect of different species and genotypes.

Because of the difficulties in screening and breeding for crop competitive ability (also see Chapter 10), few plant breeding programs have attempted to incorporate

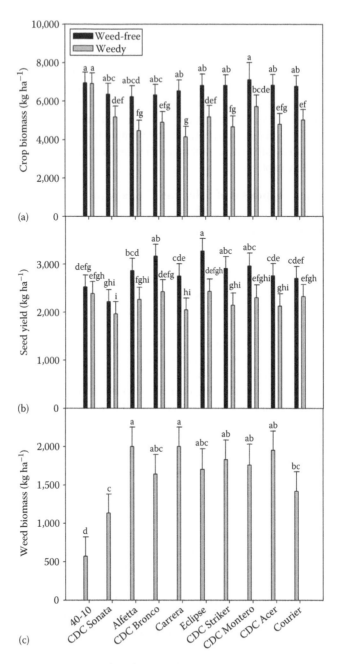

FIGURE 7.1 Field pea seed yield (a), aboveground crop biomass (b), and weed biomass (c) as affected by crop cultivar under weedy and weed-free conditions averaged over 6 site years. The varieties 40-10 and CDC Sonata are leafed and the other varieties are leafless. Bars indicate standard errors. Comparisons are made within the weed presence/absence and between cultivars with similar letters indicating no significant difference (P < 0.05) between values. (Modified from Spies, J.M. et al., *Weed Sci.*, 59(2), 218, 2011.)

competitive traits. Based on 27 wheat cultivars tested under organic conditions, Mason et al. (2007b) proposed that early flowering and maturity, increased tillering capacity, and increased plant height could be considered as desirable characteristics for organic wheat production. Other studies under nonorganic conditions have identified similar traits among wheat cultivars indicating that conventionally bred crop cultivars could be useful in organic systems. However, Reid et al. (2011) believe that breeding under organic conditions is a more viable strategy as the ranking of some cultivars for grain yield differs between the two systems. At least one study has evaluated genotypes bred for high yield, grain quality, and competitive ability. Benaragama et al. (in press) identified that some deliberately bred competitive oat cultivars were superior in competitive ability and had superior grain yield and quality. This demonstrates the possibility of breeding for competitive ability. However, Benaragama et al. (in press) also found that the differences in competitive ability and yield identified in progeny were often not substantially greater from the individual parents, indicating that the effort required to breed for competitive ability may be substantial but may not always result great returns.

Market demands may prevent farmers from choosing competitive crop varieties. When farmers grow crops for identity-preserved markets, the end user dictates the choice of crop variety or market class. Even in markets that are not identity preserved, variety traits such as disease and lodging resistance may be more important to farmers than competitive ability. For example, even though leafed peas are usually more competitive with weeds than semi-leafless, most organic growers do not grow leafed varieties because of difficulties with harvest (Spies et al., 2011). Even organic farmers should consider the weed-free yield potential of crop genotypes when selecting. Grain yield under weed-free conditions is often correlated with grain yield under weedy conditions, so choosing a variety with a high yield potential is a prudent decision. Furthermore, modern high-yielding crop varieties are often higher yielding under organic conditions than older *heritage* varieties (Shirtliffe et al., 2006; Mason et al., 2007b). Even if competitive high-yielding cultivars are available, they are not a panacea for weed management in organic systems because they cannot achieve significant weed control on their own. Rather, cultivar choice must be integrated with other cultural practices to augment overall crop competitive ability.

To grow a crop that is marketable, has relatively high yield, and has some degree of competitive ability, organic farmers should choose a crop species or genotype that meets all the following criteria:

- It has good market demand and quality parameters that buyers want.
- It has desirable agronomic traits such as disease and lodging resistance.
- It is tall and high yielding.

7.2.2 Increasing Crop Seeding Rate

7.2.2.1 Premise

Increasing the crop's seeding rate may be the most stable and reliable way to increase the crop's ability to compete with weeds. This is because increasing the seeding rate results in a larger proportion of crop plants relative to weeds in a given area (Knight et al., 2010). For example, if a crop and a weed were to have exactly the same competitive ability and if there were an equal mixture of the two, then the crop's yield would be reduced by 50%. One way of changing that yield reduction is to change the proportion of crop plants relative to weeds by increasing the seeding rate and the subsequent crop population. Using the previous example, assume that the crop and weeds were both present at 100 plants m^{-2}. Thus, the total population of plants is 200 plants m^{-2}, of which there are 100 crop plants m^{-2}. Because these hypothetical crop and weed species have the same competitive ability, the yield loss, as previously discussed, would be 50%. Increasing the crop's population to 200 plants m^{-2} would result in a greater proportion of total plants being the crop (i.e., 200 crop plants/300 total plants), thus lowering the yield loss to 33%. As this relationship is proportional and nonlinear, doubling the seeding rate does not halve the yield loss. Of course, crops and weeds often differ in relative competitive ability. Nevertheless, this relative proportional relationship is still valid.

Increasing crop seeding rate can increase the competitive ability for most crops grown in western Canada resulting in reduced weed biomass (Wax and Pendelton, 1968; Evans et al., 1991; Weiner et al., 2001; Olsen et al., 2004). Carlson and Hill (1985) used different wild oat and wheat densities to study crop density effect on weed suppression. In their study, wild oat infestation of 5.5 plants m^{-2} resulted in 20% grain yield reduction in a poor crop stand of 100 plants m^{-2}. When the crop population was 700 plants m^{-2}, 38 wild oat plants m^{-2} were needed to cause the same degree of yield reduction. Weed biomass generally decreases with increasing crop density for many weed species (Olsen et al., 2004). In a study conducted using natural weed flora, it was observed that increased wheat density affected the vegetative traits (dry weight and leaf area) and reproductive output of many weed species (Froud-Williams and Korres, 2001).

Increasing crop seeding rates can increase intraspecific competition and thereby reduce grain yield and quality. As such, high crop seeding rates can suppress weeds but reduce crop yield. Therefore, effective weed suppression occurs at a crop density increase that will not cause substantial crop yield losses (Weiner et al., 2001). In conventional systems with lower weed densities, increasing crop seeding rates may either have no impact (Spaner et al., 2001) or result in a moderate (4%–5%) yield increase (Lafond, 1994; O'Donovan et al., 2000; Lemerle et al., 2004). Under weedy conditions, crop yield can be improved with the use of higher seeding rates (Mohler, 2001b). Barton et al. (1992) found that when barley was seeded at higher rates, grain yield was greater and wild oat competition was lower. In the presence of weeds, crops generally respond to high crop density with increased grain yield. O'Donovan et al. (1999) found that barley yield decreased with increasing crop density, but this negative effect was minimized when high weed densities were present. In addition to greater weed suppression and greater yield under weed competition, increased crop seeding rate results in earlier and more even crop maturity (Blackshaw et al., 2008). However, seeding rate should be increased cautiously, as an overcrowded plant stand can reduce tillering and enhance tiller mortality due to enhanced intraspecific competition (Chen et al., 2008).

Under organic conditions, doubling the recommended seeding rate of wheat and barley enhanced weed suppression and increased grain yield by 10% (Mason et al., 2007a). In organic oat, doubling the crop density from 250 to 500 plants m^{-2} reduced weed biomass by 52% and increased grain yield by 11% (Benaragama and Shirtliffe, 2013). For organic lentils (*Lens culinaris* Medik.), increasing the crop density to 229 plants m^{-2} from the conventional recommendation of 130 plants m^{-2} reduced weed biomass by 59% and resulted in best economic returns (Baird et al., 2009b). Similarly for peas (*P. sativum* L.), increasing the crop density from the recommended 88 to 120 plants m^{-2} reduced weed biomass by 68% and maximized economic returns (Baird et al., 2009a). Due to the high seed cost of large-seeded crops, it is crucial to examine optimal seeding rates in terms of economic return. Optimal seeding rates could vary according to location (Geleta et al., 2002), type of cultivar (Weiner et al., 2001), planting date (Koscelny et al., 1991; Spink et al., 2000), and weed density (Koscelny et al., 1990). Nevertheless, the weed control effect and yield advantage of increasing seeding rates are more robust than other nonherbicidal weed management strategies. As such, general guidelines for optimum seeding rates for each agroecological region can substantially benefit organic production systems.

7.2.2.2 Critical Appraisal of Increasing Crop Seeding Rate

Increasing a crop's seeding rate may not be economical for large-seeded crops because the farmer has to cover the cost of the additional seed (Baird et al., 2009a). The expenditures on additional seed and the seed cleaning may exceed profits even with higher grain yields. Furthermore, with large-seeded crops such as field peas, typical seeders and planters may not be able to achieve the volume of seed required at high seeding rates. Seeding twice also increases costs.

Increasing seeding rate can also cause agronomic difficulties in many crops, including lodging and increased foliar disease (Mohler, 2001a). In some situations, the gains resulting from increased competitive ability may be nullified by increased losses. For example, we have observed that, in wet years, lentil responds negatively to increase seeding rate, even in the presence of weeds. In contrast, under normal rainfall, Baird et al. (2009a) found no increase in seed-borne lentil disease infection with increased seeding rates. The risk from foliar disease with a denser crop population varies depending on the type of crop and the disease risk. Considering the classic plant disease epidemic triangle may give some guidance as to when higher seeding rates should be avoided. Increased seeding rates, therefore, should be avoided for crops that are not genetically resistant to crop pathogens and are in an environment conducive to disease development. Otherwise, the yield loss from disease could be greater than the yield gain from increased competitive ability.

Increasing a crop's seeding rate is probably the most effective single-year cultural weed management technique. The increase in competitive ability is predictable, with 93% of experiments showing that higher seeding rates increased crop yield (Mohler, 2001a). Mohler (2001a) also offers some practical guidelines for increasing a crop's seeding rate when yield losses from weeds are expected. In general, for most crops that do not have a yield penalty when increasing the seeding rate under weed-free conditions, increasing the seeding rate to 1.5 times the weed-free recommendation will result in greater competitive ability. This recommendation has been borne out in western Canada. Lentil and pea both responded positively to increases up to this level and in lentil to much higher increases in seeding rate. Under weedy conditions in oat, a similar yield increase was observed (May et al., 2009). It is, therefore, recommended that, if there are no other guidelines, farmers anticipating weedy conditions should increase their seeding rate to 1.5 times the conventional recommended rate. For crops that have had guidelines developed for organic production or weedy conditions, those recommended seeding rates should be used.

7.2.3 CROP SEED SIZE

7.2.3.1 Premise

Crops sown with larger seeds have greater competitive ability than crops grown with smaller seeds, possibly because larger seeds tend to germinate quicker than smaller seeds (Boyd et al., 1971; Ries and Everson, 1973). For example, Wildeman (2004) determined that large oat seeds of the same genotype emerged sooner than small seeds. It has long been known that plants that emerge before others have a competitive advantage. This effect is known as the relative time of emergence. Its relative effect has been quantified in several weed/crop systems, with the most commonly studied weed being wild oat in western Canada (O'Donovan et al., 1985). The idea behind this principle is that plants emerging first can compete asymmetrically with other plants and, therefore, get a larger proportion of the available resources. As such, early emerging plants can shade later emerging plants and capture more sunlight relative to their size. This effect probably also holds true for roots that are able to explore the soil before others and, thus, encounter more nutrients and water than will later root systems.

Crops grown from larger seeds appear to be more competitive than those grown from smaller seeds in field studies. Xue and Stougaard (2002) found that wheat grown from large seeds reduced wild oat seed production by 25% compared to small-seeded wheat. In our studies, we have observed a similar reduction in wild oat biomass as a result of sowing large seeds (unpublished). The effect of increased seed size appears to be additive with increased seeding rate (Xue and Stougaard, 2002).

7.2.3.2 Critical Appraisal of Sowing Large Seed

In practical terms, seeding the same target population with larger seed increases the mass of seed sown per unit area. Because most grain crops are sown on a constant mass per unit area basis, increasing the individual seed size will result in a lower population of seed sown, presenting farmers with a conundrum: Should they seed a higher population of small seed per unit area or a lower population of large seed? Evidence suggests that these two mechanisms may be compensatory (Wildeman, 2004). Given that in a genetically pure crop variety, larger crop seeds do not differ genetically from smaller seeds, the difference between large and small seeds is in the amount of stored energy available to the seedling. In the absence of weed competition, the seed size of oat had no affect on the yield density relationship; small oat seeds produced the same yield at the same density as large oat seeds

(Lamb et al., 2011). Perhaps what is important is not the absolute number or size of seeds but, instead, the total mass of crop energy that is initially supplied to the crop plant's systems to compete with weeds. Of course there are biological/ecological limits to how large a seed can be or how low a seeding rate for an individual species can be, but perhaps across a narrow range of seed size and seed number compensate. This argument is similar to the longstanding seed size versus seed weight argument in ecology (Harper, 1977). Clearly, more research is needed to determine the relative effect of planted seed size versus number.

There are other practical issues that may preclude the use of larger seed as a weed control method. Individual seed size is the least plastic of crop yield components, and therefore, an individual crop genotype tends to have a narrow range of seed sizes (Hay and Walker, 1989). The cost associated with screening out larger seed may not justify the returns (Willenborg et al., 2005). Furthermore, there may not be opportunities for farmers to purchase various seed sizes. Therefore, sowing with large seed may not be a practical weed control option for most farmers.

7.2.4 CROP SPATIAL ARRANGEMENT

7.2.4.1 Premise

Plants compete for resources, including light, water, and soil nutrients. Since plants are anchored in one place and cannot move, their physical location relative to other plants determines their access to resources. The pattern in which they are planted, therefore, determines their access to resources relative to other crop plants. Ideally, the intraspecific competition between crop plants should be equivalent for all crop plants so they should occupy all available space as efficiently as possible (Weiner et al., 2001). It is a common misconception that intraspecific competition between crop plants is a negative effect. To best suppress weeds, competition between crop plants should be uniform and high. This ensures that fewer resources are available to weeds. A spatial arrangement in which crop plants are planted in an equidistant uniform pattern assures that crop plants will occupy the space as quickly as possible and therefore leave fewer resources for weeds.

Crop rows can be considered as very narrow clumps of plants (Olsen et al., 2004); plants grow within close proximity of each other, with relatively large interrow distance. Theoretically, reducing the interrow distance (narrow row planting) can make the 2D pattern less clumped (Weiner et al., 2001) and decrease the rectangularity (ratio between crop row spacing and within-row plant spacing; Fischer and Miles, 1973). For small-seeded annual grain crops, the most practical way to achieve a more equidistant plant arrangement is to plant in narrower rows. For most crops, the distance between adjacent seeds within the row is much smaller than the distance between crop rows. Thus, narrow rows will usually result in a more uniform spatial arrangement. For small-seeded grain crops, the distance between crop plants within a row is usually random, as commercial seeders do not individually plant seeds at a specific distance. For larger-seeded row crops such as corn, planters space seeds at a given distance. Such spaced planting may offer some advantages for small-seeded grain crops, although its potential has not yet been assessed due to lack of available machinery.

Crop planting pattern can substantially influence the competitive ability of a crop (Mohler, 2001a). Crops can be grown in a wide range of planting patterns with a variety of interrow and intrarow spacing. Changing the number of plants within rows and changing the distance between two crop rows can alter the planting pattern. In the absence of weeds, the morphological plasticity allows the crop to grow toward the space between rows where resource availability is high, thereby reducing intraspecific competition (Ballare et al., 1994). In weedy conditions, crop plants distributed in a clumped pattern have less capacity to suppress weeds than they would if they were in a uniform pattern (Weiner et al., 2001). In a uniform planting pattern, crop canopy cover development is faster and light penetration through the canopy is minimized (Teasdale and Frank, 1983; Murphy et al., 1996). Therefore, narrower row planting often results in decreased weed biomass and higher grain yields (Putnam et al., 1992; Teich et al., 1993; Murphy et al., 1996). Begna et al. (2001) studied corn (*Zea mays*) hybrids with two different row spacings (38 and 76 cm) and found that in both years of the experiment, the respective weed biomass was reduced by 29% and 20%, respectively. This reduction was attributed to the narrow row spacing. When Olsen et al. (2004) planted wheat in normal (the pattern achieved by a commercial seeder), mathematically random, and uniform patterns, they found that weed biomass was lower and crop biomass was higher in random and uniform patterns as compared to the normal pattern.

In the presence of weeds, narrow row spacing often increases crop yield (Mohler, 2001b). Hence, the yield advantage of narrow row planting can be due to either greater weed suppression or more efficient use of resources through greater crop ground cover and reduced weed incidence (Peters et al., 1965; Sharratt and McWilliams, 2005). In wheat, narrow row spacing results in greater yields than those obtained with wide row spacing, attributed to more resource use when the crop plants grow closer to each other (Champion et al., 1998). Similarly, high yields were observed in maize in narrow row planting, attributed to greater radiation interception (Mashingaidze et al., 2009).

The yield advantage of narrow row spacing may be greater at high within-row crop densities (Champion et al., 1998). When maize plants were grown at higher within-row density and with narrow row spacing, light interception increased from 3% to 5% (Begna et al., 2001). However, in some situations, narrow row spacing reduces yield. Fanadzo et al. (2007) found that maize yield was significantly reduced in 60 and 70 cm row spacing compared to 90 cm row spacing. Knight et al. (2010) found that, at low seeding rates in lentil, reducing row spacing had no effect on yield but, at higher seeding rates, reducing row spacing increased yield. The reason for yield reduction was mainly attributed to resource limitation in the growing environment and increased intraspecific competition.

The effect of row spacing on weed control is inconsistent. Some studies find an effect (Begna et al., 2001; Weiner et al., 2001) and others find it to be ineffective (Mohler, 2001b; Rasmussen, 2004). In a study conducted in Saskatchewan, organically grown oat was sown with row spacing at 11.5 cm instead of the usual 23 cm. The reduced row spacing was shown to have had no effect on subsequent yield or weed suppression (Benaragama and Shirtliffe, 2013). Therefore, as a cultural weed control method, creating narrow row spacing is less effective than increasing crop density. This inconsistency likely occurs because the effect of row spacing can vary depending on environment (Thill et al., 1994), crop species (Mohler, 2001b), cultivar (Marshall and Ohm, 1987; Drews et al., 2009), seeding rate (Marshall and Ohm, 1987), and the presence of weeds (Koscelny et al., 1990).

7.2.4.2   Critical Appraisal of Narrow Row Spacing

Many issues prevent farmers from targeting and obtaining an equidistant crop spatial pattern. Seeding crops with reduced interrow distance is problematic agronomically. Seeders with reduced interrow distances require more energy to pull, thereby reducing the area that the farmer can seed with the same tractor in a given time. Seeders with reduced interrow distances are also more apt to plug with residue from the previous crop (Lafond et al., 2013). In areas susceptible to erosion, crop residue retained on the soil surface reduces the erosion potential. Because seeders are very expensive, changing a seeder as a weed control measure may not be cost-effective. Furthermore, although reduced interrow distance can increase competitive ability, the range of interrow spacing available on commercial seeders is relatively small (20–35 cm). If a seeder cannot achieve a narrow row distance, farmers can compensate by increasing the seeding rate.

There is also the potential issue of soil disturbance increasing the seedling recruitment of weeds. In a reduced or no-till system, the soil and crop residue in the crop rows are disturbed and then packed in order to ensure quick and uniform crop emergence. In contrast, the space between the crop rows is undisturbed and not packed. In some cases, this lack of disturbance and packing reduces weed emergence in the interrow space. Presumably, having a greater number of crop rows would result in greater weed emergence because of greater soil disturbance, but to our knowledge, this hypothesis has not yet been tested.

Obviously, there is a limit to how wide the interrow distance can be before impacting yield. For example, seeding wheat in rows 1 m apart would result in yield reductions. This factor has been investigated extensively in relatively weed-free conditions of conventional agriculture (e.g., Lafond et al., 2013). Being unable to grow and fill in the spaces between the rows will prevent the crop from capturing all the available resources. When water and soil nutrients are not limited, competition for light will dictate the yield potential. Should the interrow distance be so large that the crop canopy does not intercept all the light, yield loss will likely occur. When weeds are present, additional yield losses can occur as weeds can now utilize these resources and compete with the crop. The most prudent recommendation for farmers is to choose a seeder with as narrow an interrow width as practical. The efficacy of space planters for small-grain crops in narrow rows has not been studied to a large extent, although initial findings by Olsen et al. (2012) suggest that uniform planting substantially increases the crop competitive ability. However, as there are currently no seeders commercially available that can individually plant small-grain seeds within a row, there are no options for farmers to adopt this technique. Furthermore, the effect of individual intrarow plant spacing needs to be evaluated in different crops and environments to determine its potential for increasing crop competitive ability in narrow row grain cropping systems.

The decision to determine row spacing in organic systems is highly dictated by the type of mechanical weed control practiced. In the Canadian prairies where mechanical weed control is mainly practiced using harrows and rotary hoes, narrow row spacing of 10–18 cm in cereals is advisable. However, in other regions where interrow cultivation is practiced, increasing the row spacing from conventional 15–20 to 25–30 cm is advocated for organic systems (Rasmussen, 2004).

7.3   IN-CROP MECHANICAL WEED CONTROL METHODS

7.3.1   Pre- and Postemergence Harrowing

Physical removal of weeds by disturbing the soil is one of the oldest methods of weed control in crops (Mohler, 2001a). Tillage, harrowing, hoeing, and hand weeding are the main mechanical weed control techniques. In organic cropping systems, the main physical weed control strategy is tertiary tillage, which uses rotary hoes, harrows (blind cultivators), and interrow cultivators (selective), either in pre- or postemergence stages. Tertiary tillage is done after the crop has been seeded and is often referred to as in-crop tillage.

In-crop tillage is most effective when crops are different than weeds in terms of growth habit, time of emergence, and time of maturity (Zimdhal, 1993). However, because of the low selectivity of the technique, mechanical weed control is often not very effective. Selectivity is the ratio between the positive weed control effect and the negative effect of crop damage by the mechanical weed control (Rasmussen, 1991). Poor selectivity often occurs with in-crop tillage because of similarities between the crop and weed such that when in-crop tillage is performed, both the weed and the crop are damaged. This is especially true for in-crop harrowing in which the damage done to the crop can often be as great as that done to the weed (Rasmussen, 1990). Reduced yield due to crop injury from harrowing has been reported for many crops (Lafond and Kattler, 1992; Kirkland, 1995; Rasmussen and Svenningsen, 1995; Jensen et al., 2004). If the weeds are larger than crop plants at the time of harrowing, selectivity is greater (Rasmussen, 1992).

Crops differ in their ability to resist (i.e., avoid soil covering) and tolerate (i.e., resist and recover from burying) postemergence harrowing. Oat was found to have a greater resistance than wheat, barley, and triticale. However, triticale was the most tolerant followed by wheat, oat, and barley (Rasmussen et al., 2009). Shirtliffe and Johnson (unpublished) found that, in western Canada, oat was the cereal most tolerant to harrowing, followed by barley and wheat. Significant cultivar differences have been observed in crop tolerance to harrowing. Hansen et al. (2007) found that among barley cultivars, difference in tolerance to harrowing was correlated to plant height at harrowing. The ability of the plant to resist bending can be vital to tolerate postemergence harrowing (Baerveldt and Ascard, 1999; Kurstjens and Perdok, 2000).

Weed control can be maximized and crop damage can be minimized by harrowing when the weeds are just emerging (i.e., at the *white thread* stage; Kurstjens et al., 2000) and by adjusting harrow settings and tractor speed (Kolb and Gallant, 2012). However, according to Rasmussen et al. (2010), adjusting harrow aggressiveness based on the crop stage is more important than time of harrowing. Adjusting the angle of the tines in relation to the soil surface controls harrow aggressiveness. Johnson (2001) found that, for peas in Canadian prairies, selectivity of flex-tine harrowing was improved by doing multiple passes with a low level of aggressiveness. Increasing the tractor driving speed can increase the soil cover percentage (Rydberg, 1994; Rasmussen and Svenningsen, 1995), but it may not increase weed kill (Rydberg, 1994) because the weed control through soil covering can be less important than uprooting of weeds (Kurstjens et al., 2000; Kurstjens and Kropff, 2001).

In organically grown spring cereals, pre- and postemergence weed harrowing are key weed control strategies (Hansen et al., 2007). Harrowing is effective because it uproots weeds and buries them in the soil, limiting their ability to regrow (Rasmussen, 1991; Kirkland, 1995; Kurstjens and Kropff, 2001). In Saskatchewan, 52% of organic farmers surveyed used postemergence harrowing to control weeds (Beckie, 2000). Cirujeda et al. (2003) reported that harrowing in winter wheat reduced weed biomass by 40%–60%. Velykis et al. (2009) showed that in organically grown spring oat and field pea, harrowing the crop at the two- to three-leaf stages resulted in a 62% reduction in weed density. Similarly, in organically grown spring barley, two to three harrowing passes reduced weed density by approximately 80% (Auskalnis and Auskalniene, 2008). In the prairies, two-pass harrowing at postemergence stage in organic oat reduced weed density by an average of 50% in 3 site years (Benaragama et al., in press). Harrowing improves yield not only because it suppresses weeds but also because it enhances soil physical and chemical properties such as soil aeration, moisture conservation, and mineralization of organic matter (Velykis et al., 2009). Nevertheless, intensive harrowing also has negative effects, such as leading to deterioration of soil structure and nitrogen leaching (Bond and Grundy, 2001; Steinmann, 2002).

Crop injury due to harrowing can be due to prevailing environmental conditions with moisture levels following harrowing, affecting crop recovery (Kirkland, 1995). Moreover, the negative impact on yield due to harrowing can depend on the time of application. Lundkvist (2009) observed yield reduction in spring wheat and oat in relation to time of harrowing. In early preemergence combined with postemergence harrowing treatments, the yield loss was 14%. This reduction was minimized when early preemergence harrowing was combined with early postemergence. In a 2-year study in an organic system, Benaragama and Shirtliffe (2013) showed that timely application of postemergence harrowing minimized crop injury and increased yield by 11%, due to greater weed control. However, harrowing cereal crops may not consistently improve grain yield (Rasmussen and Svenningsen, 1995; Velykis et al., 2009). Despite a possible yield penalty, harrowing could be a viable option to reduce weed pressure and minimize the weed seed bank in the long term.

7.3.2 Rotary Hoeing

The rotary hoe is a seldom-used weed control implement in western Canada but one that has great potential. A rotary hoe kills small, shallowly rooted weeds pre- and postcrop emergence by uprooting them with rotating tines (Figure 7.2). One advantage of the rotary hoe is that minimum-till rotary hoes can operate in high levels of crop residue (Shirtliffe and Johnson, 2012). Minimum-till rotary hoes have rotating tines

FIGURE 7.2 A minimum-till (high residue) rotary hoe showing the passively powered rotating tines.

positioned in tandem to allow crop residue to pass through the implement without plugging it. Thus, mechanical weed control can be used in cropping systems where tillage is reduced or eliminated. Using spring tine harrows under such conditions results in the machine becoming plugged with crop residue and therefore becoming ineffective. Shirtliffe and Johnson (2012) reported that multiple passes with a rotary hoe only minimally reduced the level of surface crop residue and only slightly reduced pea yield (Figure 7.3). Furthermore, they observed that the rotary hoe was able to operate within a no-till, direct-seeded system in which tillage had been completely eliminated for several years. Shirtliffe and Johnson (2012) reported that two passes of a rotary hoe at ground crack and three-node stage reduced weed biomass by 46% and increased pea yield by 42% compared to the untreated plots. Rotary hoes, therefore, can be effective in controlling annual weeds in western Canada when used at multiple passes and timings. They are, however, ineffective at controlling perennial weeds.

7.3.3 INTERROW CULTIVATION

Interrow cultivation has the advantage of physically isolating the mechanical disturbance from the crop. However, precise guidance systems are needed to navigate the narrow rows associated with small-grain production. Cultivators that can be manually steered have been available for a long time. They allow an operator to precisely guide the cultivator in the row independent of the tractor. Using a steerable interrow cultivator, Kolb et al. (2010, 2012) determined that interrow cultivation was the most effective and economical system for organic production of wheat and barley. However, Knight et al. (2010) found that interrow cultivation in field pea failed to increase weed control and crop yield. It was observed that the uncontrolled weeds within the crop row were still able to compete strongly with the crop.

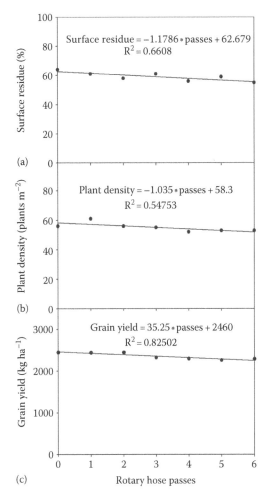

FIGURE 7.3 The effect of rotary hoe passes in a no-till cropping system on the tolerance of field pea. The panels show the effect of number of rotary hoe passes on surface crop residue (a), field pea plant density (b), and field pea yield (c). The experiment was conducted in the absence of weeds to determine crop tolerance. (Adapted from Shirtliffe, S.J. and Johnson, E.N., *Renew. Agr. Food Syst.*, 27(1), 60, 2012.)

7.3.4 Critical Appraisal of Mechanical Weed Control Methods

The efficacy of mechanical weed control often depends on the soil and environmental conditions at and following the time of treatment. In general, because mechanical weed control measures rely on burial in soil or plant uprooting to kill weeds, wet conditions, which reduce the soil tilth, tend to hamper their effectiveness. For techniques in which weeds are uprooted, the environment following the uprooting has been observed to affect the efficacy of the control measure. In general, hot dry conditions promote quick death of weeds, whereas cloudy wet conditions can allow uprooted weeds to re-root and survive (Jones et al., 1996). Uprooted plants die more quickly with dry soil conditions (Kurstjens and Kropf, 2001).

To effectively control weeds with in-crop tillage implements, multiple passes are often needed (Rasmussen, 1993; Kirkland, 1995; Shirtliffe and Johnston, 2012). Although weed control improves with multiple passes, extra passes mean more labor and equipment operating costs. An analysis of the weed control effect of multiple passes indicates that the first pass often provides more relative weed control compared to subsequent passes. In order for farmers to widely adopt this method, we believe that equipment that achieves effective control in a single pass is needed.

Mechanical weed control is unpredictable due to the aforementioned effects of environment, poor selectivity, and the narrow range of timings when implements can be used. Thus, a mechanical weed control system may require several implements in order to be effective. The rotary hoe has the advantage of operating at a high speed and causing relatively little crop damage (Bowman, 2002; Cloutier et al., 2007). However, it must take place over a very narrow time frame because it has difficulty controlling weeds that either root deeply as seedlings or are beyond the *white thread* stage. In-crop harrowing can control weeds that are somewhat larger than what a rotary hoe can manage, but harrowing can cause much greater crop damage. Finally, interrow cultivation can control large weeds and may even partially control perennial weeds. However, it may cover the crop rows if conducted at an early stage and result in root pruning at later stages. Thus, a system that uses either harrowing or rotary hoeing initially and has interrow tillage as a backup may be ideal.

7.4   CONCLUSION

While our tongue-in-cheek introduction to this chapter may seem to make light of Liebman and Gallandt (1997) and others' concepts of IWM, we fundamentally agree that IWM in its purest form must employ many diverse tactics. However, it is natural for farmers, like all people, to seek simplicity. Furthermore, although using the framework of IWM to study these weed control practices is sound, it may not be the best way to communicate these techniques. As most of these techniques are agronomic in nature, a better way to communicate them to farmers may be to present them as modifications to the agronomic practices that they already do. Thus, the implication is that IWM does not constitute an insurmountable shift in crop production practices but is instead an alteration of their current practices.

Having said that, we carried out a 2-year study (Benaragama and Shirtliffe, 2013) in organic oat crops with the objective of integrating four cultural practices (competitive genotype, increased crop density, narrow row planting, and postemergence harrowing) in order to enhance overall competitive ability of the crop agronomically coupled with mechanical weed control. Results of this study were encouraging as we found out that combined use of a competitive cultivar, doubling the seeding rate and postemergence harrowing, additively reduced weed biomass by 71%. Importantly, when comparing their individual effects, seeding rate was the biggest hammer with 52% weed suppression, competitive genotype only had 22% weed suppression, and mechanical weed control only affected weed density but not weed biomass compared to standard cultural practices. Three outcomes of this study should be highlighted. Firstly, substantial weed control in organic systems can be achieved only by integrating weed control methods. Secondly, different tactics in an IWM plan can be ranked based on their effect on weed control allowing farmers to easily pick the important practices according to their feasibility. Thirdly, farmers can achieve substantial weed control simply by altering normal cultural practices.

This review has focused on cultural and mechanical weed control methods that either increase a crop's competitive ability or selectively damage the weed. We have ignored other long-term management practices that target other aspects of a weed's life cycle (Anderson, 2005). Over time, methods that target weed seed production instead of just the yield damage that weeds cause can be most effective at reducing weed populations. Perhaps the title of Eric Gallandt's YouTube channel *zeroseedrain* best sums up his philosophy on weed control. Long-term practices that prevent weed seed, such as crop rotation, are very potent weed control methods but are beyond the scope of this review. Nevertheless, the techniques outlined in this chapter will also result in reduced weed seed rain.

This chapter has focused on organic weed control systems where synthetic herbicides are prohibited. However, nonorganic systems can also benefit from nonherbicidal weed control methods. Many cropping systems have weeds that are closely related to the crops they are grown with and as a result are impossible to control with selective herbicides. The control of wild oat in oat crops is the example that our research group is the most familiar with. To reduce yield loss due to wild oats, we recommend that growers seed their oats at approximately twice the regular seed rate (500 vs. 250 seeds m^{-2}). Herbicide-resistant weeds can also be controlled with cultural weed control techniques. In addition to controlling weeds already resistant to the herbicide group, nonherbicidal weed control also reduces the selection pressure and subsequent evolution of herbicide resistance (Beckie et al., 2006). To capitalize on these benefits of nonherbicidal weed control, we believe that many cultural and mechanical weed control techniques have applicability to nonorganic cropping systems. Given the rapid evolution of herbicide-resistant weeds (Powles and Yu et al., 2010), all farmers should be encouraged to adopt these techniques.

So of all the *hammers* (weed control methods) available to a farmer, which one should be included in their toolbox and utilized regularly? Of those methods intended to increase the crop's competitive ability, increasing the seeding rate is usually the most consistent and successful. Other methods that attempt to increase the competitive ability, such as choosing competitive crop cultivars or seeding in narrow seed rows, often are not available or if utilized are not as consistent in terms of weed control as increased seeding rate. Some methods, such as seeding large seed, may not be needed as seeding larger seed results in lower plant populations for the same weight per area seeding rate. Mechanical weed control can also offer good weed control provided that it is selective and does not damage the crop excessively. Because of this, we see promise for rotary hoeing and interrow tillage. We believe that systems that utilize a higher than normal seeding combined with timely, selective mechanical weed control are the best tools available to most organic crop farmers.

REFERENCES

Anderson, R.L. 2005. A multi-tactic approach to manage weed population dynamics in crop rotations. *Agron. J.* 97:1579–1583.

Auskalnis, A. and O. Auskalniene. 2008. Weed control in spring barley by harrowing. *Zemdirbyste (Agriculture)*. 95:388–394.

Baerveldt, S. and J. Ascard. 1999. Effect of soil cover on weeds. *Biol. Agric. Hort.* 17:101–111.

Baird, J.M., S.J. Shirtliffe, and F.L. Walley. 2009a. Optimal seeding rate for organic production of lentil in northern Great Plains. *Can. J. Plant Sci.* 89:1089–1097.

Baird, J.M., F.L. Walley, and S.J. Shirtliffe. 2009b. Optimal seeding rate for organic production of field pea in the northern Great Plains. *Can. J. Plant Sci.* 89:455–464.

Ballare, C.L., A.L. Scopel, E.T. Jordan, and R.D. Vierstra. 1994. Signaling among neighbouring plants and the development of size inequalities in plant populations. *Proc. Natl. Acad. Sci. USA.* 91:10094–10098.

Barton, D.L., D.C. Thill, and B. Shafii. 1992. Integrated wild oat (*Avena fatua*) management affects spring barley (*Hordeum vulgare*) yield and economics. *Weed Technol.* 6:129–135.

Beckie, H.J., G.S. Gill, H.P. Singh, D.R. Batish, and R.K. Kohli. 2006. Strategies for managing herbicide-resistant weeds. In H.P Singh, D.R. Batish, and R.K. Kohli, eds. *Handbook of Sustainable Weed Management.* pp. 581–625. Food Products Press, New York.

Beckie, M.A.N. 2000. Zero-tillage and organic farming in Saskatchewan: An interdisciplinary study of the development of sustainable agriculture. PhD thesis. University of Saskatchewan, Saskatoon, Saskatchewan, Canada.

Begna, S.H., R.I. Hamilton, L.M. Dwyer, D.W. Stewart, D. Cloutier, L. Assemat, K. Foroutan-Pour, and D.L. Smith. 2001. Weed biomass production response to plant spacing and corn (*Zea mays*) hybrids differing in canopy architecture. *Weed Tech.* 15:647–653.

Benaragama, D., B.G. Rossnagel, and S.J. Shirtliffe. In press. Breeding for competitive and high yielding crop cultivars (manuscript accepted for publication).

Benaragama, D. and S.J. Shirtliffe. 2013. Additive weed control by integrating cultural and mechanical weed control strategies in an organic cropping system. *Agron. J.* 105:1728–1734.

Blackshaw, R.E. 1994. Differential competitive ability of winter wheat cultivars against downy brome. *Agron. J.* 86:649–654.

Blackshaw, R.E., K.N. Harker, J.T. O'Donovan, H.J. Beckie, and E.G. Smith. 2008. Ongoing development of integrated weed management systems on the Canadian prairies. *Weed Sci.* 56:146–150.

Blackshaw, R.E., J.T. O'Donovan, K.N. Harker, and X. Li. 2002. Beyond herbicides: New approaches to managing weeds. In *Proceedings of the International Conference on Environmentally Sustainable Agriculture for Dry Areas.* Shijiazhuang, Hebei, China. Dobing Enterprises, Lethbridge, Alberta, Canada. pp. 305–312.

Bond, W. and A.C. Grundy. 2001. Non-chemical weed management in organic farming systems. *Weed Res.* 41:383–405.

Bowman, G. 2002. Steel in the field: A farmer's guide to weed management tools. Sustainable agriculture network Beltsville, MD. http://www.sare.org/Learning-Center/Books/Steel-in-the-Field (accessed November 29, 2012).

Boyd, W.J.R., A.G. Gordon, and L.J. LaCroix. 1971. Seed size, germination resistance and seedling vigor in barley. *Can. J. Plant Sci.* 51:93–99.

Boyetchko, S. and G. Peng. 2004. Challenges and strategies for development of mycoherbicides. In D.K. Arora, ed. *Fungal Biotechnology in Agricultural, Food, and Environmental Applications.* Vol. 21, pp. 111–121. Marcel Dekker Inc., New York.

Boyetchko, S. and E.N. Rosskopf. 2006. Strategies for developing bioherbicides for sustainable weed management. In H.P. Singh, D.R. Batish, and R.K. Kohli, eds. *Handbook for Sustainable Weed Management.* pp. 393–420. Haworth Press, Inc., New York.

Buhler, D.D. 2002. 50th anniversary—Invited article: Challenges and opportunities for integrated weed management. *Weed Sci.* 50:273–280.

Carlson, H.L. and J.E. Hill. 1985. Wild oat (*Avena fatua* L.) competition with spring wheat: Plant density effects. *Weed Sci.* 33:176–181.

Champion, G.T., R.J. Froud-Williams, and J.M. Holland. 1998. Interactions between wheat (*Triticum aestivum* L.) cultivar, row spacing, and density and the effect on weed suppression and crop yield. *Ann. Appl. Biol.* 133:443–453.

Chen, C., N.K. Wichman, and D.M. Westcott. 2008. Hard red spring wheat response to row spacing, seeding rate, and nitrogen. *Agron. J.* 100(5):1296–1302.

Cirujeda, A., B. Melander, K. Rasmussen, and A.I. Rasmussen. 2003. Relationship between speed, soil movement into the cereal row and intra-row weed control efficacy by weed harrowing. *Weed Res.* 43:285–296.

Cloutier, D.C., R.Y. Van der Weide, A. Peruzzi, and M.L. Leblanc. 2007. Mechanical weed management. In M.K. Upadhyaya and R.E. Blackshaw, eds. *Non-Chemical Weed Management: Principles, Concepts and Technology.* pp. 111–135. CAB International, Wallingford, U.K.

Drews, S., D. Neuhoff, and U. Köpke. 2009. Weed suppression ability of three winter wheat varieties at different row spacing under organic farming conditions. *Weed Res.* 49(5):526–533.

Entz, M.H., R. Guilford, and R. Gulden. 2001. Crop yield and soil nutrient status on 14 organic farms in the eastern portion of the northern Great Plains. *Can. J. Plant Sci.* 81:351–354.

Evans, R.M., D.C. Thill, L. Tapia, B. Shafii, and J.M. Lish. 1991. Wild oat (*Avena fatua*) and spring barley (*Hordeum vulgare*) density affect spring barley grain yield. *Weed Technol.* 5:33–39.

Fanadzo, M., A.B. Mashingaidze, and C. Nyakanda. 2007. Narrow rows and high maize densities decrease maize grain yield but suppress weeds under dry-land conditions in Zimbabwe. *Agron. J.* 6:566–570.

Fischer, R.A. and R.E. Miles. 1973. The role of spatial pattern in the competition between crop plants and weeds. A theoretical analysis. *Math. Bios.* 18:335–350.

Froud-Williams, R.J. and N. Korres. 2001. The effects of varietal selection, seed rate and weed competition on quantitative and qualitative traits of grain yield in winter wheat. *Asp. Appl. Biol.* 64:147–156.

Geleta, B., M. Atak, P.S. Baenziger, L.A. Nelson, D.D. Baltenesperger, K.M. Eskridge, M.J Shipman, and D.R. Shelton. 2002. Seeding rate and genotype effect on agronomic performance and end-use quality of winter wheat. *Crop Sci.* 42(3):827–832.

Goldberg, D.E. 1990. Components of resource competition by plants. In J.B. Grace and D. Tilman, eds. *Perspectives of Plant Competition.* pp. 22–23. Academic Press, Inc., San Diego, CA.

Hansen, P.K., I.A. Rasmussen, N. Holst, and C. Andreasen. 2007. Tolerance of four spring barley (*Hordeum vulgare*) varieties to weed harrowing. *Weed Res.* 47:241–251.

Harker, K.N., G.W. Clayton, R.E. Blackshaw, J.T. O'Donovan, and F.C. Stevenson. 2003. Seeding rate, herbicide timing and competitive hybrids contribute to integrated weed management in canola (*Brassica napus*). *Can. J. Plant Sci.* 83(2):433–440.

Harker, K.N., J.T. O'Donovan, R.E. Blackshaw, E.N. Johnson, F.A. Holm, and G.W. Clayton. 2011. Environmental effects on the relative competitive ability of canola and small-grain cereals in a direct-seeded system. *Weed Sci.* 59(3):404–415.

Harper, J.L. 1977. *The Population Biology of Plants.* Academic Press, London, U.K.

Hay, R.K.M. and A.J. Walker. 1989. *An Introduction to the Physiology of Crop Yield.* Longman Scientific & Technical, Harlow, England.

Hucl, P. 1998. Response to weed control by four spring wheat genotypes differing in competitive ability. *Can J. Plant. Sci.* 78:171–173.

Huel, D.G. and P. Hucl. 1996. Genotypic variation for competitive ability in spring wheat. *Plant Breed.* 115:325–329.

Jensen, R.K., J. Rasmussen, and B. Melander. 2004. Selectivity of weed harrowing in Lupin. *Weed Res.* 44:245–253.

Johnson, E.N. 2001. Mechanical weed control in field pea (*Pisum sativum* L.). Master's thesis. University of University of Saskatchewan, Saskatoon, Saskatchewan, Canada.

Jones, P.A., A.M. Blair, and J. Orson. 1996. Mechanical damage to kill weeds. In *Proceedings Second International Weed Control Congress*, Copenhagen, Denmark. pp. 949–954.

Jordan, N. 1993. Prospects for weed control through crop interference. *Ecol. Appl.* 3:84–91.

Kirkland, K.J. 1995. Frequency of post-emergence harrowing effects wild oat control and spring wheat yield. *Can. J. Plant Sci.* 75:163–165.

Kirkland, K.J. and J.H. Hunter. 1991. Competitiveness of Canada prairie spring wheats with wild oat (*Avena fatua* L). *Can. J. Plant Sci.* 71:1089–1092.

Knight, J.D., E. Johnson, S.S. Malhi, S.J. Shirtliffe, and B. Blackshaw. 2010. Nutrient management and weed dynamics challenges in organic farming systems in the Northern Great Plains of North America. In S.S. Malhi, Y. Gan, J.J. Schoenau, R.L. Lemke, and M.A. Liebig, eds. *Recent Trends in Soil Science and Agronomy Research in the Northern Great Plains of North America*. pp. 241–244. Research Sinpost, Trivandrum, India.

Kolb, L.N. and E.R. Gallandt. 2012. Weed management in organic cereals: Advances and opportunities. *Org. Agric.* 2:23–42.

Kolb, L.N., E.R. Gallandt, and E.B. Mallory. 2012. Impact of spring wheat planting density, row spacing, and mechanical weed control on yield, grain protein, and economic return in Maine. *Weed Sci.* 60(2):244–253.

Kolb, L.N., E.R. Gallandt, and T. Molloy. 2010. Improving weed management in organic spring barley: Physical weed control vs. interspecific competition. *Weed Res.* 50(6):597–605.

Koscelny, J.A., T.F. Peeper, J.B. Solie, and S.G. Solomon, Jr. 1990. Effect of wheat (*Triticum aestivum* L.) row spacing, seeding rate, and cultivar on yield loss from cheat (*Bromus secalinus* L.). *Weed Technol.* 4:487–492.

Koscelny, J.A., T.F. Peeper, S.B. Solie, and S.G. Solomon, Jr. 1991. Seeding date, seeding rate, and row spacing affect wheat (*Triticum aestivum*) and cheat (*Bromus secalinus*). *Weed Technol.* 5(4):707–712.

Kurstjens, D.A.G. and M.J. Kropff. 2001. The impact of uprooting and soil-covering on the effectiveness of weed harrowing. *Weed Res.* 41:211–228.

Kurstjens, D.A.G. and U.D. Perdok. 2000. The selective soil covering mechanism of weed harrows on sandy soil. *Soil Tillage Res.* 55:193–206.

Kurstjens, D.A.G., U.D. Perdok, and D. Goense. 2000. Selective uprooting by weed harrowing on sandy soils. *Weed Res.* 40(5):431–447.

Lafond, G.P. 1994. Effects of row spacing, seeding rate and nitrogen on yield of barley and wheat under zero-till management. *Can. J. Plant Sci.* 74:703–711.

Lafond, G.P. and K.H. Kattler. 1992. The tolerance of spring wheat and barley to post-emergence harrowing. *Can. J. Plant Sci.* 72:1331–1336.

Lafond, G.P., W.E. May, and C.B. Holzapfel. 2013. Row spacing and nitrogen fertilizer effect on no-till oat production. *Agron. J.* 105:1–10.

Lamb, E.G., S.J. Shirtliffe, and W.E. May. 2011. Structural equation modeling in the plant sciences: An example using yield components in oat. *Can. J. Plant Sci.* 91:603–619.

Leeson, J.Y., J.W. Sheard, and A.G. Thomas. 2000. Weed communities associated with arable Saskatchewan farm management systems. *Can. J. Plant Sci.* 80:177–185.

Lemerle, D., R.D. Cousens, G.S. Gill, S.J. Peltzer, M. Moerkerk, C.E. Murphy, D. Collins, and B.R. Cullis. 2004. Reliability of higher seeding rates of wheat for increased competitiveness with weeds in low rainfall environments. *J. Agric. Sci.* 142(4):395–409.

Lemerle, D., B. Verbeek, and N. Coombes. 1995. Losses in grain yield of winter crops from *Lolium rigidum* competition depend on crop species, cultivar, and season. *Weed Res.* 3:503–509.

Liebman, M. and E.R. Gallandt. 1997. Many little hammers: Ecological management of crop-weed interactions. In L.E. Jackson, ed. *Ecology in Agriculture*. pp. 291–343. Academic Press, San Diego, CA.

Llewellyn, R.S. 2007. Information quality and effectiveness for more rapid adoption decisions by farmers. *Field Crops Res*. 104:148–156.

Lundkvist, A. 2009. Effects of pre- and post-emergence weed harrowing on annual weeds in peas and spring cereals. *Weed Res*. 49:409–416.

Macey, A. 2010. *Certified Organic Production in Canada, 2009*. Canadian Organic Growers, Ottawa, Ontario, Canada. p. 9.

Marshall, G.C. and H.W. Ohm. 1987. Yield responses of 16 winter wheat cultivars to row spacing and seeding rate. *Agron. J*. 79(6):1027–1030.

Mashingaidze, A.B., W. Van der Werf, L.A.P. Lotz, J. Chipomho, and M.J. Kropff. 2009. Narrow rows reduce biomass and seed production of weeds and increase maize yield. *Ann. Appl. Biol*. 155:207–218.

Mason, H., A. Navabi, B. Frick, J.T. O'Donovan, and D.M. Spaner. 2007a. Cultivar and seeding rate effects on the competitive ability of spring cereals grown under organic production in northern Canada. *Agron. J*. 99(5):1199–1207.

Mason, H.E., A. Navabi, B.L. Frick, J.T. O'Donovan, and D.M. Spaner. 2007b. The weed-competitive ability of Canada western red spring wheat cultivars grown under organic management. *Crop Sci*. 47:1167–1176.

May, W.E., S.J. Shirtliffe, D.W. McAndrew, C.B. Holzapfel, and G.P. Lafond. 2009. Management of wild oat (*Avena fatua* L.) in tame oat (*Avena sativa* L.) with early seeding dates and high seeding rates. *Can. J. Plant. Sci*. 89:763–773.

Melander, B. 1993. Modeling the effects of *Elymus repens* (L.) competition on yield of cereals, peas and oilseed rape. *Weed Res*. 34:99–108.

Mohler, C.L. 2001a. Mechanical management of weeds. In M. Liebman, C.L. Mohler, and C.P. Staver, eds. *Ecological Management of Agricultural Weeds*. pp. 139–209. Cambridge University Press, New York.

Mohler, C.L. 2001b. Enhancing the competitive ability of crops. In M. Liebman, C.L. Mohler, and C.P. Staver, eds. *Ecological Management of Agricultural Weeds*. pp. 269–321. Cambridge University Press, New York.

Murphy, S.D., Y. Yakubu, S.F. Weise, and C.J. Swanton. 1996. Effect of planting patterns and inter-row cultivation on competition between corn (*Zea mays*) and late emerging weeds. *Weed Sci*. 44:865–870.

O'Donovan, J.T., R.E. Blackshaw, K.N. Harker, G.W. Clayton, and R. McKenzie. 2005. Variable plant establishment contributes to differences in competitiveness with wild oat among wheat and barley varieties. *Can. J. Plant Sci*. 85:771–776.

O'Donovan, J.T., K.N. Harker, G.W. Clayton, and L.M. Hall. 2000. Wild oat (*Avena fatua*) interference in barley (*Hordeum vulgare*) is influenced by barley variety and seeding rate. *Weed Technol*. 14:624–629.

O'Donovan, J.T., J.C. Newman, K.N. Harker, R.E. Blackshaw, and D.W. McAndrew. 1999. Effect of barley plant density on wild oat interference, shoot biomass and seed yield under zero tillage. *Can. J. Plant Sci*. 79:655–662.

O'Donovan, J.T., E.A. de. St. Renmy, P.A. Sullivan, D.A. Dew, and A.K. Sharma. 1985. Influence of the relative time of emergence of wild oat (*Avena fatua*) on yield loss of barley (*Hordeum vulgare*) and wheat (*Triticum aestivum*). *Weed Sci*. 33:498–503.

Olsen, J., L. Kristensen, J. Weiner, and H. Griepentrog. 2004. Increased density and spatial uniformity increase weed suppression by spring wheat. *Weed Res*. 45:316–321.

Olsen, J.M., H.-W. Griepentrog, J. Nielsen, and J. Weiner. 2012. How important are crop spatial pattern and density for weed suppression by spring wheat? *Weed Sci*. 60, 501–509.

Pavlychenko, T.K. and J.B. Harrington. 1934. Competitive efficiency of weeds with cereal crops. *Can. J. Res*. 10:77–94.

Peters, E.J., M.R. Gebhardt, and J.F. Stritzke. 1965. Interrelations of row spacings, cultivations and herbicides for weed control in soybeans. *Weeds*. 13:285–289.

Powles, S.B. and Q. Yu. 2010. Evolution in action: Plants resistant to herbicides. *Ann. Rev. Plant Biol.* 61:317–347.

Putnam, D.H., J. Wright, L.A. Field, and K.K. Ayisi. 1992. Seed yield and water-use efficiency of white lupin as influenced by irrigation, row spacing, and weeds. *Agron. J.* 84:557–563.

Rasmussen, I.A. 2004. The effect of sowing date, stale seedbed, row width and mechanical weed control on weeds and yields of organic winter wheat. *Weed Res.* 44:12–20.

Rasmussen, J. 1990. Selectivity-an important parameter on establishing the optimum harrowing technique for weed control in growing cereals. In *Symposium on Integrated Weed Management in Cereals*. pp. 197–204. *Proceedings of an EWRS Symposium*, Helsinki, Finland, June 4–6, 1990. European Weed Research Society, Wageningen, the Netherlands.

Rasmussen, J. 1991. A model for prediction of yield response in weed harrowing. *Weed Res.* 31:401–408.

Rasmussen, J. 1992. Testing harrows for mechanical control of annual weeds in agricultural crops. *Weed Res.* 32:267–274.

Rasmussen, J. 1993. The influence of harrowing used for post-emergence weed control on the interference between crop and weeds. In *Proceedings of the European Weed Research Society Symposium. Quantitative Approaches in Weed and Herbicide Research and Their Practical Application*. pp. 209–217. European Weed Research Society, Braunschweig, Germany.

Rasmussen, J., H. Mathiasen, and B.M. Bibby. 2010. Timing of post- emergence weed harrowing. *Weed Res.* 50(5):436–446.

Rasmussen, J., H.H. Nielsen, and H. Gundersen. 2009. Tolerance and selectivity of cereal species and cultivars to post-emergence weed harrowing. *Weed Sci.* 57:338–345.

Rasmussen, J. and T. Svenningsen. 1995. Selective weed harrowing in cereals. *Biol. Agri. Hort.* 12:29–46.

Reid, T.A., R.C. Yang, D.F. Salmon, A. Navabi, and D. Spaner. 2011. Realized gains from selection for spring wheat grain yield are different in conventional and organically managed systems. *Euphytica* 177:253–266.

Ries, S.K. and E.H. Everson. 1973. Protein content and seed size relationships with seedling vigor of wheat cultivars. *Agron. J.* 65:884–886.

Rydberg, Y. 1994. Weed harrowing-the influence of driving speed and driving directions on the degree of soil covering and the growth of the weed and crop plants. *Biol. Agri. Hort.* 10:197–205.

Seavers, G.P. and K.J. Wright. 1999. Crop canopy development and structure influence weed suppression. *Weed Res.* 39:319–328.

Seufert, V., N. Ramankutty, J.A. Foley. 2012. Comparing the yields of organic and conventional agriculture. *Nature* 485(7397):229–232.

Sharratt, B.S. and D.A. McWilliams. 2005. Microclimatic and rooting characteristics of narrow-row versus conventional-row corn. *Agron. J.* 97:1129–1135.

Shirtliffe, S.J., Brandt, S.A., and J.K. Knight. 2006. Saskatchewan organic on-farm research: Part II: Soil fertility and weed management. *Saskatchewan Agriculture Development Fund Final Report*. Project: 20020198.

Shirtliffe, S.J. and E.N. Johnson. 2012. Progress towards no-till organic weed control in western Canada. *Renew. Agric. Food Syst.* 27(1):60–67.

Spaner, D., A.G. Todd, and D.B. McKenzie. 2001. The effect of seeding rate and nitrogen fertilization on barley yield and yield components in a cool maritime climate. *J. Agron. Crop Sci.* 187(2):105–110.

Spies, J.M., T.D. Warkentin, and S.J. Shirtliffe. 2011. Variation in field pea (*Pisum sativum*) cultivars for basal branching and weed competition. *Weed Sci.* 59(2):218–223.

Spink, J.H., T. Semere, D.L. Sparkes, J.M. Whaley, M.J. Foulkes, R.W. Clare, and R.K. Scott. 2000. Effect of sowing date on the optimum plant density of winter wheat. *Ann. Appl. Biol.* 137(2):179–188.

Statistics Canada. 2012. Estimated areas, yield, production and average farm price of principal field crops, in metric units. http://www5.statcan.gc.ca/cansim/a26?lang=eng&retrLang=eng&id=0010010&tabMode=dataTable&srchLan=-1&p1=-1&p2=9 (accessed December 4, 2012).

Steinmann, H. 2002. Impact of harrowing on the nitrogen dynamics of plants and soil. *Soil Tillage Res.* 65:53–59.

Swanton, C.J. and S.D. Murphy. 1996. Weed science beyond the weeds: The role of integrated weed management (IWM) in agroecosystem health. *Weed Sci.* 44:437–445.

Teasdale, J.R. and J.R. Frank. 1983. Effect of row spacing on weed competition with snap beans (*Phaseolus vulgaris*). *Weed Sci.* 31:81–85.

Teich, A.H., A. Smid, T. Welackyl, and A. Hami. 1993. Row-spacing and seed-rate effects on winter wheat in Ontario. *Can. J. Plant Sci.* 3:31–35.

Thill, D., J.T. O'Donovan, and C.A. Mallory-Smith. 1994. Integrated weed management strategies for delaying herbicide resistance in wild oats. *Phytoprotection.* 75(4):61–70.

Van Hemmst, H.D. 1985. The influence of weed competition on crop yield. *Agric. Syst.* 18:81–93.

Velykis, A., S. Maiksteniene, A. Arlauskiene, I. Kristaponyte, and A. Satkus. 2009. Mechanical weed control in organically grown spring oat and field pea crops. *Agron. Res.* 7:542–547.

Watson, P.R., D.A. Derksen, and R.C. Van Acker. 2006. The ability of 29 barley cultivars to compete and withstand competition. *Weed Sci.* 54:783–792.

Wax, L.M. and J.W. Pendelton. 1968. Effects of row spacing on weed control in soy beans. *Weed Sci.* 15:462–465.

Weiner, J., H. Griepentrog, and L. Kristensen. 2001. Suppression of weeds by spring wheat (*Triticum aestivum*) increases with crop density and spatial uniformity. *J. Appl. Ecol.* 38:784–790.

Wildeman, J. 2004. The effect of oat (*Avena sativa* L.) genotype and plant population on wild oat (*Avena fatua* L.) competition. Master's thesis. University of Saskatchewan, Saskatoon, Saskatchewan, Canada.

Willenborg, C.J., B.G. Rossnagel, and S.J. Shirtliffe. 2005. Oat caryopsis size and genotype effects on wild oat–oat competition. *Crop Sci.* 45(4):1410–1416.

Xue, Q. and R.N. Stougaard. 2002. Spring wheat seed size and seeding rate affect wild oat demographics. *Weed Sci.* 50:312–320.

Zand, E. and H.J. Beckie. 2002. Competitive ability of hybrid and open-pollinated canola (*Brassica napus*) with wild oat (*Avena fatua*). *Can. J. Plant Sci.* 82:473–480.

Zimdhal, R.L. 1993. *Fundamentals of Weed Science*. Academic Press, San Diego, CA.

8 Insect Pest Management in Organic Cropping Systems Based on Ecological Principles

Gilles Boiteau, Tom Lowery, and Josée Boisclair

CONTENTS

8.1   INTRODUCTION

The growth of the organic sector is challenging insect pest management (IPM). An increasing demand for organic products has supported expansion of the sector but has also drawn IPM away from a primary reliance on preventive pest control methods to a dependence on curative methods, particularly insecticides, which should be employed as a last resort (Zehnder et al. 2007). Here, we state a case for the importance and value of a long-term approach to insect management based on an understanding of pest biology.

Discussions of IPM in organic crop production are most often based on comparisons between organic and conventional practices. As suggested at the 2012 First Canadian Organic Science Conference (MacKenzie and Savard 2012), the time has come to focus on the advantages, benefits, and challenges of organic production in its own right. According to organic production principles, pest management methods should be selected in a precautionary and responsible manner that emulates living ecological systems and protects environmental health (Gomiero et al. 2011). The aim is to work with natural processes in an attempt to prevent pest outbreaks. In an ideal organic production system, healthy plants resulting from the selection of appropriate crops and varieties produced within long rotations and coupled with appropriate agronomic and cultural practices help to maintain insect pest populations at low levels, reducing the need for direct curative action against insect pests (e.g., Litterick et al. 2002). Unfortunately, the reality is that insect pest pressure continues to cause significant reductions in yield and quality across a wide range of organically grown crops because preventive control methods that are effective against a particular insect species are not necessarily effective against all pests. Very few farms have had the opportunity to fully establish a prevention-based farming strategy, and integration of various methods against multiple pests remains rare and limited to single fields rather than to whole farms or entire regions (i.e., area-wide mating disruption programs). Also, farms in transition from conventional to organic benefit little from the crop tolerance to insect pests of long-established farms and must turn to reactive control methods (e.g., Boisclair and Estevez 2006; Zehnder et al. 2007; Boiteau 2008).

Farming in a manner close to nature means managing crops according to ecological principles with an emphasis on sustainable and renewable biological processes (Letourneau and van Bruggen 2006). The objective is to establish crop systems oscillating around a sustainable equilibrium point. This is a challenging task where the farm environment, a land-management system maintained to produce food, is controlled not only by the usual ecosystem feedbacks but jointly by the farmer's activities (Trewavas 2001). The equilibrium point varies according to many factors including the fluctuating pressures exerted by insects and other pests as well as variable inputs (fertilizers, irrigation, land use pattern, etc.) in response to fluctuating market demands for each crop. One of the requirements for establishing or maintaining the equilibrium is a sound IPM program. Careful planning with a strong emphasis on measures to prevent or delay the colonization of crops by pests is essential, as it is difficult to rescue a crop from insects that have established and are affecting the crop's development. Remedial actions often result in a major disruption of the agroecosystem, particularly but not limited to their negative impact on beneficial arthropods, leading to outbreaks of secondary pests and often a rebound in numbers of pest species (e.g., Rabb and Guthrie 1970; Landis et al. 2000; Gomiero et al. 2011). It means building a resilient agroecosystem with the ability to bounce back when a perturbation happens: for example, when insect pests colonize the crop or when established pests increase in numbers.

The objective of this chapter is therefore to review the management of insect pests in organic crops, its status, challenges, and promises in relation to the basic principles of organic production and ecology. Insect management plans can be considered from many vantage points, but because insect pests tend to colonize the crop from surrounding habitats, we have chosen for this chapter to apply the basic concepts of invasion ecology as suggested by Letourneau and van Bruggen (2006). This is in line with Davis and Thompson (2000) and Boiteau and Heikkila (2013) who have argued that invasion is but one type of colonization.

8.2   ESSENTIALS OF INSECT PEST MANAGEMENT

Insects are generally overlooked until damage to crop plants or reduced yields draw attention to their presence. As a result, insect pests are remembered for their population explosions rather than for their relatively stable populations oscillating around normal equilibrium levels when considered over a long period of time (Figure 8.1a). This equilibrium position is largely determined by the physical environment and

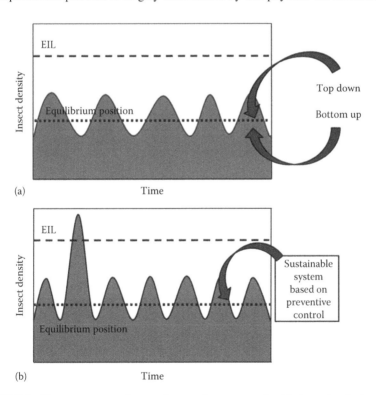

FIGURE 8.1   Pest management in organic cropping systems should aim at developing agroecosystems where the density of the insect pest species (if present) occupies an equilibrium position modulated well below the EIL by bottom-up and top-down factors (a). Occasional peaks of abundance can be easily managed (b).

(*continued*)

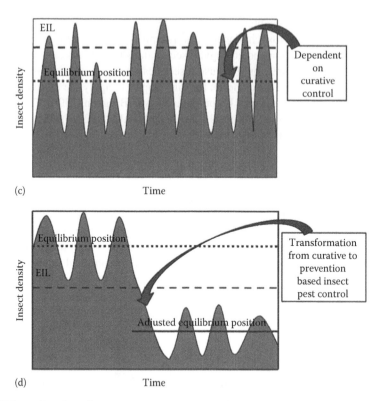

FIGURE 8.1 (continued) But repeated peaks at or above the EIL create a dependence on curative controls (c). Unless permanent changes are made to the crop production system to bring the equilibrium position back below the EIL (d). (Adapted from Luckman, W.H. and Metcalf, R.L., The pest management concept, in *Introduction to Insect Pest Management*, eds. R.L. Metcalf and W.H. Luckman, John Wiley & Sons, New York, 1994.)

feedback from density-independent and density-dependent factors. The need for insect control arises only when the density of a pest species exceeds normal abundance. The best insurance against this is a rich, biologically diverse multitrophic plant ecosystem. The integrated IPM strategy tries to take advantage of the large number of similarities between an agroecosystem and a natural ecosystem by favoring natural mortality factors and eliminating as much as possible practices that could upset these factors. Rabb (1970) defined IPM as the reduction of insect pest problems by combined actions selected after the life systems of the pests are understood and the ecological as well as economic consequences of these actions have been predicted as accurately as possible.

There are fundamental differences between natural ecosystems and agricultural ecosystems that cannot be overlooked. For example, even in organically managed grasslands, intense soil cultivation, crop harvesting, and crop rotations lead to changes in plant communities and numbers of pollinators. Self-pollinated plants that can better withstand disturbances caused by farming gain some dominance over plant species that depend on insect pollinators (Power et al. 2012).

The development of agriculture over many millennia resulted in selection of crop plants that were more palatable and productive but had fewer morphological and chemical defenses against herbivores (Risch 1987). Additionally, agriculture became ever more intensive resulting in a reduction in landscape diversity, ever larger fields, crop specialization, higher planting densities, and reliance on monocultures of similar age structure that reduced the number of beneficial insects (Estevez et al. 2000). Diverse cropping systems favor beneficial insects through the provision of alternative food sources (i.e., prey, nectar, and pollen) and refugia in hedgerows and in uncultivated or fallow areas. Little experimental data exist, but it is hypothesized that large, uniform plantings also favor pest outbreaks by increasing the lag time for colonization of large fields by natural enemies in relation to their prey (Risch 1987). These differences are often used to explain why there are more outbreaks of insects in agroecosystems than in natural forest or grassland ecosystems (Pimentel 1961; Risch 1987). Increased diversity does not however automatically lead to lower numbers of herbivores because of a greater abundance of beneficials (Risch et al. 1983). The ecological mechanisms underlying how agricultural diversification affects insect pests are numerous. In some cases, changes in within-field plant diversity may not affect the abundance of natural enemies but may, for example, decrease the ability of herbivores to colonize their host plants (Risch et al. 1983). It is generally true that diversification of organic cropping systems by planting of diverse crops and varieties or the use of trap crops or companion plantings can contribute to greater long-term stability of the agroecosystem, but a fundamental understanding of the mechanisms is critical to the selection of the diversification appropriate to each pest and each crop.

Insects are key players in the flow of energy in the organic agroecosystem as herbivores, carnivores, and detritivores. Managing insect pests in an organic production system amounts to managing herbivores competing with humans for part of the energy value of a complex integrated system made up of living (crop plants, other pests, beneficials, etc.) and nonliving elements (soil, air, water, nutrients, dead and decaying organisms, etc.). Agricultural crops are especially efficient plants capturing and storing as much as 40 times more solar energy than uncultivated plants. Much of the energy that is not used for plant metabolism is used for the production of grain, vegetable, fruit, root, or other comestible products suitable for consumption. As consumers, we are in direct competition with the other herbivores, including the insect pests, for that portion of the crop energy or biomass. Because the relationship between stored energy and biomass is fairly constant, the biomass of herbivores, predators, and parasitoids in an uncultivated system has been estimated at about 10% and 1% of the plants, respectively. The objective of ecologically based IPM is to maintain the insect pest population equilibrium level low (Figure 8.1a) or to lower it (Figure 8.1d), keeping the energy of the system balanced. IPM in organic cropping systems should aim to create conditions, through ecological means, that are not conducive to pest population colonization, establishment, or growth. Tolerable levels of pests should be defined in relation to the ecological carrying capacity, which differs for every ecosystem. Each herbivorous insect occupies an equilibrium position within the system where it shares resources with the crop and other organisms. If the population of the herbivorous insect increases beyond its equilibrium position and uses more of the available resources than allocated within the stable system, the system is destabilized and the yield of the crop will decline if the herbivorous population is not brought under control. This very sensitive equilibrium is normally established after years of oscillations around the equilibrium point.

Insect management practices should help a farm maintain a high level of sustainable productivity. For example, the use of organic mulches and amendments to improve soil health reduces density-dependent competition for nutrients between individual crop plants and maximizes plant tolerance to insect pests. The availability of resources that the insect pest needs to survive in any given crop determines its carrying capacity. The carrying capacity is determined by the interaction of "bottom-up" factors such as food and shelter as well as "top-down" factors such as natural enemies (Figure 8.1a). However, although the organic agroecosystem is ecologically based, it is important to remember that it remains a managed ecosystem and not a natural system. Flint and van den Bosch (1981) compared the agroecosystem to the early stages of a natural succession in nature. The stability of the system will increase with the incorporation of permanent density-dependent factors to compensate for the low biodiversity of the early succession or of the monoculture crops (e.g., the permanent introduction of biological control agents or the modification of the physical environment).

There are methods to determine the point at which an insect pest can no longer be tolerated on the crop. In the literature on IPM, this is usually referred to as the economic injury level (EIL), the point at which the value of the lost yield begins to exceed the cost of controlling the pest. EILs provide an estimate of the pest density at which the cost of control is less than the loss the farmer would encounter if no control was enacted. Insect species with densities frequently above their equilibrium and close to (or above) the EIL are likely to be recurrent pests (Figure 8.1c), whereas those with densities normally well below the EIL will only be occasional pests (Figure 8.1b). Crops with a preponderance of the latter are good indicators of a diversified relatively stable agroecosystem. The use of EIL is a proven and very effective method to maximize yields and control crop protection cost in conventional agriculture. The EIL, or the economic threshold (ET), a predictor of the EIL, triggers the application of reactive control measures, whereas pest abundance or injury levels below the ET do not require intervention. However, successful suppression of pests depends on the availability of reliable curative methods. In organic agriculture, these are limited to control products and a few nontoxic biological, physical, and mechanical methods. More importantly, in organic agriculture, preventive management of pests must be given priority over reactive methods and especially insecticidal products. Reliance on preventive pest management is a requirement common to all organic crop production and protection guidelines and implies a management strategy directed at maintaining the insect pest population equilibrium level lower than that of the EIL to eliminate or minimize the need for the few authorized curative controls available. Insect populations whose density frequently reaches or exceeds the EIL within a growing season, and even more critically across seasons, indicate a need to invest more resources in permanent density-dependent factors that will lower the equilibrium point and ensure overall long-term stability for the system (Figure 8.1d). In conventional agriculture, the traditional role of the EIL is to ensure the best economic return for a given crop by determining when and how often curative controls need to be applied (e.g., Luckman and Metcalf 1994). In organic agriculture, the EIL can also serve as an effective tool to monitor the efficacy of occasional corrective actions, but it is a poor tool for monitoring the effectiveness of preventive insect management practices that maintain or rebuild equilibrium.

Organic agroecosystems where the suite of agronomic practices maintains the equilibrium position of an herbivorous insect close to the EIL are the ones most likely to require occasional curative control actions. Events outside the control of the particular agroecosystem, such as the removal of a similar host crop in the adjacent areas, could force the emigration of the herbivorous species to the organic site where its population density could suddenly exceed the EIL, making the herbivorous insect an occasional pest that requires curative control. Another example might be the case of cereal crops where the tolerance for insects is very low. The effort required to manage insect pests in storage is largely determined by the abundance of the pest in the crop. Even after careful management of insect pests on the crop itself, small numbers of insects can reside within the grain and be moved to storage. The response to such situations is too often to search for the best performing and/or less costly curative treatment rather than an examination of the ecosystem and how it could be redesigned for greater sustainability.

During the transition to the equilibrium, conventional IPM monitoring, trapping, and sampling techniques (e.g., Flint and van den Bosch 1981) will continue to provide reliable measures of an insect's population growth and the associated risk for the crop. Pheromone trapping, for example, can be used to better time the release of *Trichogramma ostriniae* (Pang and Chen) against European corn borer (ECB), *Ostrinia nubilalis* (Hübner), eggs. However, these methods need to be adapted to create tools that can measure the progress accomplished toward a stable system. Year after year, the difference between the insect pest population level and the EIL could be used as an indicator of the progress accomplished in moving toward an insect density equilibrium position and away from dependence on curative control methods. Identification and quantification of pest(s) and natural enemies and analysis of the agroecosystem and landscape structure should all be considered in the development of a pest prevention management plan. There likely is a need for new tools focused on assessing the contributions of the different prevention factors or control measures to the stability of the agroecosystems rather than a single focus on pest abundance or plant injury.

There are even more challenging cases, with crops such as carrots and Brassicas, where the EIL for cabbage root fly, *Delia radicum* (L.), damage is very low and essentially below the equilibrium level of the pest. Problems can be so serious that they may limit the economic viability of the organic crops (Litterick et al. 2002) because they challenge the ability of preventive measures to adequately manage such pests.

8.3   PEST MANAGEMENT STRATEGIES

This section reviews the benefits and limitations of the tools available to (1) prevent pests from finding and invading the crop, (2) establish a stable organic crop ecosystem capable of maintaining populations of pest insects at low equilibrium levels (well below the EIL), and (3) manage those occasional pest outbreaks that exceed the EIL.

The goal is to create a strong web of bottom-up and top-down control tactics, which, in order of priority, prevents (1) colonization, (2) establishment on the crop, and (3) population growth reserving (4) curative controls for emergencies.

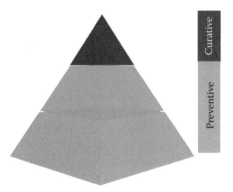

FIGURE 8.2   Pyramid of IPM in organic crop production.

It is not uncommon in IPM to logically organize control tactics into a multi-level, multicomponent pyramid with ecologically and biologically rational aspects of insect management at the base or foundation of the structure (Figure 8.2). Crop management, deep knowledge of pest biology and ecology, and area-wide management are the foundation blocks of sustainable IPM. The broader this base of tactics, the less reliant IPM is on intervention or curative tactics of the upper portions of this pyramid. However, due to the inherent complexity of the foundation level, this can be the most difficult set of practices to develop, research, and implement. The individual effects of many control tactics are weak, but their ecological interactions contribute significantly to the resilience of a preventive insect pest control program. The integration of multiple stresses and mortality factors, the "many little hammers" of Liebman and Gallandt (1997), helps prevent crop colonization and helps maintain population density at or below its equilibrium point if colonization occurs. To reinforce our emphasis on the importance of focusing the management strategy first on preventing colonization (host finding or accessing) (Section 8.3.1), then on preventing establishment (Section 8.3.2), and lastly on reducing abundance (Section 8.3.3), management methods are presented according to the role that they can play in one or more of the categories rather than in their own traditional sections as host plant resistance, biological control, semiochemicals, physical controls, etc.

8.3.1   Prevention of Crop Colonization

Preventing or reducing insect colonization of crops should be part of the core plan for the establishment or conversion to a stable organic agroecosystem. The objective is to reduce the likelihood of insect pest colonization to such a low level that most fields will remain insect pest free and that where they find or access the host, the agroecosystem will be able to tolerate or control rare colonization events. Colonization of the crop is prevented through farm location, crop isolation, crop rotation, erection of barriers, trap crops and modification of plant architecture, etc. (Letourneau and van Bruggen 2006). Although some of these preventive measures (e.g., crop rotation) are difficult to implement for perennial crops, sanitation, site selection, and the choice of appropriate cultivars are important measures to consider. They apply to the establishment of orchards and vineyards, for example. Orchards and vineyards planted on land that was previously planted to these same crops can also result in higher levels of trunk and root diseases (Westphal et al. 2002). These diseases can directly impact the health and survival of the trees and make them more prone to attack by wood-boring pests. Prompt removal of dead and dying trees and piles of wood, brush, or prunings that harbor wood-boring beetles will help reduce colonization of orchards by these pests. Biofumigation with mustard crops (Mazzola and Mullinix 2005) or a fallow period prior to planting can help reduce the incidence of disease. This approach can also prevent or reduce damage from such pests as white grubs (Noble and Sams 1999) and wireworms (Furlan et al. 2010) that develop over a period of more than 1 year and that can affect horticultural crops planted into land that was previously in pasture (Luginbill and Painter 1953).

8.3.1.1 Isolation of Farms and Fields

IPM is usually not the main consideration when the geographic location of the farm and the actual production sites are selected. Locating the farming site(s) in an area where crops can be isolated from sources of insect pest colonizers and rotated some distance away is an ideal foundation for an IPM plan. Unfortunately, other priorities or restrictions may prevent the farm from being located in areas least suitable to primary insect pests or from choosing a crop or variety that is less suitable to pest attack. Areas suitable for production of tender fruit crops and high-quality wine grapes occur in only a few locations in most countries, southern regions in Canada, for example. For this reason, these areas are intensively farmed and often have organic production in close proximity to larger acreages of conventional crops. In these situations, there will be even more pressure to develop and implement an array of agronomic and cultural practices to counteract the increased risk of insect pest colonization. In less intensively farmed regions, organic acreage for many vegetable and horticultural crops may represent a small percentage of the farm's acreage in pasture, giving farmers the flexibility to grow a given crop in field sections that are best suited agronomically and are isolated from insect pest sources. This has been the case with potato production, for example, where it is possible to select isolated sections of pasture fields that have good soil health, spatial and temporal distance from earlier potato crops, and landscape isolation (Boiteau 2008). The approach is often more inadvertent than by design and only possible as long as the organic industry remains at the seminal stage.

The role that isolation can play in protecting a crop from insect pests is both significant and well established. At the scale of continents, mountain ranges and large bodies of water continue to protect potato fields in large areas of China and in some European countries from invasive colonization by the Colorado potato beetle (CPB), *Leptinotarsa decemlineata* (Say) (Boiteau and Heikkila 2013). Seed potato production zones reserved for seed production isolated spatially from table stock and processing potato production areas and located at higher elevation have been very successful at protecting the crop from colonization by aphids that vector economically important plant virus diseases (Johnson 2008). Similarly, grapes produced in the Okanagan and Similkameen Valleys of south central British Columbia (BC) are isolated from other major production areas by mountain ranges to the east and west and unfavorable growing conditions to the north. To the south, in Washington State, grapes are grown in fragmented areas separated by regions unsuitable for commercial production due to climate or lack of irrigation. A relatively small but expanding acreage of organic wine and table grape production occurs in BC due to the low number of primary pest species that occurs there (Lowery 2010). Apples and pears produced in BC are likewise attacked by fewer major pests compared with most other production areas. For these pome fruit producers, a successful area-wide codling moth, *Cydia pomonella* (L.), program involving the use of pheromone mating disruption and releases of sterile insects has largely eliminated codling moth, a key pest of apple and pear, from many production areas of south central BC (Judd and Gardiner 2005). Growers of organic fruit crops have benefitted by establishing orchards in these valleys. A significant proportion of organic tree fruit production is located in western North America (NA) where important pests such as the plum cur-culio, *Conotrachelus nenuphar* (Herbst), and apple maggot, *Rhagoletis pomonella* (Walsh), are absent (Zehnder et al. 2007).

Isolation of organic crops also makes it possible to avoid invasion by outbreaks of secondary pests in nearby conventional crops. Recent studies conducted in BC (Lowery et al., unpublished) have demonstrated that certain commonly used fungi-cides and insecticides and the surfactant Sylgard™ (Dow Corning) used in conventional vineyards are repellent to adult leafhoppers. Applications of these pesticides can push these important pests of grapes to adjacent unsprayed areas or to neighboring vineyards.

On a different scale, shelterbelts can help protect crops from colonization by some insect pests.

There is evidence that organic farms incorporated into existing large natural eco-systems or arranged in geographical communities (such as seed production zones) not only offer direct protection against pest colonization but also provide habitat heterogeneity and biodiversity similar to or approaching that of natural systems (Rundlöf and Smith 2006; Gomiero et al. 2011). The counterpart is that small-size organic farms located among large-size conventional agricultural operations may not support high levels of heterogeneity and biodiversity.

Temporal isolation where planting time of the crop is selected to shorten the period of interaction with the pest can be very effective. For example, planting the potato crop so that it can be top-killed before peak aphid vector flights has been shown to reduce virus infection in seed potato in Canada and the United States (Johnson 2008). The development of resistance to pathogens as plants age, often termed age-related resistance, has been demonstrated for several plant/patho-gen interactions, such as the decreased susceptibility of wheat to wheat leaf rust (Kus et al. 2002). The susceptibility of rutabagas to turnip mosaic virus (TuMV) decreases considerably during the growing season (Lowery 1988); sowing early so that plants are resistant to infection when the peak of aphid flight activity occurs in late June is a recommended practice that will help reduce infection rates. As well, the crop should be isolated from fields of winter canola that serve as overwintering hosts for the disease.

8.3.1.2 Spatial and Temporal Rotation of Crops

Adequate break crop choice and rotation design may be the next (after farm site selection) most important IPM strategy in organic production. It protects against both incidence and severity of the pest infestation and is especially effective against insect pests with low mobility, a narrow host range, or those that live in the soil for part of their life cycle (Litterick et al. 2002; Younie and Litterick 2002).

According to invasion theory, organisms can be prevented from spreading into patchy islands when the distance between patches is large. Therefore, crop fields can be isolated from insect source pools such as overwintering sites by keeping large distances between fields with the same crop. Physically separating the location of the new crop from that of the previous crop can limit or prevent colonization. For example, the rotation of the current year potato crop away from the site of the previous year crop or from overwintering sites by 1.5 km decreases the probability of any CPB finding a host, while shorter distances will at least delay colonization (Weisz et al. 1996; Baker et al. 2001; Boiteau et al. 2008). Strips of natural vegetation between fields may further reduce the ability of the colonizing insects to find the host crop and perhaps reduce the required rotation distance.

Rotations create another form of temporal isolation between the insect pests and the susceptible crop that reinforces the impact of physical isolation. The 5–8 year periods of rotation between years of nutrient-demanding annual crops ensure soil health "regeneration" (e.g., Nelson et al. 2009, 2011) and also prevent the continued establishment of insect pest species. There are no data on this, but with annual crops, periodic absence of the "monoculture" host crop over a number of years might be as important as soil health regeneration in this preventive insect pest control strategy. There is evidence that eradication of remaining low-density insect pest populations during that period could be augmented and accelerated by choosing rotation crops with allelopathic properties (Formsgaard 2006). For example, wheat, maize, and white clover, often used as rotation crops, naturally produce biologically active compounds as part of their defense mechanisms against insects and other pests. This property could be exploited for the management of some insect pests by using these plants in rotations, as green manure, or by intercropping of allelopathic plants together with the crop. Allelopathy offers interesting potential in this regard, but more research will be required before it can be used systematically (Fomsgaard 2006; Keogh et al. 2009).

In some small-scale organic farms, multiple crops may be grown in small plots in the same area for consumers buying at the farm. These multiple crop areas should be rotated away over a period of years. If crops are rotated between the plots at the same site from year to year, the risk of some insect pest moving from the previous year plot to the new plots is high (Mohler and Johnson 2009). Unfortunately, in practice, the size of a majority of organic farms is insufficient to take full advantage of physically separated rotations to avoid pests. The land converted for organic production and organically certified is still limited.

8.3.1.3 Physical and Living Barriers

A number of devices and agronomic practices, including habitat manipulation, have been devised to prevent insect pests from reaching the crop (Vincent et al. 2003).

8.3.1.3.1   Physical Barriers

Barriers are structures made of materials or plants to obstruct or close passage of colonizing insects or to fence them in a space and thereby prevent them from reaching the crop (Boiteau and Vernon 2001). The structures modify the environment to interfere with the normal behavior of the target pest. For example, based on the knowledge that most overwintered CPBs walk to their spring hosts from overwintering sites, a significant reduction in numbers of beetles immigrating into potato fields was achieved by surrounding the fields with a "V"-shaped trench lined with plastic (Misener et al. 1993; Boiteau and Vernon 2001). Similarly, screen fencing applied around fields of Brassica crops prevented attack by cabbage root fly, as most females were shown to fly close to the ground in search of new hosts (Vernon and Mackenzie 1998). The use of polyethylene is usually permitted in organic farming, but some may consider it more polluting than certain insecticides. The effectiveness of these methods would be enhanced if pest pressure was reduced by rotation and isolation of susceptible crops. Physical barriers such as floating row covers can effectively control airborne insect pests such as aphids, but cost can be a significant issue and covers have to be in place before the pests enter the crop or they can increase the insect pest problem. Netting also interferes with the management of weeds, which in organic systems mostly involves hand weeding or cultivation. Netting can be used effectively when plants are small and most vulnerable to pest attack. The use of netting will allow protection for cucurbit plants when they are most vulnerable to the striped cucumber beetle (SCB), *Acalymma vittatum* (Fabricius). Delaying removal of row covers on muskmelons has been shown to minimize the impact of bacterial wilt, *Erwinia tracheiphila,* which is vectored by the cucumber beetle (Rojas et al. 2011).

Producers of organic grapes in western NA frequently string yellow sticky tape, which is particularly attractive to leafhoppers in spring, below the vine cordon (Lowery 2010). Numbers can be reduced by as much as 95% when the tape is applied to every row. In apple and pear orchards, strips of cardboard can be wrapped around the trunks of trees in summer to provide an attractive location for mature codling moth larvae moving down the trees to create their overwintering hibernacula (Judd and Gardiner 2005). The cardboard is removed and burned during the winter before larvae complete their development in spring. Unfortunately, as is the case with many physical barriers, the significant cost in materials and labor limits the use of these techniques mostly to vineyard or orchard edges and areas with high pest pressure.

Mineral oils have been widely used as both dormant and foliar sprays on a wide range of crops, both to prevent aphid transmission of nonpersistent viruses (Boiteau et al. 2009) and as physical controls for soft-bodied arthropods (e.g., mites, thrips, immature scale). Particle films, such as the kaolin-based commercial material Surround™, applied to crops prevent colonization of pome fruits and grapes by a wide array of arthropod pests through a combination of several mechanisms, such as interference with visual cues and physical interference (Vincent et al. 2003). Kaolin can also be applied to cucurbit transplants to interfere with the ability of the SCB, to recognize its host plant. Kaolin is thought to disrupt the insects' host-finding abilities by changing visual, tactile, gustative, or olfactory cues of the host plant (Puterka et al. 2005). In all cases, care must be exerted to ensure that all foliar spray materials used for the management of pests are acceptable for use under the certification guidelines.

8.3.1.3.2   Living Barriers

In a manner similar to physical barriers, plant stands grown at the perimeter of the main crop can act as trap crops or living barriers (Hokkanen 1991). Trap cropping fits within the ecological framework of habitat manipulation for the purpose of pest management, usually as one component of an IPM program (Shelton and Badenes-Perez 2006). Trap crops, which may be different species or preferred varieties of the same crop, are grown for the purpose of attracting insects or other organisms and protecting the target crop by preventing colonization or concentrating pests in certain parts of the field where they can be more economically destroyed while also causing less disruption to the ecosystem. One of the most widely cited examples of successful trap cropping that became widely used in the central valley of California in the 1960s is the use of alfalfa strips to prevent western tarnished plant bug, *Lygus hesperus* (Knight), damage to cotton (Shelton and Badenes-Perez 2006). Alfalfa trap crops can also be effective against *L. hesperus* in strawberry fields when used in combination with tractor-mounted vacuum collectors (Swezey et al. 2007). In a similar way, certain cucurbit varieties such as the blue hubbard squash can be used as trap crops for the SCB (Cavanagh et al. 2009).

Potato, being the preferred host of CPB, can be used as a trap crop to protect plantings of tomato and eggplant, which are less preferred hosts of the same plant family (Solanaceae). Hunt and Whitfield (1996) showed that rows of potato planted 16 beds apart in fields of tomato decreased beetle numbers on tomato and increased yields from 61% to 87% compared with plots that did not have the potato trap crop. In spite of these successes, Shelton and Badenes-Perez (2006) state that there have been only 10 instances of successful trap cropping at the commercial farm level. Limitations to expanded adoption include the need to dedicate land to production of the trap crop. This would be of less concern for small-scale organic farms practicing polyculture if, as in the previous examples, the trap crop was also marketable or required no additional space. Trap cropping is also knowledge intensive and there is little doubt that efforts to better understand the biology and host plant relationships of certain key pests would contribute to the development of effective trap cropping systems for other crops (Hokkanen 1991).

8.3.1.3.2.1   Case Study: Trap Crop   Two species of flea beetle, *Phyllotreta striolata* (Fabricius) and *Phyllotreta cruciferae* (Goeze), cause considerable damage to Brassica crops in Canada when adults attack seedlings or young transplants. In Ontario, southern giant curled mustard, *Brassica juncea*, also known as Indian longstem mustard, planted adjacent to broccoli and cauliflower provided almost complete protection against beetle attack (Lowery, unpublished). This information was rapidly adopted by some growers of traditional Brassica crops who plant the mustard in widely spaced drive rows to protect the main crop. Spraying the mustard is often not

necessary as it is more tolerant to flea beetle damage and hosts a large number of predators, mostly spiders, in the frilly leaf margins. Some seed catalogs now advertise southern giant curled mustard for this purpose.

8.3.2 LIMITING PEST ESTABLISHMENT

Assuming that some insect pests will occasionally find the crop, it is important to monitor for their presence and abundance. Because most insect pests are r-strategists that can rapidly increase their populations to outbreak levels that can be uncontrollable by naturally regulating factors, it is essential to design the crop system so that it will maintain the overall insect population level low from the moment of colonization (Southwood and Way 1970).

In a newly colonized crop, the number of insects invading a crop largely determines if the colonizers are likely to successfully establish the potential population size. Initially, birth and immigration are much higher than death and emigration. As the population size increases, competition for resources between individuals causes the growth rate to decrease. At some point, the growth rate approaches zero and the population reaches the carrying capacity of the environment. The crop has a specific carrying capacity and when the insect pest population density increases beyond it, the limited resources will cause a drop in abundance or crop yield (Bird et al. 2009). Hopefully, colonization occurs outside the critical time period when insect pests can most severely reduce yield or quality of the crop. Colonization inside the critical window would indicate that production management did not succeed at keeping the insect pests below damaging levels and would force a reassessment of the production practices. The possible establishment of a significant population of the pest(s) leading to a potentially large overwintering population that could threaten next year's crop may force a reconsideration of next year's crop location or the use of curative control methods against this year's pest population. Chemical control should be a last resort.

Crops new to an area may be especially at risk. The landscape in which the crop is initially introduced is itself disrupted. The insect pests locating this crop and trying to establish in the farmed area of that landscape may be new to a large proportion of the beneficial organisms already established in the disrupted area or nearby. IPM during the transition phase may therefore require more intervention. Some insect pests may also be highly specialized and prone to rapid cycles of highs and lows that are difficult to manage regardless of the production system. For each crop, there is a need to work at decreasing the degree to which the system meets or satisfies the insect pest's resource requirements while increasing the degree to which natural controls are functioning in the management system. This will largely determine the number of years required before the production system has reached long-term stability and can resist occasional attempts at colonization or can tolerate low pest densities.

8.3.2.1 Maintaining a Healthy Crop

The quality of the soil in which the crop is grown plays a role in determining whether or not the colonizing insects will establish and affect the development of the crop.

There is evidence that healthy plants living in soil with balanced nutrition can tolerate a relatively high abundance of insects. Under ideal conditions, this might be sufficient to prevent significant insect pest problems in organic systems according to some studies (e.g., Litterick et al. 2002). However, reports of severe insect pest damage are not uncommon, even in the presence of healthy soils, especially as acreages of single crops increase (Collier et al. 2001).

Phelan et al. (1996) developed the mineral balance hypothesis according to which organic matter and microbial activity associated with organically managed soils enhance the nutrient balance in plants. In turn, the healthy plants have gained increased tolerance to insect pests. The hypothesis has been developed around the ECB (Phelan et al. 1995, 1996; Keogh et al. 2009); the bean fly, *Ophiomyia phaseoli* (Tryon) (Letourneau and Msuku 1992, Letourneau 1994); the cabbage white butterfly, *Pieris rapae* (L.) (Hsu et al. 2009); and the CPB (Alyokhin et al. 2005). There is now evidence that the protection provided by these healthy soils will vary considerably between different complexes of insects and crops (e.g., Boiteau et al. 2008; Staley et al. 2010; Stafford et al. 2012), but it remains one factor to consider for integration to the others.

It is also suggested that the nitrogen concentrations in foliar tissues or phloem of organic crops and natural plants reach levels that are insufficient or unfavorable to outbreaks of aphids and whiteflies (e.g., van Bruggen 1995). Although crop N levels do tend to be low in organic fields as compared with conventional ones, a link between it and pest levels or crop damage remains difficult to confirm (e.g., Letourneau et al. 1996). Excessive vigor resulting in denser canopies is known to favor outbreaks of leafhoppers and mealybugs on grapevines through improved nutrition and a more sheltered environment (Neilsen et al. 2009). At the other extreme, vines lacking sufficient vigor are prone to attack from hard-scale and wood-boring beetles. Related to nutrient balance and plant health, a modest irrigation deficit from berry set to veraison reduced leafhopper populations on wine grapes more than 60% with only a modest reduction in yield (Dry et al. 2001).

Although it is unquestionable that nutrient sources can be manipulated to produce mineral balances that reduce the suitability of some crop plants for at least some insect pests, the relative importance of the effect among the many factors affecting insect population establishment remains poorly quantified and therefore difficult to integrate into a protection management plan. Also, it should not be forgotten that all changes to the plant's health to control insect pests can have secondary effects on yield or attractiveness to other pests including susceptibility to diseases. This is outside the scope of this chapter, but an IPM plan can never be completed without taking into consideration possible effects on other aspects of crop production.

The application of appropriate amendments, the use of sound tillage practices, and long rotation periods between crops (e.g., Nelson et al. 2009) to maintain a healthy soil will, at the very least, help sustain high levels of beneficial organisms contributing to the overall biodiversity of the crop system (Litterick et al. 2002). Unfortunately, the gains have to be balanced against the knowledge that rich enhanced soils can also encourage the buildup of populations of symphylans, cutworms, wireworms, and other pests such as the strawberry root weevil, *Otiorhynchus ovatus* (L.), and the *Lygus* bug (Letourneau and van Bruggen 2006).

8.3.2.2 Utilization of Insect-Resistant Plants

Although healthy plants may benefit from a level of tolerance to colonization, growing plant varieties with complete or even partial resistance to pest insects is probably the ideal method for preventing the establishment of colonizing (invading) insect pests. On the positive side, resistant varieties of crops tend to have been bred against primary insect pests, the ones presenting a high risk of invasion, because of the substantial investment required to develop these varieties.

On the other hand, it is not uncommon for these varieties to lack interesting flavors, colors, or culinary values in demand by a range of market segments. This is particularly true for perennial horticultural crops where enhanced flavor and appearance are of particular importance to consumers. Hybrid grape varieties are available that resist colonization by leafhoppers and other pests, but wine produced from these varieties is generally considered to be of inferior quality and of lower value. Considerable effort has been devoted to the development of grapevine rootstocks that possess resistance to grape phylloxera, *Dactylosphaera vitifolii* (Fitch), and plant parasitic nematodes, as this effort does not alter the quality of the fruit and offers other agronomic values, such as vigor control.

The value of many organic crops is not high enough to justify the development of insect-resistant varieties dedicated to the requirements of organic agroecosystems. Unfortunately, the production of varieties particularly suited to organic production systems continues to lag (Jahn 2003). Some resistance features such as leaf toughness have broad activity against many herbivorous insects feeding on many crops. For example, a comparison of hybrid wine grape varieties with and without dense hairs on the undersides of leaves showed that approximately tenfold fewer eggs were deposited by the Virginia creeper leafhopper, *Erythroneura ziczac* (Walsh), on the hairy, thick-leaved varieties compared with those having smooth leaf surfaces (McKenzie 1973). However, breeding of grape varieties resistant to leafhoppers based on leaf thickness and density of leaf hairs has not been pursued due to societal preference for wines produced from noble varieties (*V. vinifera*) and the availability of effective insecticides to control leafhoppers in conventional vineyards. The potential for the development of resistant varieties on the basis of these resistance traits exists, especially for the production of organic table and juice grapes, as these currently consist mostly of species hybrids. The development of varieties resistant to insect pests is quite complex for the organic sector where the third trophic level (predators and parasitoids) plays a major role in maintaining insect pest populations around the equilibrium point. For example, although the development of hairy varieties can reduce damage levels by some pests (e.g., whiteflies), it can also have a negative impact on the searching ability of parasitoids such as *Encarsia formosa* (Gahan) (Hua et al. 1987).

Although the use of plant varieties whose resistance is based on high levels of defensive chemicals can provide protection against many species of herbivorous insects, these compounds will not prevent feeding by specialists that are adapted to these compounds. The leaves of potato and other solanaceous crops that contain high concentrations of alkaloids and glycosides are not toxic to the CPB, for example.

Low levels of resistance based on multiple genes tend to be preferable to high resistance levels because of the associated yield stability even if a limited level of feeding occurs. It simply means that plant resistance alone will not be enough and will have to be integrated with other methods (Vaarst et al. 2003) including the level of insect resistance provided to the plant by the healthy soils. Zehnder et al. (2007) have made an interesting case for the advantages of partial host plant resistance to insect pests in organic production. As indicated earlier, economics have and continue to limit the screening effort possible across the range of crops that could benefit from it.

8.3.2.3   Biodiversity and Conservation Biological Control

8.3.2.3.1   Background

The conservation and augmentation of biodiversity, especially predators and parasitoids, can play a key role in preventing the early establishment of colonizing insect pests and subsequent population increases. Biodiversity refers to the number, variety, and variability of living organisms in a given environment. It includes diversity within species, between species, and among ecosystems. Biodiversity is one of the resilience factors that contribute to the maintenance of insect pest densities at an acceptable equilibrium position. Biodiversity of flora and fauna in organic farms should not be taken as a given but rather as something to develop. Biodiversity levels vary widely within and between the different cropping systems; for recent reviews of research on the subject (largely comparative between organic and conventional), the reader is referred to Letourneau and van Bruggen (2006) and Gomiero et al. (2011). Most studies have compared the fauna of organic and conventional farms and discussed the varied and sometimes contradictory results.

It is logical to assume that natural ecosystems with high biodiversity are less likely to be overtaken or substantially damaged by invading insects than ecosystems with low biodiversity. A crop system with high biodiversity might be presumed to also be at lower risk of successful colonization by an insect pest than a crop system with a low biodiversity. For example, the natural enemy complex has been shown to maintain many insect pests, such as spider mites in orchards, at low population levels (Letourneau and van Bruggen 2006).

The successful establishment and subsequent population growth of the colonizing pest should therefore be less likely in a crop system rich with predators, parasitoids, etc. However, there has been such a focus on attempts to demonstrate an expected pattern of higher abundance and greater diversity of beneficial arthropods on organic crops than on conventional crops that relatively little attention has been paid to the actual quantification of the impact that they have on the abundance of insect pests. The only constant from these comparative studies seems to be the great diversity of species in organic crops, which is usually attributed to the diversity of plants within the cropped area and in surrounding habitats (Booij and Noorlander 1992; Holland and Fahrig 2000; Asteraki et al. 2004).

Biodiversity by itself is not enough to alleviate insect pest problems. It is necessary to build an environment that is favorable to the natural enemies and unfavorable to the insect pests. In an evaluation of 76 separate studies, Hole et al. (2005) concluded that species diversity and abundance across a wide range of taxa tended to be higher on organic farms than on locally representative conventional farms. Few of these studies, however, had assessed the effects of increased plant biodiversity, mostly attained through a greater abundance of weeds, with numbers of pest invertebrates. As a secondary benefit to society and the environment, organic farms contributed to greater biodiversity of birds, butterflies, and mammals, potentially helping to preserve many threatened and endangered species.

The only conclusion possible so far has been that management aimed at maintaining or enhancing landscape diversity does over time create systems that are more stable and resilient and have the potential to stabilize or increase biological control. Many studies have confirmed that organic farms have a diverse arthropod fauna including an abundance of natural parasitoids and predators (e.g., Letourneau and Goldstein 2001; Hole et al. 2005). The contribution of this fauna is not limited to the cultivated field. Studies have shown that biodiversity at the landscape level has an impact on insect pest population dynamics at the farm level (Burel 1992; Polis et al. 1997). Moreover, noncultivated areas may act as reservoirs for beneficial insect populations, predators, and parasitoids, which in turn might be effective at the landscape level and consequently on parasitism of pest insects. In Germany, Thies and Tscharntke (1999) have studied the influence of landscape structure on the management of a major rapeseed pest, *Meligethes aeneus* (Fabricius) (Coleoptera: Nitidulidae). They found that in a complex landscape, parasitism was higher and damage was lower than in a more simple landscape where more intensive agriculture was practiced. According to that study, when the proportion of noncultivated land falls below 20% in a given territory, parasitism falls below 36%, a level insufficient to naturally regulate a pest population.

Habitat fragmentation affects the distribution and survival of both pests and natural enemies (Heinen et al. 1998). Spatial heterogeneity management will favor conservation of refuges for predator and parasitoid populations that could be effective at the landscape level (Ferro and McNeil 1998). Effective conservation of natural parasitoids often requires the preservation and/or restoration of late successional habitats within the agricultural landscape (Marion et al. 2006).

It is important to keep in mind when developing these conservation strategies that more biodiversity does not always imply better pest control. It is also necessary to establish the right kind of biodiversity for the agroecosystem. Nonpest prey becomes more abundant in habitats with high biodiversity, and as a result, predators may actually end up consuming fewer pests. For instance, compost has been shown to increase the number of alternate prey and reduce predation on codling moth and fruit fly pupae by carabid beetles (Mathews et al. 2004; Renkema et al. 2012).

8.3.2.3.2 Habitat Management

Crop environments rarely provide sufficient resources for natural enemies, and their populations are further harmed by frequent disturbances in the form of pesticide sprays, crop harvesting, tillage, and weed management. Manipulation of the environment to enhance the effectiveness of natural enemies by improving their survival, fecundity, and longevity or ability to locate suitable hosts is often necessary (Landis et al. 2000). Increased plant diversity within the crop (e.g., polyculture), within the farm, or at the landscape level can promote the activity of natural enemies by various means, including provision of nectar, pollen, and alternative prey and host species (Bostanian et al. 2004; Bickerton and Hamilton 2012). Understanding the biology of pests and their biocontrol agents and their relationships to specific plants is fundamental to the success of conservation biological control. Small-scale factors such as the intrafield plant community may play a critical role in determining interactions among different trophic levels such as aphids, parasitoids, and predators in cereals (Caballero-Lopez et al. 2012). It is not enough to simply increase plant diversity, as this can at times exacerbate pest problems and create weed problems (Landis et al. 2000).

From the standpoint of invasion theory, it could be argued that control tactics most useful at preventing establishment will be those that are effective at low densities (Liebhold and Tobin 2008). Generalist predators, for example, should have the most important role in preventing successful establishment by early colonizers because parasitoids and other specialized biocontrol agents present in the diverse landscape will not have time to build up the abundance required to control the colonizing insect pest species unless colonization occurs over a long period of time or at very low densities. Even then, the role of the generalist biological control agents will vary substantially between insect pests and crops. Generalist predators may have a negative impact on the ability of aphids, for example, to successfully establish on a crop such as potato (Tamaki 1981; Boiteau 1986) but are likely to have less impact on colonizing CPB. In the latter case, the unpalatability of the insect acts as a deterrent to predators even when they are present (Boiteau and McCarthy 2010). In the first case, the size of the "propagule" is the likely determinant of the success of the colonization event (Liebhold and Tobin 2008). The number of insects invading a crop largely determines the potential population size. This is where a thorough knowledge of the agroecosystem may be required to ensure that the agricultural practices do not hamper the natural processes (e.g., cutting or harvesting one crop harboring a pest and forcing it to move to a neighboring host crop thereby increasing the pest pressure on that field) but assisting it (e.g., mowing a crop thereby killing most of the aphid population and forcing the Coccinellidae over to neighboring crops with low-density aphid populations that were not attracting and retaining the predators).

Various methods can be utilized to make the habitat more favorable to beneficial insects (Gurr et al. 2012). Provision of shelter and increased complexity of plant architecture can often enhance numbers of beneficial insects and mites (Landis et al. 2000). Unsprayed or undisturbed areas within or adjacent to crops (refugia) provide a broad form of shelter for predators and parasitoids, but individual species can sometimes benefit from the physical structure of certain plants. Studies in the United Kingdom have shown that strips of uncultivated land, termed beetle banks for their support of predacious beetle populations, in fields of cereals increased predator numbers. Recommendations are to plant these strips with cocksfoot, *Dactylis glomerata*, or Yorkshire fog, *Holcus lanatus*, as these mat-forming species harbored the greatest number of predators (Landis et al. 2000).

Conditions within the crop itself can sometimes be made more favorable for ground-dwelling predators by the addition of leaf litter or mulch, through reduced or no-till practices, or by planting low-growing companion plants that are not competitive with the crop. A significant increase in numbers of spiders was achieved by providing 10–12 cm deep holes in the soil surface (Alderweireldt 1994), and shelters provided in the notches of fruit tree branches can provide housing for spiders and earwigs. Wood shavings of the type used as packaging material are most often used for this purpose. In addition to shelter, organic growers are often advised to grow plants that provide pollen or to supply homemade or commercial mixtures consisting of combinations of yeast, sugar, and, in some formulations, powdered milk diluted in water to attract and retain beneficial insects (Dufour 2002).

Noncrop vegetation within and near the crop serves to increase faunal diversity by providing alternate prey or hosts, pollen, nectar, and habitat. Fields managed with cover crops or annual weeds between crop seasons make it easier for arthropods to establish residence. Letourneau and van Bruggen (2006) have suggested that such practices may perennialize the crop habitat to allow continuity of populations of natural enemies throughout the year. Increased biodiversity augments the potential for ecosystem services to growers. So-called beetle banks consisting of boundary vegetation or grass strips will help enhance populations of natural parasitoids and predators. Intercropping with two or more crop types has been successful with some pests and some crops (e.g., Younie and Litterick 2002).

Shelterbelts and, in particular, riparian areas contribute to the management of leafhoppers on grapes and tree fruits by providing alternative hosts for mymarid (Hymenoptera: Mymaridae) egg parasitoids. The most important biological control agents for the management of leafhoppers on grapes are species of *Anagrus* egg parasitoids that require alternative leafhopper egg hosts in which to spend the winter (Lowery et al. 2007). Alternative hosts are also important in early spring when few eggs of those leafhopper species that specialize on grapes are available. In the case of *Anagrus erythroneura* (Triapitzyn and Chiappini) that parasitizes eggs of the western grape leafhopper, *Erythroneura elegantula* (Osborn), an important overwintering host are eggs of the rose leafhopper, *Edwardsiana rosae* (L.), that are deposited in fall in the stems of roses, blackberry, and related plant species. In a similar manner, parasitism of two leafroller pests of apples in western NA, *Pandemis pyrusana* (Kearfott) and *Choristoneura rosaceana* (Harris), by the introduced parasitoid *Colpoclypeus florus* (Walker) was greatly enhanced by proximity to plantings of rose and strawberry that harbored the strawberry leafroller, *Ancylis comptana* (Frolich) (Pfannenstiel and Unruh 2003). The strawberry leafroller, also an introduced species, is an important overwintering host that also helps bridge the gap between the two generations of the pest species occurring on apple. Provision of appropriate noncrop plants or location of organic vineyards and orchards near riparian areas will enhance the activity of both these beneficial insects.

8.3.2.3.2.1 Case Study: Benefits of Increased Plant Biodiversity Damage to the buds of grapevines by larvae of climbing cutworm in one particular BC vineyard regularly exceeded 20%. Introduction of white clover mixtures with grass in the drive rows and encouragement of winter annual mustards *Draba verna* and shepherd's purse, *Capsella bursa-pastoris*, in the vine rows in spring have reduced the damage to nearly zero in the past 2 years. The presence of alternate food sources in spring explains the low rates of bud damage (Mostafa et al. 2011), and because they die off very early in the season, these plants do not compete with the grapevines for nutrients or water. The results obtained for these major pests of grapevines in BC demonstrate the benefits of increased targeted plant diversity (Lowery 2010).

The balancing and managing of cropped and noncropped areas, crop species, and variety selection across time and space reviewed previously provided top-down control. This is sometimes referred to as farmscaping (Litterick et al. 2002). The concept can be broadened to include the manipulation of agricultural habitats so that they are not only more attractive to beneficial insects and other natural enemies but also less favorable for insect pests and provide bottom-up control. This way, farmscaping becomes a form of ecological engineering for pest management (Zehnder et al. 2007; Lacey and Shapiro-Ilan 2008; Gurr et al. 2012). Landscape-scale research on the importance of vegetation management and its use (e.g., a "wide-area view" sensu; Rabb 1978) to discourage insect pests and favor beneficials has been stimulated by new geographical information system capabilities (e.g., Marino and Landis 1996; Letourneau and Goldstein 2001; Thies et al. 2003), metapopulation dynamics models, and new insect-tracking methods (e.g., Gui et al. 2012). This form of ecological engineering (Lacey and Shapiro-Ilan 2008) corresponds to the "resource concentration" and "enemies" hypotheses of Root (1973) according to which insect pests are less likely to find their host plant and natural enemies should be more abundant and more effective due to the complexity of the system. It explains, for example, why insect pests tend to be less abundant in polycultures than in monocultures. Koch et al. (2012) provide a recent example of a study where an autumn-seeded winter rye, *Secale cereal* L., used as a cover crop in soybean, *Glycine max* (L.) Merr., significantly decreased insect pest pressure. Although there are many examples in conventional agriculture, this study showed that cover-cropping strategies can effectively suppress insect pests in organically managed crops (soybeans).

Practices can be as simple as using cover crops that provide supplemental food for predators and parasitoids (Olmstead et al. 2001) and also raise humidity levels and decrease light intensity at ground level compared with bare soil, which will help enhance infectivity or extend the survival of naturally occurring entomopathogenic microorganisms. Mulching and irrigation have been shown to enhance the activity of insect parasitic nematodes by improving larvicidal activity and persistence (Lacey and Shapiro-Ilan 2008). Although not without potential negative consequences, mulching increased the diversity of ground-dwelling arthropods in orchards and increased the numbers of predators. With potato and eggplant, mulch may decrease the likelihood of CPB finding the plants and increase predation (Brust 1994; Stoner 1997). Mulches can also be used to help control weeds and improve tree vigor and performance (Neilsen et al. 2009). Similarly, manure or straw mulch applied to the soil increased numbers of the carabid beetle *Bembidion lampros* (Herbst) that feeds on eggs of the cabbage root fly (Landis et al. 2000).

Management of vegetation surrounding different crop fields must be done very judiciously and take into account long-term rotation plans. Engineering the agroecosystem so that it is both favorable to the beneficials and detrimental to pests is often challenging. Plants that act as alternate hosts of crop pests (e.g., weeds as hosts of flea beetles around potato fields [Letourneau et al. 2006]; poplar trees that harbor lettuce root aphids [Phillips et al. 1999]) can be removed, but it should be remembered that noncrop vegetation also hosts natural control agents of insect pests. There is a fine balance between measures that will encourage natural enemies while at the same time not encourage insect pests. The challenge with each crop is to select plant species that are effective at reducing insect numbers while not competing for resources with the crop plant. Hence, work is needed to select background plant species that will both reduce pest insect numbers and cause the least reduction in yield to the harvestable crop (Collier et al. 2001). A thorough knowledge of the biology and ecology of both pests and beneficials is required to reach the desired equilibrium.

Genetic and phenotypic heterogeneity is a key characteristic of natural plant populations (Letourneau and van Bruggen 2006). Mixed cropping shares some of the same characteristics and helps reduce the concentration of the pests' hosts, thereby lowering the probability that pests will locate their host and increasing the probability that they will leave them. Crop mixtures could also suppress herbivores through the indirect effects of plant quality and emission of volatiles (Bukovinszky et al. 2004). This is largely based on theory and there are few fully quantified examples available. However, as cited in Letourneau and van Bruggen (2006), reviews of the literature on pest population densities in mixed cropping by Andow (1983, 1991) showed that 56% of herbivores have lower population densities in polycultures than in monocultures. Unfortunately, the searching efficiency of biological control agents may also be reduced (Bukovinszky et al. 2004) by the same factors (Letourneau and Van Bruggen 2006). Regardless, according to a review by Tooker and Frank (2012), replacing uniform plantings of a single crop variety by multiple varieties is poised to become a popular sustainable pest management strategy.

Numbers of pest insects can be considerably reduced by letting the crop become weedy, intercropping with another plant species, or undersowing with living mulch (Collier et al. 2001). Companion planting should provide additional protection against insect pests but its efficacy remains poorly documented (Collier et al. 2001; Moreau et al. 2006; Finch and Collier 2012). Intercropping is not a widespread practice in NA (Moreau et al. 2006).

The diversified landscapes of organic farming operations can only help conserve biodiversity and can perform an insect pest control function. There is no doubt that practices promoting biodiversity above- and belowground will favor natural enemies and enhance the resilience of the crop system to insect pests, but the relative roles of the different practices and their joint impact remain to be quantified for most crop systems. Our lack of knowledge of the functional response of beneficial arthropods to increasing numbers of insect pests and increasing organic crop acreages is likely a key reason for the low reliance on conservation biological control in spite of the attention given to the subject.

8.3.2.4 Disruption by Semiochemicals

Semiochemicals are compounds emitted by plants or animals that evoke a behavioral or physiological response in another organism. Research to develop semiochemicals that will prevent pest insects from finding their host crop has been largely unsuccessful. However, the use of female sex pheromone mimics is widely used and very effective for reducing the establishment of lepidopteran pests in a number of crops. Since 1992, a sterile insect release (SIR) program has been conducted in the Okanagan and Similkameen Valleys of BC for the management of codling moth, a key pest of apple and pear (Neilsen et al. 2009). The SIR program is the cornerstone of a successful area-wide integrated IPM program that also includes pheromone-based mating disruption of codling moth (Judd and Gardiner 2005). Mating disruption is more effective when used in an area-wide program, which might limit its effectiveness in small fields. Decreased effectiveness in small orchards, for example, is thought to be due to immigration of mated females from outside the treated zone. In addition to codling moth, a multispecies pheromone lure has been developed that provides control of several species of leafroller (Lepidoptera: Tortricidae) pests found on apple (Judd and Gardiner 2005). Mating disruption of lepidopteran pests of apples and pears has been widely adopted in BC and the northwestern United States partly because of the small number of major pests. In contrast, pheromones commercially available for the management of grape-berry moth are not used extensively in eastern NA due to their higher cost relative to insecticide sprays and presence of other pests that require chemical control. For a similar reason, commercial pheromone lures are effective for the control of several lepidopteran pests of vegetable crops (e.g., cabbage looper, *Trichoplusia ni* (Hübner), diamondback moth, *Plutella xylostella* (L.)), but they are not widely used.

Many predators and parasitoids use the kairomones emitted by their prey or hosts to locate them. Research conducted in hops and grapes has shown increases in numbers of predators of several families following application of herbivore-induced plant volatiles, such as methyl salicylate and (Z)-3-hexenyl acetate (James 2005). Some attractants and lures that can be used to bring predators into crops preventively are already available commercially (e.g., PredaLure™), although their efficacy requires further documentation. Additional research into methods to enhance numbers of predators and parasitoids would be of particular benefit to organic producers because of their reduced dependence on insecticides and greater reliance on biological control.

8.3.2.5 Physical Alteration of Crops

Physically altering the crop can be an active method of physical insect pest prevention (Vincent et al. 2003). Crop/plant architecture modification is often pursued for purposes other than insect control, but studies have shown that it can have a valuable impact on preventing colonization or, more often, reducing successful establishment by colonizers. For example, removal of the lower leaves of grapevines around the fruiting cluster in summer to improve fruit color and quality can help reduce colonization by overwintered leafhoppers. Removing the leaves in June after leafhoppers had mostly completed oviposition rather than in August was even more effective for the control of fruit diseases (Sholberg et al. 2008) and also reduced leafhopper populations by approximately 70% (Lowery 2010). Research to gain a better understanding of the impact of plant architecture on colonization by insects on a larger range of crops could lead to interesting applications (Simon et al. 2012).

8.3.3 CONTROL OF INSECT PESTS FOLLOWING COLONIZATION

In spite of all the efforts made to prevent colonization or successful establishment on a crop, a proportion of the dispersing insects will circumvent all barriers, and the insect density will rise above its normal abundance, and the economy of the farm or the quantity or quality of the crop will be substantially affected. Curative control measures will then be required to re-establish equilibrium. Control measures may also be required where pest abundance remains close to the EIL regardless of preventive measures and pest outbreaks are frequent. In these instances, the preventive methods described in the previous sections should be better integrated into the failing strategy. The requirement for curative controls should be occasional, and repeated use would suggest that the control strategy has drifted from a prevention-based one to a curative-based one (Figure 8.3). Consideration should be given to a redesign of the protection and production system in which the insect pests are below equilibrium density.

Curative control measures include inundative releases of biological control agents, use of insecticides, and mechanical removal of pests. Within the different types of

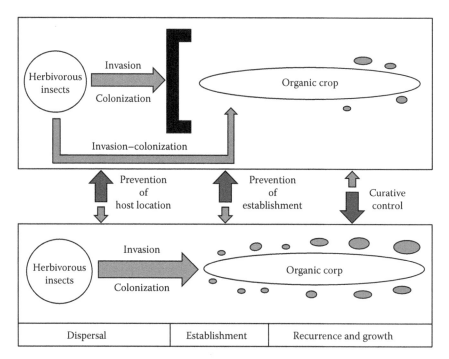

FIGURE 8.3 Diagrammatic comparison of an IPM strategy based on curative control (lower box) and on preventive methods (upper box). In a preventive approach, host crop finding by dispersing insect pests can be largely prevented by paying attention to the location of farms and crops, managing the landscape and the crop, and introducing physical or living barriers. The establishment on the crop by colonizers that successfully dispersed to the crop can be reduced by maintaining high soil quality, healthy plants, conserving biodiversity and biological control agents, as well as using host plant resistance. Occasional insect outbreaks may require inundative biological control or insecticides, but repeated use would suggest that the control strategy has drifted from a prevention-based one to a curative-based one. Consideration should be given to a redesign of the protection and production system in which the insect pests are below equilibrium density.

insect control practices available to organic producers, curative control methods most often challenge the limits of eligibility set by the various certification bodies. Propane flamers for the control of herbivorous pests such as the CPB, for example, are usually permitted in organic farming (more frequently as weeders), but reliance on petroleum fuels can be considered more polluting and energetically more costly than some insecticides (Wu and Sardo 2010). In the case of biological insecticides, their effectiveness and simplicity of use make them very attractive, and as a result, discussions often arise as to their suitability for inclusion in an ecologically based protection system. It has also been argued that organic insecticides do not necessarily mitigate environmental risks (e.g., Bahlai et al. 2010).

8.3.3.1 Inundative Releases of Biological Control Agents

In spite of the frequent reference to biological control as a mainstay of organic farming, surveys indicate that biological control agents are rarely utilized on organic farms (Langer 1995). This is likely because the introduction of new biocontrol agents to fight field pests has often not been as effective as expected, although some have demonstrated potential (Cloutier et al. 2002). This is the case with *Podisus maculiventris* (Say) and *Perillus bioculatus* (F.), two species of stinkbugs that are effective predators of CPB larvae but are not available in sufficient numbers for inundative releases. Inundative releases have become a method of choice in greenhouse vegetable production where confinement of the biocontrol agents ensures higher efficacy at reduced costs (Van Lenteren and Woets 1988).

Attempts to maintain populations of certain introduced pests below the EIL can be difficult, often due to the absence of specific biological control agents. In these situations, a better long-term approach might include studies to determine the most important predators and parasitoids that help maintain the pest at low population levels in its native area. Classical biological control, the discovery, importation, and establishment of exotic natural enemies, has been utilized successfully to reduce numbers of several important pests of horticultural crops (e.g., larval parasitoids *Cotesia glomerata* and *Cotesia rubecula* from Europe/Asia to control cabbage white butterfly on Brassica crops in Canada), but there is no guarantee that the introduced predator or parasitoid will perform well in its new environment. In the case of *Anagrus daanei* that parasitizes eggs of the Virginia creeper leafhopper mentioned previously, a lack of suitable overwintering hosts limits the effectiveness of this species in BC (Lowery et al. 2007). Although a serious pest of grapes in western NA, the Virginia creeper leafhopper is uncommon on grapes in eastern NA, most likely due to greater control by *A. daanei* and another native species, *Anagrus tretiakovae*, in that region.

Classical biological control cannot be carried out by individual growers, but farmers and industry groups can provide information that assists in the decision-making process and are often able to aid in the establishment of the introduced organisms. Particularly for parasitoids that tend to have restricted host ranges, the introduction of exotic biological control agents can contribute to greater sustainability through reduced costs and less disruption to the environment. Care must be taken, however, to ensure that the introduced species does not disrupt the ecosystem or cause unforeseen difficulties, as in the case of the multicolored Asian lady beetle, *Harmonia axyridis* (Pallas), which has negatively affected populations of native coccinellids and become a vineyard pest.

8.3.3.1.1 Case Study: Curative Biological Control

For more than 15 years, releases of *Trichogramma* wasps to control the ECB have been used successfully in conventional sweet corn production in Québec (Yu and Byers 1994, Conseil des Productions Végétales du Québec 2000) and the United States (Gardner et al. 2011). Sweet corn growers now have access to a brochure to guide them in their *Trichogramma* releases (Jean 2008). Based on regional phero-mone trapping and coordinated through collaboration between public and private agencies, multiple releases of this ECB egg parasitoid are conducted in sweet corn fields. The use of *Trichogramma* against ECB ensures that there is no detrimental effect from sprays to natural enemies present in the crop, which minimizes the need to intervene against aphids that can interfere with pollination and cause aesthetic damage to the cob due to sooty mold. This is an example of a curative biologi-cal approach successful under conventional production that could be transferred to organic sweet corn crop production.

8.3.3.2 Application of Insecticides

In principle, organic production should rarely require the use of insecticides unless preventive methods and biological control have failed to prevent the buildup of insect pest populations beyond acceptable levels. In practice, however, organic farms, like conventional farms, have access to a number of mineral and natural insecticides (Avery 2006; Bahlai et al. 2010). There is a growing industry dedicated to the devel-opment, production, and marketing of these materials or equipment related to their use and application. The inventory of authorized substances for insect control on each crop remains relatively small but shows signs of increasing. Organic products include plant-derived pesticides, microbial agents, and other naturally available materials. Registered products differ between countries and perhaps regions and must meet the approval of local or national organic certification agency guidelines. In Canada, the Organic Production Systems–Permitted Substances Lists and General Principles and Management Standards are available at http://www.tpsgc-pwgsc.gc.ca/ongc-cgsb/programme-program/normes-standards/comm/32-20-agriculture-eng.html and in the United States at the Organic Material Review Institute https://www.omri.org/ Products are normally applied only after pest monitoring has indicated that curative control measures are needed to prevent an unacceptable loss of yield or quality. Restrictions on some products have increased but others have become a mainstay of organic production for certain crops (Bahlai et al. 2010).

It is not within the scope of this chapter to outline specific organic materials available to producers of organic crops, and production guides and other sources of information are available for this purpose. Within the context of an ecological approach to pest management, certain insecticides are favored over others because of their reduced disturbance to the agroecosystem. Entomopathogenic bacteria, fungi, viruses, and nematodes are inherently selective and safe. They pose less threat to applicators and cause minimal disruption to populations of natural enemies and the environment providing a sustainable approach to insect pest control (Lacey and

Shapiro-Ilan 2008). The bacterium *Bacillus thuringiensis* (Bt) is the most widely used microbial insecticide in both organic and conventional agriculture (Lacey and Siegel 2000). It is used mostly to control a wide range of lepidopteran pests, including the complex of leafroller species that affect orchard crops and also for lepidopteran pests of vegetable crops, particularly Brassica crops for the control of imported cabbageworm, diamondback moth, and cabbage looper. The ease of handling and cost competitiveness of commercial preparations of Bt have facilitated their success (Lacey and Siegel 2000).

Spinosad, a commercially available microbial insecticide derived from the metabolites of the bacteria *Saccharopolyspora spinosa*, can be used in a manner similar to that of Bt. It is active against a wider range of insect pests, including many Lepidoptera, Diptera, Hymenoptera, Thysanoptera, and some Coleoptera, including the CPB (Kawalska 2009). It is still considered relatively benign toward beneficial insects, although there is increasing evidence of toxicity to pollinators and parasitoids (Williams et al. 2003; Besard et al. 2011). As an example of the complex interactions between pests and their natural control agents, although Spinosad is considered to be "soft" on beneficial insects, it is damaging to several species of predacious thrips (Thysanoptera) that help control spider mites and phytophagous thrips on grapevines and tree fruits in BC. Elsewhere, activity against plant-feeding thrips is considered advantageous.

With the exception of Bt and Spinosad, commercialization and widespread use of microbial insecticides are limited by their high cost, selectivity, reduced efficacy, and limited stability under field conditions compared with synthetic, physical (e.g., oils), and plant-based (e.g., pyrethrum) pest control products (Lacey and Shapiro-Ilan 2008). Entomopathogenic nematodes and codling moth granulosis virus have been registered more recently for the control of codling moth on pome fruits in some countries, but with minimal uptake by the organic industry. Regardless of efficacy, the higher cost of most microbial insecticides will prevent their widespread incorporation into mainstream pest management programs (Lacey and Shapiro-Ilan 2008). Investigation of methods to reduce the cost of microbial control agents and improve efficacy through strain selection, incorporation into attractive insect baits, or through the concomitant use of mulches or other strategies would contribute greatly to the commercial use of these control agents. Oils and other materials (e.g., kaolin-based Surround) that control pests through physical means are broadly active against small, soft-bodied insects of several taxa (e.g., mites, scale, aphids) as well as the eggs of some pests such as codling moth (Vincent et al. 2003) and are useful in organic programs as curative sprays in addition to their preventive role. Highly purified petroleum oils can be used both as dormant sprays for the control of overwintering eggs and mites and as foliar sprays during the growing season.

Plant-based botanical insecticides are often more disruptive of the ecosystem than the previous materials, mostly due to their broad spectrum of activity and damage to populations of nontarget organisms. Although it affords rapid control of pests, natural pyrethrum derived from the flowers of *Chrysanthemum cinerariaefolium* is broadly active and considered very harmful to beneficial insects, while the use of rotenone, another plant-derived botanical, is banned or highly restricted by most organic certification organizations due to its toxicity and negative effects on nontarget arthropods, including those living in marine environments (Zehnder et al. 2007). Commercial extracts derived from seeds of the Indian neem tree, *Azadirachta indica*, are the most widely used botanical insecticides (Zehnder et al. 2007). Neem insecticides are somewhat selective toward herbivorous pests (Lowery and Isman 1995), are relatively persistent, and are less expensive than many other organic insecticides. Research is ongoing to discover additional botanical insecticides with antifeedant or repellent properties (e.g., Gökçe et al. 2012). Frequent reliance on these and other insecticides for the management of pests suggests the need to review the pest management program and consider possible methods to enhance the activity of beneficial insects or incorporate other preventive measures.

8.3.3.3   Mechanical Removal

Insect pests can also be controlled using machinery specifically developed for pest management purposes. Tractor-mounted vacuum collectors (Vincent and Boiteau 2001) can successfully remove insect pests from a range of crops, and propane flamers can significantly reduce insect pest populations (Vincent et al. 2003). Propane flamers for the control of herbivorous pests such as the CPB are usually permitted in organic farming (more frequently as weeders). Mechanical control remains an interesting alternative for the future with recent research suggesting that lighter tractor-mounted insect collectors could be effective against such insect pests as the CPB (Vincent et al. 2003). Research on the design of mechanical distributors of insect predators could even make it possible to "spray" biological control agents across the surface of crop fields (de Ladurantaye et al. 2010).

8.4   ECONOMIC CHALLENGES TO PREVENTIVE IPM

The long-term environmental and societal benefits of organic insect crop protection are most often discussed within the general concept of "pest" control, perhaps because it makes it possible to pool together success stories for various crops and different types of pests (weeds, fungi, insects). It has the disadvantage of simplifying the challenge presented by insects and of hiding the diversity of issues that exist in organic crop protection. In parallel, the concept of IPM, originally developed for insect pests, drifted toward weed and disease management. As a result, the impact of insect pests on the yield of organic crops has often been left unquantified and even occasionally dismissed with statements such as "insect pests are not a problem because of the resistance of healthy plants produced under organic principles" (e.g., Litterick et al. 2002). More often, the economic value of the loss caused by insect pests comes from comparative trials of organic and conventional agriculture, but these studies have mostly been conducted on relatively small plots that may have constrained the benefits expected from biodiversity and landscape heterogeneity. Of course, a valid assessment would have to be done in relation to the carrying capacity and resilience of the given crop ecosystem—this is the only way to determine if the production system is truly sustainable—one of the key fundamental objectives of organic farming. One has to be realistic. The level of insect pest control possible under organic crop production will vary according to the crop and the associated environment.

Insect pests can indeed be a major constraint to growing organic field crops and cause important reductions in crop yield and quality (Collier et al. 2001). Although insect pest control can be very effective, yields, at least for some crops, will be affected, and suitable agricultural policies that take into account the contributions of organic systems to general environmental health must be developed (Gomiero et al. 2011). There are more than one set of practices for any given crop that is capable of producing good crop yields and good financial returns. This means flexibility and adaptability to a range of circumstances, but it also means a lack of "ready for use" schemes that could easily be recommended. As pointed out by Letourneau and van Bruggen (2006), further development and increased adoption of organic production systems will require dedicated research, price supports, agricultural policies, and land use practices. For example, it is interesting that organic production of cereal crops in Europe became more frequent only after the removal of price support made it unprofitable to grow cereals intensively (Letourneau and van Bruggen 2006). Organic cereal crop production was able to fill the vacuum because the low-cost production methods of organic farms leave room for a profit.

As organic growers increase their acreage in response to greater demand, pressure from agri-food corporations seems to be encouraging producers to become more dependent on curative methods that are more responsive and allow shorter rotation periods or "save" more of the yield at a particular location and time. Agri-food corporations that buy and distribute organic products will have to shift their commercial model if organic production is to remain close to its founding principle of preventive management.

Gomiero et al. (2011) have shown a strong belief among specialists that the benefits of landscape heterogeneity and biodiversity to insect pest control and crop productivity to organic production will not be fully realized until the total area of organic farmland increases substantially. Policies promoting the contiguity of organic farms and landscape heterogeneity in general might be more beneficial in terms of biodiversity and of insect pest biological control than any particular farm practice. The lack of data on the relationship between the cost of such policies and the value of the resulting ecosystem services limits the willingness of decision-makers to move ahead. As a starting point, on the basis of the premium required by organic production calculated by Reganol et al. (2001), the policy incentives could approach 12%.

8.5   CONCLUSION

It is tempting and productive, to a point, to compare organic agroecosystems to natural systems (e.g., Letourneau and van Bruggen 2006), but it may be more realistic to consider natural systems as a source of inspiration always keeping in mind that they are intrinsically different; the ecological principles can be borrowed but have to be deeply adapted to meet the constraints of the organic agricultural systems.

The application of ecological principles to pest management in organic agroecosystems remains very tentative for most crops. Detailed ecological studies of plant–arthropod and predator–prey interactions leading to specific applications for the management of key pests in particular crops are required. For example, recent work by Batary et al. (2012) highlighted how the response of insects to local or landscape-scale farm management practices can differ between cereal crops and grasslands. There are unfortunately still many recommendations issued from ecological theory that are entirely based on assumptions and whose actual potential remains to be assessed before realistic IPM practices for organic producers can be established. For example, the general recommendation to conserve toads because of their great appetite for insects and other arthropods has undeniable merit, but research is showing that the impact of toads will vary considerably between crop/pest systems. The ability of toads to learn to avoid larvae and adult stages of the CPB because of their toxicity will limit their consumption and therefore their role in the management of that particular insect pest but perhaps not of other pests (Boiteau and McCarthy 2010). Wood ash often recommended as a general barrier to pests based on its success with slugs and its use in early history will have no or little impact on some pests. Wood ash is toxic to CPB under dry conditions but becomes ineffective outdoors because it is too hydrophilic (Boiteau et al. 2012). Ash could however be considered against stored insects or as soil amendment.

The benefits of an area-wide approach to insect pest control are well demonstrated and should encourage increased cooperation between organic and conventional growers. The two production systems share the same environment and influence each other (Boiteau 2010). The landscape of conventional farms influences the role that conservation biological control can play at reducing insect pest populations in neighboring organic farms, for example. The use of inundative releases of *Trichogramma* in fields of conventional sweet corn could easily be transferred to organic fields of corn. The benefits are not limited to the farm but extend beyond to the health of the environment.

The occasional failure of general insect control recommendations against particular cases can and has been used to support the view that organic farming is an ideology rather than a scientific approach to agriculture (e.g., Kirchmann and Thorvaldsson 2000; Rigby and Càceres 2001; Trewavas 2001). It might be more profitable to look at it as an indicator of the great need for more in-depth research to address the diverse requirements of specific crops. The repeated need to turn to curative insect pest control methods, which some consider to have the potential to be harmful to the environment (e.g., Elliot and Mumford 2002), should be considered an indicator of a need for increased research on a better preventive strategy rather than a search for additional curative methods. Faced with an increasing demand for its products, the organic industry is having to rapidly adapt its production methods to increase yield, acreage, or growing season. These changes are having inevitable impacts on the insect communities of these crops and their agroecosystems and creating interesting challenges to maintain insect pests at an equilibrium position. Because insect pest protection in an organic production system relies essentially on ecological/behavioral modification methods, there will be an increased need for a better understanding of insect pest behavior as it relates to dispersal and host plant selection as well as parasitoid and predator behavior. Organic production is knowledge intensive; the research activity level during the next decade will play an important role in ensuring its continued growth. Mostly, the integration of these prevention-based insect pest control systems into the production system will require patience and persistence.

REFERENCES

Alderweireldt, M. 1994. Habitat manipulations increasing spider densities in agroecosystems—Possibilities for biological control. *Journal of Applied Entomology* 118 (1):10–16.

Alyokhin, A., G. Porter, E. Groden, and F. Drummond. 2005. Colorado potato beetle response to soil amendments: A case in support of the mineral balance hypothesis? *Agricultural Ecosystems and Environment* 109:234–244.

Andow, D.A. 1983. Effect of agricultural diversity on insect populations. In *Environmentally Sound Agriculture: Selected Papers, 4th Conference, International Federation of Organic Agriculture Movements*, Cambridge, MA, August 18–20, 1982, ed. W. Lockeretz, pp. 91–115. New York: Praeger.

Andow, D.A. 1991. Vegetational diversity and arthropod population response. *Annual Review of Entomology* 36:561–586.

Asteraki, E.J., B.J. Hart, T.C. Ings, and W.J. Manley. 2004. Factors influencing the plant and invertebrate diversity of arable field margins. *Agriculture, Ecosystems and Environment* 102 (2):219–231.

Avery, A.A. 2006. Organic pesticide use: What we know and don't know about use, toxicity, and environmental impacts. In *Crop Protection Products for Organic Agriculture, ACS Symposium Series*, Chapter 5, pp. 58–77. Washington, DC: American Chemical Society.

Bahlai, C.A., Y. Xue, C.M. McCreary, A.W. Schaafsma, and R.H. Hallett. 2010. Choosing organic pesticides over synthetic pesticides may not effectively mitigate environmental risk in soybeans. *PLoS One* 5 (6).

Baker, M.B., D.N. Ferro, and A.H. Porter. 2001. Invasions on large and small scales: Management of a well-established crop pest, the Colorado potato beetle. *Biological Invasions* 3 (3):295–306.

Batáry, P., A. Holzschuh, K.M. Orci, F. Samu, and T. Tscharntke. 2012. Responses of plant, insect and spider biodiversity to local and landscape scale management intensity in cereal crops and grasslands. *Agriculture, Ecosystems and Environment* 146(1):130–136.

Besard, L., V. Mommaerts, G. Abdu-Alla, and G. Smagghe. 2011. Lethal and sublethal side-effect assessment supports a more benign profile of spinetoram compared with spinosad in the bumblebee *Bombus terrestris*. *Pest Management Science* 67 (5):541–547.

Bickerton, M.W. and G.C. Hamilton. 2012. Effects of intercropping with flowering plants on predation of *Ostrinia nubilalis* (Lepidoptera: Crambidae) eggs by generalist predators in bell peppers. *Environmental Entomology* 41 (3):612–620.

Bird, G.W., M. Grieshop, P. Hepperly, and J. Moyer. 2009. Climbing Mt. Organic: An ecosystem approach to pest management. In *Organic Farming: The Ecological System*, ed. C. Francis, pp. 191–208. Madison, WI: ASA/CSSA/SSA/ASF Publishing.

Boisclair, J. and B. Estevez. 2006. Lutter contre les insectes nuisibles en agriculture biologique: intervenir en harmonie face à la complexité. *Phytoprotection* 87:83–90.

Boiteau, G. 1986. Native predators and the control of potato aphids. *Canadian Entomologist* 118:1177–1183.

Boiteau, G. 2008. État de la lutte dirigée contre les insectes ravageurs en production biologique de pommes de terre. *Agricultures* 17 (4):382–387.

Boiteau, G. 2010. Insect pest control on potato: Harmonization of alternative and conventional control methods. *American Journal of Potato Research* 87 (5):412–419.

Boiteau, G. and J. Heikkila. 2013. Successional and invasive colonization of the potato crop by the colorado potato beetle: Managing spread. In *Insect Pests of Potato. Global Perspectives on Biology and Management*, eds. P. Giordanengo, A. Alyokhin, and C. Vincent, pp. 339–371. San Diego, CA: Elsevier.

Boiteau, G., D.H. Lynch, and R.C. Martin. 2008. Influence of fertilization on the Colorado potato beetle, *Leptinotarsa decemlineata*, in organic potato production. *Environmental Entomology* 37 (2):575–585.

Boiteau, G. and P.C. McCarthy. 2010. Is there a role for stripes of adults and colour of larvae in determining the avoidance of the Colorado potato beetle by the American toad? *Canadian Journal of Zoology* 88 (5):468–478.

Boiteau, G., J.D. Picka, and J. Watmough. 2008. Potato field colonization by low-density populations of Colorado potato beetle as a function of crop rotation distance. *Journal of Economic Entomology* 101 (5):1575–1583.

Boiteau, G., M. Singh, and J. Lavoie. 2009. Crop border and mineral oil sprays used in combination as physical control methods of the aphid-transmitted potato virus Y in potato. *Pest Management Science* 65 (3):255–259.

Boiteau, G., R.P. Singh, P.C. McCarthy, and P.D. MacKinley. 2012. Wood ash potential for Colorado potato beetle control. *American Journal of Potato Research,* 89 (2):129–135.

Boiteau, G. and R. Vernon. 2001. Physical barriers for the control of insect pests. In *Physical Control Methods in Plant Protection*, eds. C. Vincent, B. Panneton and F. Fleurat-Lessard, pp. 224–247. New York: Springer-Verlag.

Booij, C.J.H. and J. Noorlander. 1992. Farming systems and insect predators. *Agriculture, Ecosystems and Environment* 40 (1–4):125–135.

Bostanian, N.J., H. Goulet, J. O'Hara, L. Masner, and G. Racette. 2004. Towards insecticide free apple orchards: Flowering plants to attract beneficial arthropods. *Biocontrol Science and Technology* 14 (1):25–37.

Brust, G.E. 1994. Natural enemies in straw-mulch reduce Colorado potato beetle populations and damage in potato. *Biological Control* 4:163–169.

Bukovinszky, T., H. Trefas, J.C. van Lenteren, L.E.M. Vet, and J. Fremont. 2004. Plant competition in pest-suppressive intercropping systems complicates evaluation of herbivore responses. *Agriculture Ecosystems and Environment* 102:185–196.

Burel, F. 1992. Effect of landscape structure and dynamics on species diversity in hedgerow networks. *Landscape Ecology* 6 (3):161–174.

Caballero-López, B., J.M. Blanco-Moreno, N. Pérez-Hidalgo, J.M. Michelena-Saval, J. Pujade-Villar, E. Guerrieri, J.A. Sánchez-Espigares, and F.X. Sans. 2012. Weeds, aphids, and specialist parasitoids and predators benefit differently from organic and conventional cropping of winter cereals. *Journal of Pest Science* 85 (1):81–88.

Cavanagh, A., R. Hazzard, L.S. Adler, and J. Boucher. 2009. Using trap crops for control of *Acalymma vittatum* (Coleoptera: Chrysomelidae) reduces insecticide use in butternut squash. *Journal of Economic Entomology* 102 (3):1101–1107.

Cloutier, C., G. Boiteau, and M.S. Goettel. 2002. *Leptinotarsa decemlineata* (Say), Colorado potato beetle (Coleoptera: Chrysomelidae). In *Biological Control Programmes in Canada 1981–2000*, eds. P.G. Mason and J.T. Huber, pp. 145–152. Wallingford, U.K.: CABI Publishing.

Collier, R.H., S. Finch, and G. Davies. 2001. Pest insect control in organically-produced crops of field vegetables. *Mededelingen (Rijksuniversiteit Te Gent. Fakulteit Van De Landbouwkundige En Toegepaste Biologische Wetenschappen)* 66 (2):259.

Conseil des Productions Végétales du Québec. 2000. Lutte biologique contre la pyrale du maïs à l'aide de trichogrammes dans la culture du maïs sucré. In *Fiche technique*: Conseil des Productions Végétales du Québec.

Davis, M.A. and K. Thompson. 2000. Eight ways to be a colonizer; two ways to be an invader. *Bulletin of the Ecological Society of America* 81:226–230.

De Ladurantaye, Y., M. Khelifi, C. Cloutier, and T.A. Coudron. 2010. Short-term storage conditions for transport and farm delivery of the stink bug *Perillus bioculatus* for the biological control of the Colorado potato beetle. *Canadian Biosystems Engineering/Le Genie des Biosystems au Canada* 52:4.1–4.7.

Dry, P.R., B.R. Loveys, M.G. McCarthy, and M. Stoll. 2001. Strategic irrigation management in Australian vineyards. *Journal International des Sciences de la Vigne et du Vin* 35 (3):129–139.

Dufour, R. 2002. Farmscaping to enhance biological control. NCAT/USDA 2002 [cited October 2012]. Available from www.attra.org.

Elliot, S.L. and J.D. Mumford. 2002. Organic, integrated and conventional apple production: Why not consider the middle ground? *Crop Protection* 21 (5):427–429.

Estevez, B., G. Domon, and É. Lucas. 2000. Use of landscape ecology in agroecosystem diversification towards phytoprotection. *Phytoprotection* 81 (1):1–11.

Ferro, D.N. and J.N. McNeil. 1998. Habitat management and conservation of natural enemies of insects. In *Conservation Biological Control*, ed. P. Barbosa, pp. 123–132. San Diego, CA: Academic Press.

Finch, S. and R.H. Collier. 2012. The influence of host and non-host companion plants on the behaviour of pest insects in field crops. *Entomologia Experimentalis et Applicata* 142 (2):87–96.

Flint, M.L. and R. van den Bosch. 1981. *Introduction to Integrated Pest Management*. New York: Plenum Press.

Fomsgaard, I.S. 2006. Chemical ecology in wheat plant-pest interactions. How the use of modern techniques and a multidisciplinary approach can throw new light on a well-known phenomenon: Allelopathy. *Journal of Agricultural and Food Chemistry* 54 (4):987–990.

Furlan, L., C. Bonetto, A. Finotto, L. Lazzeri, L. Malaguti, G. Patalano, and W. Parker. 2010. The efficacy of biofumigant meals and plants to control wireworm populations. *Industrial Crops and Products* 31 (2):245–254.

Gardner, J., M.P. Hoffmann, S.A. Pitcher, and J.K. Harper. 2011. Integrating insecticides and *Trichogramma ostriniae* to control European corn borer in sweet corn: Economic analysis. *Biological Control* 56 (1):9–16.

Gökçe, A., R. Isaacs, and M.E. Whalon. 2012. Dose-response relationships for the antifeedant effects of *Humulus lupulus* extracts against larvae and adults of the Colorado potato beetle. *Pest Management Science* 68 (3):476–481.

Gomiero, T., D. Pimentel, and M.G. Paoletti. 2011. Environmental impact of different agricultural management practices: Conventional vs. Organic agriculture. *Critical Reviews in Plant Sciences* 30 (1–2):95–124.

Gui, L.Y., G. Boiteau, B.G. Colpitts, P. MacKinley, and P.C. McCarthy. 2012. Random movement pattern of fed and unfed adult Colorado potato beetles in bare-ground habitat. *Agricultural and Forest Entomology* 14 (1):59–68.

Gurr, G.M., S.D. Wratten, and W.E. Snyder, eds. 2012. *Biodiversity and Insect Pests: Key Issues for Sustainable Management*. New York: Wiley-Blackwell.

Henein, K., J. Wegner, and G. Merriam. 1998. Population effects of landscape model manipulation on two behaviourally different woodland small mammals. *Oikos* 81 (1):168–186.

Hokkanen, H.M.T. 1991. Trap cropping in pest management. *Annual Review of Entomology* 36:119–138.

Hole, D.G., A.J. Perkins, J.D. Wilson, I.H. Alexander, P.V. Grice, and A.D. Evans. 2005. Does organic farming benefit biodiversity? *Biological Conservation* 122 (1):113–130.

Holland, J. and L. Fahrig. 2000. Effect of woody borders on insect density and diversity in crop fields: A landscape-scale analysis. *Agriculture, Ecosystems and Environment* 78 (2):115–122.

Hsu, Y.T., T.C. Shen, and S.Y. Hwang. 2009. Soil fertility management and pest responses: A comparison of organic and synthetic fertilization. *Journal of Economic Entomology* 102 (1):160–169.

Hua, L.Z., F. Lammes, J.C. van Lenteren, P.W.T. Huisman, A. van Vianen, and O.M.B. de Ponti. 1987. The parasite-host relationship between *Encarsia formosa* Gahan (Hymenoptera, Aphelinidae) and *Trialeurodes vaporariorum* (Westwood) (Homoptera, Aleyrodidae). *Journal of Applied Entomology* 104 (1–5):297–304.

Hunt, D.W.A. and G. Whitfield. 1996. Potato trap crops for control of Colorado potato beetle (Coleoptera: Chrysomelidae) in tomatoes. *Canadian Entomologist* 128 (3):407–412.

Jahn, M. 2003. On-farm and centralized approaches to breeding vegetables for organic farming systems. *Hortscience* 38:866.

James, D.G. 2005. Further field evaluation of synthetic herbivore-induced plant volatiles as attractants for beneficial insects. *Journal of Chemical Ecology* 31:481–495.

Jean, C. 2008. Les Trichogrammes dans le maïs sucré: lutte contre la pyrale du maïs. Saint-Augustin-de-Desmaures, Quebec, Canada: Para-Bio.

Johnson, D.A., ed. 2008. *Potato Health Management*, 2nd edn., Plant Health Management Series. Saint Paul, MN: American Phytopathological Society.

Judd, G.J.R. and M.G.T. Gardiner. 2005. Towards eradication of codling moth in British Columbia by complimentary actions of mating disruption, tree banding and sterile insect technique: Five-year study in organic orchards. *Crop Protection* 24 (8):718–733.

Kasirajan, S. and M. Ngouajio. 2012. Polyethylene and biodegradable mulches for agricultural applications: A review. *Agronomy for Sustainable Development* 32 (2):501–529.

Keogh, B., J. Humphreys, P. Phelan, M. Necpalova, I.A. Casey, and E. Fitzgerald. 2009. Organic management strategies and its effect on clover-based grassland production. *Irish Journal of Agricultural and Food Research* 48 (2):267.

Kirchmann, H. and D. Thorvaldsson. 2000. Challenging targets for future agriculture. *European Journal of Agronomy* 12:145–161.

Koch, R.L., P.M. Porter, M.M. Harbur, M.D. Abrahamson, K.A.G. Wyckhuys, D.W. Ragsdale, K. Buckman, Z. Sezen, and G.E. Heimpel. 2012. Response of soybean insects to an autumn-seeded rye cover crop. *Environmental Entomology* 41 (4):750–760.

Kowalska, J. 2009. Spinosad effectively controls Colorado potato beetle, *Leptinotarsa decemlineata* (Coleoptera: Chrysomelidae) in organic potato. *Acta Agriculturae Scandinavica, Section B—Soil & Plant Science* 60 (3):283–286.

Kus, J.V., K. Zaton, R. Sarkar, and R.K. Cameron. 2002. Age-related resistance in *Arabidopsis* is a developmentally regulated defense response to *Pseudomonas syringae*. *Plant Cell* 14 (2):479–490.

Lacey, L.A. and D.I. Shapiro-Ilan. 2008. Microbial control of insect pests in temperate orchard systems: Potential for incorporation into IPM. *Annual Review of Entomology* 53:121–144.

Lacey, L.A. and J.P. Siegel. 2000. Safety and ecotoxicology of entomopathogenic bacteria. In *Entomopathogenic Bacteria: From Laboratory to Field Application*, eds. J.F. Charles, A. Delecluse, and C. Nielsen-LeRoux, pp. 253–273. Dordrecht, the Netherlands: Kluwer Academic.

Landis, D.A., S.D. Wratten, and G.M. Gurr. 2000. Habitat management to conserve natural enemies of arthropod pests in agriculture. *Annual Review of Entomology* 45:175–201.

Langer, V. 1995. Pests and diseases in organically grown vegetables in Denmark: A survey of problems and use of control methods. *Biological Agriculture and Horticulture* 12 (2):151–171.

Letourneau, D.K. 1994. Bean fly, management practices, and biological control in Malawian subsistence agriculture. *Agriculture, Ecosystems and Environment* 50:103–111.

Letourneau, D.K., L.E. Drinkwater, and C. Shennan. 1996. Effects of soil management on crop nitrogen and insect damage in organic vs. conventional tomato fields *Agriculture, Ecosystems and Environment* 57:179–187.

Letourneau, D.K. and B. Goldstein. 2001. Pest damage and arthropod community structure in organic vs. conventional tomato production in California. *Journal of Applied Ecology* 38:557–570.

Letourneau, D.K. and M.A.B. Msuku. 1992. Enhanced *Fusarium solani f.* sp. *phaseoli* infection by bean fl y in Malawi. *Plant Disease* 76:1253–1255.

Letourneau, D.K. and A.H.C. Van Bruggen. 2006. Crop protection in organic agriculture. In *Organic Agriculture: A Global Perspective*, eds. P. Kristiansen, A.M. Taji, and J.P. Reganold, pp. 93–120. Clayton, Victoria, Australia: CSIRO.

Liebhold, A.M. and P.C. Tobin. 2008. Population ecology of insect invasions and their management. *Annual Review of Entomology* 53:387–408.

Liebman, M. and E.R. Gallandt. 1997. Many little hammers: Ecological management of crop–weed interactions. In *Ecology in Agriculture*, ed. L.E. Jackson, pp. 287–339. San Diego, CA: Academic Press.

Litterick, A., C.A. Watson, and D. Atkinson. 2002. Crop protection in organic agriculture—A simple matter? In *Proceedings of the UK Organic Research 2002 Conference*, ed. J. Powell, 203–206. Aberystwyth, U.K.: University of Wales.

Lowery, D.T. 1988. *Turnip Mosaic Virus (TuMV) of Rutabaga*. Guelph, Ontario, Canada: Ontario Ministry of Agriculture and Food.

Lowery, D.T. 2010. Insect and mite pests of grape. In *Best Practices Guide for Grapes for British Columbia Growers*. Victoria, B.C., Canada: British Columbia Ministry of Agriculture and Lands.

Lowery, D.T. and M.B. Isman. 1995. Toxicity of neem to natural enemies of aphids. *Phytoparasitica* 23 (4):297–306.

Lowery, D.T., S.V. Triapitsyn, and G.J.R. Judd. 2007. Leafhopper plant associations for *Anagrus* parasitoids (Hymenoptera: Mymaridae) in the Okanagan valley, British Columbia. *Journal of the Entomological Society of British Columbia* 104:9–15.

Luckman, W.H. and R.L. Metcalf. 1994. The pest management concept. In *Introduction to Insect Pest Management*, eds. R.L. Metcalf and W.H. Luckman, pp. 1–31. New York: John Wiley & Sons.

Luginbill, P. and R. Painter. 1953. *May Beetles of the United States and Canada*. Washington, DC: U.S. Department of Agriculture.

MacKenzie, J. and M. Savard, eds. 2012. *Proceedings of the 2012 Canadian Organic Science Conference, Canadian Organic Science Conference and Organic Science Cluster Strategic Meetings*. Winnipeg, Manitoba, Canada. Truro, Nova Scotia, Canada: Organic Agriculture Centre of Canada.

Marino, P.C., and D.A. Landis. 1996. Effect of landscape structure on parasitoid diversity and parasitism in agroecosystems. *Ecological Applications* 6(1):276–284.

Mathews, C.R., D.G. Bottrell, and M.W. Brown. 2004. Habitat manipulation of the apple orchard floor to increase ground-dwelling predators and predation of *Cydia pomonella* (L.) (Lepidoptera: Tortricidae). *Biological Control* 30:265–273.

Mazzola, M. and K. Mullinix. 2005. Comparative field efficacy of management strategies containing *Brassica napus* seed meal or green manure for the control of apple replant disease. *Plant Disease* 89 (11):1207–1213.

McKenzie, L.M. 1973. *The Grape Leafhopper Erythroneura ziczac (Hymenoptera: Cicadellidae) and Its Mymarid (Hymenoptera) Egg-Parasite in the Okanagan Valley, British Columbia*. Burnaby, British Columbia, Canada: Simon Fraser University.

Misener, G.C., G. Boiteau, and L.P. McMillan. 1993. A plastic-lining trenching device for the control of Colorado potato beetle: Beetle excluder. *American Potato Journal* 70 (12):903–908.

Mohler, C.L. and S.E. Johnson, eds. 2009. *Crop Rotation on Organic Farms: A Planning Manual*. 156p. Ithaca, NY: Natural Resources, Agriculture and Economics Service (NRAES), Cooperative Extension.

Moreau, T.L., P.R. Warman, and J. Hoyle. 2006. An evaluation of companion planting and botanical extracts as alternative pest controls for the Colorado potato beetle. *Biological Agriculture and Horticulture* 23 (4):351–370.

Mostafa, A.M., D.T. Lowery, L.B.M. Jensen, and E.K. Deglow. 2011. Host plant suitability and feeding preferences of the grapevine pest *Abagrotis orbis* (Lepidoptera: Noctuidae). *Environmental Entomology* 40 (6):1458–1464.

Neilsen, G.H., D.T. Lowery, T.A. Forge, and D. Neilsen. 2009. Organic fruit production in British Columbia. *Canadian Journal of Plant Science* 89 (4):677–692.

Nelson, K.L., G. Boiteau, D.H. Lynch, R.D. Peters, and S. Fillmore. 2011. Influence of agricultural soils on the growth and reproduction of the bio-indicator *Folsomia candida*. *Pedobiologia* 54 (2):79–86.

Nelson, K.L., D.H. Lynch, and G. Boiteau. 2009. Assessment of changes in soil health throughout organic potato rotation sequences. *Agriculture, Ecosystems and Environment* 131 (3–4):220–228.

Noble, R.R.P. and C.E. Sams. 1999. Biofumigation as an alternative to methyl bromide for control of white grub larvae. In *Annual International Research Conference on Methyl Bromide Alternatives and Emissions Reductions*, San Diego, CA, November 1–4, eds. G.L. Obenauf and A. Williams, pp. 92-1–92-3.

Olmstead, M.A., R.L. Wample, S.L. Greene, and J.M. Tarara. 2001. Evaluation of potential cover crops for Inland Pacific Northwest vineyards. *American Journal of Enology and Viticulture* 54:292–303.

Pfannenstiel, R.S. and T.R. Unruh. 2003. Conservation of leafroller parasitoids through provision of alternate hosts in near-orchard habitats. Paper read at First *International Symposium on Biological Control of Arthropods*, Honolulu, HI, January 14–18, 2002.

Phelan, P.L., J.F. Mason, and B.R. Stinner. 1995. Soil-fertility management and host preference by European corn borer, *Ostrinia nubilalis* (Hubner), on *Zea mays* L.: A comparison of organic and conventional chemical farming. *Agriculture, Ecosystems and Environment* 56 (1):1–8.

Phelan, P.L., K.H. Norris, and J.F. Mason. 1996. Soil-management history and host preference by *Ostrinia nubilalis*: Evidence for plant mineral balance mediating insect-plant interactions. *Environmental Entomology* 25 (6):1329–1336.

Phillips, S.W., J.S. Bale, and G.M. Tatchell. 1999. Escaping an ecological dead-end: Asexual overwintering and morph determination in the lettuce root aphid *Pemphigus bursarius* L. *Ecological Entomology* 24: 336–344.

Pimentel, D. 1961. Species diversity and insect population outbreaks. *Annals of the Entomological Society of America* 54:76–86.

Polis, G.A., W.B. Anderson, and R.D. Holt. 1997. Toward an integration of landscape and food web ecology: The dynamics of spatially subsidized food webs. *Annual Review of Ecology and Systematics* 28:289–316.

Power, E.F., D.L. Kelly, and J.C. Stout. 2012. Organic farming and landscape structure: Effects on insect-pollinated plant diversity in intensively managed grasslands. *PLoS One* 7 (5).

Puterka, G.J., D.M. Glenn, and R.C. Pluta. 2005. Action of particle films on the biology and behavior of pear psylla (Homoptera: Psyllidae). *Journal of Economic Entomology* 98 (6):2079–2088.

Rabb, R.L. 1978. A sharp focus on insect populations and pest management from a wide-area view. *Bulletin of the Entomological Society of America* 24:55–61.

Rabb, R.L. and F.E. Guthrie, eds. 1970. *Concepts of Pest Management. Proceedings of a Conference*. North Carolina State University of Raleigh, North Carolina, March 25–27, 1970. Raleigh, NC: North Carolina State University.

Reganold, J.P., J.D. Glover, P.K. Andrews, and H.R. Hinman. 2001. Sustainability of three apple production systems. *Nature* 410 (6831):926–930.

Renkema, J.M., D.H. Lynch, G.C. Cutler, K. MacKenzie, and S.J. Walde. 2012. Ground and rove beetles (Coleoptera: Carabidae and Staphylinidae) are affected by mulches and weeds in highbush blueberries. *Environmental Entomology* 41 (5):1097–1106.

Rigby, D. and D. Càceres. 2001. Organic farming and the sustainability of agricultural systems. *Agricultural Systems* 68:21–40.

Risch, S.J. 1987. Agricultural ecology and insect outbreaks. In *Insect Outbreaks*, eds. P. Barbosa and J.C. Schultz, pp. 625–629. Toronto, Ontario, Canada: Academic Press, Inc.

Risch, S.J., D. Andow, and M. Altieri. 1983. Agroecosystem diversity and pest control: Data, tentative conclusions and new research directions *Environmental Entomology* 12:625–629.

Rojas, E.S., M.L. Gleason, J.C. Batzer, and M. Duffy. 2011. Feasibility of delaying removal of row covers to suppress bacterial wilt of muskmelon (*Cucumis melo*). *Plant Disease* 95 (6):729–734.

Root, R.B. 1973. Organization of a plant-arthropod association in simple and diverse habitats: The fauna of collards (*Brassica oleracea*). *Ecological Monographs* 43:95–124.

Rundlöf, M., and H.G. Smith. 2006. The effect of organic farming on butterfly diversity depends on landscape context. *Journal of Applied Ecology* 43:1121–1127.

Shelton, A.M. and F.R. Badenes-Perez. 2006. Concepts and applications of trap cropping in pest management. *Annual Review of Entomology* 51:285–308.

Sholberg, P.L., T. Lowery, and P. Bowen. 2008. Effect of early leaf stripping on bunch rot, powdery mildew and sour rot of wine grapes. In *Crop Protection Research Advances*, eds. E.N. Burton and P.V. Williams, pp. 199–212. Hauppauge, NY: Nova Science Publishers, Inc.

Simon, S., K. Morel, E. Durand, G. Brevalle, T. Girard, and P.E. Lauri. 2012. Aphids at crossroads: When branch architecture alters aphid infestation patterns in the apple tree. *Trees—Structure and Function* 26 (1):273–282.

Southwood, T.R.E. and M.J. Way. 1970. Ecological background to pest management. In *Concepts of Pest Management*. eds. R.L. Rabb and F.E. Guthrie, pp. 6–29. Raleigh, NC: NC State University.

Stafford, D.B., M. Tariq, D.J. Wright, J.T. Rossiter, E. Kazana, S.R. Leather, M. Ali, and J.T. Staley. 2012. Opposing effects of organic and conventional fertilizers on the performance of a generalist and a specialist aphid species. *Agricultural and Forest Entomology* 14 (3):270–275.

Staley, J.T., A. Stewart-Jones, T.W. Pope, D.J. Wright, S.R. Leather, P. Hadley, J.T. Rossiter, H.F. Van Emden, and G.M. Poppy. 2010. Varying responses of insect herbivores to altered plant chemistry under organic and conventional treatments. *Proceedings of the Royal Society B: Biological Sciences* 277 (1682):779–786.

Stoner, K.A. 1997. Influence of mulches on the colonization by adults and survival of larvae of the Colorado potato beetle (Coleoptera: Chrysomelidae) in eggplant. *Journal of Entomological Science* 32:7–16.

Swezey, S.L., D.J. Nieto, and J.A. Bryer. 2007. Control of western tarnished plant bug *Lygus hesperus* Knight (Hemiptera: Miridae) in California organic strawberries using alfalfa trap crops and tractor-mounted vacuums. *Environmental Entomology* 36 (6):1457–1465.

Tamaki, G. 1981. Biological control of potato pests. In *Advances in Potato Pest Management*, eds. J.H. Lashomb and R. Casagrande, pp. 178–192. Stroudsburg, PA: Hutchinson Ross Publ. Co.

Thies, C., I. Steffen-Dewenter, and T. Tscharntke. 2003. Effects of landscape context on herbivory and parasitism at different spatial scales. *Oikos* 101:18–25.

Thies, C. and T. Tscharntke. 1999. Landscape structure and biological control in agroecosystems. *Science* 285 (5429):893–895.

Tooker, J.F. and S.D. Frank. 2012. Genotypically diverse cultivar mixtures for insect pest management and increased crop yields. *Journal of Applied Ecology* 49 (5):974–985.

Trewavas, A. 2001. Urban myths of organic farming. *Nature* 410:409–410.

Vaarst, M., S. Roderick, V. Lund, W. Lockeretz, and M. Hovi. 2003. Organic principles and values: The framework for organic animal husbandry. In *Animal Health and Welfare in Organic Agriculture*, eds. M. Vaarst, S. Roderick, V. Lund and W. Lockeretz, pp. 1–12. Wallingford, CT: CABI.

van Bruggen, A.H.C. 1995. Plant-disease severity in high-input compared to reduced-input and organic farming systems. *Plant Disease* 79:976–984.

Van Lenteren, J.C. and J. Woets. 1988. Biological and integrated pest control in greenhouses. *Annual Review of Entomology* 33:239–269.

Vernon, R.S. and J.R. Mackenzie. 1998. The effect of exclusion fences on the colonization of rutabagas by cabbage flies (Diptera: Anthomyiidae). *Canadian Entomologist* 130 (2):153–162.

Vincent, C. and G. Boiteau. 2001. Pneumatic control of agricultural insect pests. In *Physical Control Methods in Plant Protection*, eds. C. Vincent, B. Panneton, and F. Fleurat-Lessard, pp. 270–281. New York: Springer-Verlag.

Vincent, C., G. Hallman, B. Panneton, and F. Fleurat-Lessard. 2003. Management of agricultural insects with physical control methods. *Annual Review of Entomology* 48:261–281.

Weisz, R, Z. Smilowitz, and S. Fleischer. 1996. Evaluating risk of Colorado potato beetle (Coleoptera: Chrysomelidae) infestation as a function of migratory distance. *Journal of Economic Entomology* 89:435–441.

Westphal, A., G.T. Browne, and S. Schneider. 2002. Evidence for biological nature of the grape replant problem in California. *Plant and Soil* 242 (2):197–203.

Williams, T., J. Valle, and E. Viñuela. 2003. Is the naturally derived insecticide Spinosad® compatible with insect natural enemies? *Biocontrol Science and Technology* 13 (5):459–475.

Wu, J.Y. and Sardo, V. 2010. Sustainable versus organic agriculture. In *Sociology, Organic Farming, Climate Change and Soil Science*, ed. E. Lichtfouse, pp. 41–76. Dordrecht, the Netherlands: Springer.

Younie, D. and A. Litterick. 2002. Crop protection in organic farming. *Pesticide Outlook* 13 (4):158–161.

Yu, D.S. and J.R. Byers. 1994. Inundative release of *Trichogramma brassicae* Bezdenko (Hymenoptera: Trichogrammatidae) for control of European corn borer in sweet corn. *Canadian Entomologist* 126 (2):291–301.

Zehnder, G., G.M. Gurr, S. Kühne, M.R. Wade, S.D. Wratten, and E. Wyss. 2007. Arthropod pest management in organic crops. *Annual Review of Entomology* 52:57–80.

Section III

Integrating Approaches

9 Glenlea Organic Rotation
A Long-Term Systems Analysis

*Martin H. Entz, Cathy Welsh, Shauna Mellish,
Yuying Shen, Sarah Braman, Mario Tenuta,
Marie-Soleil Turmel, Katherine Buckley,
Keith C. Bamford, and Neil Holliday*

CONTENTS

The years teach much which the days never know

Ralph Waldo Emerson
Essays, Second Series: Experience 1844

9.1 INTRODUCTION

Canada has a history of embracing long-term crop rotation studies, the first of which was started in 1910 (Campbell et al., 1990). Some of the original studies still exist (Janzen, 2001). Rotation studies that include grain and forage phases were started in the 1930s. Three of these studies are still operating, two at Agriculture and Agri-Food's research centers at Indian Head, Saskatchewan and Lethbridge, Alberta, and one at the University of Alberta—The Breton Plots. All have provided important information for organic farm management.

As organic agriculture grew, however, so did the need for dedicated long-term *organic* studies. The oldest among the organic rotation studies in Canada is the Glenlea plots, located south of Winnipeg, in Manitoba's Red River Valley. This chapter outlines some of the highlights from the Glenlea study over its 20-year history.

9.2 THE FIRST 12 YEARS

Glenlea has multiple goals. Planning took place soon after *Our Common Future* (Brundtland, 1987) was published, a book that introduced the term sustainable development. We asked questions about how sustainable fertilizer and pesticide use was. Could proper rotation reduce the need for these external inputs? Could we reduce the inputs to the point where we were farming organically? Entomologists involved in the planning process had a special request—large plot size. They were interested in trapping ground beetles, where large plots are better.

The original design tried to satisfy all of us. Agronomists got three different crop rotations. The entomologists got their large plot size (1/4 ha per subplot). The crop ecologists and weed scientists got subplots that included different fertilizer and pesticide input combinations (Figure 9.1).

There was, however, a cost. We felt it was not possible to include all these components and maintain all rotation phases each year—that would make the study too large and too expensive to operate. So we decided to have each 4-year rotation run in sequence. The three rotations were as follows: (1) wheat–pea–wheat–flax (grain only); (2) wheat–clover green manure–wheat–flax (green manure–grain), and (3) wheat–alfalfa–alfalfa–flax (forage–grain). We used the flax *test* crop, which is common to all rotations, to evaluate some of the agronomic outcomes.

Half-hectare native prairie grass restorations were established—one in each replicate (Figure 9.2). These perennial systems were designed to mimic the original prairie and are viewed as benchmark treatments. Species include the native perennial grasses *Andropogon gerardii* Vitman var. (L.), *Sorghastrum nutans* (L.), *Panicum virgatum* (L.), *Agropyron smithii* (Rydb.), *Elymus lanceolatus* (Scribn. and Smith) Gould., and *Elymus trachycaulus* (Link) Gould ex Shinners. The experimental design was a factorial randomized complete block design in a split plot arrangement and included three replications.

Synthetic fertilizer — no Pesticides — no Organic	Synthetic fertilizer—yes Pesticides—no
Synthetic fertilizer — no Pesticides — yes	Synthetic fertilizer — yes Pesticides — yes Conventional

FIGURE 9.1 Subplot scheme for Glenlea, 1992–2003. Each 1 ha main plot was divided systematically into four subplots with different fertilizer and pesticide inputs.

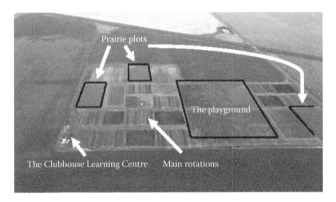

FIGURE 9.2 Aerial view of the Glenlea plots showing different study components. This picture was taken in 2012, after changes were made to the study. However, from the outset, *playground* areas were included at the site. The playgrounds occupy land that is not sufficiently uniform to be in the study, but provide an opportunity to conduct short-term studies on organically managed lands. (Photo courtesy of Gary Martens.)

9.2.1 BIOMASS AND ENERGY PRODUCTION

The first measure of productivity was grain and forage crop yields. In general, flax *test crop* yields were affected more by input management than by rotation. During the first 8 years of the study, crop yields in the organic system were 10%–50% lower than in the conventional system with limited influence of rotation type (Table 9.1). The main reason for the yield decline was weed interference. By year 12, however, the rotation effect was very strong. The yield decrease due to organic was lower in the forage–grain rotation compared with the grain-only system (Table 9.1). The mechanism for superior yield performance in the forage–grain system is related in great part to better weed control.

TABLE 9.1

Grain Yield (kg/ha) for Flax *Test Crops* at Glenlea for 1995, 1999, and 2003

Rotation	Inputs	Yield (kg/ha)		
		1995	1999	2003
Grain-only rotation	+f+h	1876	1378	1693
	+f−h	975	600	81
	−f+h	1312	1050	964
	−f−h	960	605	231
Green manure–grain rotation	+f+h	1808	1826	971
	+f−h	1233	1099	77
	−f+h	1109	1584	670
	−f−h	1022	993	170
Forage–grain rotation	+f+h	1712	1453	1328
	+f−h	1291	998	94
	−f+h	1550	1531	1287
	−f−h	1373	1378	482
Rotation (R)		0.0747	0.0915	0.0010
Management (M)		0.0001	0.0001	0.0010
R×M		0.1575	0.1922	0.0010
SEM		117.01	156.12	615.00

Statistical analysis for yield performed on log-transformed data. Means considered significant when $P < 0.05$.

f, fertilizer added based on soil test recommendation; h, recommended herbicides used for weed control; +/− refers to either addition (+) or omission (−) of fertilizer or herbicide.

An important observation was the decline in organic forage yield over time. Total hay production, from two cuts per year, in the conventional and organic forage–grain systems was 9,093 and 9,277 kg/ha, respectively, in 1993–1994; 12,936 and 10,785 kg/ha, respectively, in 1997–1998; and 12,752 and 8,955 kg/ha, respectively, in 2001–2002. This was a signal that limitations were increasing in the organic system. Could they be nutrient related?

The second measure of productivity involved total caloric energy production (EP) per unit area of land. An energy balance was conducted for the 1992–2003 period by Hoeppner et al. (2006). As expected, organic systems used less energy than conventional systems. Annual energy use dropped from 5708 to 2019 MJ/ha/year due to organic in the grain-only system and from 4104 to 1848 MJ/ha/year in the forage–grain rotation (Table 9.2). Reduction in energy use for organic systems was primarily due to reduction in N fertilizer use (Hoeppner et al., 2006).

Organic systems produced fewer calories than conventional systems, 46% fewer in the grain-only system but only 21% fewer in the forage–grain system (Table 9.2). Lower energy yield sacrifice in the forage–grain system highlights

TABLE 9.2

Rotational Energy Production and Consumption and Energy Efficiency at the Glenlea Long-Term Cropping Systems Study, 1992–2003

Rotation	Treatment (MJ/ha)	Energy Production (EP) (MJ/ha)	Energy Consumption (EC) (MJ/ha)	Energy Efficiency (EP/EC)	Amended Energy Efficiency[a] (EP/EC)
Grain only	Organic	252,054	24,233	10.4	10.4
Grain only	Conventional	465,841	68,489	6.8	6.8
Forage–grain	Organic	742,229	22,180	33.5	11.9
Forage–grain	Conventional	937,344	49,256	19.0	7.4
Significance (P>F)					
Rotation (R)		<0.0001	—	<0.0001	0.2281 ns
Management (M)		0.0012	—	0.0006	0.0018
R×M		0.8207 ns	—	0.0102	0.5657 ns
SEM		42,053	—	1.6	0.9

[a] Assuming alfalfa production/meat production conversion factor of 9:1.

the importance of good crop rotations in reduced external input farming systems. When the best organic system (forage–grain) was compared with the predominant farming system in the area (conventional grain-only production), the organic system produced more energy (61,000 MJ/ha/year compared with only 39,000 in the conventional grain-only system). The highest EP, however, was in the conventional forage–grain system.

Of course, not all calories produced in the forage–grain rotation are human useable. Human useable calories after forage conversion (9:1) to animal products were 20% lower in the organic forage–grain system compared with the conventional grain-only system. In Canada, a high proportion of grains are fed to ruminant and monogastric livestock. Therefore, future research should focus on EP of integrated versus grain-based rotations where different proportions of the grain crops are used for livestock feed. Energy efficiency (energy output/energy input) and amended energy efficiency (after 9:1 ruminant conversion) were affected mostly by management system. In both rotations, the organic systems had 35% higher energy efficiency than their conventional counterparts.

9.2.2 SOIL NUTRIENTS

Soil nutrient status in year 12 showed significant rotation effects for all four macronutrients (Table 9.3). The forage–grain rotation had lower levels of available P, K, and S than the grain-only rotation.

Phosphorous deficiencies have been identified by several workers (e.g., Entz et al., 2001; Knight et al., 2010) as a threat to the long-term sustainability of organic farming. Results from 2003 showed that only the organic forage–grain system was in any danger of a P deficiency. Only a small decline in available P due to organic

TABLE 9.3

Soil Nutrient Status (kg/ha) for the Glenlea Long-Term Cropping Systems Study Flax Test Crop in 2003

Rotation	Inputs	NO₃–Nᵃ	P₂O₅ᵇ	Kᵇ	Sᵇ
Annual	f+h+	32	46	1316	141
	f–h–	22	33	1312	86
Green manure	f+h+	29	24	1169	87
	f–h–	31	37	1116	76
Forage	f+h+	81	24	1140	63
	f–h–	37	11	1073	26
Rotation (R)		0.0024	0.0020	0.0002	0.0535
Inputs (I)		0.0093	0.1899	0.0620	0.1029
R×I		0.0158	0.0153	0.3866	0.6219
SEM		5.40	3.30	30.98	10.64

ᵃ Sampling depth 0–60 cm.

ᵇ Sampling depth 0–15 cm.

f, fertilizer added based on soil test recommendation; h, recommended herbicides used for weed control. +/– refers to either addition (+) or omission (–) of fertilizer or herbicide.

was observed in the grain-only rotation, while available P was actually higher in the organic compared with the conventional green manure system. This was an interesting observation; one possible explanation is that green manure crops solubilize P making it more plant available.

Within organic systems, NO₃–N was highest in the forage–grain rotation. However, a rotation by management interaction showed that the reduction in NO₃–N due to organic management was much greater for the forage–grain compared with the other two rotations (Table 9.3). Perhaps the P deficiency in the organic forage–grain system was limiting N assimilation by the perennial legume phase due to poor fixation or simply low growth.

9.2.3 WEEDS

Weed measurements were recorded in flax years. Results show an increase in weed density over time. This was partly attributed to above average precipitation from 1997 to 2003, which among other problems, limited fall tillage opportunities. Among the rotations, total weed density was consistently lower in the forage–grain compared with the other two rotations (Table 9.4). For wild oat, green foxtail, and redroot pigweed, weed densities in the organic forage–grain system were always less than the conventional grain-only system.

A detailed weed community analysis was conducted in 1999 by Humble (2001). Redundancy analysis ordination showed that the grain-only rotation was

TABLE 9.4
Total In-Crop Weed Population Density in Flax at Glenlea, MB. Statistical Analysis Performed on Log-Transformed Data

Rotation and Inputs	Total Weed Density (Plants/m^2)		
	1995	1999	2003
W-P-W-F			
+f+h	16	1889	1379
+f–h	16	942	3204
–f+h	7	98	707
–f–h	12	532	2041
W-Cl-W-F			
+f+h	15	91	1738
+f–h	11	154	1812
–f+h	10	64	1502
–f–h	6	128	2235
W-A-A-F			
+f+h	4	40	594
+f–h	7	148	1307
–f+h	6	42	680
–f–h	6	110	1338
ANOVA (P-value)			
Rotation (R)	0.03	0.001	0.01
Management (M)	0.27	0.0001	0.07
R×M	0.21	0.06	0.18

f, fertilizer added based on soil test recommendation; h, recommended herbicides used for weed control. +/– refers to either addition (+) or omission (–) of fertilizer or herbicide.

associated ($P < 0.05$) with green foxtail and wild oat, the green manure–grain rotation was associated with stinkweed and Canada thistle, and the forage–grain rotation was associated with dandelion.

Weed population analysis 4 years later (in 2003) showed an abundance of wild oat, green foxtail, and thistle in the grain-only and green manure–grain rotations (Table 9.5). In contrast, the forage–grain rotation had few of these weed species. In a Manitoba farm field survey, Ominski et al. (1999) observed few wild oat, green foxtail, or Canada thistle plants following forage crops, thereby supporting the Glenlea data.

We were interested in the hypothesis that adding fertilizer would increase weed populations; the subplot arrangement at Glenlea allowed this hypothesis to be tested in different crop rotations. Significant rotation by input interactions was

TABLE 9.5

Population Density of Major Weed Species Prior to In-Crop Spraying in Flax at Glenlea, MB, in 2003

Rotation and Inputs	Weed Seedling Density (Plants/m²)					
	Green Foxtail	Wild Oat	Wild Mustard	Redroot Pigweed	Stinkweed	Canada Thistle
W-P-W-F						
+f+h	1212	60	5	42	21	2
+f–h	1773	754	212	2	43	8
–f+h	633	3	4	28	9	0
–f–h	1731	55	126	1	24	13
W-Cl-W-F						
+f+h	728	4	114	86	758	2
+f–h	729	63	911	4	60	4
–f+h	800	10	68	181	352	1
–f–h	1413	6	566	6	178	5
W-A-A-F						
+f+h	21	0	185	3	353	0
+f–h	14	4	1203	3	26	2
–f+h	61	1	246	3	313	0
–f–h	50	1	1201	0	19	5
ANOVA (P-value)						
Rotation (R)	0.002	0.003	0.009	0.38	0.13	0.005
Inputs (I)	0.06	0.005	<0.0001	<0.0001	0.02	0.0002
R×I	0.64	0.05	0.01	0.007	0.05	0.02

Statistical analysis performed on log-transformed data.

f, fertilizer added based on soil test recommendation; h, recommended herbicides used for weed control; +/– refers to either addition (+) or omission (–) of fertilizer or herbicide.

observed for five of the six weed species tested (Table 9.5). Results for wild oat and green foxtail showed that as nutrient level increased (+f (fertilizer) +h (herbicide) vs. –f+h subplot), populations of these weeds did in fact increase (Table 9.5). However, the increase was observed only in the *poor* (grain-only) rotation and not in the green manure–grain or forage–grain rotations. Under superior rotation conditions (e.g., forage–grain rotation), an increase in fertilizer did not increase weed densities, presumably because the rotation provided sufficient weed control benefits. Results for Canada thistle mirror those for the grassy weeds. Therefore, for three of the most problematic weeds in organic farming on the prairies (Entz et al., 2001), raising soil fertility did not aggravate the problem—provided suitable rotations were used. Results may have been different if manures were used to fertilize the conventional systems (Dyck and Liebman, 1994).

We did observe that rotations that suppressed wild oat and green foxtail often favored wild mustard. For example, wild mustard populations were much higher in the forage–grain compared with the other organic rotations (Table 9.5). Farmers practicing forage rotation to manage grassy weeds and Canada thistle will need to use alternative strategies such as harrowing in order to keep wild mustard populations low. Unfortunately, the high clay content at Glenlea (66% clay) severely reduced both the opportunity (too wet) and effectiveness (lumpy soil) of pre- or postemergence harrowing.

An absence of redroot pigweed in the system with the lowest level of plant available P (organic forage–grain) was observed in 2003 (Table 9.5) and continues to be observed (data not shown). Blackshaw and Molnar (2009) reported that redroot pigweed is highly responsive to available soil P status. Therefore, low redroot pigweed populations in the organic forage–grain rotation may be related to low available soil P status.

9.2.4 BIODIVERSITY: GROUND BEETLES

Carabid beetle populations act as indicators of ecosystem health (Carcamo, 1995) plus some beetles consume weed seeds as part of their diverse diet (Lund and Turpin, 1977). A detailed assessment of carabid beetle populations in 1995 and 1999 indicated that total beetle capture tended to be greatest in grain-only > green manure–grain > forage–grain rotations (significant only in 1999) and f+h– > –f–h > f+h+ > f–h+ (Humble, 2001; Table 9.6). The redundancy analysis indicated that total beetle capture was most influenced by weeds; more weeds, more beetles. For example, the f+h– systems had the most weeds (Table 9.4) and the most beetles (Table 9.6). These observations identify an important effect of plant diversity on beetle populations.

Humble (2001) used multivariate analysis to test for associations between weed and carabid beetle species. Four consistent associations were observed in 1995 and 1999. These were *Harpalus pensylvanicus* and redroot pigweed, *Amara carinata* and stinkweed, and *Amara placidum* and *Convolvulus calidum* and wild mustard (Humble, 2001). These data lead to interesting questions about harnessing the weed seed consumption abilities of carabid beetles to manage weeds in organic systems. According to Cromar et al. (1999), up to 82% of produced seeds may be consumed by seed predators, most of which would be ground beetles. Cardina et al. (1996) indicated that ground beetles could account for half of the post-dispersal seed predation.

No significant effects of rotation or management system were observed for beetle diversity (Humble, 2001). The main difference in diversity was between the arable systems and the restored prairie. Shannon diversity index and evenness values averaged 1.6 and 0.6, respectively, for the arable systems compared with 2.2 and 0.85, respectively, for the prairie (Humble, 2001).

9.2.5 SUMMARY

The Glenlea study was initiated at a time when little scientific information existed on organic farming systems in Canada. Many of the results were not surprising and had been articulated by organic farmers in the region. However, by being able to compare performance of organic systems with conventional ones—and over a longer time period in the same study—we have increased our understanding of (1) how adoption

TABLE 9.6

Average Ground Beetle Capture at the Glenlea Long-Term Cropping Systems Study for 1995 and 1999, as Influenced by Crop Rotation Type and Crop Input Management

Rotation and Inputs	Average Ground Beetle Capture	
	1995	1999
Rotation 1		
+f+h	19.3	63.0
+f−h	30.0	122.0
−f+h	14.7	42.0
−f−h	18.7	76.3
Rotation 2		
+f+h	18.7	32.0
+f−h	17.3	83.3
−f+h	10.3	33.3
−f−h	18.7	75.3
Rotation 3		
+f+h	12.7	33.0
+f−h	21.7	47.3
−f+h	16.7	34.0
−f−h	17.7	37.0
Prairie	18.7	15.0
ANOVA (P-value)		
Prairie vs. rotations	0.3763	0.002
Rotation (R)	0.3997	0.0327
Treatment (T)	0.014	0.0001
R × T	0.1401	0.0087
SEM	2.88	11.67

Statistical analysis performed on log-transformed data, excluding prairie. Prairie statistically compared with +f+h subplots of rotations 1, 2, and 3. Means considered significant when $P < 0.05$.

f, fertilizer added based on soil test recommendation; h, recommended herbicides used for weed control; +/− refers to either addition (+) or omission (−) of fertilizer or herbicide.

of organic farming changes the system, (2) opportunities in the organic system, and (3) future problems long-term organic farming systems may encounter.

Under the conditions of this study, the organic system that appeared most sustainable was the forage–grain rotation. The decline in available soil P is a troubling development in this system but has since led us to new and interesting questions.

9.3 PHASE II: 2004–2012

9.3.1 ADJUSTMENTS TO THE STUDY

Changes were implemented after the first 12 years. In 2005, all phases of each rotation were included each year. This was accomplished by splitting the large plots into 4 m × 28 m plots, one for each crop in the rotation (see Figure 9.2). The variable crop input treatments were dropped, leaving just the organic and conventional comparison. Further, the study was reduced to two farming systems, the forage–grain and the grain-only systems.

In the forage–grain rotation, we switched the positions of wheat and flax (now forage–forage–wheat–flax). In the grain-only conventional rotation, genetically modified (GM) soybean was substituted for faba bean (now soybean–wheat–flax–oat) to assist in control of herbicide-tolerant weeds that had developed in that rotation. Also, the second wheat crop in the rotation was switched to oats in order to increase crop diversity. Starting in 2005, faba bean in the organic grain-only rotation was green manured instead of harvested for seed. Therefore, the grain-only organic system was transformed to a rotation that included a green manure every fourth year. Starting in 2008, the green manure in this rotation was switched from faba bean to a plant mixture (forage peas, non-GMO soybeans, and oats).

In 2002, a separate portion of each large plot in the forage–grain systems was manured with 10 Mg/ha composted beef cattle manure. This compost plot is described by Welsh et al. (2009) and was used exclusively in that study (no longer in system). In 2007, all organic plots were split in half; one randomly selected half of each plot (a 4 m × 12 m area) was treated with composted beef cattle manure. The manure source in all cases was Agriculture and Agri-Food Canada, Brandon Research Centre.

9.3.2 SOIL PHOSPHOROUS STUDIES

The 2003 observation of low available P in certain systems (Table 9.3) prompted a more detailed analysis of the P question. A simple P balance for the 1992 to 2004 period showed large deficits for the organic forage–grain system (−117 kg/ha) compared with the organic grain-only system (−52 kg/ha) and the conventional grain-only system (+32 kg/ha) (Welsh et al., 2009).

Earlier work by Oberson et al. (1993) showed little difference in unavailable or recalcitrant P on soils collected from conventional, biodynamic, and bioorganic cropping systems in the Swiss DOK trial. This was despite the observation that biodynamic and bioorganic systems had lower available P concentrations than soils from the conventional system (similar to Glenlea results). They concluded that a large reserve of P still existed in organic systems. In 2004, Welsh et al. (2009) used the modified Hedley sequential extraction procedure to extract different P pools from readily plant available, to moderately available, to unavailable P. The organic systems had lower concentrations of readily plant available P fractions than conventional, but recalcitrant fractions were not significantly different ($P < 0.05$). In this way, our results mirrored those of Oberson et al. (1993).

The organic forage–grain system displayed a reduction in recalcitrant P. Total P in this system was 259 mg/kg compared with 316 mg/kg in the organic grain-only system and 344 mg/kg in the conventional grain-only system (Welsh et al., 2009). Therefore, in the organic hay export (forage–grain) system, more than just the labile P fractions decreased over 14 years of organic management. The Welsh study demonstrated that only high-yield, high-P export organic systems are a concern for developing plant available P deficiency in the long term. It was interesting that the manure (applied in 2002) only raised the level of total P to 286 mg/kg in the organic forage–grain system. Oberson et al. (1993) also observed only a limited ability of compost to raise available P levels in their organic production systems. However, in both studies, additions of compost restored the system yield potential.

A 2009 assessment of available P (Bell et al., 2012) demonstrated that the P mining effect of the organic forage–grain rotation at Glenlea extended beyond the 0–15 cm soil depth. Available P in the 15–30 cm depth was 1.73 mg/kg in the organic forage–grain system compared with 2.8 mg/kg in the conventional forage–grain system ($P<0.05$). The P mining effect of the forage system extended to 60 cm. Stocks of plant available P in the 0–60 soil depth were 13.5 kg/ha in the forage–grain organic system compared with an average 40 kg/ha in the other arable systems and 48 kg/ha in the prairie (Bell et al., 2012). This observation suggests that limiting soil sampling for available P to the top 15 cm of soil may be inadequate in organic systems.

Microbial biomass P (MBP) was measured at Glenlea in 2011 (Braman, 2012). MBP levels were higher in organic systems, especially in autumn after soil rewetting following a prolonged drought. Autumn MBP measurements averaged 155 µg/g for organic systems compared with 90 µg/g for conventional systems and 200 µg/g for the prairie. Because microbial biomass C (MBC) and MBP tend to increase and decrease together (Brookes et al., 1982), low microbial P in the conventional system is likely due to low MBC (data not shown). These observations suggest that biological forms of P may be important in organic systems and require further research.

Another P fraction of interest is water-extractable P; these compounds tend to show decreased sorption and increased mobility in soils in comparison to ortho-phosphate (Bolan et al., 2011). At Glenlea, water-extractable P averaged 12 mg/kg in conventional systems compared with 7.8 mg/kg in organic systems (Welsh et al., 2009). Campbell et al. (1996) reported that rotations containing sweetclover green manure and bromegrass hay have higher rates of P removal and lower concentration of water-extractable organic P. Similar results were reported by Xu et al. (2013) working at Glenlea. These observations show that organic systems have a reduced risk of P runoff.

9.3.3 Arbuscular Mycorrhizal Fungi

Several previous studies have observed higher arbuscular mycorrhizal fungi (AMF) colonization in organic compared with conventional systems (see Chapter 3). In the most recent assessment of AMF at Glenlea, total colonization in flax averaged 30% for conventional compared with 47% in organic systems (Welsh et al., 2012). One reason for higher colonization in the organic system was low soil P. Mäder et al. (2000) and Oehl et al. (2002) attributed higher AMF in organic systems to a lack of available P. One other possible reason for differences in AMF is that the pH of soil at Glenlea is significantly lower in conventional than in organic systems (Welsh et al., 2009).

AMF spore abundance was assessed in 2005 (Welsh et al., 2012). Spore abundance was 78, 135, 98, and 130 spore/g dry soil for grain-only conventional, grain-only organic, forage–grain conventional, and forage–grain organic systems, respectively. The increases in spore populations due to organic management were correlated to increases in total AMF colonization (Welsh et al., 2012). The lower available P in organic systems was likely the greatest contributor to the larger density of spores. In a greenhouse experiment, Douds and Schenck (1990) observed decreased sporulation of different AMF species in the presence of high soil P.

Correspondence analysis showed differences in AMF spore community structure with some treatments (Welsh et al., 2012). The prairie plots were separated (on the biplot) from all other management systems (data not shown). This was expected since the management of soil in the prairie was profoundly different from the other systems. A shift in diversity between conventional and organic systems was also observed, but only in the grain-only rotation. Since differences in plant available P levels in the soil were more extreme between organic and conventional management in forage–grain compared with the grain-only rotation, the shift in diversity between management systems in the grain-only rotation may have been due to selection of AMF species by crop plants (or weeds) rather than a shift in diversity caused by P limitation. Crop rotation therefore may have had more of an impact on AMF diversity than management system. AMF spore density and diversity are shown in Figure 9.3.

9.3.4 SOIL NITROGEN

Measurements in 2003 showed lower available N in organic compared with conventional systems (Table 9.3). A follow-up study was conducted in 2006 to better understand the N-supplying power of the soils. Greenhouse and laboratory incubations were used to measure the N-supplying power of surface soils collected in autumn, 2006. Results of the 32-week greenhouse incubation showed slightly higher N supply due to organic in the grain-only rotation (103.4 vs. 94.4 kg/ha) compared with a much larger increase with organic management in the forage–grain rotation (conventional at 108.9 kg/ha vs. organic at 132.4). One reason for this trend was higher microbial biomass in the organic system, especially for the forage–grain rotation (Braman, 2012). This observation was supported by a laboratory incubation for NO_3 using the Campbell et al. (1993a) method. Therefore, while both organic systems had higher N mineralization than their conventional counterparts, the effect was once again greater in the forage–grain rotation.

Nitrate–N release curves for the 100-day incubation were as follows:

Grain-only conventional $= 0.1419x + 3.6067 r^2 = 0.83$
Grain-only organic $= 0.1683x + 4.1321 r^2 = 0.82$
Forage–grain conventional $= 0.1926x + 4.579 r^2 = 0.83$
Forage–grain organic $= 0.212x + 4.837 r^2 = 0.85$

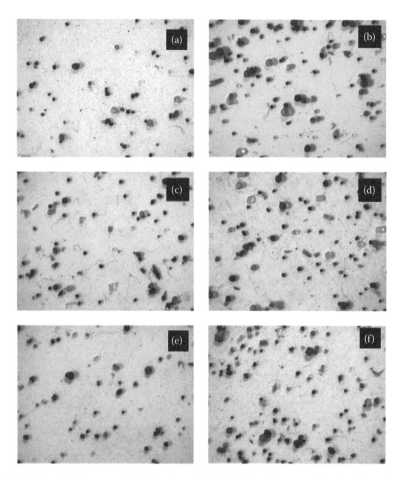

FIGURE 9.3 AMF spores in 100 g dry soil. (a) Grain-only conventional; (b) grain-only organic; (c) green manure–grain conventional; (d) green manure–grain organic; (e) forage–grain conventional; (f) forage–grain organic.

All of the N results here point to a higher N-supplying power of organic versus conventional soils, with the forage–grain system showing a greater benefit of organic management compared with the grain-only system. The results also suggest that low levels of available P in the organic forage–grain system did not appear to be limiting soil N supply power in this rotation.

Total soil N was assessed in 2009 by Bell et al. (2012). A significant rotation by management interaction for N in the 0–30 cm soil zone showed higher N stocks under organic management in the forage–grain rotation (increase from 6.4 to 7.2 tons/ha due to organic), but lower N stocks under organic management in the grain-only rotation (decrease from 7.7 to 6.8 tons/ha). A similar trend was observed in the 30–60 cm soil zone (Bell et al., 2012). These results confirm higher N in the organic system in the forage–grain rotation. On the other hand, the organic system in the grain-only rotation appears to be N limited.

9.3.5 Soil Carbon

Soil C was measured directly and indirectly. Nelson (2005) observed lower C in organic compared with conventional systems (4.8 vs. 5.1; P=0.01). Surprisingly, the organic systems had better soil stability. Organic systems had higher mean weight diameter (2.83 vs. 2.65, P=0.17) and wet aggregate stability (1.20 vs. 1.10, P=0.03) compared with the conventional system. One explanation for better soil stability at slightly lower C levels in the organic system may be higher AMF populations producing more glomalin than in the conventional system. Glomalin measurements at Glenlea would be a fascinating future research project.

MBC was measured in 2011 (Braman, 2012). A strong rotation by management interaction showed that MBC was higher in the organic (1648 µg/g) compared with the conventional (1476 µg/g) forage–grain system but that MBC was lower in the grain-only organic (1080 µg/g) compared with the grain-only conventional (1179 µg/g) system. Most literature supports the view that all organic systems have higher MBC (Fleissbach et al., 2007; Joergenson et al., 2010). Perhaps, the grain-only organic system in the present study would have fallen in line with the literature had the plots been regularly manured (which is the case for many studies in the literature).

A detailed assessment of surface and subsoil C was conducted on soil samples taken in the spring of 2009 (Bell et al., 2012). There was less C in both the surface, and especially the subsoil (between 60 and 120 cm), in the organic than the conventional systems. The authors concluded that results may be different if organic plots had been manured more regularly. Unfortunately, the Bell study did not include the manured forage–grain plots at Glenlea. No previous studies have measured subsoil C in organic field plots.

Plant residue C returned to the soil after harvest for the period 1992–2006 was also estimated (Bell et al., 2012). C return was 55 and 31 tons C/ha for the conventional and organic grain-only rotations, respectively, compared with 72 and 46 tons C/ha for the conventional and organic forage–grain rotations, respectively. It was interesting to note that the forage–grain organic system returned 83% as much C as the standard farming system in the area—the conventional grain-only system. The prairie returned 120 tons C/ha over the same period.

Xu et al. (2013) observed higher water-extractable organic C in the forage–grain conventional system; forage-cropping effects on water-extractable organic C may be due to the greater release of root exudates and return of residues to the soil (as seen in the present study—Bell et al., 2012), differences in rooting mechanisms, and higher fungal (Welsh et al., 2012) and bacterial biomass (Gulser, 2006).

The C research at Glenlea shows that while organic systems have slightly lower C stocks compared with conventional systems, they appear to have a more active C pool—at least in the forage–grain rotations. The observation of lower subsoil C in all organic systems is a worrying trend and requires further attention.

9.3.6 Some Organic Systems Mimic Prairie

The prairie plots allow us to test the question "Which arable farming systems best mimic the prairie?" Braman (2012) observed that two farming systems had MBC,

CO_2 respiration, and metabolic quotient (qCO_2) values statistically similar to the prairie. These two systems were the manured and unmanured organic forage–grain systems. Braman (2012) suggested that one reason for the comparatively poorer performance of the conventional systems was that synthetic fertilizer may be depleting some C fractions. This theory has been proposed by others (Mulvaney et al., 2009) but requires further validation.

It is important to recognize that the prairie in this study is restored grassland, not a natural prairie. Briar et al. (2012), working on nematode communities at Glenlea, concluded that while the prairie had superior nematode assemblages compared with the arable systems, it still lacks many of the characteristics of an undisturbed prairie. Nevertheless, the prairie plots are an important resource for future studies that address questions of sustainability and resilience of soil systems.

9.3.7 Grain Yield Performance 2005–2012

Yields from the 1996 to 2012 wheat crops are shown in Figure 9.4. One striking observation was the switch in which rotation produced the highest organic wheat yields. For the first 15 years, the organic forage–grain system produced more wheat than the organic grain-only system (and also the original green manure–grain system). However, by 2006, the organic grain-only system was producing higher wheat yields than the organic forage–grain system. Campbell et al. (1993b) observed

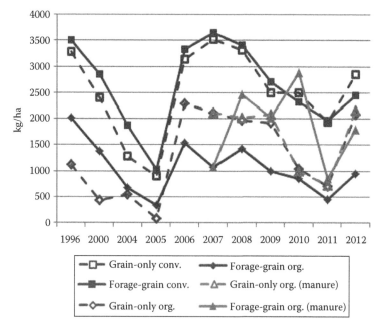

FIGURE 9.4 Wheat yields at Glenlea for the period 1996–2012. Annual data points for wheat yield available after 2005, when all phases appeared each year.

a similar trend at Indian Head, Saskatchewan, where an unfertilized forage–grain rotation produced wheat yields comparable to those in a fertilized grain-only rotation for the first 20 years of the study. After 20 years, however, wheat yields in the forage–grain system dropped below those from the fertilizer grain system—the reason was low soil P. At Indian Head, wheat productivity was maintained in the low-input system for 20 years; at Glenlea, a similar decrease in wheat yield was observed after 15 years. These two observations provide important guidelines for nutrient management in organic farming systems.

The other factor that may have contributed to improved wheat yields in the grain-only system was the introduction of green manuring. Starting in 2005, instead of growing peas or faba bean for seed, the faba bean was grown as a green manure. The effect of this rotational change was immediate in terms of raising organic wheat yields (Figure 9.4). This observation demonstrates that farmers who have ignored using a green manure crop may be able to restore productivity by introducing one. It appears it is never too late to start green manuring.

In response to P deficiencies in certain rotations at Glenlea, all organic plots were split in the fall of 2007, and one half of each plot was treated with composted beef cattle manure (10 tons/ha in grain-only system and 20 tons/ha in the forage–grain rotation). In the forage–grain system, the effects of compost on yield were immediate. Wheat yield in the manured system increased by 1000 kg/ha over the unmanured plots in 2008, and for the next 3 years (2008, 2009, and 2010), wheat yields averaged 85% as high in the organic forage–grain system compared with the conventional forage–grain systems (Figure 9.4). These results demonstrate the rapid restoration of productivity in a P-depleted organically managed clay soil.

A curious decline in performance of manured organic wheat in the forage–grain system was observed in 2011. This may be partly due to very wet soil conditions in early spring, which appeared to affect the organic system more than conventional systems. P may also be involved as available P was once again low after three post-manuring crops. Starting in the autumn of 2011, a 2 tons/ha compost maintenance rate has been added during the forage phase.

Manure additions had no effect on yield performance in the grain-only rotation (Figure 9.4). Soil test P results had consistently shown no P deficiency in this system (Welsh et al., 2009; Bell et al., 2012). This observation suggests that soil testing for P provides adequate guidance for P management.

Only infrequently did wheat in the forage–grain rotations yield less than the grain-only system (Figure 9.4), and it only happened under dry conditions. The greatest effect was in 2012, after the consecutive droughts of 2011 and 2012. Under dry conditions, the less water-intensive grain-only system appears to have conserved more water for wheat production.

Wild mustard remains one of the most abundant weeds at Glenlea. One of the most interesting observations at Glenlea is the near absence of wild mustard in the unmanured organic forage–grain system. It appears that the P deficiency in this system limits growth of this nonmycorrhizal weed. Flax, on the other hand, grows very well in this P-deficient system (the unmanured organic forage–grain). Hence, this is a story of flax having a unique competitive advantage over wild mustard—and it is all due to soil biology.

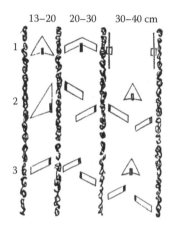

FIGURE 9.5   Tools for weeding in 13–40 cm row spacing. Translation from German "The dark spots on the tools represent the attachment points to the cultivator." (From Krafft, G., *Lehrbuch der Landwirtschaft: Pflanzenbau*, Verlagsbuchhandlung Paul Parey, Berlin, Germany, 1918.)

A final factor contributing to grain crop yield in 2012 was improved weed control in organic systems. Starting in 2012, interrow tillage was performed on the organic wheat and flax crops. The implement uses backswept knives that traveled approximately 1 cm below the soil surface between the narrow-rowed crops. Interrow tillage in small grains is a historical practice in Europe (Krafft, 1918; H.R. Entz, pers comm.). Figure 9.5 shows knife options for undercutting weeds in narrow-row grain production systems. High organic wheat yields, especially in the grain-only organic system at Glenlea in 2012, were attributed partly to weed control offered by this interrow cultivation.

9.3.8 PULLING IT ALL TOGETHER: A VISUAL SUMMARY

Spider analysis provided a holistic comparison of the four main rotation systems. In general, organic systems performed better for environmental health. For example, insect diversity, AMF colonization, MBC, and energy efficiency were typically higher in organic systems, though in some cases, the organic forage–grain system performed better than the grain-only system (Figure 9.6). Others (e.g., Hole et al., 2005) have also observed higher biodiversity in organic farming systems. Organic systems also performed better by having lower levels of water-soluble P, thus reducing risk of water pollution and eutrophication.

Conventional systems performed better for wheat yield and produced more human useable energy, though some interaction with crop rotation was observed. Other agronomic parameters such as weed control, P balance, and level of available N were superior in the conventional systems.

The spider analysis clearly showed how the addition of manure in the organic forage–grain rotation improved the overall sustainability of that system; P balance, wheat yield, and EP were all increased with manure (Figure 9.6).

When the performance of the most popular farming system in the Glenlea area (conventional grain-only production) was compared with the best organic system (forage–grain with manure added), advantages of organic farming become very clear. The organic system produced about 85% as much human useable energy, but performed better than the conventional system in almost every other category. This clearly shows the potential of organic agriculture. Higher energy yield in the conventional system also assumes that none of the grain crops, including soybean, were fed to livestock. The forage–grain conventional system had the highest production of human useable energy, but did not perform nearly as well as the organic system on environmental parameters.

9.4   CARMAN ORGANIC FIELD LABORATORY: A COMPLEMENTARY STUDY

One problem with static, long-term studies is that the rotations decided on at the beginning may not end up being the best agronomic cropping systems. This has been

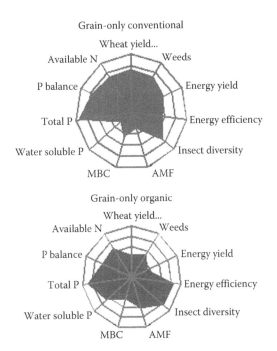

FIGURE 9.6   Effect of crop rotation and management system on wheat yield (max. 3500 kg/ha), weed density (max. 3700 plants/m²), human useable energy yield (max. 60,000 MJ/ha), energy efficiency (max. output/input ratio 12), carabid beetle diversity (max. 2.5 Shannon index), AMF (max. 7 cm/mL), MBC (max. 2500 µg/g), water-soluble P (max. 10 mg/kg), total soil-extractable P (max. 350 mg/kg), P balance (max. 150), and available N (max. 25 mg/kg). In all cases, the center of the graph is zero. Summary of data collected between 1992 and 2012.

(*continued*)

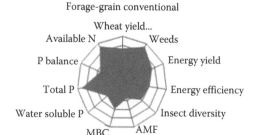

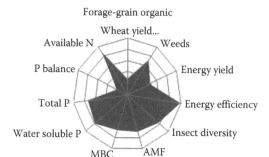

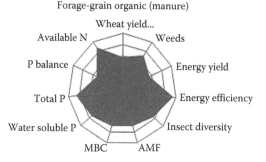

FIGURE 9.6 (continued)   Effect of crop rotation and management system on wheat yield (max. 3500 kg/ha), weed density (max. 3700 plants/m²), human useable energy yield (max. 60,000 MJ/ha), energy efficiency (max. output/input ratio 12), carabid beetle diversity (max. 2.5 Shannon index), AMF (max. 7 cm/mL), MBC (max. 2500 µg/g), water-soluble P (max. 10 mg/kg), total soil-extractable P (max. 350 mg/kg), P balance (max. 150), and available N (max. 25 mg/kg). In all cases, the center of the graph is zero. Summary of data collected between 1992 and 2012.

the case with the organic grain-only system at Glenlea; this system lacks N due to insufficient green manures in the rotation.

The Carman organic field laboratory was established in 2003 to allow an optimum organic cropping system to be tested. The rotation at Carman includes a green manure crop every third year. Plots are larger (0.7 ha each) that allow more typical tillage regimes. Finally, one of the two green manures in the 6-year rotation is grazed (with sheep) that improved the economics of this system.

TABLE 9.7

Yield of Green Manure (kg/ha Dry Matter) and Grain Crops (kg/ha), plus Net Returns for an 8-Year Period at the Ian N, Morrison Research Farm, Carman, Manitoba

	Year 1	Year 2	Year 3	Year 4	Year 5	Year 6
Year	Green Manure[a] (kg/ha)	Spring Wheat (kg/ha)	Soybean (kg/ha)	Green Manure[b] (kg/ha)	Flax (kg/ha)	Oat[c] (kg/ha)
2004	5397					
2005	4171	2470	2232	2729	1251	709
2006	3483	4148	1558	1994	1621	2612
2007	8480	3056	2128	7902	1188	3568
2008	5800	3625	1839	7593	1189	1838
2009	6165	4070	1919	7905	2265	3831
2010	3930	3310	2157	9058	1804	2633
2011	3352	1803	1862	8921	684	3404
Average	5097	3212	1956	6586	1429	2656
Net return ($ per acre)	−168 (30)[d]	215	317	−166	357	111

The 6-year rotation is green manure (grazed)–wheat–soybean–green manure–flax–oat.

[a] Year 1 green manure crops were pea/oat in 2004 and 2007–2011 and chickling vetch in 2005 and 2006. Green manure biomass weights include weed biomass, which was less than 10%.

[b] Year 4 green manure crops were berseem clover in 2005–2006, hairy vetch/oat in 2007, and hairy vetch/barley in 2008–2011. Green manure biomass weights include weed biomass (less than 10%).

[c] Fall rye was grown in place of oat in 2007.

[d] Value in parentheses when pea/oat green manure is grazed with sheep. The net return of negative $168 when green manure not grazed.

Yields from this unreplicated study are given in Table 9.7. Results indicate much higher yields than at Glenlea, though they are still 5%–20% lower than conventionally managed crops on the research station. The average annual net return across the entire rotation, which includes both fixed and variable costs, is $144 per acre for the organic rotation compared with $153 per acre for a comparable conventional system at the research center (Fryza and Carlberg, 2012, Dept. of Agribusiness, U of M, unpublished). We conclude that under these optimum management conditions, the organic system is a viable option for farmers.

9.5 CONCLUSIONS

Glenlea has produced practical results for the organic farming community. We have been able to identify strengths and weaknesses of organic systems. We have concluded that integrated systems that include perennial and annual crop phases along with some nutrient recycling are the most sustainable systems. We have provided a strong case for continued expansion of the organic sector.

The Glenlea study has demonstrated that organic systems offer more than just food—they also offer better environmental services.

ACKNOWLEDGMENTS

We are grateful to the staff at the Glenlea Research Station, in particular, Aime Chaput for fixing broken equipment and Mike Stuski for field operations. Thanks to the people of Manitoba and Canada, who supported the Glenlea study through their taxes. Thanks to the many farm writers for taking an interest in Glenlea and promoting the study—a special mention to Laura Rance. Special thanks to all the summer students who participated in field data collection since 1992. Finally, thanks to James B. Frey and his band of high school students for the construction of the Clubhouse Learning Centre.

REFERENCES

Bell, L., B. Sparling, M. Tenuta, and M.H. Entz. 2012. Soil profile carbon and nutrient stocks under long-term conventional and organic crop and alfalfa-crop and re-established grassland. *Agric. Ecosyst. Environ.* 158:156–163.

Blackshaw, R.E. and L.J. Molnar. 2009. Phosphorus fertilizer application method affects weed growth and competition with wheat. *Weed Sci.* 57:311–318.

Bolan, N.S., D.C. Adriano, A. Kunhikrishnan, T. James, R. McDowell, and N. Senesi. 2011. Dissolved organic matter: Biogeochemistry, dynamics, and environmental significance in soils. In: D.L. Sparks, ed., *Advances in Agronomy*, Vol. 110. Elsevier Academic Press Inc., San Diego, CA, pp. 1–75.

Braman, S. 2012. Soil health investigations at the Glenlea long-term crop rotation study. MSc thesis, Department of Plant Science, University of Manitoba, Winnipeg, Manitoba, Canada.

Briar, S.S., C. Barker, M. Tenuta, and M.H. Entz. 2012. Soil nematode responses to crop management and conversion to native grasses. *J. Nematol.* 44(3):245–254.

Brookes, P.C., D.S. Powlson, and D.S. Jenkinson. 1982. Measurement of microbial biomass phosphorus in soil. *Soil Biol. Biochem.* 14:319–329.

Brundtland, G. 1987. *Our Common Future: World Commission on Environment and Development.* Oxford University Press, Oxford, U.K.

Campbell, C.A., B.H. Ellert, and Y.W. Jame. 1993a. Nitrogen mineralization potential in soils. In: M.R. Carter, ed., *Soil Sampling and Methods of Analysis.* Canadian Society of Soil Sciences/Lewis Publishers, Boca Raton, FL, pp. 341–349.

Campbell, C.A., G.P. Lafond, J.T. Harapiak, and F. Selles. 1996. Relative cost to soil fertility of long-term crop production without fertilization. *Can. J. Plant Sci.* 76:401–406.

Campbell, C.A., G.P. Lafond, and R.P. Zentner. 1993b. Spring wheat yield trends as influenced by fertilizer and legumes. *J. Prod. Agric.* 6:564–568.

Campbell, C.A., R.P. Zentner, H.H. Janzen, and K.E. Bowren. 1990. Crop rotation studies on the Canadian prairies. Research Branch, Agriculture Canada Publication 1841/E, Ottawa, Ontario, Canada.

Carcamo, H.A. 1995. Effect of tillage on ground beetles (Coleoptera: Carabidae): A farm-scale study in central Alberta. *Can. Entomol.* 127:631–639.

Cardina, J., H.M. Norquay, B.R. Stinner, and D.A. McCartney. 1996. Post-dispersal predation of velvetleaf (*Abutilon theophrasti*) seeds. *Weed Sci.* 44:534–539.

Cromar, H.E., S.D. Murphy, and C.J. Swanton. 1999. Influence of tillage and crop residue on postdispersal predation of weed seeds. *Weed Sci.* 47:184–194.

Douds, D.D., Jr. and N.C. Schenck. 1990. Increased sporulation of vesicular-arbuscular mycorrhizal fungi by manipulation of nutrient regimes. *Appl. Environ. Microbiol.* 56:413–418.

Emerson, R.W. Essays. *Second Series: Experience 1844.* http://www.emersoncentral.com/experience.htm (accessed November 11, 2013).

Entz, M.H., R. Guilford, and R. Gulden. 2001. Crop yield and nutrient status on 14 organic farms in the eastern Northern Great Plains. *Can. J. Plant Sci.* 81:351–354.

Fleissbach, A., H. Oberholzer, L. Gunst, and P. Mäder. 2007. Soil organic matter and biological soil quality indicators after 21 years of organic and conventional farming. *Agric. Ecosyst. Environ.* 118:273–284.

Gulser, C. 2006. Effect of forage cropping treatments on soil structure and relationships with fractal dimensions. *Geoderma* 131:33–44.

Hoeppner, J.W., M.H. Entz, B. McConkey, B. Zentner, and C. Nagy. 2006. Energy use and efficiency in two Canadian organic and conventional crop production systems. *Renew. Agric. Food Syst.* 21:60–67.

Hole, D.G., A.J. Perkins, J.D. Wilson, I.H. Alexander, P.V. Grice, and A.D. Evans. 2005. Does organic farming benefit biodiversity? *Biol. Conserv.* 122:113–130.

Humble, S.M. 2001. Weeds and ground beetles (Coleoptera: Carabidae) as influenced by crop rotation type and crop input management. MSc thesis, Department of Plant Science, University of Manitoba, Winnipeg, Manitoba, Canada.

Janzen, H.H. 2001. Soil science on the Canadian prairies-peering into the future from a century ago. *Can. J. Soil Sci.* 81:489–503.

Joergenson, R.G., M. Mäder, and A. Fliessbach. 2010. Long-term effect of organic farming on fungal and bacterial residues on relation to microbial energy metabolism. *Biol. Fertil. Soils* 46:303–307.

Knight, J.D., R. Buhler, J.Y. Leeson, and S.J. Shirtliffe. 2010. Classification and fertility status of organically managed fields across Saskatchewan, Canada. *Can. J. Soil Sci.* 90:667–678.

Krafft, G. 1918. *Lehrbuch der Landwirtschaft: Pflanzenbau.* Verlagsbuchhandlung Paul Parey, Berlin, Germany.

Lund, R.D. and F.T. Turpin. 1977. Carabid damage to weed seeds found in Indiana corn fields. *Environ. Entomol.* 6:695–698.

Mäder, P., S. Edenhofer, T. Boller, A. Wiemken, and U. Niggli. 2000. Arbuscular mycorrhizae in a long-term field trial comparing low-input (organic, biological) and high-input (conventional) farming systems in a crop rotation. *Biol. Fertil. Soils* 31:150–156.

Mulvaney, R.L., S.A. Khan, and T.R. Ellsworth. 2009. Synthetic nitrogen fertilizers deplete soil nitrogen: A global dilemma for sustainable cereal production. *J. Environ. Qual.* 38:2295–2314.

Nelson, A.G. 2005. Soil erosion risk and mitigation through crop rotation on organic and conventional cropping systems. MSc thesis, Department of Plant Science, University of Manitoba, Winnipeg, Manitoba, Canada.

Oberson, A., J.C. Fardeau, J.M. Besson, and H. Sticher. 1993. Soil phosphorous dynamics in cropping systems managed according to conventional and biological agricultural methods. *Biol. Fertil. Soils* 16:111–117.

Oehl, F., A. Oberson, H.U. Tagmann, J.M. Besson, D. Dubois, P. Mäder, H. Roth, and E. Frossard. 2002. Phosphorus budget and phosphorus availability in soils under organic and conventional farming. *Nutr. Cycl. Agroecosyst.* 62:25–35.

Ominski, P.D., M.H. Entz, and N. Kenkel. 1999. The influence of alfalfa (*Medicago sativa*) on weeds in subsequent crops: A comparative survey. *Weed Sci.* 47:282–290.

Welsh, C., M. Tenuta, D.N. Flaten, J.R. Thiessen-Martens, and M.H. Entz. 2009. High yielding organic crop management decreases plant-available but not recalcitrant soil phosphorus. *Agron. J.* 101:1027–1035.

Welsh, C.M., M. Tenuta, T.P. McGonigle, Z. Gao, and M.H. Entz. 2012. Increased mycorrhizal colonization of flax unable to provide phosphorus nutrition to sustain yield in very P deficient soil as a result of long-term cropping and organic management. *Can. J. Plant Sci.* (in review).

Xu, N., H.F. Wilson, J.E. Saiers, and M.H. Entz. 2013. Effects of crop rotation and management system on water extractable organic matter concentration, structure, and bioavailability in a chernozemic agricultural soil. *J. Environ. Qual.* 42(1):179–190. doi: 10.2134/jeq2012.0126.

10 Breeding Cereals for Organic Soil Properties, Plant Nutrition, and Weed Control

Hakunawadi Alexander Pswarayi, Stephen Fox, Pierre Hucl, and Dean Spaner

CONTENTS

10.1 BACKGROUND

Cereal yields in organic systems are often lower than in conventional systems. This is because most cultivars deployed in organic systems have been developed in and for conventional systems (Lammerts van Bueren and Myers 2012). There are differences in soil fertility, weed, pest, and disease management (Lammerts van Bueren et al. 2002; Murphy et al. 2007; Wolfe et al. 2008) between organic and conventional systems, with organic systems generally poorer in nutrients and more challenged by pests and diseases. Cultivar adaptation is driven by the environment of selection. During breeding and selection, the testing environments serve to screen out those cultivars that are unadapted to the environments. Thus, selecting in an environment is essentially matching genes to the testing environment. Conventional cultivars are selected in systems of nutrient abundance and are also selected for high responsiveness to highly soluble fertilizers, provided in adequate quantities (Guppy and McLaughlin 2009; Lynch 1998; Marschner and Rengel 2003). Hence, in organic systems, conventional cultivars are highly likely to yield below optimum due to poor adaptation.

Poor adaptation, which translates to lower than optimal yields from conventional cultivars in organic systems, is because conventional cultivars generally lack important traits required under organic conditions (Lammerts van Bueren et al. 2002; Murphy et al. 2007; Wolfe et al. 2008). Selection under conventional systems tends to eliminate traits not suitable for conventional systems. Specifically, traits important for independence to external inputs (Lammerts van Bueren and Myers 2012), which are necessary for adaptation to organic environments, may be screened during selection in conventional systems. These include traits such as those essential for competitiveness against weeds, adaptation to arbuscular mycorrhizal fungi (AMF) for enhanced nutrient uptake, and disease and pest resistance and early plant vigor (Lammerts van Bueren et al. 2002).

In most cases, cereals have greater N demand during the later growth stages, which cannot be supplied from mineralization (Panga and Leteya 2000). There is thus a general lack of control of N supply to crops in organic systems (Mäder et al. 2002) to satisfy the different specific crop stage demands. Breeding for enhanced efficiencies in N and P acquisition and utilization is considered an important long-term strategy in organic systems (Wolfe et al. 2008). Against weeds, breeding for competitive wheat cultivars is considered an efficient alternative.

10.2 BREEDING PHOSPHORUS-EFFICIENT CULTIVARS

Breeding for phosphorus efficiency (PE) is possible due to availability of genotypic variation. PE is defined as grain yield per unit of nutrient P supplied from the soil (Moll et al. 1982). Breeders can select for two broad traits that determine P efficiency, P uptake or extraction, and P utilization. Uptake efficiency (UptE) is the ability of the plant to extract nutrients from the soil and is governed by root morphological and architectural characteristics or traits (Reynolds et al. 2001). Utilization efficiency (UtLE) is the ability of the plant to convert the absorbed nutrient into grain yield (Reynolds et al. 2001). Genetically, both P extraction (acquisition) and utilization efficiencies are polygenic traits that are independently inherited and can be independently selected (Gerath 1993). Selecting for P efficiency (uptake or utilization) does pose some problems. Firstly, there is the large amount of chemical analyses required to compare genotypes for either UptE or UtLE (Batten 1992), which might translate to delayed cultivar release and costly seed development. Secondly, it is difficult to choose which trait, UptE or UtLE, due to their environmental sensitivities that have given rise to variable responses in different environments (Reynolds et al. 2001).

For example, some of the environmental conditions that influence uptake and utilization of phosphorus in the field have been reported to be factors affecting water use efficiency (French and Schultz 1984), soil temperature (Barrow 1974), soil acidity (Reynolds et al. 2001), and planting date (Batten and Khan 1987; Smith and Batten 1976). Generally, conflicting results from different experiments can be inferred as evidence of environmental influence. For instance, some CIMMYT trials in acid Andisols with no aluminum (Al) toxicity showed that uptake was more important than utilization in explaining P use efficiency (Reynolds et al. 2001). In contrast, and using the same group of genotypes, UtLE was more important when evaluated in an alkaline vertisol (Manske et al. 2000a). Thirdly, selecting for P efficiency can be difficult because some traits are inefficient to measure, especially root traits related to UptE or extraction efficiency.

There are distinct traits to select for UptE/extraction efficiency. These traits are increasing the root length by increasing the root surface/soil contact area, increasing nutrient availability through rhizosphere modification, and increasing the effective root area (Reynolds et al. 2001). Likewise, different traits are involved in P UtLE. They include PE ratio = grain yield/unit P uptake (mg g^{-1}); phosphorus harvest index (PHI) = grain P content/unit P uptake $\times$ 100 (the proportion of shoot phosphorus in the grain at maturity); grain yield per unit of phosphorus translocated to the grain; and the proportion of shoot phosphorus in the grain at maturity or PHI (Batten 1992). Secondary traits can also be selected for P UtLE. The secondary traits are biomass, leaf area at flowering and kernels per head at serious P-deficient conditions (Olsen P = 2.4 mg kg^{-1}) or number of heads at slight P deficiency (Olsen 6.6 mg kg^{-1}) (Wang Lan-zhen et al. 2004), tallness, and delayed heading due to P deficiency.

10.2.1   BREEDING FOR P UPTAKE/EXTRACTION EFFICIENCY

10.2.1.1   Increasing the Root Length

Two factors make breeding for increased root length in organic systems logical. The first is that phosphates are generally immobile in soil and a long root system enables the plant to scavenge and acquire P. Also, phosphate concentrations in soil solutions are very small (<0.05 μg^{-1}), resulting in very little phosphate movement to roots by capillary action (Reynolds et al. 2001).

Several studies have reported that there is variation for root length among wheat genotypes (Gahoonia and Nielsen 2004; Gahoonia et al. 1997, 1999) to enable selection. Additional factors make selecting for root length even more compelling. They are (1) the high (0.14–0.51) heritabilities for root length (Porceddu et al. 1978), (2) high correlation values (0.50–0.60) between root length density and P uptake or grain yield under low P supply (Reynolds et al. 2001), (3) high relative root growth rates by genotypes tolerant to low soil P (Leon and Schwang 1992; Wissuwa et al. 2009), (4) higher P uptake (Manske et al. 2000b), and (5) better yields (Barraclough and Leigh 1984) in wheat genotypes with greater root length in low P organic soils.

Two ways are available to select for increased root length: selection for modified root morphology or selecting for increased root hair development. Selecting for modified root morphology involves selecting for roots with higher specific root length, that is, roots with smaller diameter (finer roots). Smaller root diameters are ideal as they enable a larger soil volume per unit root surface area to deliver nutrient P to the plant (Gahoonia and Nielsen 2004). In wheat, root fineness has been reported to be an important determinant of P UptE (Jones et al. 1989). Larger root diameters cover less surface areas for a given constant level of root biomass compared to smaller root diameters, which is less efficient for the same amount of maintenance photosynthates. However, the fact that finer roots have to be replaced more frequently, a process requiring more carbon cost (Persson 1982), may cancel the advantage of genotypes with finer roots. In contrast, selecting for root hair development as a way of increasing root length is less costly metabolically and also relatively easier to assess compared to selecting for finer roots (Gahoonia et al. 2004; Hetrick 1991; Röhm and Wener 1987). Obviously, it is relatively easier, less time consuming, and accurate to measure the biomass of root hairs than the diameter of a fine root. Root hairs are known to enhance P uptake due to their increased effective root surface area. In turn, an increased effective root surface area increases the effectiveness of roots to exploit the rhizosphere soil for nutrients (Fohse et al. 1991; Gahoonia et al. 1997).

While it is logical and possible to breed for root length, challenges exist that make this trait less favorable with breeders. Firstly, the large number of genotypes that need screening in breeding programs is both labor intensive and time consuming (Reynolds et al. 2001). Typically, preliminary selection trials in breeding programs are made of thousands of genotypes. Secondly, the screening methodologies used for measuring root lengths are prone to errors (Gahoonia and Nielsen 2004). To avoid these errors, model simulation has been suggested. However, model simulations are problematic in that they need field experiments for validation. Such field experiments are notorious for being labor intensive (Gahoonia and Nielsen 2004; Hoad et al. 2001; Lynch 1995). Hence, root traits have not been popular with breeders. We, however, make mention of them in this chapter because the traits have been studied and selections attempted by some breeders.

10.2.1.2　Rhizosphere Modification

Essentially, rhizosphere modification aims at exploiting the unavailable P bound on soil particles in the rhizosphere. It is estimated that 40%–50% of a plant's total P supply comes from poorly available organic P through solubilization by root exudates (Randall 1995). There are genotypic differences for protons and complex organic molecules that are capable of solubilizing soil-bound P making it available for uptake, to enable selecting for root exudates (Lambers et al. 2006; McLachlan 1980). In wheat, cultivars growing in an Andisol soil showed an association between acid phosphatases and P uptake (Portilla-Cruz et al. 1998). Examples of exudates that can be manipulated to enhance phosphate solubilization are phosphatases that transform poorly available organic P into inorganic forms available to the plant (Randall 1995). However, selecting for enhanced excretion of phosphatases to access organic P forms in soils is less promising because of complex sorption/desorption characteristics in soil (George et al. 2004).

10.2.1.3   Increasing Effective Root Area Selecting for AMF

The effective root area is increased by a symbiotic association between plants and AMF. Hence, breeding and selecting for increased effective root area are essentially selecting for symbiotic associations between host plants and AMF. Many crop species partially depend on AMF for uptake of immobile nutrients (Smith and Read 1997). Although mycorrhizal fungi can take up a number of nutrients, including nitrogen, phosphorus, potassium, calcium, magnesium, sulfur, iron, manganese, copper, and zinc (Al-Karaki et al. 2004; Cruz et al. 2004; Mohammad et al. 2003, 2005; Ryan et al. 2004), AMF are most beneficial in the uptake of relatively immobile nutrients such as phosphorus, copper, and zinc (Habte 2006). It is reported that the influx of P in roots colonized by AMF can be up to three or five times higher than in nonmycorrhizal roots (Smith and Read 1997). This influx is aided by an extensive hyphal network that attaches to host plant roots and ramifies through the soil matrix, outside of the rhizosphere (Harrier and Watson 2003), to increase the capability of plants to take up mineralized soil nutrients present at low concentrations outside the rhizosphere (Mader et al. 2000). Increased uptake of P through mycorrhizal colonization can significantly increase P concentrations in wheat grains. However, the intensity of the effect can be altered by wheat cultivar, mycorrhizal species, or the soil environment (Al-Karaki et al. 2004).

Mycorrhizae can generate a number of additional benefits to the soil system and the plant such as stabilization of soil aggregates, suppression of plant fungal pathogens, reduction of plant parasitic infection by nematodes, protection of plants from drought and saline conditions, and protection of plants from heavy metals (Habte 2006). In addition, mycorrhizae can positively affect the community structure of other soil microorganisms, by contributing carbon compounds to the soil system as well as influencing good soil structure formation (Hamel 2004; Hamel and Strullu 2006). The numerous benefits from AMF are reported to be critical to the development of sustainable agricultural systems (Douds et al. 1997; Hamel 2004; Plenchette et al. 2005; Rabatin and Stinner 1989). There is, however, a cost to these numerous benefits. In particular, the host plant must supply photosynthetically fixed carbon to the AMF in exchange for nutrients. Generally, this translates to a cost of 4%–20% of the plant's photosynthetically fixed carbon to the AMF or fungal partner (Bago et al. 2000; Wright et al. 1998).

There is genotypic variation in plant colonization by AMF and plant responsiveness to AMF (Baon et al. 1993; Hetrick et al. 1993; Manske et al. 1995); hence, breeding and selection for the trait are possible. There are, however, some conflicting reports about genetic variation, especially among wheat genotypes. While some studies suggested differences in mycorrhizal association among wheat cultivars (Vlek et al. 1996), others, in particular CIMMYT's extensive screening of spring wheat cultivars for mycorrhizal association, found very small differences (Manske et al. 2000b).

Conflicting reports can indicate differences in the materials tested or environments under which the materials were tested. The influence of the environment on AMF associations has been observed in some studies. One example is Janos (2007) who reported that the relative performance of colonized plants (i.e., responsiveness) will change depending on environmental conditions. Given that most mycorrhizal associations with plants have been shown to be quantitatively governed (Smith and Goodman 1999), the environmental influence on AMF associations would be expected. If a trait is environmentally influenced, then cultivars should be tested in different environments during development.

Genotypic variation in colonization and responsiveness (benefitting) seems to be based on the ability of plants to regulate the photosynthetic carbon cost in the AMF/plant relationship (Graham and Eissenstat 1994; Javot et al. 2007; Schaarschmidt et al. 2007). The ability to regulate the photosynthetic carbon cost can be successfully bred with two alternate routes. One way is to manipulate the plant genetic factors that impact the cost–benefit balance of an AMF symbiosis, in particular the rate at which resources are diverted to the fungus, in return for a given level of enhanced performance or reduction in production costs (Ruairidh et al. 2007). Ideally, one would select genotypes with the least rate of diversion to fungi and maximum benefit to the plant host. The alternative is to select for alterations in plant regulation of fungal growth (Ruairidh et al. 2007). An ideal situation would be one in which fungal growth would not be the excessive and therefore not costly, in terms of the fixed carbon needed to support the extensive hyphae of the AMF.

Two traits, root colonization and responsiveness to AMF colonization, are equally important in AMF root symbiosis and therefore should be selected simultaneously. This was implied in one study of wild ancestors of wheat, where it was observed that although the roots of all accessions tested were colonized by mycorrhizal symbionts, not all of the accessions showed positive growth response (Hetrick et al. 1993). From the same foregoing study, it is speculated that two sets of genes are involved, one set for the process of colonization and the other for nutrient acquisition by the plant (Hetrick et al. 1993).

Over the years, breeding and selection appear to have resulted in selecting for genes for colonization at the expense of responsiveness. This derives from observations made from several studies where it was reported that the degree of colonization was greater in modern cultivars than older cultivars (Hetrick et al. 1992; Kirk et al. 2011; Zhu et al. 2001). This was supported by a study comparing allelic diversity between landraces and modern cultivars, where it was observed that allele richness had declined in modern cultivars (Fu et al. 2005), which could imply selection favoring one set of genes, the genes for colonization. In another related study, alleles contributing to beneficial plant–microbe interactions that are associated with efficient nutrient acquisition strategies were reported to have been possibly lost, compromising yield (Lambers et al. 2006), which further confirmed the possible loss of one set of genes, those relevant for responsiveness to AMF.

Two theories exist as to why it is responsiveness to AMF associations that has suffered during the years of selection. The first is related to the environment of selection used for modern cultivars. In general, low soil P concentrations found in natural ecosystems favor plant responsiveness to AMF. However, modern cultivars are typically selected in standardized, high-fertility soil conditions with a primary focus on yield (Wissuwa et al. 2009). Under such conditions, rhizosphere microbial communities are faced with an environment that differs substantially from the one in which plant–microbial interactions originally evolved from (Drinkwater and Snapp 2007), resulting in genes relevant for symbiotic associations being screened out during breeding and selection. The second theory postulates that, by breeding and selecting for resistance to fungal pathogens, genes important in supporting interactions with AMF have been lost (Smith and Goodman 1999). Therefore, susceptibility to fungal pathogens can be inferred to have been an adaptive strategy to allow AMF associations with plants (Wissuwa et al. 2009). If this is true, breeding is faced with a conundrum of how to simultaneously select for both AMF associations for P UptE and resistance to pathogenic fungi in organic systems. It must be pointed out that there are, however, conflicting reports as to which genotypes, landraces, or modern cultivars are more responsive to AMF associations. Some studies suggested that modern wheat cultivars were more responsive under conventional management (Nelson et al. 2011).

10.2.2 BREEDING FOR NUTRIENT P UTILIZATION EFFICIENCY

There are several traits that can be possibly selected for P UtLE: PE ratio, PHI, grain yield per unit of P translocated to the grain, the proportion of shoot P in the grain at maturity or PHI, crop biomass production efficiency (BPE), and secondary traits (tallness, biomass production, delayed heading, and number of heads). Secondary traits (tallness, biomass production, delayed heading, and number of heads) used to select for P UtLE are relatively easier to measure and select for and are more useful for breeding.

Generally, the traits selected for P UtLE increase the ability of the plant to convert as much of the absorbed P as is possible into grain. This implies that not all of the P absorbed from the soil by the plant is used to produce the grain. Phosphorus UtLE is related to improving BPE (Reynolds et al. 2001). Biomass production is a stable trait across environments (Schulthess et al. 1997). Stability across environments implies less testing sites required during selection and the possibility of selecting cultivars for broad adaptation.

There is a negative correlation between increased biomass and P concentration in grain, complicating selection for this trait. An excessively low value of P concentration in the grain derives from P dilution in the plant. Such low P concentrations are known to affect seed vigor (Reynolds et al. 2001). P dilution in the plant in turn derives from two possible scenarios: increasing biomass production with the current levels of P in the plant or the maintenance of biomass production with a lower concentration of P in the plant (Reynolds et al. 2001). Essentially, selecting for increased biomass production with the same levels of P in the plant or to maintaining the same biomass production with lower P in the plant is actually selecting cultivars that extract less P from the soil. This is ideal for sustainable land use (Schulthess et al. 1997), especially in organic systems where low soil P is typical.

10.3 BREEDING FOR NITROGEN USE-EFFICIENT WHEAT CULTIVARS

Nitrogen use efficiency (NUE) is defined as NUE = grain yield/N applied (with or without top dressing, i.e., all available N) (Moll et al. 1982). In the case of organic systems, N-efficient cultivars produce grain from organic N fertilizers ploughed into the soil. NUE is determined by two broad traits, N UptE and N UtLE. UptE, defined as total plant N/N applied, describes the ability of the plant to extract N from the soil. UtLE, as grain yield/total plant N, describes the ability of the plant to convert the absorbed nutrient into grain yield (Huggins and Pan 2003; Moll et al. 1982). Both UptE and UtLE have been credited with genetic gains in NUE in cultivars released under low N environments (Ortiz-Monasterio et al. 1997a). However, UptE and UtLE can be simultaneously or independently selected for in cultivars. There is adequate variability among wheat genotypes (Barraclough et al. 2010; Laperche et al. 2006; Le Gouis et al. 2000; Ortiz-Monasterio et al. 1997a; van Ginkel et al. 2001) and fairly high trait heritabilities (Coque and Gallais 2006; Presterl et al. 2002) to enable selection for both UptE and UtLE. Within the UptE and UtLE, broad traits are specific traits that can be broadly classified into physiological and root traits.

10.3.1 Physiological Traits

There are several physiological traits that determine NUE in wheat genotypes. These traits include harvest index (HI) and biomass production (Reynolds et al. 2001); the ability to take up more N at low concentrations (Chapin 1980; Jiang and Hull 1998); reduced plant N requirements (Cassman et al. 2002); N conservation, internal N cycling, and adaptation to low N growing conditions (Dawson et al. 2008); conservation of N within plant tissues rather than maximum biomass production per unit N (Silla and Escudero 2004; Vazquez de Aldana and Berendse 1997); rapid early-season N uptake, efficient remobilization and complete leaf senescence, and late-season N uptake; and the ability of leaves to continue photosynthesis under N stress (Dawson et al. 2008). The ability of leaves to continue photosynthesis under N stress is also known as *stay green*. Rapid early-season N uptake, efficient remobilization, and complete leaf senescence are traits better suited for environments with severe N stress, especially late in the season when grain-filling demands for N are high. The rapid early-season uptake stocks up on the abundant N typically found during early season from mineralization in most organic conditions. Remobilization and complete senescence are important in ensuring that most of the N in the plant is channeled to the grain. Late-season N uptake is ideal in environmental conditions where soil N is available late in the season (Dawson et al. 2008), which is, generally, not the case in organic systems. Given that variable weather conditions can precipitate N shortages early in the season or late in the season, the best strategy for breeders would be to select for both early- and late-season uptake traits in cultivars destined for organic systems.

HI is one of the most exploited traits in selecting for NUE that is associated with NUE in most old cultivars (Reynolds et al. 2001). However, further improvements in NUE through breeding seem to be limited (Calderini et al. 1995; Fischer 1981), most probably due to a selection plateau. As a result, two alternatives have been suggested: either to increase grain yield, while maintaining and/or reducing nutrient concentration in the plant, or to reduce the total nutrient concentration in the plant, while increasing or maintaining grain yield. However, these two alternatives have a major problem in that they lower grain protein, therefore wheat end-use quality (Weegels et al. 1996). The reduction in grain protein concentration as grain yield increases occurs due to the negative correlation between grain protein and grain yield

(Anderson and Hoyle 1999 in wheat; Calderini et al. 1995; Dawson et al. 2008; Ortiz-Monasterio et al. 1997b; Sinebo et al. 2004 in Ethiopian barley; Witcombe et al. 2008).

Careful selection strategies can break the linkage between NUE and grain quality (e.g., Dawson et al. 2008). The strategies include selecting for increased N storage capacity of the leaves and stems, as well as selecting for increased allocation of N to nonstructural proteins (Martre et al. 2007). In addition, there are other traits that can help break the negative correlation between NUE and quality. They are (1) traits for the internal N requirements for growth and development (Singh and Buresh 1994); (2) traits giving the ability to translocate, distribute, and remobilize absorbed N (Ladha et al. 1993); (3) traits for flag leaf N import/export and leaf senescence patterns; and (4) traits for NUE in converting CO_2 to carbohydrates.

The ability of leaves to continue photosynthesis under N stress, also known as stay green, is one of the NUE traits that is easy to evaluate (visually) in a breeding program, without any need for complicated equipment. The stay green trait enables the plant to retain their green color and continue photosynthesizing for much longer than is normal. Generally, genotypes that stay green for much longer produce greater biomass and have higher grain yield potential. In some extreme cases, the stay green trait may delay plant leaf senescence until the grain is completely mature (Dawson et al. 2008). In such cases, the stay green trait gives plants greater capacity to take up N during grain filling due to continued leaf activity that promotes the uptake of soil N (Woodruff 1972). The stay green trait is especially crucial in annual grains such as wheat (*Triticum aestivum* L.) because the size and duration of active leaf tissue directly bear on N uptake and the amount of plant N available for remobilization (Habash et al. 2006; Palta and Fillery 1995). It is estimated that 60%–95% of the grain N comes from the remobilization of N stored in roots and shoots before anthesis (Habash et al. 2006; Palta and Fillery 1995).

10.4 BREEDING FOR WEED CONTROL IN ORGANIC SYSTEMS

The traits that can be selected for weed control in organic systems are tolerance to damage by mechanical weed control machinery, especially tine weeders (Donner and Osman 2006; Lammerts van Bueren et al. 2010; Murphy and Jones 2008), some root characteristics, and cultivar competitiveness against weeds (Mason and Spaner 2006; Mason et al. 2007a,b,c). Tolerance to mechanical weed control damage is an essential trait since mechanical weed control is practiced at higher than normal frequencies in most organic systems (Lammerts van Bueren et al. 2010) due to higher weed pressures. While there is evidence of genotypic differences in root characteristics among spring wheat cultivars (e.g., O'Brien 1979), many root characteristics are of less practical use by plant breeders, given the difficulties associated with assessing root traits (Wu et al. 2000).

There is variation for cultivar competitiveness against weeds among wheat genotypes (Huel and Hucl 1996; Lemerle et al. 1996; Wicks et al. 1986) to enable breeding and selection. As a result, some competitive ideotypes have been developed for a number of geographic regions (Lemerle et al. 2006; Mason et al. 2007b). However, the variation in competitive ability among wheat genotypes has not been confirmed in some studies (Cosser et al. 1997; Rasmussen 2004). Others (Cousens and Mokhtari 1998;

Didon and Bostrom 2003; Lemerle et al. 2001b) reported that cultivar competitiveness against weeds varied with environments. Hence, conflicting reports can be an indication of environmental influence on a trait, among other factors. Sensitivity to different environments makes it imperative to test cultivar competitiveness against weeds over a wide range of environments that include different agronomic conditions (Cousens and Mokhtari 1998; Lemerle et al. 2001b) in order to make efficient selections. As well, sensitivity to different environments may mean having to breed and select for narrow and very specific locations or systems, instead of broad-based weed competiveness.

Cultivar competitive ability against weeds is contributed by several specific traits (Lemerle et al. 2006; Mason et al. 2007b) that work together (Lemerle et al. 1996) or interact (Mason et al. 2007a). However, it is generally difficult to identify traits that consistently increase competitive ability against weeds (Mason et al. 2007a). Specific traits contributing to competitiveness against weeds can be divided into two broad groups, depending on the mode of action against weeds. They are the ability to tolerate weed pressure while maintaining grain yield and the ability to suppress weed growth and weed seed production (Coleman et al. 2001). With the ability to tolerate weeds, cultivars can grow and be productive side by side with weeds. Cultivars with an ability to suppress weeds will actively impede weed growth and reproduction. The two broad traits may or may not occur together in a single individual genotype (Lemerle et al. 2001a) and can be described by above- and belowground plant morphological and physiological characteristics (Lemerle et al. 2001a). Most studies have, however, concentrated on examining aboveground morphological and physiological traits (Mason et al. 2007a) because they are easy to measure and assess.

10.4.1 Morphological Characteristics

Morphological characteristics are related to the physical characteristics, especially of the plant canopy structure. Canopy structure determines the crop's ability to shade weeds (Mason et al. 2007a); hence, shading ability is used as a measure of the overall competitive ability of a genotype against weeds (Eisele and Köpke 1997). Several parameters that describe the ability of the canopy structure to shade weeds that can be selected include leaf area index (LAI), flag leaf length and angle, and canopy tallness or plat height. LAI is negatively correlated with weed seed yield (Huel and Hucl 1996), that is, a large LAI implies better weed suppression. Flag leaf length has a strong negative correlation with wheat yield loss (Huel and Hucl 1996; Lemerle et al. 1996), implying shorter leaves incur higher yield losses due to weed competition. Leaf angle describes the erectness of leaves. Higher leaf angles result in less erect or planophile leaves that in turn makes genotypes more competitive against weeds (Huel and Hucl 1996; Fischer et al. 2000; Richards and Whytock 1993) by increasing the shading and light interception abilities of the crop canopy. However, some studies have shown that flag leaf angle is positively correlated with wheat yield reduction (Huel and Hucl 1996), which means, the higher the flag leaf angle, the higher the yield losses. Hence, an ideal ideotype may have planophile nonflag leaves in order to suppress weeds and erect flag leaves to generate high yields.

Plant height influences competitiveness against weeds by influencing the amount of solar radiation the canopy structure intercepts. Taller wheat cultivars were observed to be more yield stable across weed environments than semidwarf cultivars that are more sensitive to changes in weed level (Huel and Hucl 1996; Mason et al. 2007a). Tall plants intercept more solar radiation (photosynthetically active radiation [PAR]) above the weed layer, enabling the crop to escape heavy competition for soil moisture and nutrients (Mason et al. 2007a,b,c; Richards and Whytock 1993) and suppress weeds (Champion et al. 1998). Solar interception during early (Lemerle et al. 1996) and late (Wicks et al. 1986) growth stages is associated with wheat grain yield maintenance under competition from weeds.

There are conflicting reports from studies on the importance of tallness to competitiveness against weeds. Some studies have reported that shorter cultivars were good competitors against weeds (Wicks et al. 1986), while other studies have reported variability in suppressing weeds by some taller cultivars compared to shorter cultivars (Cosser et al. 1997). Crop researchers seem to agree that selection should combine tall cultivars with high LAI for good ground cover (Champion et al. 1998; Cosser et al. 1997; Fischer et al. 2000; Gooding et al. 1993; Huel and Hucl 1996; Hucl 1998; Korres and Froud-Williams 2002; Lemerle et al. 1996; Wicks et al. 1986) and rapid canopy development in order to be more competitive against weeds.

10.4.2 PHYSIOLOGICAL TRAITS

A number of easily assessable physiological traits can be selected to enhance the competitive ability of cultivars against weeds. Most of these physiological traits are also selected to assess yield potential during cultivar development, meaning no extra work in terms of data recording or sample processing in a breeding program. The traits include heading date, days to maturity, tillering capacity, and number of fertile tillers, greater spikes m^{-2}, strong early-season vigor, and greater early biomass accumulation (Drews et al. 2009; Lemerle et al. 1996). Strong early-season vigor is related to increased yield, increased spikes m^{-2}, and reduced weed biomass (Mason et al. 2007a). Early-season ground cover also reduces subsequent weed biomass (Huel and Hucl 1996). Fast early-season growth, early maturity, and elevated fertile tiller number are particularly important traits for a competitive spring wheat ideotype organically grown in the northern growing regions of the Canadian Prairies (Drews et al. 2009; Mason et al. 2007a). Greater tiller numbers have generally been identified as an important trait in most competitive genotypes in a number of studies of mainly Australian wheat genotypes (Lemerle et al. 1996).

10.5 SELECTION ENVIRONMENT

10.5.1 DIRECT OR INDIRECT SELECTION?

Adaptation is a measure of genotypic performance in an environment. Testing for adaptation is essentially testing if a cultivar has genes suitable to the demands of a particular environment. The best way to know if a cultivar carries genes suitable for a particular environment is to test the cultivar in that particular or target environment. Testing a cultivar in a target environment is called direct selection.

The decision to use direct selection in breeding programs is based on the occurrence of significant genotype by environment interaction (GE). In organic systems, significant GE is common due to the extremely variable conditions (Wolfe et al. 2008). Variability in organic conditions is governed by biodiversity at soil, crop, field, whole rotation or polyculture, and landscape levels and in the way they integrate crops and livestock production systems on the farm (Mäder et al. 2002). This variability in conditions in organic systems supports the use of direct selection in breeding cultivars targeted for organic systems (e.g., Burger et al. 2008; Messmer et al. 2009; Reid et al. 2009). In addition, given the inherent stresses that characterize organic environments (Lammerts van Bueren and Myres 2012), direct selection under representative stress conditions may be the most efficient form of selection (Banziger et al. 2006; Lafitte et al. 2006; Reid et al. 2009, 2011). Indirect selection would result in cultivars characterized by irrelevant traits not suitable for the stressed conditions. However, selecting solely in the representative stress conditions translates to costly breeding programs given the number of different stress environments in organic systems.

Currently, indirect selection under conventional conditions is the preferred method for selecting cultivars destined for organic systems. It is envisaged that such indirect selection for organic cultivars will continue for the next two decades, due to the financial constraints (Osman et al. 2007) brought about by direct selection. However, modifications that make conventional systems somewhat resemble organic systems may enhance the selection efficiency of indirect selection. In particular, if conventional conditions are modified to low-input (LI) conditions by significantly reducing the rates of external inputs such as fertilizers, then LI conditions are used for early-generation selections (Oberforster 2006; Oberforster et al. 2000; Przystalski et al. 2008). Later-generation selections are then run as parallel replicated yield trials under both organic and LI conditions in order to separate breeding material for organic and conventional farming systems (Löschenberger et al. 2008). The organic systems used in the later-generation trials are the different target organic farms where the cultivars are destined to be grown; hence, farmers participate in the development of cultivars suitable for their particular environments. The by-product of combining both direct and indirect selections is collaborative breeding and selection between scientists and farmers, which is termed participatory plant breeding (PPB).

10.5.2   PPB: Combining Direct and Indirect Selection

The concept of PPB exploits and compliments different skills and knowledge possessed by farmers and scientists (Desclaux 2005; Wolfe et al. 2008). Scientists may perform a number of functions such as (1) the provision of relevant information to enable the choosing genotypic materials that carry desired traits, (2) carrying out early-generation selections for highly heritable traits such as dry matter content under LI conditions, and (3) selecting for traits that might not be easily measurable at the farmer level, such as those traits that may need laboratory facilities (examples in Table 10.1). In addition, scientists may help with analyses that are useful for documenting populations or genotypes developed (Dawson et al. 2011).

TABLE 10.1
Traits for Organic Soil Properties, Nutrition, and Weed Control

Trait	Remarks
P extraction efficiency and UtLE	
Higher biomass production	Labor intensive to assess
AMF colonization	Sophisticated to measure
AMF responsiveness	Sophisticated to measure
Finer roots	Labor intensive to assess
Increased root hair development	Labor intensive to assess
Enhanced root exudates (phosphatases)	Sophisticated to measure
Nitrogen UptE and UtLE	
Ability to take N at low N concentration	Sophisticated to measure
Reduced plant N requirements	Sophisticated to measure
N conservation within plants	Sophisticated to measure
Leaf stay green under N stress	Sophisticated to measure
Plant internal N cycling	Sophisticated to measure
Low biomass production/unit N	Sophisticated to measure
Low internal N turnover	Sophisticated to measure
Lower relative growth rate	Sophisticated to measure
Increased root affinity for soil N at low soil N levels	Sophisticated to measure
Late-season N uptake	Sophisticated to measure
Rapid early-season N uptake	Sophisticated to measure
Efficient remobilization with complete leaf senescence	Sophisticated to measure
Higher root to shoot ratio	Labor intensive to assess
Better root growth	Labor intensive to assess
Root distribution	Labor intensive to assess
Root architecture	Labor intensive to assess
Mycorrhizal connections	Labor intensive to assess
Genotype competitiveness against weeds	
Tall plants	Labor intensive to assess
Planophile leaves	Labor intensive to assess
Elevated PAR interception	Sophisticated to measure
Greater early biomass accumulation	Labor intensive to assess
Rapid leaf canopy development	Labor intensive to assess
High LAI	Labor intensive to assess
Early-season vigor	Subjective
Early maturity	
High tillering capacity	Labor intensive to assess
High fertile tillers/unit area	Labor intensive to assess
Tolerance to mechanical damage	Labor intensive to assess

P, phosphorus; N, nitrogen; AMF, arbuscular mycorrhizal fungi; PAR, photosynthetically active radiation.

In PPB programs, where farmers use their farms to further develop cultivars, partially developed seed from scientists is given out to farmers for further selection, especially for yield under their individual organic farms, which may result in different cultivars being developed from farm to farm. Some PPB programs involve farmers in selecting genotypes at central research stations, not their individual farms. Farmers in PPB programs contribute in-depth knowledge of environmental conditions and plant traits that are adaptive to their specific farm conditions (Dawson et al. 2011). Involving local farmers in PPB projects also helps in the maintenance and conservation of genetic diversity through attention to on-farm management of genetic resources (Enjalbert et al. 2011; Goldringer et al. 2001). Genetic diversity is important in providing base genotypes for future organic systems' plant breeding programs.

10.6 GENETIC DIVERSITY: THE USE OF MIXTURES AND THEIR RELATED PROBLEMS

Cultivars for organic systems should be adaptive, that is, flexible, robust, yield stable, and capable of compensating for the unfavorable and variable conditions (Lammerts van Bueren and Myres 2012; Lammerts van Bueren et al. 2007). Such adaptive cultivars may include heterogeneous cultivar mixtures that carry diverse alleles. Cultivar mixtures have several advantages. Firstly, mixtures have been reported to have inherent phenotypic plasticity and the ability to produce more than one phenotype, depending on the environmental conditions (Byers 2005; Górny 2001). For example, spring wheat variety mixtures were observed to provide greater stability with little or no reduction in yield, while providing greater competitive ability (Kaut et al. 2009) under variable conditions. Secondly, mixtures are said to be resilient materials that allow for more buffering capacity at both the spatial and the temporal levels (Finckh et al. 2000; Wolfe 1985). The buffering capacity of mixtures derives from complementation and compensation among different plant neighbors growing in a mixture (Finckh et al. 2000; Wolfe 1985). Thirdly, mixtures in organic systems may help with disease control, which is crucial given the non-usage of synthetic pesticides. Studies from several developing countries have reported the capacity for disease control attributed to mixtures (e.g., Almekinders and Hardon 2006). In these studies, the genetic diversity in mixtures was inferred to generate genetic mosaics that were helpful in delaying the development of epidemics and plagues (Finckh et al. 2000; Wolfe 1985). The use of heterogeneous cultivar mixtures is therefore regarded as logical (Witcombe and Virck 2001) and insurance against the impact of biotic and abiotic stress factors on crop yield and quality (Finckh 2008) in organic systems. However, there are several problems related to the use of mixtures.

The first problem is that they are inconsistent in productivity as yield has been observed to vary unpredictably. Productivity inconsistencies may be attributed to a probable lack of compatibility among component cultivars (Phillips et al. 2005). Using breeding populations derived from well-designed composite crosses or from consecutive harvest and resowing of genotype mixtures may overcome inconsistencies in cultivar mixtures (Allard 1988, 1990; Harlan and Martini 1929; Murphy et al. 2005; Phillips and Wolfe 2005; Suneson 1960). Using resowings or consecutive harvests in the same environment enables the mixtures to respond continuously to environmental changes over time (Allard 1988, 1990; Harlan and Martini 1929; Suneson 1960). When using composite crosses, component genotypes must be used. They should be adapted to specific regions, such as modern landraces that are farmer selected, resilient to heterogeneous environmental conditions, and have the capacity to continually evolve (Dawson et al. 2011). Such component genotypes will in turn positively influence the yield performance of cultivar mixtures by increasing the average yield of the different populations over time as a result of natural selection and competition among plants (Allard 1988, 1990; Harlan and Martini 1929; Suneson 1960). If the composite crosses come from populations grown in different environments, they will have an added advantage of being able to differentiate adaptive traits such as earliness components and powdery mildew resistance in wheatmade crosses (Goldringer et al. 2006; Paillard et al. 2000).

The second problem related to cultivar mixtures is their usefulness. In particular, some of the heterogeneous crops from cultivar mixtures are not suitable for milling, baking, and other processing (Wolfe et al. 2008). Careful selection for similarity in quality traits among component cultivars may be a solution to the problem. Thirdly, legislation for seed regulation is one of the biggest problems encountered with cultivar seed mixtures. No seed can be sold in most industrialized countries of the west without legislation. Legislation for seed regulation was put into place to verify claims about cultivar performances and to ensure that cultivars met the internationally recognized definition of a cultivar. Before the advent of seed regulation, it was common to have different companies marketing a popular cultivar under different names to create brand names (Chablé et al. 2012), which brought confusion in the market. The problem with legislation, established in 1940, lies in the requirements that define a cultivar (Chablé et al. 2012). Specifically, a cultivar is defined in industrialized countries for arable species as distinct, uniform, and stable (DUS). Any material that does not meet these three cultivar requirements cannot be marketed (Desclaux et al. 2008). Given the heterogeneity of cultivar mixtures, these three cultivar requirements exclude cultivar mixtures from qualifying for registration, especially the requirement for uniformity. Thus, the DUS criterion is a stumbling block to organic agriculture progress because the conventional seed regulations used to register all cultivars do not optimally encompass characteristics of mixtures and populations needed for organic crop production (Chablé et al. 2012). However, in some countries (e.g., Denmark), the DUS criterion is only applied to component cultivars constituting a cultivar mixture, which enables mixtures to be registered and marketed. It is important to note that the DUS criterion does not apply to the United States.

Additional seed regulation legislation called the value for cultivation (VCU) in other countries (Austria, Denmark, Germany, and Switzerland) is an additional stumbling block to the marketing of cultivar mixtures. The VCU is a criterion that tests new cultivars for progress in agronomic and product quality characteristics (Chablé et al. 2012). According to the VCU criterion, a new cultivar will only pass and be registered if it has better agronomic performance or quality than existing cultivars. This is difficult to achieve with cultivar mixtures; hence, most fail the VCU tests. There are, however, flexibilities in the application of the VCU criterion that may allow cultivar mixtures to pass and therefore be registered. Specifically, the European Union (EU), in theory, allows member states to choose the agronomic conditions for VCU testing, which allows members to determine how to assess traits and characteristics important to organic agriculture (Chablé et al. 2012). In addition, efforts are being made in Europe to facilitate the registration and marketing of cultivar mixtures and populations without the DUS and VCU criteria. VCU trials are not compulsory in the United States and Canada.

10.7 BREEDING TECHNIQUES

In general, breeding techniques/methodologies that encompass philosophies underlying organic agriculture are recommended for use in developing cultivars intended for organic systems. Broadly, the philosophies underlying organic agriculture advocate for a natural way (be it breeding and selection or production) (Lammerts van Bueren et al. 1999). Such natural breeding techniques have an added advantage of allowing farmers to participate in breeding activities on their farms and in their own regions. Farmer involvement in breeding frees farmers from dependence on multicontinental seed houses for their seed needs. Sophisticated techniques such as in vitro techniques that require laboratories and highly trained professionals will exclude farmers from participatory breeding activities.

There are also speculative fears of disease problems such as cancers as a result of employing techniques with in vitro procedures. In addition, in vitro techniques infringe on the aspect of naturalness, which is one of the pillars of organic varieties (Lammerts van Bueren and Struik 2004). However, in vitro techniques are still used, for most organic seed. A total ban is likely to affect the availability of cultivars suitable for organic systems, given that the current wheat breeding programs use embryo or microspore culture together with colchicine application for doubling haploids. It is envisaged that an immediate banning of the in vitro techniques will set back development at least 30 years as many modern varieties have been bred or multiplied using a technology involving in vitro culture. It is impossible to identify cultivars with a history of embryo culture in their origins (Lammerts van Bueren et al. 2010). There is a need to develop alternative methods for breeding organic varieties since the use of material produced with current in vitro techniques is temporarily accepted (Lammerts van Bueren and Struik 2004). There are efforts to distinguish the cultivars produced using in vitro techniques in some countries (e.g., Hungary and Switzerland have designed a certification system for specific organic breeding programs, where no in vitro techniques are applied) (Lammerts van Bueren et al. 2010). Organic farming regulations in the United States and Europe also prohibit the use of genetically engineered crops (Lammerts van Bueren et al. 2010) on the grounds that they violate the natural constitution of crops.

10.8 SUMMARY

Breeding cereals for organic soil properties, plant nutrition, and weed control is generally breeding cultivars tolerant to low N, low P, and high weed pressure. The availability of genetic variability for tolerance to these stresses found among genotypes makes breeding and selection possible. Both direct and indirect selections are recommended in breeding cultivars adapted to organic systems. Indirect selection is carried out in the initial stages of cultivar development under LI conditions to assess the availability of required traits and for traits that require specialist processing by researchers at research stations. Direct selection carried out on organic farms, by participating farmers, is the final stage in which adaptation to specific farm environments is screened for using yield, a trait easily evaluated at the farm level. The collaboration between scientists and farmers in developing cultivars is called PPB. Both pure and cultivar mixtures should be developed for organic systems. Cultivar mixtures are recommended for they are flexible and thus can continually evolve and adapt to the ever-changing organic conditions. The use of cultivar mixtures has given rise to the need to loosen current legislation that defines cultivars in order to accommodate registration and marketing of cultivar mixtures.

REFERENCES

Al-Karaki, G., B. McMichael, and J. Zak. 2004. Field response of wheat to arbuscular mycorrhizal fungi and drought stress. *Mycorrhiza* 14:263–269. doi:10.1007/s00572-003-0265-2.

Allard, R.W. 1988. Genetic changes associated with the evolution of adaptedness in cultivated plants and their wild progenitors. *J. Hered.* 79:235–238.

Allard, R.W. 1990. The genetics of host-pathogen coevolution. Implication for genetic resources conservation. *J. Hered.* 81:1–6.

Almekinders, C. and J. Hardon (eds.). 2006. *Bringing Farmers Back to into Breeding, Experiences with Participatory Plant Breeding and Challenges for Institutionalisation.* Agromisa special 5. Agromisa, Wageningen, the Netherlands.

Anderson, W.K. and F.C. Hoyle. 1999. Nitrogen efficiency of wheat cultivars in a Mediterranean environment. *Aust. J. Exp. Agric.* 39:957–965. doi:10.1071/EA98045.

Bago, B. et al. 2000. Carbon metabolism and transport in arbuscular mycorrhizas. *Plant Physiol.* 124:949–958.

Bänziger, M., P.S. Setimela, D. Hodson, and B. Vivek. 2006. Breeding for improved drought tolerance in maize adapted to southern Africa. *Agric. Water Manage.* 80:212–224.

Baon, J.B., S.E. Smith, and A.M. Alston. 1993. Mycorrhizal responses of barley cultivars differing in P efficiency. *Plant Soil* 157:97–105.

Barraclough, P.B., J.R. Howarth, J. Jones, R. Lopez-Bellido, S. Parmar, C.E. Shepherd, and M.J. Hawkesford. 2010. Nitrogen efficiency of wheat: Genotypic and environmental variation and prospects for improvement. *Eur. J. Agron.* 33:1–11.

Barraclough, P.B. and R.A. Leigh. 1984. The growth and activity of winter wheat roots in the field: The effects of sowing date and soil type on root growth of high-yielding crops. *J. Agric. Sci. (Camb.)* 103:69–74.

Barrow, N.J. 1974. Factors affecting the long-term effectiveness of phosphate and molybdenum fertilizer. *Commun. Soil Sci. Plant Anal.* 5:355–387.

Batten, G.D. 1992. A review of phosphorus efficiency in wheat. *Plant Soil* 146:163–168.

Batten, G.D. and M.A. Khan. 1987. Effect of time of sowing on grain yield, and nutrient uptake of wheats with contrasting phenology. *Aust. J. Agric. Res.* 27:881–887.

Burger, H., M. Schloen, W. Schmidt, and H.H. Geiger. 2008. Quantitative genetic studies on breeding maize for adaptation to organic farming. *Euphytica* 163:501–510.

Byers, D.L. 2005. Evolution in heterogeneous environments and the potential of maintenance of genetic variation in traits of adaptive significance. *Genetica* 123:107–124.

Calderini, D.F., S. Torres-Leon, and G.A. Slafer. 1995. Consequences of wheat breeding on nitrogen and phosphorus yield, grain nitrogen and phosphorus concentration and associated traits. *Ann. Bot.* 76:315–322.

Cassman, K.G., A. Dobermann, and D.T. Walters. 2002. Agroecosystems, nitrogen- use efficiency and nitrogen management. *Ambio* 31 (2):132–140.

Chablé, V., N. Louwaars, K. Hubbard, B. Baker, and R. Bocci. 2012. Plant breeding, cultivar release, and seed commercialization: Laws and policies applied to the organic sector. In Lammerts van Bueren, E.T. and J.R. Myers (eds.). *Organic Crop Breeding*. John Wiley & Sons Ltd, The Atrium, Southern Gate, Chichester, U.K, pp. 139–160.

Champion, G.T., R.J. Froud-Williams, and J.M. Holland. 1998. Interactions between wheat (*Triticum aestivum* L.) cultivar, row spacing and density and the effect on weed suppression and crop yield. *Ann. Appl. Biol.* 133:443–453.

Chapin III, F.S. 1980. The mineral nutrition of wild plants. *Ann. Rev. Ecol. System.* 11:233–260.

Coleman, R.D., G.S. Gill, and G.J. Rebetzke. 2001. Identification of quantitative trait loci for traits conferring weed competitiveness in wheat (*Triticum aestivum* L.). *Aust. J. Agric. Res.* 52:1235–1246.

Coque, M. and A. Gallais. 2006. Genomic regions involved in response to grain yield selection at high and low nitrogen fertilization in maize. *Theor. Appl. Genet.* 112:1205–1220.

Cosser, N.D., M.J. Gooding, A.J. Thompson, and R.J. Froud-Williams. 1997. Competitive ability and tolerance of organically grown wheat cultivars to natural weed infestations. *Ann. Appl. Biol* 130:523–535.

Cousens, R.D. and S. Mokhtari. 1998. Seasonal and site variability in the tolerance of wheat cultivars to interference with *Lolium rigidum*. *Weed Res.* 38:301–307.

Cruz, C., J.J. Green, C.A. Watson, F. Wilson, and M.A. Martins-Louçâo. 2004. Functional aspects of root architecture and mycorrhizal inoculation with respect to nutrient uptake capacity. *Mycorrhiza* 14:177–184. doi:10.1007/s00572-003-0254-5.

Dawson, J.C., D.R. Huggins, and S.S. Jones. 2008. Characterizing nitrogen use efficiency in natural and agricultural ecosystems to improve the performance of cereal crops in low-input and organic agricultural systems. *Field Crops Res.* 107:89–101.

Dawson, J.C., P. Rivière, J.-F. Berthellot, F. Mercier, P. de Kochko, N. Galic, S. Pin, E. Serpolay, M. Thomas, S. Giuliano, and I. Goldringer. 2011. Collaborative plant breeding for organic agricultural systems in developed countries. *Sustainability* 3:1206–1223; doi:10.3390/su3081206.

Desclaux, D. 2005. Participatory plant breeding for organic cereals. In: *Proceedings of the Eco-Pb Workshop on Organic Plant Breeding Strategies and the Use of Molecular Markers*. Driebergen, the Netherlands, 2006. pp. 17–23.

Desclaux, D., J.M. Nolot, Y. Chiffoleau, E. Gozé, and C. Leclerc. 2008. Changes in the concept of genotype environment interactions to fit agriculture diversification and decentralized participatory plant breeding. Pluridisclipinary point of view. *Euphytica* 163:533–546.

Donner, D. and A. Osman (eds.). 2006. *Handbook Cereal Cultivar Testing for Organic and Low-Input*. Available through COST SUSVAR http://www.cost860.dk/publications/.

Drews, S., D. Neuhoff, and U. Kopke. 2009. Weed suppression ability of three winter wheat cultivars at different row spacing under organic farming conditions. *Weed Res.* 49:526–533.

Drinkwater, L.E. and S.S. Snapp. 2007. Understanding and managing the rhizosphere in agro-ecosystems. In Cardon, Z.G. and J.L. Whitbeck (eds.). *The Rhizosphere—An Ecological Perspective*. Elsevier, New York, pp. 155–178.

Douds, D.D., Jr., L. Galvez, M. Franke-Snyder, C. Reider, and L.E. Drinkwater. 1997. Effect of compost addition and crop rotation point upon VAM fungi. *Agric. Ecosyst. Environ.* 65:257–266. doi:10.1016/S0167-8809(97)00075-3.

Eisele, J.A. and U. Kopke. 1997. Choice of cultivar in organic farming: New criteria for winter wheat ideotypes. *Pflanzenbauwissen-schaften* 1(1):5, 19–24.

Enjalbert, J., J.C. Dawson, S. Paillard, B. Rhoné, Y. Rousselle, M. Thomas, and I. Goldringer. 2011. Dynamic management of crop diversity: From an experimental approach to on-farm conservation. *C. R. Biol.* 334:458–468.

Finckh, M.R. 2008. Integration of breeding and technology into diversification strategies for disease control in modern agriculture. *Eur. J. Plant Pathol.* 121:399–409.

Finckh, M.R., E.S. Gacek, H. Goyeau, C. Lannou, U. Merz, C.C. Mundt, L. Munk, J. Nadziak, A.C. Newton, C. de Vallavieille-Pope, and M.S. Wolfe. 2000. Cereal cultivar and species mixtures in practice, with emphasis on disease resistance. *Agronomie* 20:813–837.

Fischer, A.J., C.G. Messersmith, J.D. Nalewaja, and M.E. Duysen. 2000. Interference between spring cereals and Kochia scoparia related to environment and photosynthetic pathways. *Agron. J.* 92:173–181.

Fischer, R.A.1981. Optimizing the use of water and nitrogen through breeding of crops. *Plant Soil* 58:249–278.

Fohse D., N. Classen, and A. Jungk. 1991. Phosphorus efficiency of plants. II. Significance of root radius, root hairs and cation–anion balance for phosphorus influx in seven plant species. *Plant Soil* 132:261–272.

French, R.J. and J.E. Schultz. 1984. Water use efficiency of wheat in a Mediterranean environment II. Some limitations to efficiency. *Aust. J. Agric. Res.* 35:765–775.

Fu, Y.B., G.W. Peterson, K.W. Richards, D. Somers, R.M. DePauw, and J.M. Clarke. 2005. Allelic reduction and genetic shift in the Canadian hard red spring wheat germplasm released from 1845 to 2004. *Theor. Appl. Genet.* 110:1505–1516.

Gahoonia, T.S., D. Care, and N.E. Nielsen. 1997. Root hairs and phosphorus 533 acquisition of wheat and barley cultivars. *Plant Soil* 191:181–188.

Gahoonia, T.S. and N.E. Nielsen. 2004. Root traits as tools for creating phosphorus efficient crop cultivars. *Plant Soil* 260:47–57.

Gahoonia, T.S., N.E. Nielsen, and O.B. Lyshede. 1999. Phosphorus (P) acquisition of cereal cultivars in the field at three levels of P fertilization. *Plant Soil* 211:269–281.

George, T.S., A.E. Richardson, P. Hadobas, and R.J. Simpson. 2004. Characterization of transgenic *Trifolium subterraneum* L. which expresses phyA and releases extracellular phytase: Growth and P nutrition in laboratory media and soil. *Plant Cell Environ.* 27:1351–1361.

Gerath, H. 1993. Entwicklung von Stickstoffeffizientem Winterraps. *BML-Forschung* 8:9–11.

Goldringer, I., J. Enjalbert, J. David, S. Paillard, J.L. Pham, and P. Brabant. 2001. Dynamic management of genetic resources: A 13 year experiment on wheat. In Cooper, H.D., Spillane, C., Hodgkin, T., eds. *Broadening the Genetic Base of Crop Production*; IPGRI/ FAO, Rome, Italy, pp. 245–260.

Goldringer, I., C. Prouin, M. Rousset, N. Galic, and I. Bonnin. 2006. Rapid differentiation of experimental populations of wheat for heading-time in response to local climatic conditions. *Ann. Bot. Lond.* 98:805–817.

Gooding, M.J., A.J. Thompson, and W.P. Davies. 1993. Interception of photosynthetically active radiation, competitive ability and yield of organically grown wheat cultivars. *Asp. Appl. Biol. Physiol. Cultivars* 34:355–362.

Górny, A.G. 2001. Photosynthetic activity of flag leaves in diallel crosses of spring barley under varied nutrition and soil moisture. *Cereal Res. Commun.* 29/1–2:159–166.

Graham, J.H. and D.M. Eissenstat. 1994. Host genotype and the formation and function of VA mycorrhizae. *Plant Soil* 159:179–185.

Graham, J.H. and D.M. Eissenstat. 1998. Field evidence for the carbon cost of citrus mycorrhizas. *New Phytol.* 140:103–110.

Guppy, C.N. and M.J. McLaughlin. 2009. Options for increasing the biological cycling of phosphorus in low-input and organic agricultural systems. *Crop Pasture Sci.* 60:116–172.

Habash, D.Z., S. Bernard, J. Shondelmaier, Y. Weyen, and S.A. Quarrie. 2006. The genetics of nitrogen use on hexaploid wheat: N utilization, development and yield. *Theor. Appl. Genet.* 114:403–419.

Habte, M. 2006. The roles of arbuscular mycorrhizas in plant and soil health. In Uphoff, N., A.S. Ball, E.C.M. Fernandes, H. Herren, O. Husson, M. Laing, C. Palm, J. Pretty, and P. Sanchez (eds.). *Biological Approaches to Sustainable Soil Systems*. Taylor & Francis Group LLC, Boca Raton, FL, pp. 129–148.

Hamel, C. 2004. Impact of arbuscular mycorrhizal fungi on N and P cycling in the root zone. *Can. J. Soil Sci.* 84:383–395.

Hamel, C. and Strullu, D.G. 2006. Arbuscular mycorrhizal fungi in field crop production: Potential and new direction. *Can. J. Plant Sci.* 86:941–950.

Harlan, H.V. and Martini, M.L. 1929. A composite hybrid mixture. *J. Am. Soc. Agron.* 21:487–490.

Harrier, L.A. and C.A. Watson. 2003. The role of arbuscular mycorrhizal fungi in sustainable cropping systems. *Adv. Agron.* 79:185–225.

Hetrick, B.A.D. 1991. Mycorrhizas and root architecture. *Experientia* 47:355–362.

Hetrick, B.A.D., G.W.T. Wilson, and T.S. Cox. 1992. Mycorrhizal dependence of modern wheat cultivars, landraces, and ancestors. *Can. J. Bot.* 70:2032–2040.

Hetrick, B.A.D., G.W.T. Wilson, and T.S. Cox. 1993. Mycorrhizal dependence of modern wheat cultivars and ancestors: A synthesis. *Can. J. Bot.* 71:512–518.

Hoad, S.P., G. Russell, M.E. Lucas, and I.J. Bingham. 2001. The management of wheat, barley, and oat root systems. *Adv. Agron.* 74:193–254.

Hucl, P. 1998. Response to weed control by four spring wheat genotypes differing in competitive ability. *Can. J. Plant Sci.* 78:171–173.

Huel, D.G. and P. Hucl. 1996. Genotypic variation for competitive ability in spring wheat. *Plant Breed.* 115:325–329.

Huggins, D.R. and W.L. Pan. 2003. Key indicators for assessing nitrogen use efficiency in cereal-based agroecosystems. *J. Crop Prod.* 8:157–185.

Janos, D.P. 2007. Plant responsiveness to mycorrhizas differs from dependence upon mycorrhizas. *Mycorrhiza* 17:75–91.

Javot, H. et al. 2007. A Medicago truncatula phosphate transporter indispensable for the arbuscular mycorrhizal symbiosis. *Proc. Natl. Acad. Sci. USA* 104:1720–1725.

Jiang, Z and R.J. Hull. 1998. Interrelationships of nitrate uptake, nitrate reductase and nitrogen use efficiency in selected Kentucky bluegrass cultivars. *Crop Sci.* 38:1623–1632.

Jones, G. P.D., G. J. Blair, and R. S. Jessop. 1989. Phosphorus efficiency in wheat-a useful selection criterion? *Field Crops Res.* 21:257–264.

Kaut, A.H.E.E., H.E. Mason, A. Navabi, J.T. O'Donovan, and D. Spaner. 2009. Performance and stability of performance of spring wheat variety mixtures in organic and conventional management systems in western Canada. *J. Agric. Sci.* 147:141–153, doi:10.1017/S0021859608008319.

Kirk, A.P., M.H. Entz, S.L. Fox, and M. Tenuta. 2011. Mycorrhizal colonization, P uptake and yield of older and modern wheats under organic management. *Can. J. Plant Sci.* 91:663–667.

Korres, N.E. and R.J. Froud-Williams. 2002. Effects of winter wheat cultivars and seed rate on the biological characteristics of naturally occurring weed flora. *Weed Res.* 42:417–428.

Ladha, J.K., A. Tirol-Padre, K. Reddy, and W. Venture. 1993. Prospects and problems of biological nitrogen fixation in rice production: A critical assessment. In Palacios, R., J. Mora, and W.E. Newton (eds.). *New Horizons in Nitrogen Fixation*. Kluwer Academic Publishers, Dordrecht, the Netherlands, pp. 677–682.

Lafitte, H. et al. 2006. Improvement of rice drought tolerance through backcross breeding: Evaluation of donors and selection in drought nurseries. *Field Crops Res.* 97:77–96.

Lambers, H., M.W. Shan, M.D. Cramer, S.J. Pearse, and E.J. Veneklaas. 2006. Root structure and functioning for efficient acquisition of phosphorus: Matching morphological and physiological traits. *Ann. Bot.* 98:693–713.

Lammerts van Bueren, E.T., M. Husscher, M. Haring, J. Jonggerden, J.D. van Mansvelt, A.P.M. den Nijs, and G.T.P. Ruivenkamp. 1999. Sustainable organic plant breeding. Final report: A vision, choices, consequences and steps. *Louis Bolk Instutuut natuurwetenschappelijk onderzoek.* G24:1–60.

Lammerts van Bueren, E.T. and J.R. Myers. 2012. Organic crop breeding: Integrating organic agricultural approaches and traditional and modern plant breeding methods. In Lammerts van Bueren, E.T. and J.R. Myers (eds.). *Organic Crop Breeding.* John Wiley & Sons Ltd, The Atrium, Southern Gate, Chichester, U.K.

Lammerts van Bueren, E.T. and Struik, P.C. 2004. The consequences of the concept of naturalness for organic plant breeding and propagation. *NJAS-Wagen J. Life Sci.* 52:85–95.

Lammerts van Bueren, E.T., P.C. Struik, and E. Jacobsen. 2002. Ecological concepts in organic farming and their consequences for an organic crop ideotype, *Netherlands J. Agric. Sci.* 50:1–26.

Lammerts van Bueren, E.T., P.C. Struik, M. Tiemens-Hulscher, and E. Jacobsen. 2003. The concepts of intrinsic value and integrity of plants in organic plant breeding and propagation. *Crop Sci.* 43:1922–1929.

Lammerts van Bueren, E.T., H. Verhoog, M. Tiemens-Hulscher, P.C. Struik, and M.A. Haring. 2007. Organic agriculture requires process rather than product evaluation of novel breeding techniques. *NJAS Wagen J. Life Sci.* 54:401–412.

Lammerts van Bueren, E.T. et al. 2010. The need to breed crop cultivars suitable for organic farming, using wheat, tomato and broccoli as examples: A review. *NJAS-Wagen. J. Life Sci.* 58:193–205. doi:10.1016/j.njas.2010.04.001.

Laperche, A., M. Brancourt-Hulmel, E. Heumez, O. Gardet, and J. Le Gouis. 2006. Estimation of genetic parameters of a DH wheat population grown at different N stress levels characterized by probe genotypes. *Theor. Appl. Genet.* 112:797–807.

Le Gouis, J., D. Beghin, E. Heumez, and P. Pluchard. 2000. Genetic differences for nitrogen uptake and nitrogen utilization efficiencies in winter wheat. *Eur. J. Agron.* 12:163–173.

Lemerle, D., G.S. Gill, C.E. Murphy, S.R. Walker, R.D. Cousens, S. Mokhtari, S.J. Peltzer, R. Coleman, and D.J. Luckett. 2001a. Genetic improvement and agronomy for enhanced wheat competitiveness with weeds. *Aust. J. Agric. Res.* 52:527–548.

Lemerle, D., A. Smith, B. Verbeek, E. Koetz, P. Lockley, and P. Martin. 2006. Incremental crop tolerance to weeds: A measure for selecting competitive ability in Australian wheats. *Euphytica* 149:85–95.

Lemerle, D., B. Verbeek, R.D. Cousens, and N.E. Coombes. 1996. The potential for selecting wheat cultivars strongly competitive against weeds. *Weed Res.* 36:505–513.

Lemerle, D., B. Verbeek, and B. Orchard. 2001b. Ranking the ability of wheat cultivars to compete with *Lolium rigidum. Weed Res.* 41:197–209.

Leon, J. and K.U. Schwang. 1992. Description and application of a screening method to determine root morphology traits of cereals cultivars. *Z. Acker. Pflanzenbau* 169:128–134.

Loschenberger, F., A. Fleck, H. Grausgruber, H. Hetxendorfer, G. Hof, J. Lafferty, M. Marn, A. Neumayer, G. Pfaffinger, and J. Birschitzhy. 2008. Breeding for organic agriculture: The example of winter wheat in Austria. *Euphytica.* 163:469–480.

Lynch, J. 1998. The role of nutrient efficient crops in modern agriculture. In Rengel, Z. (ed.). *Nutrient Use in Crop Production.* The Haworth Press, New York, pp. 241–264.

Lynch, J. and K.L. Nielsen. 1995. Simulation of root system architecture. In *Plant Roots: The Hidden Half*, Y. Waisel, A. Eshel, and U. Kafkafi, eds., second edn., Marcel Dekker, New York.

Mäder, P., S. Edenhofer, T. Boller, A. Wiemken, and U. Niggli. 2000. Arbuscular mycorrhizae in a long-term field trial comparing low-input (organic, biological) and high-input (conventional) farming systems in a crop rotation. *Biol. Fertil. Soils.* 31:150–156. doi:10.1007/s003740050638.

Mäder, P, D. Fliessbach, D. Dubois, L. Gunst, P. Fried, and U. Niggli. 2002. Soil fertility and biodiversity in organic farming. *Science* 296:1694–1697.

Manske, G.G.B., A.B. Luttger, R.K. Behl, and P.L.G. Vlek. 1995. Nutrient efficiency based on VA mycorrhizae (AMF) and total root length of wheat cultivars grown in India. *Angew. Bot.* 69:108–110.

Manske, G.G.B., J.I. Ortiz-Monasterio R.M. van Ginkel, R.M. Gonzalez, R.A. Fischer, S. Rajaram, and P. Vlek. 2000a. Importance of P-uptake efficiency vs. P-utilization for wheat yield in acid and calcareous soils in Mexico. *Eur. J. Agron.* 14(4):261–274.

Manske, G.G.B., J.I. Ortiz-Monasterio, M. van Ginkel, R.M. Gonzalez, S. Rajaram, E. Molina, and P.L.G. Vlek. 2000b. Traits associated with improved P-uptake efficiency in CIMMYT's semidwarf spring bread wheat grown on an acid andisol in Mexico. *Plant Soil* 22(1):189–204.

Marschner, P. and Z. Rengel. 2003. Contributions of rhizosphere interactions to soil biological fertility. In: *Soil Biological Fertility: A Key to Sustainable Land Use in Agriculture.* Kluwer Academic Publishers, Dordrecht, the Netherlands.

Martre, P., J.R. Porter, P.D. Jamieson, and E. Triboï. 2007. Modeling grain nitrogen accumulation and protein composition to understand the sink/source regulations of nitrogen utilization in wheat. *Plant Physiol.* 133:1959–1967.

Mason, H.E., A. Navabi, B.L. Frick, J.T. O'Donovan, and D.M. Spaner. 2007a. The weed-competitive ability of Canada Western Red Spring wheat cultivars grown under organic management. *Crop Sci.* 47:1167–1176.

Mason, H.E., A. Navabi, B.L. Frick, J.T. O'Donovan, and D.M. Spaner. 2007b. Cultivar and seeding rate effects on the competitive ability of spring cereals grown under organic production in northern Canada. *Agron. J.* 99:1199–1207. doi: 10.2134/agronj2006.0262.

Mason, H.E. and D. Spaner. 2006. Competitive ability of wheat in conventional and organic management systems; a review of the literature. *Can. J. Plant Sci.* 86:333–343.

Mason, H.E. et al. 2007c. The weed competitive ability of Canada Western Red Spring Wheat cultivars grown under organic management. *Crop Sci.* 47:1167–1176.

McLachlan, K.D. 1980. Acid phosphatase activity of intact roots and phosphorus nutrition of plants. II. Variation among wheat roots. *Aust. J. Agric. Res.* 31:441–448.

Messmer R., Y. Fracheboud, M.Banziger, M.Vargas, P.Stamp, and J.M. Ribaut. 2009. Drought stress and tropical maize: QTL–by–environment interactions and stability of QTLs across environments for yield components and secondary traits. *Theor. Appl. Genet.* 119:913–930. doi: 10.1007/s00122-009-1099-x.

Mohammad, M.J., H.I. Malkawi, and R. Shibli. 2003. Effects of arbuscular mycorrhizal fungi and phosphorus fertilization on growth and nutrient uptake of barley grown on soils with different levels of salts. *J. Plant Nutr.* 26:125–137. doi:10.1081/PLN-120016500.

Mohammad, M.J., W.L. Pan, and A.C. Kennedy. 2005. Chemical alteration of the rhizosphere of the mycorrhizal-colonized wheat root. *Mycorrhiza* 15:259–266.

Moll, R.H., E.J. Kamprath, and W.A. Jackson. 1982. Analysis and interpretation of factors which contribute to efficiency of nitrogen utilization. *Agron. J.* 74:562–564.

Murphy, K. and S.S. Jones. 2008. Genetic assessment of the role of breeding wheat for organic systems. In *Wheat Production in Stressed Environments,* H.T. Buck, J.E. Nisi, and N. Salomon, eds. Springer, the Netherlands. Vol. 12, pp. 217–222.

Murphy, K., D. Lammer, S. Lyon, B. Carter, and S.S. Jones. 2005. Breeding for organic and low-input farming systems: An evolutionary-participatory breeding method for inbred cereal grains. *Renew. Agric. Food Syst.* 20:48–55.

Murphy, K.M., K.G. Campbell, S.R. Lyon, and S.S. Jones. 2007. Evidence of varietal adaptation to organic farming systems. *Field Crops Res.* 102:172–177.

Nelson, A., S. Quideau, B. Frick, P. Hucl, D. Thavarajah, J. Clapperton, and D. Spaner. 2011. The soil microbial community and grain micronutrient concentration of historical and modern hard red spring wheat cultivars grown organically and conventionally in the black soil zone of the Canadian prairies. *Sustainability* 3:500–517; doi:10.3390/su3030500.

Oberforster, M. 2006. Ist die Sortenzulassungsprüfung biogerecht? Österreichische Fachtagung für biologische Landwirtschaft, March 21–22. *HBLFA für Landwirtschaft Raumberg-Gumpenstein*, Irdning, Austria. pp. 15–20.

Oberforster, M., G. Plakolm, J. Sollinger, and M. Werteker. 2000. Are descriptions of conventional cultivar testing suitable for organic farming? In Alföldi, T., W. Lockeretz, and U. Niggli (eds.). *The World Grows Organic. Proceedings of the 13th International IFOAM Scientific Conference*. Research Institute of Organic Agriculture (FiBL), Frick, Switzerland, p. 242.

O'Brien, L. 1979. Genetic variability of root growth in wheat (*Triticum aestivum* L.). *Aust. J. Agric. Res.* 30:587–595.

Ortiz-Monasterio, J.I., K.D. Sayre, S. Rajaram, and M. McMahon. 1997a. Genetic progress in wheat yield and nitrogen use efficiency under four nitrogen rates. *Crop Sci.* 37:898–904.

Ortiz-Monasterio, R.J.I., R.J. Peña, K.D. Sayre, and S. Rajaram. 1997b. CIMMYT's genetic progress in wheat grain quality under four N rates. *Crop Sci.* 37(3):892–898.

Osman, A.M., E.T. Lammerts van Bueren, and C. Almekinders. 2007. Mobilising chains to stimulate spring wheat breeding for organic farming. In Osman, A.M., K.J. Müller, and K.P. Wilbois (eds.), *Different Models to Finance Plant Breeding, European Consortium for Organic Plant Breeding*, Frankfurt, Germany. pp. 27–30.

Paillard, S., I. Goldringer, J. Enjalbert, G. Doussinault, C. de Vallavieille-Pope, and P. Brabant. 2000. Evolution of resistance against powdery mildew in winter wheat populations conducted under dynamic management. I. Is specific seedling resistance selected? *Theor. Appl. Genet.* 101:449–456.

Palta, J.A. and I.R.P. Fillery. 1995. N application increases pre-anthesis contribution of dry matter to grain yield in wheat grown on a duplex soil. *Aust. J. Agric. Res.* 46:507–518.

Panga, X. P. and J. Leteya. 2000. Challenge of timing nitrogen availability to crop nitrogen requirements. *Soil Sci. Soc. Am. J.* 64:247–253.

Persson, H. 1982. The importance of fine roots in boreal forests. In *Root Ecology and Its Practical Applications. International Symposium at Gumpenstein*, Austria. pp. 595–608.

Phillips, S.L., M. Shaw, and M.S. Wolfe. 2005. The effect of potato cultivar mixtures on epidemics of late blight in relation to plot size and level of resistance. *Ann. Appl. Biol.* 147:245–252.

Phillips, S.L. and M.S. Wolfe. 2005. Evolutionary plant breeding for low-input systems. *J. Agric. Sci.* 143:245–254.

Plenchette, C., C. Clermont-Dauphin, J.M. Meynard, and J.A. Fortin. 2005. Managing arbuscular mycorrhizal fungi in cropping systems. *Can. J. Plant Sci.* 85:31–40.

Porceddu, E., F. Martignano, and G.T. Scaracia Mugnozza. 1978. Interpopulation variation in coleoptile and primary root length in durum wheat. *In Proceedings of the Fifth International Wheat Genetics Symposium*, Indian Society of Genetics and Plant breeding, New Delhi, India, pp. 852–860.

Portilla-Cruz, I., E. Molina Gayosso, G. Cruz-Flores, I. Ortiz-Monasterio, and G.G.B. Manske. 1998. Colonización micorrízica arbuscular, actividad fosfatásica y longitud radical como respuesta a estrés de fósforo en trigo y triticale cultivados en un andisol. *Terra* 16(1):55–61.

Presterl, T., S. Groh, M. Landbeck, G. Seitz, W. Schmidt, and H.H. Geiger. 2002. Nitrogen uptake and utilization of European maize hybrids developed under conditions of low and high nitrogen input. *Plant Breed.* 121:480–486.

Przystalski, M. et al. 2008. Comparing the performance of cereal cultivars in organic and nonorganic cropping systems in different European countries. *Euphytica* 163:417–433.

Rabatin, S.C. and B.R. Stinner. 1989 The significance of vesicular–arbuscular mycorrhizal fungal-soil macroinvertebrate interactions in agroecosystems. *Agric. Ecosyst. Environ.* 27:195–204. doi:10.1016/0167-8809(89)90085-6.

Randall, P.J. 1995. Genotypic differences in phosphate uptake. In Johansen, C., K.K. Lee, K.K. Sharma, G.V. Subbarao, and E.A. Kueneman (eds.). *Genetic Manipulation of Crop Plants to Enhance Integrated Management in Cropping Systems. 1. Phosphorus: Proceedings of an FAO/ICRISAT Expert Consultancy Workshop*. International Crops Research Institute for the Semi-Arid Tropics, Andhra Pradesh, India. pp. 31–47.

Rasmussen, A. 2004. The effect of sowing date, stale seedbed, row width and mechanical weed control on weeds and yields of organic winter wheat. *Weed Res.* 44:12–20.

Reid, T.A., R.-C. Yang, D.F. Salmon, A. Navabi, and D. Spaner. 2011. Realized gains from selection for spring wheat grain yield are different in conventional and organically managed systems. *Euphytica.* 177:253–266. doi: 10.1007/s10681-010-0257-1.

Reid, T.A., R.-C. Yang, D.F. Salmon, and D. Spaner. 2009. Should spring wheat breeding for organically managed systems be conducted on organically managed land? *Euphytica.* 169:239–252.

Reynolds, M.P., J.I. Ortiz-Monasterio, and A. McNab (eds.). 2001. *Application of Physiology in Wheat Breeding.* CIMMYT, Mexico, DF.

Richards, M.C. and G.P. Whytock. 1993. Varietal competitiveness with weeds. *Aspect. Appl. Biol.* 34:345–354.

Röhm, M. and D. Werner. 1987. Isolation of root hairs from seedlings of *Pisum sativum.* Identification of root hair specific proteins by in situ labelling. *Physiol. Plant.* 69:129–136.

Ruairidh, J.H., S.C. Gutjahr, and U. Paszkowski. 2007. Cereal mycorrhiza: An ancient symbiosis in modern agriculture. *Trends Plant Sci.* 13(2):93–97.

Ryan, M.H., J.W. Derrick, and P.R. Dann. 2004. Grain mineral concentrations and yield of wheat grown under organic and conventional management. *J. Sci. Food Agric.* 84:207–216.

Schaarschmidt, S. et al. 2007. Regulation of arbuscular mycorrhization by carbon. The symbiotic interaction cannot be improved by increased carbon availability accomplished by root specifically enhanced invertase activity. *Plant Physiol.* 143:1827–1840.

Schulthess, U., B. Feil, and S.C. Jutzi. 1997. Yield-independent variation in grain nitrogen and phosphorus concentration among Ethiopian wheats. *Agron. J.* 89:497–506.

Silla, F. and A. Escudero. 2004. Nitrogen-use efficiency: Trade-offs between N productivity and mean residence time at organ, plant and population levels. *Funct. Ecol.* 18:511–521.

Sinebo, W., R. Gretzmacher, and A. Edelbauer. 2004. Genotypic variation for nitrogen use efficiency in Ethiopian barley. *Field Crop. Res.* 85:43–60. doi:10.1016/S0378-4290(03)00135-7.

Singh, U. and R.R. Buresh. 1994. Fertilizer technology for increased fertilizer efficiency in paddy rice fields. *15th World Congress of Soil Science*, Acapulco, Mexico. Vol. 5a, pp. 643–653.

Smith, A.N. and G.D. Batten. 1976. The efficiency, accuracy and reliability of soil testing for phosphorus in wheat: A study in the south west wheat belt. *Tech. Bull.* 11, Department of Agriculture, New South Wales, Australia.

Smith, K.P. and R.M. Goodman. 1999. Host variation for interactions with beneficial plant-associated microbes. *Annu. Rev. Phytopathol.* 37:473–491.

Smith, S.E. and D.J. Read. 1997. *Mycorrhizal Symbiosis.* Academic Press, San Diego, CA, p. 605.

Suneson, C.A. 1960. Genetic diversity—A protection against plant diseases and insects. *Agron. J.* 52:319–321.

van Ginkel, M., I. Ortiz-Monasterio, R. Trethowan, and E. Hernandez. 2001. Methodology for selecting segregating populations for improved N-use efficiency in bread wheat. *Euphytica.* 119:223. doi:10.1023/A:1017533619566.

Vazquez de Aldana, B. and F. Berendse. 1997. Nitrogen-use efficiency in six perennial grasses from contrasting habitats. *Funct. Ecol.* 11:619–626.

Vlek, P.L.G., A.B. Lüttger, and G.G.B. Manske. 1996. The potential contribution of arbuscular mycorrhiza to the development of nutrient and water efficient wheat. In Tanner, D.G., T.S. Payne, and O.S. Abdalla, (eds.). *The Ninth Regional Wheat Workshop for Eastern, Central and Southern Africa.* CIMMYT, Addis Ababa, Ethiopia. pp. 28–46.

Wang Lan-zhen, M.I., C.F.-J. Guo-hua-J, and F.-S. Zhang. 2004. Study on the relationship between phosphorus efficiency and agronomic traits of winter wheat. *Plant Nutr. Fertil. Sci.* 10(4):355–360.

Weegels, P.L., R.J. Hamer, and J.D. Schofield. 1996. Critical review: Functional properties of wheat glutenin. *J. Cereal Sci.* 23:1–18.

Wicks, G.A., R.E. Ramsel, P.T. Nordquist, J.W. Schmidt, and Challaiah. 1986. Impact of wheat cultivars on establishment and suppression of summer annual weeds. *Agron. J.* 78:59–62.

Wissuwa, M., M. Mazzola, and C. Picard. 2009. Novel approaches in plant breeding for rhizo-sphere-related traits. *Plant Soil* 321:409–430.

Witcombe, J.R., P.A. Hollington, C.J. Howarth, S. Reader, and K.A. Steele. 2008. Breeding for abiotic stresses for sustainable agriculture. *Philos. T. R. Soc. B.* 363:703–716.

Witcombe, J.R. and D.S. Virck. 2001. Number of crosses and population size for participatory and classical plant breeding. *Euphytica* 122:451–462.

Wolfe, M.S. 1985. The current status and prospects of multiline cultivars and cultivar mixtures for disease control. *Ann. Rev. Phytopathol.* 23:251–273.

Wolfe, M.S., J.P. Baresel, D. Desclaux, I. Goldringer, S. Hoad, G. Kovacs, F. Löschenberger, T. Miedaner, H. Østergård, and E.T. Lammerts van Bueren. 2008. Developments in breeding cereals for organic agriculture. *Euphytica* 163:323–346.

Woodruff, D.R. 1972. Cultivar variation in nitrogen uptake and distribution in wheat. *Aust. J. Exp. Agric. Anim. Husb.* 12:511–516.

Wright, D.P. et al. 1998. Mycorrhizal sink strength influences whole plant carbon balance of *Trifolium repens* L. *Plant Cell Environ.* 21:881–891.

Wu, H.W., T. Haig, J. Pratley, D. Lemerle, and M. An. 2000. Allelochemicals in wheat (*Triticum aestivum* L.): Variation of phenolic acids in root tissues. *J. Agric. Food Chem.* 48:5321–5325.

Zhu, Y.-G. et al. 2001. Phosphorus (P) efficiencies and mycorrhizal responsiveness of old and modern wheat cultivars. *Plant Soil* 237:249–255.

11 Organic Voices
Agronomy, Economics, and Knowledge on 10 Canadian Organic Farms

James B. Frey and Martin H. Entz

CONTENTS

11.1 INTRODUCTION: VIEWING ORGANIC AGRICULTURE THROUGH A SOCIAL AGROECOLOGICAL LENS

The purpose of this study is to contextualize agronomic, economic, and knowledge-related issues relevant to organic agriculture by examining the experiences of 10 Canadian organic farming households. Attention is also given to the broader social and ecological landscape on which the organic farming system exists, resulting in a more holistic understanding of how organic farmers navigate the complexities of this system. The research objectives are (1) to understand the participants' rationale for farming organically, (2) to identify the challenges and new approaches used, and (3) to examine the types of sources of knowledge that are important to participants, as well as how that knowledge is shared across networks. Finally, building on Holling and Gunderson's (2002) adaptive cycle, the research findings are presented as comprising an adaptive cycle for organic agriculture.

In adopting a broader, case-study approach, the research provides depth (greater context and personal narrative) to previous work examining organic farming systems (cf. Knight and Shirtliffe 2003; OACC 2008) that used a predominantly agronomic approach.

11.2   RESEARCH SETTING AND DESIGN

The study data were obtained during semistructured interviews with 10 farming households. The characteristics of the households are presented in Table 11.1. All farmers in this study had experience with organic farming methods and organic certification. One household had been certified organic, but discontinued that certification, while continuing to farm organically. Another farmer had been certified organic, but returned to conventional farming practices.

The study participants were selected from a database of organic farmers and are primarily located in Manitoba, as well as Saskatchewan and Ontario. Participants were chosen to represent a broad cross section of agricultural types, including grain, forages, livestock, and horticulture, or some combination thereof. Consequently, farm size was not a selection criterion. However, participants were selected according to the number of years they had farmed organically.

TABLE 11.1
Characteristics of Case Study Organic Farmers

Farmers[a]	Type of Operation	Total Acres (Acres Organic)[b]	Certified	Continuing to Farm Organically
1. Just starting (4–7 years)				
Phillip	Grain and livestock	1200 (1200)	Yes	Yes
Katie, Al, and Tim	Grain	1600 (270)	Yes	Yes
2. It's been a while (8–15 years)				
Peter	Grain and livestock	350 (350)	Yes	Yes
Olivier	Grain and forages	1400 (1400)	Yes	Yes
Matt	Grain, forages, and livestock	500 (500)	Yes	Yes
Doug	Grain	650 (0)	No	No
3. Done it forever (16+ years)				
Carlos	Grain, forages, and livestock	500 (500)	Yes	Yes
Janet and Will	Vegetables	15 (15)	No	Yes
Nicole and Hugh	Grain	3500 (3500)	Yes	Yes
Gerhard	Grain and livestock	1100 (1100)	Yes	Yes

[a] Names have been changed to provide anonymity.
[b] The number in parenthesis indicates the acres farmed organically.

A minimum of 4 years of organic farming experience was considered sufficient to account for the transition period during which farmers are working toward certification, as well as learning essential skills. Participants were divided into three categories (Table 11.1):

- 4–7 years (*just starting*)
- 8–15 years (*it's been a while*)
- 16+ years (*done it forever*)

Individuals were first contacted by e-mail with an explanation of the study's goals and to establish interview times. Semistructured interviews occurred in person (on-farm) or over the phone and lasted approximately 45–60 min. One follow-up question was asked to all participants in a subsequent e-mail. To achieve a less formal interview setting, as well as to minimize the amount of time required for transcription, participants' responses were recorded by hand (i.e., without the use of an electronic recorder) and later copied to a digital computer file. Participants' responses presented in this chapter are therefore not meant to represent exact (verbatim) speech; nevertheless, they reflect the meaning of the response as closely as possible.

In addition to learning about farm characteristics (such as size and crop rotation), the interview schedule was designed to obtain information about the following three themes:

1. Participants' rationale for farming organically
2. Challenges and new approaches (to agronomy, farming systems, and marketing)
3. Knowledge and knowledge networks

Working with these themes allowed for a more holistic perspective of farmers' experiences and viewpoints by acknowledging the broader social agroecological context in which they function. Similar themes were explored by Duram (1999), resulting in *thick* (i.e., data-rich) case studies of organic farmers.

11.3   WHY ORGANIC?

Within the literature, the question of why farmers might choose to use organic farming practices is asked and answered in different ways. One common way to explore this question is to examine conventional farmers' motives for potentially converting to organic farming (sometimes examined in tandem with reasons for not converting to organic farming).

For example, Khaledi et al.'s (2007) survey of 23 Saskatchewan conventional farmers determined that their interest in organic farming stems primarily from the potential for higher prices for organic commodities (related to higher income), reduced input costs, reduced reliance on agrichemicals, consumer demand for organic products, and environmental and consumer health benefits. Although these features represent conventional farmers' perceptions, and not necessarily the actual experiences of organic farmers, they are nevertheless important for their potential to motivate a farmer to *go organic.*

Attitude appears to play an important role in farmers' decision-making regarding production practices. Egri (1999) states that organic farmers tend to hold more extreme levels of environmental concern, as well as generally negative views of agrichemicals. Organic farmers may also apply multiple or symbolic meanings to agricultural practices (Egri 1997). These meanings include food and environmental safety, stewardship, craftsmanship, and legitimacy (of organic practices), among others.

One of this case study's participants, Olivier, says, "I never liked the chemical side of farming. It just doesn't seem right to put chemicals on food and then onto the kitchen table. I especially feel that way about preharvest applications, because those chemicals will still be on the food." Some farmers, such as Will, count environmental costs, such as energy consumption and *carbon footprint*, among their reasons for farming organically. "Do you realize how much energy goes into the Haber–Bosch process [for making nitrogen fertilizer]?" asks Will. "Tons!"

Health and food safety also emerged as important themes among participants. For Will, the decision to eschew agrichemicals is, in part, a practical one: "It's not just that it doesn't seem appropriate to use poisons on food. I think I'd probably end up poisoning myself accidentally." For Carlos, interest in organic production began as a way to address family health issues. "My brother suffered from severe food allergies as an infant," he says. "My parents began growing organic food for him, and his allergies cleared up completely. They decided that if it had that effect on him, it must also be good for everyone else. So they switched the whole farm over."

However, some organic farmers may attach more moderate environmental or ethical significance to organic practices, moved instead by primarily economic interests. Some participants in the current study cited financial incentives and *competitiveness* as reasons to explain why they chose to farm organically. Peter notes: "Premiums were a big incentive for getting in when we did, but my motivation was really to differentiate our product and to not compete with other producers." Similarly, Doug says, "I went organic at a time when the prices for agricultural commodities were really low. My initial interest came from the potential to make more money through organic premiums. I saw this as a way of being [economically] competitive with my relatively smaller farm."

Organic farmers' varying attitudes toward their own farming practices result in what Fairweather (1999) calls *committed organic* and *pragmatic organic* farmers, with the latter expressing a higher willingness to return to conventional practices if organic premiums were to decrease. Doug, who no longer farms organically, describes the struggle of a *pragmatic* farmer: "Basically, I needed to make things work financially. I have a wife and two kids, and I had to pay the bills. When the prices for conventional commodities started going up again, it was just too much to resist, so I went back to conventional."

The temptation to go back to conventional practices makes sense to Olivier, a committed organic farmer, who says,

> You've got to believe that organic is something good and that there is something better about the system than farming conventionally. If you're only in it for the money, you won't last. Don't even bother. You really have to want it, because the money is there in conventional. As a conventional farmer, you also have access to all the easy fixes, like sprays and fertilizers. Organic farming is more long term, and problems are not fixable in one year. So you need a different type of commitment.

Another farmer, Phillip, sums it up more succinctly: "Organic farming can look a lot like conventional farming if you only care about the money."

Organic farmers may also experience a shifting of perspectives and motivations over time. Peter, for whom premiums and competitiveness were initially important incentives to farm organically, experienced a shift toward a more *committed organic* perspective. "My thinking changed over time," he says. "Now it's more philosophical. Money is still important, but I am focused on organic farming as a way to address soil health, biodiversity, food health, and sustainability."

Succession planning and preparedness also play an important role in some farmers' decision to farm organically. Katie, Al, and their son, Tim, are still transitioning their farm from conventional to organic. "Al and I won't be farming forever," says Katie. "Maybe ten more years. Tim wants to farm organically, and we figure that it's easier for him to get into that type of farming than high-risk, conventional farming. When conventional agriculture works, it works well. But the costs are high, and so the risk is high." Katie also describes the uncertainty she feels regarding the future use of agrichemicals: "We see organic farming as a way to stay ahead of the game and to avoid the problems that will likely come in the future with chemical regulations. It's already happening for chemicals on lawns. It will probably happen for agriculture, too."

Whatever reasons a farmer has for farming organically, they will inevitably encounter situations that challenge their production practices. The following section will examine the major challenges to farming organically faced by the participants in this study, as well as some of the new approaches taken to overcome those challenges.

11.4 CHALLENGES AND NEW APPROACHES

Farming organically is not for the faint of heart. Organic practices are accompanied by a host of challenges. Some are common to farming in general: severe or unpredictable weather, energy costs, and market fluctuations, to name a few. Other challenges are unique to organic systems: weed and other pest pressure resulting from the nonuse of pesticides, fertility management issues resulting from the nonuse of synthetic fertilizers, the use of crops not common in conventional systems, and emerging or unpredictable markets for organic goods, among others.

For the purposes of this study, the challenges to organic farming are separated into two broad, and often interrelated, categories:

1. Agronomy and farming systems
2. Finances and markets

11.4.1 Agronomy and Farming Systems

The term *agronomy* is used here to describe a broad range of farm attributes, including soil fertility, crop rotation, timing of seeding and harvest, tillage and other forms of weed control, and animal management. The sum of these parts and their application to producing useful crops or other agricultural outputs is the farming system.

Not surprisingly, many of the farmers in this study pointed to weed control as one of the biggest challenges facing their farms. Without the use of herbicides, they rely on tillage and crop rotation, as well as livestock, when present, to manage weeds. Consequently, weed pressure in organic systems has the potential to result in high crop yield losses (Nelson et al. 2012; Rasmussen and Ascard 1995; Teasdale and Cavigelli 2010). During drought years, weeds have the potential to make moisture unavailable for crops, thereby reducing yields. This effect can be reinforced by the presence of a moisture-absorbing cover crop (Lotter et al. 2003).

Conversely, high moisture levels have the potential to play a confounding effect on weed control, making it difficult to perform tillage and other mechanical operations. "In wet years, weeds can be a huge challenge," says Will. "Moving any equipment on our clay soils is a big mess."

In addition to the impact that weeds have on crop yields, presence of weeds in fields has the potential to provoke neighbors who farm conventionally. Tim, who is still bringing conventional land into organic production, explains: "We purposefully started on a piece of land that is shielded on all sides by trees and a road. It's not near other farmers' fields. If they start to see fields of weeds growing next to theirs, they might have a problem with that."

Farmers in this study have pursued a number of strategies to manage weeds on their farms. Two major weapons in their arsenal are tillage and crop rotation. Farmers who have livestock also use animal grazing as a form of weed control. For example, Olivier, who does not own livestock, says, "Our wheat this year was very clean, just about perfect, because it was coming after a 4-year alfalfa break. But I also rely on tillage to control weeds. In general, I till two or three times over the year, maybe more."

Tillage is also important for Katie, Al, and Tim, but they have developed other strategies for weed control as well. Al explains: "We tend to use heavier seeding rates, which makes the crop a little more competitive. Sometimes we'll double seed the crop, if necessary. We've also played with the timing of our operations. We tend to seed later in spring, after we'd had a chance to cultivate." However, the latter strategy can result in problems later in the season, says Al. "If you get a cool fall, seeding late can be bad for yield."

Crop rotation is an important tool for longer-term weed control, providing spatial, architectural, and biological competition (Davies and Welsh 2002). Hugh, a grain producer, has developed a complex rotation that is designed not only to provide maximal economic returns but also to combat weeds. This rotation includes cereals, such as wheat, durum, and spelt, as well as oilseeds (flax and sometimes hemp), pulses (lentils), and an alfalfa seed crop. These crops are followed by a green manure fallow period lasting 2 years or more. Green manure crops include alfalfa, black lentil, and forage pea.

Gerhard, who now raises livestock, grows oats, barley, and peas, as well as a forage mix of sainfoin, alfalfa, sweet clover, and timothy. Before livestock were incorporated into his farming system, Gerhard says, "Quackgrass and other weeds were becoming a big problem. Even with tillage, they spread in the fall and early spring. Wild oats and thistles also got out of hand in many places." The situation improved, however, when livestock were brought in. "Now we just silage the patches with weeds, or we make bales. That way, the seeds don't get deposited into the weed bank, and we turn the plants into something nutritious for the animals. Our logic is that if something has a use, it's useful. Weeds don't always have to be villains or demons."

Soil fertility and soil health also emerged as an important challenge for many of the participants in this study. Maintaining adequate levels of important nutrients—chiefly, nitrogen and phosphorus—was seen as critical to the success of organic farming. Nutrient management in organic systems occurs in a radically different manner than in conventional systems: nutrients are introduced into the system in the form of green manures and other crop residues, animal manures, and compost, as well as slow-release sources, such as rock phosphate (Stockdale et al. 2002). Furthermore, these inputs typically rely on soil-related processes to make nutrients available for cropping, whereas conventional fertilizer inputs tend to be more immediately available in soluble, plant-available forms. Managing the quantity and quality (and possibly also timing) of organic inputs has important consequences for soil biota, nutrient transformation processes, and the accumulation of soil organic matter (Bunch 2002; Stockdale et al. 2002).

The greater reliance on soil processes (biological and chemical) for maintaining soil fertility is important for producers and results in different management decisions than those faced by conventional producers. Gerhard suggests that, in some ways, conventional producers have an easier task in managing soil fertility. "It's pretty formulaic. You just add the right chemicals at the recommended rates."

Organic farmers, on the other hand, must balance the desire to produce a marketable crop with the need to employ soil-improving methods. Some producers struggle more than others. For example, Doug says,

> I was used to continuous cropping [in a conventional system], and suddenly I was using green manures in my rotation. On paper, they sound great, and I'm sure they're good. But in my experience, I didn't see a significant nutrient boost to my crops the following year. I would produce a big crop of sweet clover, and I'd plow it down, and then the following year, I would see almost no improvement in my yields. It was hard for my land to get 'refilled' agronomically.

Similarly, Gerhard says that the long-term effect of removing forages and straw for bedding for livestock was to "really degrade the soil structure and fertility of my fields. We found out that if we didn't return nutrients to the soil in the form of composted manure or a green manure, the whole system suffered. So we haul manure, and we use hairy vetch and rye for fertility and to reduce erosion."

Having animals in the system appears to be an important feature toward improving fertility and cycling nutrients. Phillip, who produces grain and livestock, says, "Every third year, sometimes every year, I plant a green manure, and I graze my animals on it. That way, I get animal manure on my land as well, and the two combined allow for a slow release of nutrients. It's like compounding interest." He goes on to describe the long-term effect of failing to manage soil fertility: "It's like withdrawing from the piggy bank without putting anything into it. I've thought about how moving grain off my farm is essentially taking away nutrients from my farm. I'd like to sell animals exclusively and conserve the nutrients and manure on my farm."

Developing a crop rotation that satisfies a farmer's need to control weeds, manage soil fertility, and produce a marketable crop takes time, commitment, and the willingness to experiment. Hugh, who set aside a quarter section on which to test organic methods before choosing to go organic, looks back at the experience as an important learning opportunity. "We didn't perfect the system by any means," he says. "But we learned the basic concepts of crop rotation and using green manures. Since then, we've been continually improving on the system."

All participants in this study see healthy soil as a central feature of successful organic farming. Healthy soil is described in nuanced terms: it is a function of fertility, organic matter, soil biota, moisture management, and other features. Phillip states: "I'm always trying to improve the soil. I treat the soil as a living organism and an ecosystem." Organic practices, according to farmers in this study, are uniquely equipped to address issues in soil health. As Katie says, "We see organic farming as a way to connect farming directly with soil health. We're focused on learning about the soil and how to build it up. It takes time. There are no quick fixes as far as soil is concerned."

Nevertheless, improvements in soil quality are already apparent to some farmers. Olivier, for example, notes: "The soil smells better, and it's softer. It's not as hard-packed as it was before, even though I move more equipment across it. I think that's because the alfalfa really loosens the soil and because there's more organic matter."

11.4.2 Finances and Markets

Most farmers, whether organic or conventional, wear two hats: they must be skillful managers not only of their natural resources but of their financial resources as well. To be successful, they must spend wisely and have a strong understanding of markets. Balancing the dual role of producer and marketer comes naturally to some, but for others, it can be a tremendous challenge. Moreover, whereas conventional farmers have the benefit of established markets for their produce, the market for organic products is a smaller, more recent one (Lockeretz 2007). Organic farmers must learn to identify new markets, small wholesalers, and direct marketing opportunities (Duram 1999). Consequently, organic farmers may potentially experience a greater challenge in effectively connecting with markets and therefore in balancing their dual role.

The current study asked participants to describe the economic challenges they face in farming organically and in marketing their produce. The range of responses will be examined in detail in this section.

Some financial pressures were experienced by farmers during their transition to organic farming. For example, Peter, a grain and livestock producer, experienced higher costs during this transition period. "With only a portion of our land certified organic, I couldn't produce enough to feed my own animals," he says. "I had to buy grain and forages, and that was a huge expense. Thankfully, we didn't have to do that for too long."

Ongoing operational costs represent an important financial hurdle for some participants. Activities such as tillage and green manure management can result in higher fuel and other costs, as Olivier notes:

> My fuel costs are higher [than when I farmed conventionally], because I rely more on tillage to control weeds. I also need to control my green manure crops, so I have a Schulte mower. There's wear and tear of my equipment to consider as well. It all adds up, but I would say that it's still worth it if I don't have to deal with chemicals.

When asked whether he adds composted manure to his fields, Olivier says, "We don't have our own animals, so we don't have an on-farm source for manure. I'm very reluctant to purchase it, because it's another cost and another way that we become reliant on an outside company."

A strategy taken by some farmers in this study to minimize operational costs is to strive for high levels of self-sufficiency—that is, producing as much of their own inputs as possible. Carlos, who raises both livestock and crops, says,

> One of the important ways in which we keep our costs down is to produce almost everything we need for on-farm operations. We only buy minerals for the animals and grass seed for the pasture, which is difficult to produce in our area. Otherwise, we produce all of our own nutrients, seed, and feed. This is another way in which we can be holistic and self-sustaining.

Improved rotations can provide farmers with a way to reduce costs associated with tillage for weed control. "My rotation, over time, has allowed me to reduce the amount of tillage necessary to control weeds," says Hugh. "Having alfalfa [a perennial with strong weed-suppressing characteristics] in the rotation helps as well." Gerhard points out that reducing tillage can have other benefits: "Our soil is sandy, so there is a risk of erosion if we're not careful. We keep residue or a cover crop on the soil."

An improved rotation has provided other economic opportunities for farmers like Hugh. "We used to have a shorter rotation, and that was partly a reflection of the market," he says.

> Back in the 1980s, there wasn't much of a market for lentils. But we experimented with different crops over time, and that has opened up new opportunities. If we were just producing cereals and then growing a green manure, we'd only be producing a marketable crop half the time. Extending our rotation increases our diversity and lets us produce a range of crops, like alfalfa seeds and lentils. Our income goes up and our risk goes down.

For some participants, marketing their product does not present a major challenge. Will, who farms close to an urban center, says, "We're probably unique in that we've never really had any trouble marketing, even from day one. We were kind of the 'first kids on the block' to be doing this, and we've enjoyed a very good reputation at the farmers' market. Being close to the city helps."

For Doug, marketing activities provided a welcome challenge and, following his return to conventional farming, were the aspect of organic farming that he misses the most: "Marketing was one of my favorite parts of the whole enterprise. I liked connecting with buyers and different places where I could sell my grain. You just don't have that aspect in conventional farming, and that is disappointing. I'm an entrepreneur by nature, so I enjoyed it quite a bit."

Nevertheless, some producers state that simply accessing markets represents their greatest challenge. This may be especially true for producers of animal and dairy products, says Hugh. "Canadian processors of organic goods are often poorly positioned to accommodate producers, compared to processors in the United States. It's easier with grains, because they are not perishable, but marketing products like milk can be very tough." For example, Matt, who produces grain and livestock, says, "We've put out some feelers to a local meat buyer who has said that they could take some of our animals. But whatever they don't take, we have to send out east, and the returns are much lower."

Distance to markets can be a confounding factor for producers. "It's tough," says Olivier. "I hate being on the phone all the time, but I have to keep phoning certified buyers to move my produce. Basically, every place I sell to is at least 300 kilometers away. Thankfully, it's usually the buyer who pays for transportation." Producers of perishable goods, such as vegetables, face a similar challenge. "We're close to the city," says Will, "but some farmers have to transport their veggies a long way to get to markets. If they're not very well known on top of that, then it's really challenging."

Receiving help with the marketing aspects of farming has been important for some participants. Katie, Al, and Tim have participated in marketing training sessions provided by Manitoba Agriculture. Additionally, they recently had an opportunity to meet with potential buyers. "It was a bit like speed dating," says Katie. "We sat down with marketers and buyers and told them about our farm and our product. It gave us a chance to talk about ourselves in a real human way. That has been pretty successful, and we've already had a few people contact us because of that."

Phillip has also received help marketing his produce. In his case, he cooperates with a friend who also farms organically. "Together, we're able to pool our produce to gain marketing strength. He has a lot of experience, and that has made it easier for me, but I also see it as a bit of weakness, because I'm not learning to do it on my own. That might hurt me in the long run."

The challenges faced by organic farmers in marketing their produce are summed up by Gerhard, who says, "It can already be very difficult to deal with issues like soil fertility, crop rotation, and equipment. But organic farmers also have to deal with some very difficult markets. Having some market security and a fair, reasonable return for our product would make the job much easier."

11.5 KNOWLEDGE AND KNOWLEDGE NETWORKS

The third theme explored during interviews with the study participants was knowledge, as well as networks for the transmission of knowledge. Organic farming can generally be described as a knowledge-intensive form of agriculture, relying less on essential technologies (the *hardware*) than on an understanding of how to use the technologies (the *software*) (Rogers 1995). Following Padel (2001), this study asserts that organic farming is an *information-based* innovation that requires farmers to seek knowledge outside the agricultural mainstream and to network with other organic farmers.

11.5.1 SOURCES OF KNOWLEDGE

All of the participants in this study identified books, magazines, brochures, or other published information as an important source of knowledge about organic farming. For some farmers, one or more books played an important role in their introduction to organic farming, as Hugh explains: "I had just started farming organically. I found a book by Saskatchewan author Paul Hanley. I learned a lot of fundamental information about how to apply organic thinking to my farm."

Published information plays an ongoing role in helping organic farmers to stay abreast of developments and information. Peter says, "I read lots of books and have subscriptions to quite a few publications. My bookshelf is full. I didn't grow up reading much, but it's emerged as something that's really important for me. It's a great way to keep informed."

Four of the participants stated that experiences at university (completing individual courses, a portion of a degree, or a full degree) were instrumental in shaping their thinking about organic agriculture. University courses for all four participants pertained to agriculture, although not explicitly to organic production. An additional participant expressed a desire to receive training in holistic farm management practices.

Other academic sources of information and expertise are important also. Katie explains that researchers' knowledge has been vital to the success of their activities, primarily because of their distance from other sources of information. "We've received most of our information from academic sources, such as researchers and books, as well as field days and conferences," she says. "In our area, we just don't have any farmers doing what we're doing, so there's no one else we can call up to ask a question."

Conferences and field days have provided some of the participants with opportunities to network with other organic farmers and to form valuable knowledge partnerships. Matt recalls the formative role that an organic conference had in his decision to farm organically:

> I really only went to learn about new production practices. I wasn't necessarily interested in becoming an organic farmer. But I had a great experience, and I left with a strong sense of conviction and purpose and a passion that led me to go organic. Over the years, I have been able to build up a network of organic farmers through conferences and field days. It has allowed me to develop a community of likeminded individuals, and that's been an important way to learn about what others are trying on their farms.

Friendships and acquaintances between farmers are essential sources of knowledge for some of the study participants, especially for newer farmers. The experience of other farmers can assist in decision-making and help to prevent costly errors. Olivier describes one such friend:

> He's been farming for a long time, more than twenty years, and I'm able to turn to him for a lot of advice. He doesn't live in my area, so some of the ways he does things are particular to his area, but I still learn a lot. I also learn from the mistakes he's made in the past, and that's really important, because sometimes mistakes can be very costly.

Another important role that friendships between organic producers can play is to reduce farmers' feelings of isolation and provide emotional support. Although the general perception of organic farming among conventional producers has improved gradually over time (Padel 2001), there is evidence that organic producers still feel isolated by their production practices and perspectives toward farming. "I don't get any hostility from my conventional neighbors," explains Phillip, "but I do feel like organic farming is something that just doesn't make sense to them. You tell a conventional producer about it and he just looks at you like you're from the moon." Matt agrees: "In some ways, my decision to farm organically has made me feel less a part of my own [geographic] farming community, because of the huge distance between our farming practices. It's easy to feel isolated as an organic farmer, and so those contacts with likeminded individuals are very important."

In that regard, Peter describes a relationship with a nearby farming household as an important way to overcome feelings of isolation:

> They used to farm organically, and we helped each other out a lot with equipment and even land. Unfortunately, they couldn't make it work financially, and went back to conventional. It was a major blow for us, but they are still committed to organic in a philosophical sense, so I have someone to share ideas with and a place to turn to for moral support.

More formal acquaintances can also function as important sources of knowledge for organic farmers. One participant mentioned an important contact with an organic veterinarian, and several participants also referred to their organic inspector as a valuable source of information. As someone who visits a number of organic farms and sees a variety of approaches to production, an organic inspector can help to steer farmers toward new ideas and away from potential pitfalls. Katie says, "Our inspector is great. We are able to ask questions, and so in some ways, it's like having a teacher. And the inspections are done with a lot of common sense, so we don't feel as if we need to defend ourselves. We've also learned a lot about how to take care of the documentation required."

Additionally, one participant previously worked as an organic inspector and considers that experience to be important in shaping their own farming practices. "I went to many different farms, and I had the opportunity to see their fields, their rotations, their soil, and their weeds. I learned a lot about the challenges they face, and that has helped me in some ways."

Two other important sources of information include government agencies (as in the example provided by Katie regarding *speed marketing*) and producer associations.

The most frequently mentioned association among participants was the Organic Producers Association of Manitoba (OPAM), which provides inspection services to members, as well as information and networking opportunities (OPAM 2012).

A need for improved extension and agronomist services was identified by two participants. "There is a need for ag reps and agronomists that understand organic agriculture enough to give solid advice," says Hugh. "You see that sort of thing more in Quebec, where they have a well-developed organic sector. An agronomist without any organic training isn't very helpful."

In addition to being a source of information on an almost infinite number of topics, the Internet may provide farmers with the means to network and discuss issues in a cost-effective and timely manner. "It might be possible for organic farmers to network with each other over the Internet on a forum like Facebook," says Gerhard. "They could talk about whatever matters to them. For example, they might talk about the impact that genetically modified alfalfa would have on organic producers. It would let that kind of discussion happen in a public, involved way."

Olivier also sees the value in forming an online organic producer group. "We could pool our knowledge," he says.

> There are guys with a lot of experience, and even if they just share a few key points, for example, saying, 'Last year I did this on my land,' that is the type of information that can save farmers from having to do it all again on their own. There's no reason to want to keep that type of information from others, and for younger or less experienced organic farmers, it would really speed up the learning process.

11.5.2 GEOGRAPHY OF KNOWLEDGE: ORGANIC FARMERS AS HOLDERS OF INDIGENOUS KNOWLEDGE

On-farm trial-and-error processes and long-term observation provide organic farmers with valuable opportunities not only to determine which forms of exogenous (off-farm) knowledge are relevant under local conditions but also to develop their own forms of indigenous knowledge. The term *indigenous knowledge* is used here in the sense of McClure (1989 in DeWalt 1994), who writes:

> Indigenous knowledge systems are learned ways of knowing and looking at the world. They have evolved from years of experience and trial-and-error problem solving by groups of people working to meet the challenges they face in their local environments, drawing upon the resources they have at hand.

Although the term *indigenous* is often used in association with people groups (cf. Semali and Kincheloe 1999), the previous definition demonstrates the potential for knowledge to be intrinsically linked to geography. When such localized knowledge is taken as part of the broader agroecosystem, it becomes a valuable natural resource (Warren 1991 in Agrawal 1995) and can be a tool for management.

An intimate understanding of location is crucial in organic farming systems, in which fewer external inputs are available to mask local conditions. Understanding the unique agroecological features of a specific geographic location (e.g., soil type and fertility, microclimate, topography, moisture patterns, landscape interactions) enables organic farmers to make effective use of outside knowledge. Several of the farmers who participated in this study emphasized the need to contextualize outside knowledge in this manner. Tim explains,

> We have unique conditions on our farm—soil and weather, and such—and our farming circumstances are unique as well. We've seen that organic methods can work on the small-scale university plots, and now we're interested in applying them on a much bigger scale on our farm. That might also allow us to be an example for other local farmers who are interested.

Beyond simply contextualizing and "scaling up" outside knowledge, some participants saw themselves as having a role in producing new knowledge. Phillip says,

> As an organic farmer, a person really needs to be a pioneer. Farming this way is a lost art, and we [organic farmers] are all rediscovering it. There might still be room for improvement after 300 years of trying. No books will ever say everything, so you need to understand your own farm. And that's the fun part.

Organic farmers' role as holders and pursuers of indigenous knowledge raises a number of interesting questions. Firstly, how do organic farmers view themselves in relation to the farming system? Indigenous knowledge typically de-emphasizes (or does not recognize) the subject–object dichotomy that dominates western scientific thinking (Banuri and Apffel-Marglin 1993). The individual is viewed as part of the whole, and social–cultural life is not distinct from the physical environment in which it occurs. It is possible that organic farmers' perceptions of *holistic* approaches to farming influence their positioning on their own mental landscapes.

A second question that arises in connection to indigenous knowledge relates to the broader social context in which organic farmers exist. As the agri-food system becomes increasingly globalized, organic farmers must adapt their livelihood strategies and ways of life to survive in an increasingly unbounded social agroecological system. How do organic farmers navigate this system and approach complex interactions? It may be farmers' indigenous knowledge that enables them to access or gather resources across the system (Figure 11.1). These may be physical resources such as seeds, equipment, and markets, or they may be intangible, such as information, relationships, and moral support. Furthermore, resource-gathering activities may occur on a local level (on-farm or in cooperation with neighbors), as well as on a broader, more global scale. Examples of the latter type of activities may include learning to use the Internet to access information, visiting farms in other parts or outside of Canada, or establishing unique economic agreements with international buyers.

Although it is beyond the scope of this study to answer the previous questions directly, they may provide a starting point for future research. Viewing organic farmers as holders of indigenous knowledge fosters improved, two-way systems for knowledge transfer between farmers and researchers (Parrish 1999) and enables both groups to take advantage of the other's creativity and innovativeness (DeWalt 1994).

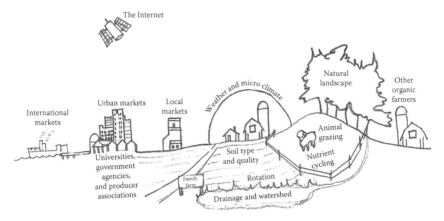

FIGURE 11.1 Conceptual social agroecological landscape. Organic farmers may rely on unique, indigenous knowledge to navigate and interact with different features of this landscape and access resources across scales.

11.6 ORGANIC AGRICULTURE AS AN ADAPTIVE CYCLE

This study has examined the experiences of 10 Canadian organic farming households around three themes: (1) participants' rationale for farming organically, (2) challenges and new approaches (to agronomy, farming systems, and marketing), and (3) knowledge and knowledge networks. Taken as a whole, participants' responses to these themes indicate that a high degree of adaptiveness is a necessary characteristic of successful organic farmers. Whereas conventional agriculture seeks to minimize and control variables within the farming system, organic agriculture acknowledges and accounts for unpredictable variables to achieve resilience. Folke et al. (2003, p. 352) define resilience as "the capacity to lead a continued existence by incorporating change." Following those authors, organic agriculture can be characterized as a system that assumes change and explains stability, instead of assuming stability and explaining change.

Building on the adaptive cycle concept developed by Holling and Gunderson (2002), this section will provide a conceptual adaptive cycle for organic agriculture (Figure 11.2). The majority (9 out of 10) of participants in this study came out of the conventional farming system. Farmers were motivated to enter the organic system by a combination of factors, including low conventional commodity prices, high input costs, a dislike for agrichemicals, and the potential for organic premiums. Two participants had established long-term on-farm trials (16 and 160 ac) to evaluate organic methods prior to committing to changing the remainder of their farms.

The adaptive cycle comprises of four phases—namely, the growth, conservation, release, and renewal phases (Holling and Gunderson 2002). Figure 11.2 shows recently certified organic farmers entering into the adaptive cycle during the growth phase, when they are chiefly concerned with applying new knowledge, adapting to their new constraints, and gaining skills.

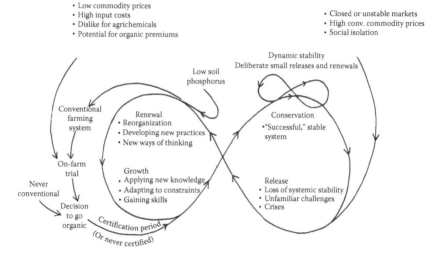

FIGURE 11.2 Conceptual adaptive cycle for organic agriculture. Individuals enter the organic system in response to different stimuli. Organization (connectedness and potential) within the system increases in the conservation phase. Instability and new stimuli trigger the release phase, causing some farmers to return to the conventional farming system. Others reorganize and re-enter the growth phase. (Adapted from Holling, C.S. and Gunderson, L.H. in: L.H. Gunderson and C.S. Holling (eds.) *Panarchy: Understanding Transformations in Human and Natural Systems*, Island Press, Washington, D.C., 2002, pp. 25–62.)

As their understanding grows and their proficiency increases, farmers enter into the conservation phase of the adaptive cycle, characterized by a *successful*, stable farming system. It is during this phase that organic farming seems to *work* for the farmer.

However, as Folke et al. (2003, p. 356) point out, "It is impossible to lock a system in a steady state for eternity or to manage it for stability and security in a command-and-control fashion." Attempts to maintain a system in a constant state often lead to organizational and ecological *brittleness*. Consequently, when new stressors and uncertainties emerge, the system becomes destabilized. In the context of organic farming, these stressors might include a host of agronomic and financial challenges, as well as closed or unstable markets, high prices for conventional commodities, and feelings of social isolation. When subjected to these unfamiliar challenges (or crises), organic farmers are thrust into the release phase of the adaptive cycle and experience a loss of systemic stability.

At this point in the cycle, a farmer may choose to exit the organic farm system. For Doug, the decision to return to farming conventionally stemmed from more than just financial challenges. "My soil phosphorus levels were going down," he says. "I had to change the way I was doing things, so I got out." Returning to the conventional farming system may be a long-term decision, but some farmers may choose to try farming organically again at a later date. Doug says, "I would definitely consider going back. I enjoyed it. But this time, I would plan ahead carefully. I thought I could just figure things out as I went along, but I see now that you ought to develop a strategy ahead of time to make things work."

Organic farmers in the release phase who do not exit the organic system are forced to re-examine their farming practices, bringing them into the renewal phase. It is during this phase that farmers reorganize, develop new practices, and acquire new ways of thinking. A key aspect of this phase is what Folke et al. (2003) refer to as *memory*—in this context, the learning that occurs as a result of crises within the social agroecological system. This memory builds over time and allows organic farmers to make decisions that increase systemic resilience and deal with ongoing change.

Creativity is also an essential component to navigating this phase of the adaptive cycle. Carlos describes the role of creativity on his farm:

> I have worked on different farms across Canada and in Europe, and from those kinds of experiences, a person can pull together some ideas and try to make them work on their own farm. But to really succeed, you need imagination. Creativity is an essential part of dealing with the challenges that will certainly come.

Organic farmers who have experienced the renewal phase will once more enter into the growth phase, applying the lessons learned and, inevitably, facing new challenges. The time frame for each phase is dependent on external factors, such as markets and natural conditions, as well as farmers' knowledge (memory) and other individual characteristics.

It may also be possible for organic farmers who are still in the conservation phase to reduce the *brittleness* of their farming system by deliberately triggering small releases of stability, leading to more rapid and less drastic renewal phases. Examples of such intentional triggers can include plowing up a multiyear alfalfa field to release stored fertility or using mixed or relay intercropping, which increases the unpredictability of interactions between species. In this manner, organic farmers may be able to remain in a state of high potential and connectedness and achieve a type of dynamic stability (Figure 11.2).

Finally, it is likely that a similar adaptive cycle could be developed for conventional agriculture. Farmers within that system must also learn to adapt and respond to agronomic and economic stimuli. However, the system can also be characterized as one that seeks to minimize variations and control change. As was observed by this study's participants, conventional agriculture also tends to occur in a more standardized, formulaic manner, with more resources available to assist farmers in decision-making. Consequently, an adaptive cycle for conventional agriculture would likely have different characteristics and function on a different time scale, at the risk of becoming more brittle in the conservation stage.

11.7 CONCLUDING REMARKS

Taking a close look at the experiences of 10 Canadian organic farming households provides a detailed perspective of the challenges facing the organic farming system. Agronomic issues, such as soil fertility and weed control, were identified as important. Crop rotation and tillage were seen as important tools available to organic farmers for addressing agronomic challenges. Financial challenges were also recognized, including fuel and other input costs. Research participants identified market access and stability, as well as distance to markets, as factors with the potential to negatively affect their financial health. Partnerships and other forms of assistance aided some farmers in overcoming marketing challenges.

Knowledge and networks for the transmission of knowledge emerged as important resources for participants in this research, enabling them to better approach both agronomic and marketing challenges. This finding supports Padel's (2001) assertion that organic farming is an *information-based* innovation.

Adaptiveness and creativity were also seen as essential components to successful, long-term organic farming. As farmers learn to anticipate challenges and uncertainties, their farming systems gain resilience, and they are better equipped to farm in a manner that is agronomically, financially, ecologically, and socially sound.

REFERENCES

Agrawal, A. (1995). Dismantling the divide between indigenous and scientific knowledge. *Development and Change*, 26(3), 413–429.

Banuri, T. and Apffel-Marglin, F. (eds.). (1993). *Who Will Save the Forests? Knowledge, Power and Environmental Destruction*. London, U.K.: Zed.

Bunch, R. (2002). Nutrient quantity or nutrient access? A new understanding of how to maintain soil fertility in the tropics. Unpublished manuscript. Retrieved May 18, 2012, from http://sustainablegrowthtexas.com/resource/Roland.pdf.

Davies, D. and Welsh, J. (2002). Weed control in organic cereals and pulses. In: D. Younie, B.R. Taylor, J.M. Welch, and J.M. Wilkinson (eds.) *Organic Cereals and Pulses*, pp. 77–114. Papers presented at conferences held at the Heriot-Watt University, Edinburgh, and at Cranfield University Silsoe Campus, Bedfordshire, U.K., November 6 and 9, 2001. Southampton, U.K.: Chalcombe Publications.

DeWalt, B. (1994). Using indigenous knowledge to improve agriculture and natural resource management. *Human Organization*, 53(2), 123–131.

Duram, L. (1999). Factors in organic farmer's decisionmaking: Diversity, challenge, and obstacles. *American Journal of Alternative Agriculture*, 14(1), 2–10.

Egri, C. (1997). War and peace on the land: An analysis of the symbolism of organic farming. *Studies in Cultures, Organizations, and Societies*, 3(1), 17–40.

Egri, C. (1999). Attitudes, backgrounds and information preferences of Canadian organic and conventional farmers: Implications for organic farming advocacy and extension. *Journal of Sustainable Agriculture*, 13(3), 45–72.

Fairweather, J. (1999). Understanding how farmers choose between organic and conventional production: Results from New Zealand and policy implications. *Agriculture and Human Values*, 16, 51–63.

Folke, C., Colding, J., and Berkes, F. (2003). Synthesis: Building resilience and adaptive capacity in social-ecological systems. In: F. Berkes, J. Colding, and C. Folke (eds.) *Navigating Social-Ecological Systems: Building Resilience for Complexity and Change*. Cambridge, U.K.: Cambridge University Press, pp. 352–356.

Holling, C.S. and L.H. Gunderson. (2002). Resilience and adaptive cycles. In: L.H. Gunderson and C.S. Holling (eds.) *Panarchy: Understanding Transformations in Human and Natural Systems*. Washington, D.C.: Island Press, pp. 25–62.

Khaledi, M., Gray, R., Weseen, S., and Sawyer, S. (2007). Assessing barriers to conversion to organic farming: An institutional analysis. Submitted to: Advancing Canadian Agriculture and Agri-Food Saskatchewan (ACAAFS). University of Saskatchewan, Saskatoon, Saskatchewan, Canada.

Knight, J. and Shirtliffe, S. (2003). Saskatchewan organic on-farm research: Part I: Farm surety and establishment of on-farm research infrastructure. University of Saskatchewan, Saskatoon, Saskatchewan, Canada.

Lockeretz, W. (2007). What explains the rise of organic farming? In: W. Lockeretz (ed.) *Organic Farming: An International History*. Cambridge, MA: CABI, p. 130.

Lotter, D., Seidel, R., and Liebhardt, W. (2003). The performance of organic and conventional cropping systems in an extreme climate year. *American Journal of Alternative Agriculture*, 18(2), 1–9.

Nelson, A., Pswarayi, A., Quideau, S., Frick, B., and Spaner, D. (2012). Yield and weed suppression of crop mixtures in organic and conventional systems of the western Canadian Prairie. *Agronomy Journal*, 104(3), 756–762.

OACC. (2008). *Research Needs Assessment of Manitoba Organic Farmers*. Organic Agriculture Centre of Canada, University of Manitoba, Winnipeg, Manitoba, Canada.

OPAM. (2012). Home. In: *Organic Producers Association of Manitoba*. http://www.opam-mb.com/index.html (accessed October 20, 2012).

Padel, S. (2001). Conversion to organic farming: A typical example of diffusion of an innovation? *Sociologia Ruralis*, 41(1), 40–61.

Parrish, A. (1999). Agricultural extension education and the transfer of knowledge in an Egyptian oasis. In: L. Semali and J. Kincheloe (eds.) *What Is Indigenous Knowledge?* New York: Falmer Press, pp. 269–283.

Rasmussen, J. and Ascard, J. (1995). Weed control in organic farming systems. In: D.M. Glen, M.P. Greaves, and H.M. Anderson (eds.) *Ecology and Integrated Farming Systems. Proceedings of the 13th Long Ashton International Symposium on Arable Ecosystems for the 21st Century*, Bristol, U.K., September 14–16, 1993, pp. 49–67. Abstract retrieved October 26, 2012 from http://www.cabdirect.org.

Rasmussen, J. and J. Ascard 1995. Weed control in organic farming systems. In: M. Glen, M.P. Greavers, and H.M. Anderson (eds.) *Ecology and Integrated Farming Systems*, Chichester, UK; Wiley Publishers, pp. 49–67.

Rogers, E. (1995). *Diffusion of Innovations*. New York: The Free Press.

Semali, L. and Kincheloe, J. (1999). Introduction: What is indigenous knowledge and why should we study it? In: L. Semali and J. Kincheloe (eds.) *What is Indigenous Knowledge?*. New York: Falmer Press, p. 69.

Stockdale, E.A., Shepherd, M.A., Fortune, S., Cuttle, S.P., 2002. Soil fertility in organic farming systems-fundame ntally different? *Soil Use Manage*. 18, 301–308.

Teasdale, J. and Cavigelli, M. (2010). Subplots facilitate assessment of corn yield losses from weed competition in a long-term experiment. *Agronomy for Sustainable Development*, 30(2), 445–453.

Section IV

Economics, Energy, and Policy

12 Economics of Energy Use in Organic Agriculture

Emmanuel K. Yiridoe

CONTENTS

12.1 INTRODUCTION

Although agriculture uses energy in various forms or types, agriculture also produces energy. One of the reasons why energy use in agriculture is important is because of the increasing cost of direct and indirect energy, linked to depletion of energy from fossil fuel sources, and increasing use in intensive agricultural systems. The concern has prompted a heightened interest in, and a growing need to develop, more sustainable agricultural production systems (Zinck et al. 2004). The agri-food sector (including on-farm production, processing and packaging of agri-food products, and distribution and marketing subsectors) accounts for an estimated 19% of national fossil fuel energy use in the United States (Pimentel 2006).

In Canada, direct energy use in agriculture accounts for about 5.5% of total energy consumption in the country (Canadian Agricultural Energy End-Use Data Analysis Center 1998). Consumption of energy from fossil fuels also accounts for significant greenhouse gas (GHG) emissions. For example, the Canadian agricultural sector accounts for about 10% of total GHG emissions in the country (Zentner et al. 2005). Agriculture could, and should be expected to, contribute to Canada's energy conservation and GHG reduction targets. Compared with conventional agriculture, organic farming systems have a potential to improve agricultural energy conservation and GHG emissions reduction (Gomiero et al. 2008; Lynch et al. 2011; MacRae et al. 2010), mainly because the latter tends to use more natural (as opposed to manufactured) inputs, particularly for nitrogen supplementation. A significant component of the energy inputs used in conventional agriculture involves off-farm activities, such as manufactured fertilizers, pesticides, animal feed, and veterinary supplies.

The importance of energy use in agriculture is also linked to the importance of food in meeting human dietary energy requirements and other indirect energy outputs from agriculture. Some of the forms of such energy include (1) energy-conserving products used to produce clothing and shelter, (2) energy for agricultural work and related tasks, (3) energy in biofuels (e.g., forest biomass and animal biomass) and potential energy in agricultural by-products, and (4) human and domestic animal dietary energy requirements (Spedding and Walsingham 1976).

The various types of energy commonly used in agriculture may be classified in terms of (1) whether they are renewable or nonrenewable and (2) direct or indirect use in agriculture. Energy from renewable sources generally has lower CO_2 emissions (e.g., solar, biogas, biofuels, and wind), compared with energy from fossil-based fuels or other nonrenewable sources. Direct energy used in agriculture reflects energy typically used on-farm. Indirect energy consumption in agriculture includes energy typically embodied in products off-farm, including production of farm inputs such as fertilizer and farm equipment and machinery. Other sources of energy include human labor and animal draft power. Energy is also used in important farm operations such as seedbed preparation and cultivation, planting, and harvesting of crop and animal products. Beyond the farm gate, energy is used in producing, transporting, and distributing farm inputs (e.g., seeds, pesticides, manure and fertilizer, and farm machinery and equipment), as well as in farm product processing, packaging, and distribution.

The global oil crisis of the 1970s and associated energy price hikes, especially during the 1980s, heightened interest in the impacts of declining fossil fuel energy reserves and increasing energy prices on agriculture and future food production (Cleveland 1995; Kendall and Pimentel 1994), as well as agricultural producers' response to changing energy situations over time (OECD 1982).

The agricultural energy knowledge gap requires a better understanding of specific stages of the agricultural production processes in which energy is used (Spedding and Walsingham 1976; Williams et al. 2006), what quantities are used (Cleveland 1995; Gomiero et al. 2008), and efficiencies associated with using such energy (Greening et al. 2000; Herring 1999, 2006). In addition, conservationists are interested in (potential) impacts and implications of where and how agricultural energy use can be improved (through, e.g., potential reduction in energy inputs and increased energy output in products produced) (Hanley et al. 2009; Weber and Matthews 2008). There is also a general lack of technical information on how organic agricultural systems affect farm energy budgets in Canada (Hoeppner et al. 2006; MacRae et al. 2010), and broader energy dynamics issues (Hoeppner et al. 2006). Gomiero et al. (2008) also noted a need for more research to better understand the energy dynamics and environmental impacts of organic farming practices.

12.2   TRENDS IN ENERGY USE IN AGRICULTURE: CANADA

Studies suggest continuing growth in nonrenewable energy use in Canadian agriculture (see, e.g., Canadian Agricultural Energy End-Use Data Analysis Center [CAEEDA] 1998, 2000; Eco-Energy Carbon Capture and Storage Task Force 2008) linked largely to increasing substitution of energy for human labor, continuing efforts to increase agricultural productivity, and intensification and mechanization of farming systems and associated changes in farm input requirements (Zentner et al. 2005). Various aspects of the agri-food sector account for increased energy use in Canadian agriculture. For example, nonrenewable energy used in spring wheat production in the prairie region of Canada increased from an average of about 2500 MJ ha^{-1} in 1948 to 8400 MJ ha^{-1} by 1981 (Hopper 1984). In addition, in Saskatchewan alone, energy input use in agriculture increased by 61% between 1961 and 1976 (Stirling 1979) and by 11% between 1990 and 1996 (Coxworth 1997). The general trend in energy input use in agriculture for the country as a whole mirrors the patterns reported for Saskatchewan.

Data for total energy use in Canada as a whole suggest a rising trend from 1990 to 2009 (Figure 12.1). However, there are differences in the general trend and other attributes, when total energy use is disaggregated into component sectors. For example, Figure 12.2 indicates that energy use is considerably higher in Canada's residential sector than in the agriculture sector. In addition, the energy data for the agricultural sector suggest a generally flat trend during the 19-year period, from 1990 to 2009. On-farm energy use, specifically for crop and livestock production, during the same period, indicates a peak in 1996, followed by a steady decline over the next decade, to levels similar to the early 1990s (Figure 12.3). The reasons for the decline from the

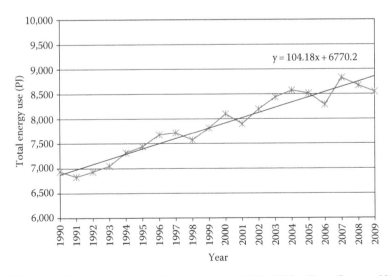

FIGURE 12.1   Total energy use (PJ) in Canada, 1990–2009. (Data Source: Natural Resources Canada, 2011.)

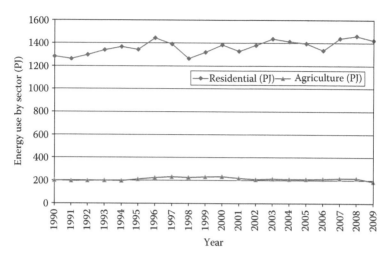

FIGURE 12.2 Comparison of energy use in Canada's residential and agricultural sectors, 1990–2009. (Data Source: Natural Resources Canada, 2011.)

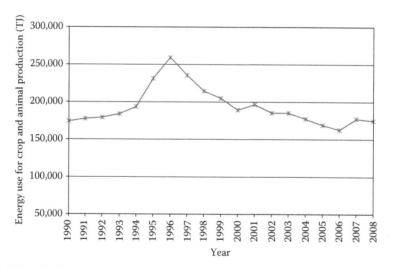

FIGURE 12.3 Energy use (TJ) for crop and livestock production in Canada, 1990–2008. (Data Source: Statistics Canada, 2012.)

1996 level include technological advances and introduction of energy efficiency and conservation programs such as the Energy Star® program (CAEEDA 2000).

Besides the general trend in energy input use in Canadian agriculture, overall energy use efficiency in dominant agricultural regions in Canada suggests a declining trend in recent decades, due largely to energy input use increasing at a faster pace relative to energy outputs produced (Zentner et al. 2005). One of the consequences of the intensification of agriculture in industrialized countries such as Canada is a corresponding increased use of fertilizers, the production and distribution of which require substantial energy. Nitrogen fertilizer use is the leading single largest energy use in nonorganic agriculture and accounts for about 37% of total energy use in industrialized countries (Azeez and Hewlett 2008). As with other OECD countries, chemical fertilizer consumption in Canada increased substantially, especially since the 1960s, both on an aggregate and a per capita basis (Boyd 2001). Total Nitrogen-Phosphorus-Potassium (NPK) fertilizer consumption (Figure 12.4) and fertilizer consumption per unit arable land area (Figure 12.5)

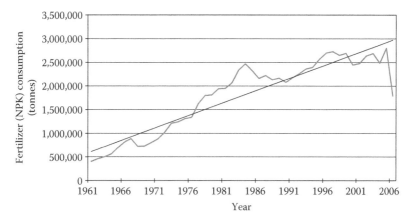

FIGURE 12.4 Fertilizer (NPK) consumption in Canada, 1961–2006. (Data Source: Food and Agriculture Organization of the United Nations (FAO), *FAOSTAT-Fertilizers*, FAO, Rome, Italy, available online at http://faostat.fao.org/site/575/default.aspx#ancor, accessed on August 10, 2012.)

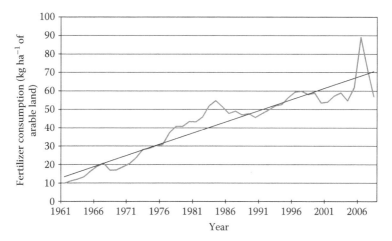

FIGURE 12.5 Fertilizer consumption per unit area in Canada, 1961–2008. *Notes*: Fertilizer products include nitrogenous, potash, and phosphate fertilizers (including ground rock phosphate). Animal and plant manures are not included. (Data Source: World Bank, Fertilizer consumption per unit and area in Canada, 1961–2008, available at http://data.worldbank.org/indicator/AG.CON.FERT.ZS, accessed on July 2011.)

in Canada increased substantially during 1961–2006. The data suggest that total (NPK) fertilizer consumption increased by 597% between 1961 and 2005. By comparison, the percentage increase per unit of arable land was 805% during the same period, due to increased fertilizer use in the (arable) farm than fertilizer use in the nonfarming activities (e.g., in golf course management).

12.3 NATURE AND CHARACTERISTICS OF ENERGY IN AGRICULTURE

Most agricultural inputs and outputs from production can be expressed in terms of energy used or produced. Thus, energy input and output are important attributes of assessing the energy efficiency and environmental impacts of agricultural production. Important components of energy balance on farms include (1) total energy consumption, (2) energy output (i.e., stored in products produced), and (3) energy efficiency (Hoeppner et al. 2006). Energy consumption estimates commonly quantify all energy inputs into a production process, including energy use in manufacturing farm machinery and equipment, fertilizer, and other chemical inputs and in the production of manure and natural pesticides.

Energy output in agri-food products reflects the energy values of agricultural outputs and food produced. Some analysts have noted that energy output in food will likely have to increase in order to meet the food energy requirements of the ever-growing world population (e.g., Smil 2000), which is projected to peak at 9.22 billion in 2075 (United Nations Department of Economic and Social Affairs 2004).

The third dimension of the on-farm energy balance, energy use efficiency, is the component commonly considered in energy assessments and comparisons (such as between organic and conventional farming systems). In general, energy use efficiency is measured in terms of energy in agricultural output produced per unit of farm input energy used. Concerns with the long-term sustainability of energy from fossil fuel sources suggest that improving efficiency in energy use in food production has potential as an important conservation strategy (Geller and Attali 2005). Partly because of this importance, analysts tend to be more interested in energy use efficiency in producing particular agricultural products than with energy generated per unit of energy used (Spedding and Walsingham 1976; Zinck et al. 2004). Although the efficiency with which energy is used in agriculture is important, evaluation of the overall performance and sustainability of agricultural production systems should not focus on energy balance or dynamics alone but should also consider the efficiency with which nonenergy inputs are used, as well as other dimensions of agricultural system performance and sustainability (Spedding and Walsingham 1976; Zinck et al. 2004).

The OECD (1997) endorsed the use of energy intensity and efficiency as relevant indicators for quantifying and assessing energy use in agriculture. Energy use in agricultural production is commonly expressed in terms of consumption per unit land area (for crop products) or per unit output produced, while energy efficiency reflects an input/output ratio (Stolze et al. 2000). This study distinguishes between *energy efficiency* and *energy conservation*, consistent with Herring (2006).

Energy efficiency is used in this study to mean optimizing energy services and uses, while the latter concept is operationalized to mean reducing energy by using *less* energy services.

12.4 ECONOMICS OF ENERGY USE IN AGRI-FOOD SECTOR: ARE THERE OPPORTUNITIES FOR SAVINGS IN ORGANIC AGRICULTURE?

Energy use in agriculture affects the overall development of the agricultural sector and, indirectly, the rest of the economy through effects on input costs and agricultural output prices. In addition, the benefits from improving energy efficiency in agriculture extend beyond the agri-food sector to the larger economy. Energy efficiency can also generate savings to individual farmers (as energy consumers) and increased agricultural productivity (Herring 2006). In addition, savings and increased profitability from energy efficiency gains can be used to support and stimulate financing of a switch to low-input agriculture or alternative energy systems with lower CO_2 emissions (Gomiero et al. 2008).

Proponents of "Factor Four" (Von Weizsacker 1998) and environmentalists who promote energy efficiency and conservation strategies hold onto the premise that improvements in energy efficiency will result in an overall reduction in energy consumption in the economy (Herring 1999, 2006). This premise is central to many national energy efficiency programs, which are promoted by various levels of government as a cost-effective strategy to address climate change and global warming. However, there is debate about whether microlevel improvements in energy efficiency will result in a reduction in energy consumption at the macro or national scale (Berkhout et al. 2000; Herring 1999). The debate is linked to concerns that a *rebound* or *take-back* effect of microlevel savings from individual investments in energy efficiency may be offset by higher overall national energy consumption. Technological advancements and improvements that result in more energy-efficient equipment and appliances can translate into less energy used in producing the same level of output, *ceteris paribus*. The energy efficiency further leads to a cost reduction per unit of services from such equipment. The input cost and, ultimately, output price reduction may then translate into increased overall energy consumption. In other words, improving energy efficiency may not necessarily reduce overall energy consumption, with implications for increasing GHG emissions and a reduction in environmental quality.

Is there a counterintuition connected with the rebound effect? Economic theory postulates that an improvement in the efficiency of the use of a good (such as energy) leads to a fall in the implicit price of the product, thereby stimulating (increased) consumption of that product in a closed economy. The resulting rebound effect can lead to (1) a substitution effect in which, because energy becomes cheaper, individuals such as farmers end up using more of it (e.g., to drive farm machines and cool or heat farm housing systems), and (2) an income effect in which the lower energy cost increases disposable income and, ultimately, more spending (Berkhout et al. 2000; Geller and Attali 2005). Laitner (2000) reported evidence that energy efficiency investments improve national income and reduce energy prices. However, the overall economy-wide rebound effect tends to result in a very small effect (i.e., about 0.5%) in reducing energy savings (Geller and Attali 2005).

Empirical studies suggest some evidence of a rebound effect at the national or macroeconomy level (Berkhout et al. 2000). However, the rebound effect is generally small in magnitude (Geller and Attali 2005). There is a general lack of empirical studies on the rebound effect specifically for energy use in the agriculture sector. In a comprehensive review of empirical studies for various economic sectors and energy end uses in the United States, Greening et al. (2000) concluded that the rebound effect for all industrial processes in the country was less than 20% for a 100% increase in energy efficiency. By comparison, Haas and Biermayr (2000) reported a rebound effect of 20%–30% in an analysis of residential building retrofits in Austria. Overall, the size of the take-back or rebound effect depends on the ability of individual firms to substitute the energy input for other inputs.

On the other hand, even proponents of the rebound effect note economic benefits from improving energy efficiency in agriculture (Greening et al. 2000), especially at the microlevel. Financial and economic impacts of improving energy use efficiency in agriculture are linked primarily to the *economical use* of farm energy. Both organic and conventional agriculture farmers are concerned with the cost of farm inputs that contain substantial energy. Increased cost of production tends to increase the price of outputs produced and, ultimately, decrease farm profits, *ceteris paribus*.

An improvement in energy efficiency would result in overall improvement in general welfare through reduced energy cost to consumers (Geller and Attali 2005). Developing and promoting more widespread use of alternative energy security and conservation strategies (as opposed to reliance on energy efficiency options alone) have the potential to address the threats to a long-term sustainable agriculture (Herring 2006; Howard et al. 2000). Organic agriculture, with its emphasis on low or reduced input use, has the potential to contribute to reducing energy consumption in agriculture (Gomiero et al. 2008; Kasperczyk and Knickel 2006). A logical question then relates to the potential of organic agriculture to improve energy use and conservation, compared with nonorganic agricultural production alternatives. This is addressed in the following section through a review of empirical studies that compared energy performance of organic versus conventional production systems.

12.5 COMPARISON OF ENERGY USE IN ORGANIC VERSUS CONVENTIONAL AGRICULTURE

Studies have compared the energy performance of organic with conventional agricultural production. An important reason for the growing interest in the effect of organic agriculture on energy conservation and long-term sustainability is primarily because organic methods of farming tend to use holistic systems to produce and, in some cases, deliver the products to final consumers, as opposed to merely being concerned with producing the individual product. In addition, organic production commonly uses natural inputs and processes, compared with many manufactured inputs (e.g., N fertilizer) used in conventional production. This potential prompted the FAO to note that more widespread conversion from conventional to organic agriculture is a major source of energy savings in agriculture (FAO 2002).

It is important to acknowledge that although there are important differences between organic and conventional agriculture, there are also many similarities, including aspects of farm energy budgets per unit of production. For example, both farming systems depend on solar energy for processes linked to photosynthesis. In addition, similar energy cost is incurred in manufacturing common farm inputs such as tractors. Furthermore, costs per unit distance traveled to market farm produce are similar, while energy tied to animal breeding stock and crop seeds does not differ by much between the two farming systems (Pimentel 2006).

On the other hand, there are considerable differences in energy use, linked primarily to the dependence on chemical fertilizer, especially N fertilizer, and pesticide use in nonorganic farming systems. According to Gomiero et al. (2008), synthetic N fertilizer use in organic production can account for over 50% of total farm energy input in conventional production.

Most studies comparing energy efficiency (and other environmental impacts) of organic and conventional production systems commonly involve specific enterprises or whole farming systems for individual products (e.g., MacRae et al. 2010; Lynch et al. 2011). On the other hand, inherent complexities associated with agricultural production systems suggest that whole farming system–level comparisons are more relevant than analysis involving individual crop or livestock production activities (MacRae et al. 2010).

In the following sections, findings from empirical studies that compared organic and conventional whole farming systems are reviewed and discussed first, followed by empirical findings for selected individual crop and livestock production. The review focuses on energy input use, energy output, and energy use efficiency comparisons between organic and conventional production. Many of the original studies initially considered had important technical problems especially pertaining to the design of the field experiments and production systems compared. For example, in several studies, crops considered in rotation systems were not quite similar for organic and conventional alternatives. Such differences can complicate inferences from the findings. In addition, for some studies, agricultural energy comparisons between organic and conventional farming systems were based on simulation modeling, as opposed to actual field experimental studies. Although organic and conventional crop and livestock production studies based on applications of decision support tools are useful in many ways, such studies were not included in this review, in part, because there are often differences in calibrating (actual agricultural production conditions) and validating simulation model performance and usefulness as assessment tools.

12.5.1 ORGANIC VERSUS CONVENTIONAL WHOLE FARMING SYSTEMS COMPARISON

In a recent comprehensive review of selected studies for the EU and North America that compared energy use (and other environmental impacts) between organic and nonorganic agricultural systems, Gomiero et al. (2008) concluded that, overall, organic systems tend to use less energy than conventional or other nonorganic intensive agricultural production systems. The comparative review reported that, overall, organic systems generated energy savings, both in terms of energy use per unit land area (i.e., GJ ha^{-1}) ranging from 10% to 70% and in terms of energy use per crop and livestock output produced (i.e., GJ tonne^{-1} product) between 15% and 45%, compared with conventional or other nonorganic alternatives (Gomiero et al. 2008).

In a similar comparison, Kustermann and Hülsbergen (2008) assessed energy inputs and outputs for 33 organic and 48 conventional agriculture operations in Germany (Table 12.1). Average energy use for the 33 organic farms was 5.6 GJ ha^{-1}, compared with 12.6 GJ ha^{-1} for the 48 conventional agriculture farms. On the other hand, energy efficiency per unit of output (estimated as ratio of energy output to energy input) was about 27% higher for the organic (i.e., 12.6 GJ GJ^{-1}) than the conventional systems (i.e., 9.9 GJ GJ^{-1}) (Kustermann and Hülsbergen 2008). Nutrient N added to crops was a major factor accounting for the differences in energy use between the two types of farming systems. Nutrient N input, averaged for all farms considered in each group, was lower for the organic farming systems (149 kg N ha^{-1}) than for conventional production (236 kg N ha^{-1}) by 37% (Kustermann and Hülsbergen 2008).

Some studies also compared energy use and outputs for various crop rotation systems. However, the actual crops or enterprises considered in some comparative studies tend to differ between the organic and nonorganic management systems and therefore complicate evaluation of the two production systems. One exception is a well-designed Canadian study by Hoeppner et al. (2006). In a 12-year field experimental study in Manitoba (see Chapter 9 for more details on the Glenlea rotation), involving a grain-based crop production system (i.e., wheat–pea–wheat–flax) and an integrated forage–grain-based system (i.e., wheat–alfalfa–alfalfa–flax) managed under organic and conventional systems, Hoeppner et al. (2006) reported that estimated total on-farm energy consumption was lower for organic systems than for the conventional alternatives evaluated. The results were similar for both grain-based and integrated forage–grain cropping systems (Table 12.1). Hoeppner et al. (2006) also found that differences in energy input use due to organic versus conventional production were higher than differences due to crop rotation type. Chemical N fertilizer use accounted for the main difference between the organic and conventional production systems.

12.5.2 Comparison for Individual Crop and Livestock Production

In general, energy use efficiency is higher for crop than for animal production (see Tables 12.2 and 12.3), primarily because most farm animal products are produced from conversion of plants or plant-based products, with attendant energy losses (and inefficiencies) and additional nonfossil energy inputs (Ziesemer 2007). On the other hand, Holmes (1974) noted that, on average, the pecuniary value of energy in animal products or outputs is about 30 times the cost of the associated inputs, compared with four times for crop products. In addition, for a given commodity, a particular species may generate more of a specific nutrient (e.g., protein) and energy than other

TABLE 12.1
Comparison of Energy Use between Organic versus Conventional Agricultural Systems: Whole Farm Analysis

Study	Farming Systems Considered	Measure or Energy Units Estimated	Production Systems		
			Organic	Conventional	% Change[a]
Hoeppner et al. (2006)	Wheat–pea–wheat–flax rotation	Total energy consumption (MJ ha^{-1})	24,233	68,498	−65
		Energy output (MJ ha^{-1})	252,054	465,841	−46
		Energy efficiency (MJ ha^{-1})[b]	10.4	6.8	53
	Wheat–alfalfa–alfalfa– flax rotation	Total energy consumption (MJ ha^{-1})	22,181	49,255	−55
		Energy output (MJ ha^{-1})	742,229	937,344	−21
		Energy efficiency (MJ ha^{-1})	33.5	19.0	76
Bos et al. (2007)	i. Potato/sugar beet/ wheat/carrot/onion/ peas ii. Leek/bean/carrot/ strawberry/lettuce/ cabbage	Energy use per ha[c]	—	—	−39
Kustermann and Hülsbergen (2008)	33 organic versus 48 conventional farming systems (estimates represent means across all farms)	Energy input (GJ ha^{-1})	5.6	12.6	−56
		Energy output/input ratio (GJ GJ^{-1})	12.6	9.9	27
		N input (kg N ha^{-1})	149	236	−37

[a] Change is expressed as a percentage of conventional production system.
[b] Energy efficiency estimated in terms of ratio of energy output to energy input or consumption.
[c] Separate values for two cropping systems not reported in original study.

species. Indeed, this consideration undergirds most modern breeding and genetic trait selection efforts.

In a study involving several organic and conventional specialized model dairy farms in the Netherlands, Bos et al. (2007) reported that average energy use per unit land area for dairy farms (which included feed crops such as grass/clover mixtures and corn) was higher for conventional (75 GJ ha^{-1}) than for organic farms (39 GJ ha^{-1}). In addition, energy use per Mg of milk produced ranged from 4.3 to 5.5 GJ for

TABLE 12.2

Comparison of Energy Use between Organic versus Conventional Livestock and Poultry Production Systems

Animal/Poultry Product and Study	Energy Use (GJ Tonne⁻¹)		
	Organic	Conventional	% Change[a]
Chicken			
Azeez and Hewlett (2008)	16.89	15.17	11
Milk (unit = 1 m³)			
Haas et al. (2001)	1.2	2.7	−54
Cederberg and Mattsson (1998)	2.41	2.85	−15
Refsgaard et al. (1998)	3.34	2.88	−13
Beef			
Azeez and Hewlett (2008)	15.56	26.54	−59
Sheep			
Azeez and Hewlett (2008)	10.79	24.99	−43
Pork			
Azeez and Hewlett (2008)	14.28	21.97	−65

[a] Change is expressed as a percentage of conventional production system.

conventional farms, compared with 3.6–4.5 GJ for organic farms. Concentrates fed to conventional dairy accounted for the highest proportion (i.e., 30%) of total energy use. Azeez and Hewlett (2008) noted that although milk production in the United Kingdom requires relatively low energy use per unit volume compared with meat and egg production, which uses about 10–30 GJ tonne⁻¹ of output, the large total quantity of milk produced in the UK agri-food sector results in milk being the largest single commodity user of energy.

Within the crop subsector, the least energy-intensive crops generally include green vegetable crops (e.g., cabbage and lettuce), followed by field or arable crops (Table 12.3). By comparison, high-value horticultural crops grown in temperature-controlled greenhouses tend to require intensive management and energy use in both organic and conventional systems (Azeez and Hewlett 2008). Bos et al. (2007) reported that energy use per unit land area is generally higher for root crops (e.g., carrots and potatoes) than for other vegetable–horticulture crops (e.g., lettuce, cabbage, peas) grown in the field. Bos et al. (2007) also found that energy use in organic carrot production was double the amount in conventional production, primarily because of high energy use in flaming weeds in the organic carrot system. Flame weeders, however, are not used extensively in Canadian organic vegetable production.

12.5.3 Factors Influencing Differences in Energy Use

Energy use and output tend to differ according to production system type (e.g., organic versus nonorganic), commodity type (i.e., crop versus animal production), and intensity of management practices (e.g., conventional tillage versus no till)

TABLE 12.3
Comparison of Energy Use between Organic versus Conventional Crop and Fruit Production Systems

Commodity/Study	Energy Use (GJ Tonne⁻¹)			Energy Use (GJ ha⁻¹)		
	Organic	Conventional	% Change[a]	Organic	Conventional	% Change[a]
a. Field crops						
Potatoes						
Mäder (2002) and Mäder et al. (2002)	3.98	3.70	–7	40.69	28.42	–30
Alfoldi et al. (1995)	0.08	0.07	7	27.5	38.2	–28
Azeez and Hewlett (2008)	1.71	1.49	14	n.a.	n.a.	n.a.
Reitmayr (1995)	0.07	0.05	29	14.3	19.7	–27
Haas and Köpke (1994)	0.07	0.80	–18	13.1	24.0	–46
Winter wheat						
Alfoldi et al. (1995)	2.84	4.21	–33	10.8	18.3	–41
Reitmayr (1995)	1.89	2.38	–21	8.2	16.5	–51
Haas and Köpke (1994)	1.52	2.70	–43	6.1	17.2	–65
Carrots						
Azeez and Hewlett (2008)	0.45	0.60	–75	n.a.	n.a.	n.a.
Onion						
Azeez and Hewlett (2008)	1.05	1.25	–84	n.a.	n.a.	n.a.
b. Fruits						
Apples						
Geier et al. (2001)	2.13	1.73	23	33.8	37.35	–9.5
Citrus						
Barbera and La Mantia (1995)	0.83	1.24	–33	24.9	43.3	–43
Olives						
Barbera and La Mantia (1995)	13.0	23.8	–45	10.4	23.8	–56

[a] Change is expressed as a percentage of conventional production system.

(Rathke et al. 2007). In addition, changes in agricultural technology over time can affect the amount of energy used. Differences in experimental design and analytical methods tend to complicate energy performance comparisons between the two production systems (Gomiero et al. 2008). Overall, however, the majority of studies report that organic systems tend to be more energy efficient than conventional or other nonorganic agriculture alternatives (Tables 12.1 and 12.2).

A trade-off related to energy savings from reduced input use in organic agriculture is that organic production typically requires more human labor/energy than its conventional counterpart. The human energy requirements for the two production systems are influenced by various factors, including commodity type, size of the operation, and climatic conditions (Ziesemer 2007). In a Danish study, Barthelemy (1999) reported a 35% increase in labor requirements for conversion from conventional to organic production. Similarly, Green and Maynard (2006) reported that, compared with conventional agriculture, UK organic agriculture provided 32% more employment per farm enterprise. In a comprehensive review of more than 40 studies for the EU, Offerman and Nieberg (2000) reported that organic production provided about 10%–20% more jobs per hectare, on average, than nonorganic farming systems.

Although organic agricultural production systems are typically more energy efficient than their conventional alternatives, potential benefits from energy savings and conservation will be linked to economical or reduced energy uses. Policy advisers can help by promoting widespread adoption of organic farming systems (see Chapter 15 for possible policy supports). On the other hand, consumers of agri-food products can enhance the agriculture energy conservation efforts through increased climate- and energy-friendly food choices, including buying organic products, sometimes locally produced products (see MacRae et al. [2010] for a discussion of some of the parameters for energy-efficient local production), foods requiring little or no processing, and food choices that emphasize less (more) animal (crop) products.

12.6 METHODOLOGICAL ISSUES AND CHALLENGES

Energy in agriculture is commonly analyzed in terms of energy consumption (associated with direct and indirect uses) and energy efficiency (Ziesemer 2007). Economic studies on energy use in agriculture commonly involve (1) comparison of energy cost associated with producing various animal and crop products for a given time period (Leach 1976; Pimentel and Pimentel 1979), (2) assessment of farm energy use across countries for a particular time period (e.g., OECD, 1982), and (3) energy performance of alternative agricultural production systems and methods (e.g., Bayliss-Smith 1984; Smil 1991). Other studies have estimated aggregate (direct and indirect) energy use in agriculture for particular countries (e.g., Cleveland 1995; Steinhart and Steinhart 1974) and energy productivity in agriculture (e.g., Panesar and Fluck 1993).

Energy consumption is commonly measured in terms of total energy used per unit farm input (e.g., 1 ha) during a specified time period. Some analysts have quantified fossil fuels and human labor use in such units as calories or joules ha^{-1} year^{-1}

(see, e.g., Bos et al. 2007; Mendoza 2002; Williams et al. 2006). As noted earlier, energy efficiency is commonly expressed in terms of energy use per unit of crop/livestock output or per calorie produced (in tonnes or joules per calories). One advantage of energy efficiency evaluations is that they allow for transforming and normalizing comparisons among alternative production systems or for various crop or livestock products (Refsgaard et al. 1998). Applied economists commonly estimate energy use indexes using energy price and production cost information (Cleveland 1995) and applying various modeling techniques (e.g., translog cost functional form) (Capalbo et al. 1985).

An advantage of quantifying energy associated with physical inputs used in agriculture is that the approach allows for use of detailed technical engineering data to evaluate specific agricultural processes or outputs (e.g., production of fertilizer or farm equipment) (Cleveland 1995). The cost of energy used in producing farm inputs tends to change with time. Yet, there is a general lack of time series agricultural statistical data (in physical metrics) for a diverse range of farm inputs (Cleveland 1995).

A second methodological challenge associated with estimating energy consumption based on physical inputs used is that energy cost associated with important services (e.g., veterinary services, agricultural insurance, irrigation services, farm financial services) is inherently not denominated in physical metrics and, consequently, is often not included in agricultural energy cost estimations. In addition, there are no standard methods for quantifying the energy in various inputs and other materials used in agriculture. Fluck and Baird (1980), for example, identified more than 20 studies that used different methods to quantify energy use in producing a tonne of chemical fertilizer. Thus, comparative studies for individual crop and animal products tend to use values for energy cost generated using different methods and sometimes covering different time periods.

Perhaps the major methodological challenge with analyzing energy use in agriculture is linked to the observation that farms tend to differ in many respects, including management style and practices, geographical location, commodities produced, size or scale, and climatic conditions. Kasperczyk and Knickel (2006), for example, noted that there is "no clearly defined system of conventional farming" because there are variants that range from *high-input intensive systems to near-organic* production systems. Similarly, although certified organic systems have to meet minimum industry standards and requirements, actual implementation of the practices may differ due, for example, to differences in scale or size of the operation (Ziesemer 2007). Spedding and Walsingham (1976) put it more succinctly when the authors noted that energy analysis and comparisons across regions may have limited relevance primarily because of regional or site-specific differences. Furthermore, crop varieties and livestock species can differ within and across farming systems.

The social implications of energy use in agriculture, especially the implications of changes in human labor (i.e., human forms of energy) in organic and nonorganic farming systems, although important, are largely unexplored (Ziesemer 2007). Changes in labor demand (and supply) requirements in both organic and nonorganically managed systems present not only opportunities but also various challenges, including impacts on (un)employment and migration, and regional and national economic development. In many OECD countries, organic agriculture tends to attract more *hobby farmers* than in intensive conventional agriculture (Green and Maynard 2006) and, therefore, has potential for revitalizing and preserving rural communities (Ziesemer 2007). The impact of organic production to changes in employment in agricultural and rural regions will become important if there is substantial growth in the organic sector (see also Chapter 14 of this volume).

Although energy comparison of agricultural production systems is important, used alone, the information does not provide a comprehensive and complete understanding of the multiple dimensions and characteristics of different farming systems. Thus, energy analyses need to be complemented with a consideration of other (especially) nonmarket benefits and costs associated with each agri-food system. For example, it is not possible to assess some costs and benefits (e.g., improved community relationships) associated with organic versus conventional (or so-called factory) farms in pecuniary units or other quantitative measures. Chapters 13 and 14 in this volume explore these themes more fully.

12.7   SUMMARY AND CONCLUSIONS

Agriculture is both a sink and source of energy. Although most agricultural inputs used and outputs produced can be quantified in terms of energy, there are also important agricultural services that cannot be estimated in such physical units. In addition, besides energy, there are other important agricultural system indicators that need to be evaluated for a more comprehensive assessment of agricultural system sustainability and performance. Energy use in organic and nonorganic agriculture is inherently connected with overall agricultural development and human welfare and, therefore, the larger economy (through impacts from farm input costs and commodity prices). Energy efficiency and intensity are important indicators of energy use in agriculture and are commonly used to assess the performance of agricultural production systems.

Economic assessments of the energy performance of agricultural production systems commonly involve the cost of energy used in alternative production systems, or estimation of energy used in producing particular agricultural outputs, or for the agricultural sector across countries. Other studies have assessed productivity per unit of energy use. Intertemporal effects of energy use in agriculture are important and need to be considered in agricultural energy use analysis and comparisons. Differences in energy analysis methods and animal species and crop variety differences, as well as variability in climatic and other growing conditions, can complicate economic comparisons of energy use in agriculture.

Although technology and innovation are sources of energy savings, the key to long-term energy efficiency and conservation centers on energy security and conservation through reduced and economical use. In this regard, organic agriculture has greater potential than conventional agricultural production, because of its emphasis on low and/or reduced input use. Increasing the energy efficiency impacts from sustained growth in organic agriculture will require increased support and promotion by government authorities and agricultural administrators. Within the organic agriculture sector itself, greater energy efficiency gains will derive from more emphasis on vegetable crops and field crops and less so on temperature-controlled greenhouse horticulture and production (and consumption) of livestock products.

REFERENCES

Alfoldi, T., E. Spiess, U. Niggli, J.M. Besson, H.F. Cook, and H.C. Lee. 1995. Energy input and output for winter wheat in biodynamic, bio-organic and conventional production systems. In *Soil Management in Sustainable Agriculture*, H.F. Cook and H.C. Lee, eds. *Proceedings of the Third International Conference on Sustainable Agriculture*, pp. 574–578. Wye College Press, Wye College, University of London, U.K.

Azeez, G.S.E. and K.L. Hewlett. 2008. The comparative energy efficiency of organic farming. *Proceedings, 2nd Conference of the International Society of Organic Agriculture Research (ISOFAR)*, June 18–20, 2008, Modena, Italy.

Barbera, G. and T. La Mantia. 1995. Analisi agronomica energetica. Filiere atte allo sviluppo di aree collinari e montane: il caso dell'agricoltura biologica. Chironi G Vo. 1. RAISA-University of Palermo, Palermo, Italy.

Barthelemy, P.A. 1999. Changes in agricultural employment. In *Agriculture, Employment, Rural Development: Facts and Figures—A Challenge for Agriculture*, European Commission, pp. 40–53. EUROSTAT. Available online at http://ec.europa.eu/agriculture/envir/report/en/emplo_en/report.htm, accessed on August 20, 2012.

Bayliss-Smith, T.P. 1984. Energy flows and agrarian change in Karnataka: The Green revolution at micro-scale. In *Understanding Green Revolutions*, T.P. Bayliss-Smith, ed. Cambridge, MA: Cambridge University Press, pp. 153–172.

Berkhout, P.H.G., J.C. Muskens, and J.W. Velthijisen. 2000. Defining the rebound effect. *Energy Policy* 28: 425–432.

Bos, J., J. Haan, W. Sukkel, and R. Schils. 2007. Comparing energy use and greenhouse gas emissions in organic and conventional farming systems in the Netherlands. *Proceedings of the Third QLIF Congress: Improving Sustainability in Organic and Low Input Food Production Systems*, University of Hohenheim, Stuttgart, Germany. Available online at: http://orgprints.org/view/projects/int_conf_qlif2007.html, accessed on October 29, 2012.

Boyd, D.R. 2001. *Canada vs. the OECD: An Environmental Comparison*. University of Victoria, Victoria, British Columbia, Australia.

Canadian Agricultural Energy End-Use Data Analysis Center (CAEEDA). 1998. Energy consumption in the Canadian agricultural and food sector. Technical Report. CAEEDA, University of Saskatchewan, Saskatoon, Saskatchewan, Canada.

Canadian Agricultural Energy End-Use Data Analysis Center (CAEEDA). 2000. Direct energy use in agriculture and the food sectors: Separation by farm type and location. Technical Report. CAEEDA, University of Saskatchewan, Saskatoon, Saskatchewan, Canada.

Capalbo, S.M., T. Vo, and J. Wade. 1985. An Econometric Database for Measuring Agricultural Productivity and Characterizing the Structure of US Agriculture. Resources for the Future, Washington, DC.

Cederberg, C. and B. Mattsson. 1998. Life cycle assessment of Swedish milk production: A comparison of conventional and organic farming. In *Proceedings of the International Conference on Lyfe Cycle Assessment in Agriculture, Agro-industry and Forestry*, D. Ceuterick, ed., Brussels, Belgium, pp. 161–170.

Cleveland, C.J. 1995. The direct and indirect use of fossil fuels and electricity in USA agriculture, 1910–1990. *Agriculture, Ecosystems and Environment* 55: 111–121.

Coxworth, E. 1997. Energy use trends in Canadian agriculture: 1990–1996. Report to Canadian Agricultural Energy Use Data Analysis Center, University of Saskatchewan, Saskatoon, Saskatchewan, Canada.

Eco-Energy Carbon Capture and Storage Task Force. 2008. *Canada's Fossil Energy Future: The Way Forward on Carbon Capture and Storage.* Natural Resources Canada, Ottawa, Ontario, Canada.

Fluck, R.C. and C.D. Bird. 1980. *Agricultural Energetics.* The AVI Publishing Company, Wesport, VT.

Food and Agriculture Organization of the United Nations (FAO). 2002. Organic agriculture, environment and food security. Environment and Natural Resources Series No. 4. Environment and Natural Resources Service Sustainable Development Department. http://www.fao.org/DOCREP/005/Y4137E/y4137e00.htm#TopOfPage, accessed on June 20, 2011.

Food and Agriculture Organization of the United Nations (FAO). 2012. *FAOSTAT-Fertilizers.* FAO, Rome, Italy. Available online at http://faostat.fao.org/site/575/default.aspx#ancor, accessed on August 10, 2012.

Geier, U., B. Frieben, Gutsche, and U. Kopke. 2001. *Okobilanz des Apfelerzeugung in Hamburg: Vergleich intergrieter und okologischer Bewirtschaftung.* Schriftenreihe Institut fur Organischen Landbau, Verlag Dr. Koster, Berlin, S. 130.

Geller, H. and S. Attali. 2005. The experience with the energy efficiency policies and programmes in IEA countries: Learning from the critics. International Energy Agency (IEA) Information Paper. IEA, Paris, France.

Gomiero, T., M.G. Paoletti, and D. Pimentel. 2008. Energy and environmental issues in organic and conventional agriculture. *Critical Reviews in Plant Sciences* 27: 239–254.

Green, M. and R. Maynard. 2006. The employment benefits of organic farming. *Aspects of Applied Biology* 79: 51–55.

Greening, L.A., D.J. Greene, and C. Difiglio. 2000. Energy efficiency and consumption—The rebound effect: A survey. *Energy Policy* 28: 389–401.

Haas, G. and U. Köpke. 1994. *Vergleich der Klimarelevanz Ökologischer und Konventioneller Landbewirtschaftung.* Studie (Studie H) im Auftrag der Enquetekommission des Deutschen Bundestages "Schutz der Erdatmosphäre". Economia Verlag, Karlsruhe, Germany.

Haas, G., F. Wetterich, and U. Kopke. 2001. Comparing intensive, extensified and organic grassland farming in southern Germany by process life cycle assessment. *Agriculture, Ecosystem and Environment* 83: 43–53.

Haas, R. and P. Biermayr. 2000. The rebound effect for space heating: Empirical evidence from Austria. *Energy Policy* 28(6–7): 403–410.

Hanley, N., P.G. McGregor, J.K. Swales, and K. Turner. 2009. Do increases in energy efficiency improve environmental quality and sustainability. *Ecological Economics* 68: 692–709.

Herring, H. 1999. Does energy efficiency save energy? The debate and its consequences. *Applied Energy* 63: 209–226.

Herring, H. 2006. Energy efficiency: A critical review. *Energy* 31: 10–20.

Hoeppner, J.W., M.H. Entz, B.G. McConkey, R.P. Zentner, and C.N. Nagy. 2006. Energy use and efficiency in two Canadian organic and conventional crop production systems. *Renewable Agriculture and Food Systems* 21(1): 60–67.

Holmes, W. 1974. Assessment of Alternative nutrient sources. In, Proceedings of the 21st Easter School in Agricultural Science - Meat. D.J.A. Cole, and R.A. Lawrie, eds. Butterwork, London, UK, pp. 535–553.

Hopper, G.R. 1984. The energetics of Canadian Prairie wheat production: 1948–1981. *Energy Developments: New Forms, Renewables, and Conservation. Proceedings of Energex 1984*, Regina, Saskatchewan, Canada, pp. 677–683.

Howard, J., D. Mitchell, D. Spennemann, and M. Webster-Mannison. 2000. Is today shapping tomorrow for tertiary eduction in Australia? A comparison of policy and practice. *International Journal of Sustainability in Higher Education.* 1: 83–96.

Huber, P.W. and M.P. Mills. 2005. *The Bottomless Well: The Twilight of Fuel, the Virtue of Waste, and Why We Will Never Run Out of Energy*. Basic Books, New York.

Kasperczyk, N. and K. Knickel. 2006. Environmental impacts of organic farming. In *Organic Agriculture: A Global Perspective*, P. Kristiansen, A. Taji, and J. Reganold, eds. Cornell University Press, Ithaca, New York.

Kendall, H.W. and D. Pimentel. 1994. Constraints on the expansion of the global food supply. *Ambio* 23: 198–205.

Kustermann, B. and K.J. Hülsbergen. 2008. Emission of climate relevant gases in organic and conventional cropping systems. Paper presented at the *16th IFOAM Organic World Congress*, Modena, Italy.

Laitner, J.A. 2000. Energy efficiency: Rebounding to a sound analytical perspective. *Energy Policy* 28: 471–475.

Leach, G. 1976. *Energy and Food Production*. IPC Science and Technology Press, Guilfoord, Surrey, UK.

Lynch, H.D., R. MacRae, and R.C. Martin. 2011. The carbon and global warming potential impacts of organic farming: Does it have a significant role in an energy constrained world? *Sustainability*. 3: 322–362.

MacRae, R.J., D. Lynch, and R.C. Martin. 2010. Improving energy efficiency and GHG mitigation potentials in Canadian organic agriculture. *Journal of Sustainable Agriculture* 34: 549–580.

Mäder, P. 2002. The ins and outs of organic farming. *Science* 298: 1889–1890.

Mäder, P., A. Fließbach, D. Dubois, L. Gunst, P. Fried, and U. Niggli. 2002. Soil fertility and biodiversity in organic farming. *Science* 296: 1694–1697.

Mendoza, T.C. 2002. Comparative productivity, profitability and energy use in Organic, LEISA and conventional rice production in the Philippines. *Livestock Research for Rural Development* 14(6): Online. Available online at http://www.cipav.org.co/lrrd/lrrd14/6/mend146.htm, accessed on August 20, 2012.

Natural Resources Canada. 2011. Total end-use sector—Energy use analysis. Available online at: http://oee.nrcan.gc.ca/corporate/statistics/neud/dpa/tablesanalysis2/res_00_1_e_1_4.cfm? Accessed May 16, 2012.

Offerman, F. and H. Nieberg. 2000. Economic performance of organic farms in Europe. In *Organic Farming in Europe: Economics and Policy*, S. Dabbert, N. Lampkin, J. Michelsen, H. Nieberg, and R. Zanoli, eds., Vol. 5, pp. 1–220. University of Hohenheim, Stuttgart, Germany. Available online at https://www.uni-hohenheim.de/i410a/ofeurope/organicfarmingineurope-vol5.pdf, accessed on August 15, 2012.

Organization for Economic Cooperation and Development (OECD). 1982. *The Energy Problem and the Agro-Food Sector*. Organization of Economic Cooperation and Development, Paris, France.

Organization for Economic Cooperation and Development (OECD). 1997. *Environmental Indicators for Agriculture*. OECD, Paris, France.

Panesar, B.S. and R.C. Fluck. 1993. Energy productivity of a production system: Analysis and measurement. *Agricultural Systems* 43: 415–437.

Pimentel, D. 2006. *Impacts of Organic Farming on the Efficiency of Energy Use in Agriculture*. The Organic Center, Cornell University, Ithaca, New York.

Pimentel, D., and M. Pimentel. 1979. Food, Energy and Society. Edward Arnold: London, UK.

Rathke, G.-W., B.J. Wienhold, W.W. Wilhelm, and W. Diepenbrock. 2007. Tillage and rotation effect on corn-soybean energy balances in eastern Nebraska. *Soil and Tillage Research* 97: 60–70.

Refsgaard, K., N. Halberg, and E.S. Kristensen. 1998. Energy utilization in crop and dairy production in organic and conventional livestock production systems. *Agricultural Systems* 57: 599–630.

Reitmayr, T. 1995. Entwicklungen eines rechnergestützten Kennzahlensystems zur ökonomischen und ökologischen Beurteilung von agrarischen Bewirtschaftungsformen— dargestellt an einem Beispiel. *Agrarwirtschaft Sonderheft* 147: 255–272.

Smil, V. 1991. *General Energetics: Energy in the Biosphere and Civilization*. Wiley & Sons, New York.

Smil, V. 2000. *Feeding the World: A Challenge for the Twenty-First Century*. The MIT Press, Cambridge, MA.

Spedding, C.R.W. and J.M. Walsingham. 1976. The production and use of energy in agriculture. *Journal of Agricultural Economics* 27(1): 19–30.

Statistics Canada. 2012. 2011 Census of Agriculture. Available online at: http://www29.statcan. gc.ca/ceag-web/eng/community-agriculture-profile-profil-agricole?geold=470000000 &selectedVarlds=350%2C. Accessed January 24, 2013.

Steinhart, J.S., and C.E. Stainhart. 1974. Energy use in the United States food system. *Science* 184: 307–316.

Stirling, B. 1979. *Use of Non-Renewable Energy on Saskatchewan Farms: A Preliminary Study*. Saskatchewan Research Council, Regina, Saskatchewan, Canada.

Stolze, M., A.H. Piorr, and S. Dabbert. 2000. The environmental impact of organic farming in Europe. In *Organic Farming in Europe: Economics and Policy*, M. Stolze, A. Piorr, A. Haring, and S. Dabbert, eds., Volume 6, pp. 1–38. University of Hohenheim, Hohenheim, Germany. Available online at https://www.uni-hohenheim.de/i410a/ ofeurope/organicfarmingineurope-vol6.pdf, accessed on July 2012.

United Nations Department of Economic and Social Affairs. 2004. *World Population to 2300*. UN Publication No. ST/ESA/SER.A/236, New York.

Von Weizsacker, E.U. 1998. *Factor Four: Doubling Wealth, Halving Resource Use*. Earthscan, London, U.K.

Weber, C.L. and H.S. Matthews. 2008. Food-miles and the relative climate impacts of food choices in the United States. *Environmental Science and Technology* 42: 3508–3513.

Williams, A., E. Audsley, and D. Sandars. 2006. Determining the environmental burdens and resource use in the production of agricultural and horticultural commodities. Technical Report. DEFRA Research Project IS0205. Cranfield University and DEFRA, Bedford, U.K. Available online at www.silsoe.cranfield.ac.uk and www.defra.gov.uk, accessed on August 20, 2012.

World Bank. 2011. Fertilizer consumption per unit and area in Canada, 1961–2008. Available online at http://data.worldbank.org/indicator/AG.CON.FERT.ZS, accessed on July 2011.

Zentner, R.P., S.A. Brandt, C.N. Nagy, and B. Frick. 2009. Economics and energy use efficiency of alternative cropping strategies for the dark brown soil zone of Saskatchewan. Project No. 20070029. Final Technical Report prepared for Saskatchewan Agriculture Development Fund, Saskatoon, Saskatchewan, Canada.

Zentner, R.P., C.N. Nagy, G.P. Lafond, B.G. McConkey, and A.M. Johnson. 2005. Energy performance of alternative cropping systems and tillage methods in Saskatchewan. Selected paper presented at the *ASA-CSSA-SSSA International Conference*, Salt Lake City, UT.

Ziesemer, J. 2007. *Energy Use in Organic Food Systems*. Natural Resources Management and Environment Department, Food and Agriculture Organization of the United Nations, Rome, Italy.

Zinck, J.A., J.L. Berroteràn, A. Farshad, A. Moameni, S. Wokabi, and E. Van Ranst. 2004. Approaches to assessing sustainable agriculture. *Journal of Sustainable Agriculture* 23(4): 87–109.

13 Will More Organic Food and Farming Solve Food System Problems? Part I: Environment*

Rod MacRae, Derek H. Lynch, and Ralph C. Martin

CONTENTS

---

* This chapter is based on Lynch (2009), Lynch et al. (2011, 2012a), and MacRae et al. (2004).

13.1 INTRODUCTION

The Canadian food and agriculture sector has for some time faced significant environmental, food safety, and financial difficulties. Canadian agri-environmental indicators continue to lag behind most other OECD nations (OECD 2008), consumer confidence in many food system processes remains weak (Bonti-Ankomah and Yiridoe 2006), and farm incomes continue to be problematic (Wiebe 2012). Policy responses have to date generated only limited improvements (MacRae 2011).

Most Canadian government jurisdictions have treated organic food and farming as a niche market to be supported in limited ways, but rapid growth rates in the past decade suggest that restricting policy supports on this basis is misplaced. In several European countries, such as Austria and Denmark, the sector has grown to become a significant percentage of the agrifood economy and rural landscape, with attendant environmental, economic, and social benefits (Willer et al. 2013).

This chapter presents data and analysis to support the position that organic food and farming are more than simply a niche market opportunity. Given relatively low adoption levels to date in North America, the extensive potential benefits of organic farming systems are not yet very visible. However, there is growing evidence that adoption of such systems produces multiple environmental, social, and financial benefits that can solve pressing food system problems. Although not all studies reviewed address just field crops, they represent the bulk of organic hectares in North America, so the analysis offered here addresses primarily organic field cropping.

This chapter reports primarily on peer-reviewed literature and governmental and paragovernmental documents. It focuses particularly on literature employing an agroecological analytical framework (cf. Altieri 1987; Gliessman 1997; Chapter 1). We also focus on systems comparisons. Many of the studies reported compare, directly or indirectly, organic and nonorganic farming systems (comprising typically a range of practices from conventional to integrated farming). To produce useful comparisons, it is important to focus on the entire farming system or larger food system dynamics as opposed to examining specific elements outside of their larger operating context. It is also important to compare systems that have common components, including comparable management capacities. Clearly, poorly managed systems, whether organic or conventional, generate problems. A poorly managed organic system compared to a well-managed conventional one may reveal more about the management capacity of the farmer than the way the farming system behaves. We are interested in structural comparisons, so we assume good management in systems being compared. In doing so, we are attempting to analyze how the structure of organic farming offers benefits that are not necessarily associated with conventional farming. It is not meant to be an exhaustive review, but to include key recent findings across various indicators.

In each section, we discuss the current state of evidence of organic system benefits relative to conventional production and identify areas where organic farming could be improved. Other chapters in this volume address issues of soil organic carbon storage, soil quality and soil biological diversity (Chapters 4 and 5), and energy efficiency (Chapter 12), so here we focus on aboveground biodiversity, reduced pollution, and greenhouse gas (GHG) emissions. We focus particularly on studies examining GHG emissions, as fewer meta-analyses across a range of commodities have been conducted in this domain relative to biodiversity.

13.2   PLANT AND WILDLIFE BIODIVERSITY

Intensive arable cropping, extensive pesticide use, and simplified rotations have strongly affected the composition, heterogeneity, and interspersion of wildlife habitat in agricultural landscapes. In general, conventional farm fields have lower within-field and between-field variabilities, and field margins and other noncrop habitats have been reduced or eliminated. In contrast, a growing number of meta reviews of the literature (Bartram and Perkins 2003; Bengtsson et al. 2005; Hole et al. 2005; Birkhofer et al. 2008; Gomiero et al. 2011; Lynch et al. 2012a) suggest species abundance and richness, across a wide range of taxa (including arable flora, birds, mammals, and invertebrates), benefit from organic management. Fuller et al. (2005) found that organic fields had 68%–105% more plant species and 74%–153% greater abundance of weeds than nonorganic fields, 5%–48% more spiders in preharvest crops, 16%–62% more birds in the first winter, and 6%–75% more bats. Most studies concluded that results are strongest for plant diversity and extend to other taxa depending on interaction with noncrop habitat and landscape complexity. These benefits result from differences in cropping, landscape management, nutrient use, soil management, and pesticide use between organic and conventional systems. In the Canadian context, the widespread adoption and use of genetically engineered (GE) crops (primarily canola, corn, and soybeans) have created further negative impacts on on-farm biodiversity (Morandin and Winston 2005; Lynch et al. 2012a). In the following, we provide some indicative results focusing on vegetative, bird, and insect biodiversity effects. Increasingly, research is expanding our understanding of the specific ecosystem services (including conservation, biological pest control, and pollination) linked to such biodiversity. The reader is referred to Gomiero et al. (2011) and Lynch et al. (2012a) for more detailed discussion of this emerging area of research.

13.2.1   Vegetative Diversity

Different measures of vegetative diversity have been used in comparative studies—native and exotic plant species, rangeland vegetative diversity, and weed and weed seed abundance. This last measure, frequently seen as problematic in conventional production, is generally viewed as desirable in organic systems because of the roles low-level weed populations play in soil cover, nutrient cycling, and food for beneficial insects.

In Ontario, Canada, Boutin et al. (2008) conducted a study comparing 16 conventional and 14 organic farms, examining plant species in working landscapes and woody hedgerows. They found more native and exotic plant species on organic sites. Winqvist et al. (2011), examining 153 farms in Europe, found that the abundance and species richness of wild plants were enhanced by both organic farming practices and more complex landscapes. As noted in Section 13.2.3, GE canola production in Alberta greatly reduced vegetative diversity when compared to organic management (Morandin and Winston 2005).

Excess N and P application to temperate grasslands is considered a major cause of biodiversity loss (UNEP 2010). Few studies have examined the impact of organic management of rangelands and grasslands, with no applications of synthetic N and P, on vegetative biodiversity, but as reported from Iowa (Wiltshire et al. 2010), the avoidance of such fertilizer applications and reduced overgrazing in organic management may promote more diverse and multifunctional swards. There is a need for additional studies under the semiarid conditions common to much of North America because results from more temperate regions may not be comparable.

In Nebraska, aboveground weed biomass (primarily grass species) and weed seed bank diversity, evenness, and richness were greater after 12 years of organic compared with conventional grain rotations, particularly in the green manure (alfalfa) compared to the manure-based organic system (Wortman et al. 2010). In dryland conditions in Spain, comparing 28 pairs (organic and conventional) of farms, the organic farms displayed significantly greater differences in abundance, richness, and diversity in weed communities than the conventional farms. These dryland results were more favorable for organic production than in temperate zone trials in Northern Europe, the result of richer weed flora and weed seed availability under these dryland conditions and management (Ponce et al. 2011).

13.2.2 Farmland Birds

Landscape heterogeneity, including noncrop habitats, is likely as important as farming system in determining the distribution, composition, abundance, and richness of different organisms on agricultural landscapes, including birds (Purtauf et al. 2005; Gabriel et al. 2010; Flohre et al. 2011; Gomiero et al. 2011; Winqvist et al. 2011). Working in southern Ontario, Freemark and Kirk (2001) counted birds over a breeding season on 72 field sites on 10 organic and conventional farms. Sites were matched for crop type, adjacent noncrop habitat, and, when possible, field size and shape to detect effects of agricultural practices. Of 68 bird species recorded, species richness, abundance, and frequency were significantly higher on the organic sites. Local habitat and agricultural practices were roughly equal contributors to the variation in bird species, with differences most likely to be the result of fewer nesting sites and food supply, because of lower plant species richness, cover, seed availability, and soil invertebrates on conventional farms. These results are roughly comparable to European studies where agricultural intensification is likely the main reason for a decline of 52% in farmland bird populations (Pan-European Common Bird Monitoring Scheme 2011). Winqvist et al. (2011) found breeding bird populations were more likely to be found on organic farms and in more complex landscapes.

13.2.3 Insects

Impacts of organic farming on insect biodiversity are less clear than other organisms, with results highly dependent on species studied, habitat scale, region, and biodiversity measure employed. *Lepidoptera*, primarily butterflies, have been examined extensively in Europe, but less so in North America. Results are inconclusive regarding the effects of farming systems on their biodiversity. Boutin et al. (2011) examined the contribution of crop fields and structurally similar woody hedgerow habitats to regional moth diversity within eight pairs of farmlands managed organically or conventionally near Peterborough, Ontario. The study found no difference in moth assemblages except for greater species richness of the *Notodontidae* family on organic farms. In general, across all farms, hedgerows harbored more species than fields did.

Crowder et al. (2010) argued that organic farming promotes evenness of species, or communities, of natural predators. A high level of evenness of natural predator species, rather than species richness alone, provided the strongest degree of pest control and crop growth in organic potato production systems in the United States. After 30 years of the Swiss DOK trial, Birkhofer et al. (2008) also found organic management fostered natural enemies and biological control of aphids. Garratt et al. (2011) conducted a meta-analysis of pests and natural enemies and concluded that performance and abundance of all groups of natural enemies, except the coleopterans, consistently showed a positive response to organic agriculture. As the effect was more pronounced at the farm than field scale, they concluded that larger-scale characteristics, including habitat heterogeneity on organic farms, were encouraging natural enemies, a conclusion consistent with Krauss et al. (2011). In their comprehensive study on 153 European farms, Winqvist et al. (2011) examined predation on aphids in cereal fields and found it to be greatest in organic fields within complex landscapes. The opposite was true for conventional fields.

When organic farms are *islands* among large areas of nonorganic farms, there may be a *dilution* of potential organic benefits. Only with policy supports to encourage organic farming within regions, such as a sub-watershed, may this dilution effect be avoided (Lynch et al. in press).

In northern Alberta, Canada, Morandin and Winston (2005) examined native wild bee abundance and pollination deficit (the difference between potential and actual pollination) in organic, conventional (non-GE), and GE herbicide-resistant canola fields. At bloom, there was no pollination deficit in organic fields, moderate (−16%) in conventional fields, and the greatest deficit (−22%) in GE fields, attributed to markedly reduced weed diversity and abundance (and thus forage for bees). In Germany (Gabriel and Tscharntke 2007; Krauss et al. 2011) and Ireland (Power and Stout 2011), pollinators and insect-pollinated arable crops and grasslands benefited from organic farming.

13.3   NUTRIENT LOADING AND POLLUTION RISKS

Leached nutrients present significant pollution problems in many agricultural landscapes. On organic field crop operations, reduced off-farm nutrient impacts are closely linked to the use of more complex rotations, legume biological nitrogen fixation and organic matter inputs, and reduced overall nutrient intensity (Lynch 2009; Griffiths et al. 2010; Lynch et al. 2012b). Extended rotations in organic systems can enhance soil biological pools, which may contribute to nutrient dynamics in these systems in ways not fully yet appreciated (Mader et al. 2002; Stockdale et al. 2002; Nelson et al. 2009). Legume crops also act as an *N buffer*, reducing the likelihood of excesses or deficits of N. By helping to lower application rates of organic amendments, they also help reduce soil P and K accumulation (Lynch et al. 2004). For more specific details on nutrient dynamics in soils, see Chapters 3 through 6 of this volume.

Only a few comparative studies have taken a wide angle view of nutrient pollution. An extensive earlier European comparison of organic and conventional farming systems and their environmental impacts found that organic farming was the same or better than conventional on all environmental indicators, except there was some potential for more soil erosion in some cases (although most comparisons were more positive for organic) and some possibilities for more nitrate leaching (Stolze et al. 2000). A study of 15 organic dairy farms in Ontario, Canada, found NPK loading (imports–exports) was greatly reduced under commercial organic dairy production compared with more intensive confinement-based systems (Roberts et al. 2008). Interestingly, organic farm nutrient surpluses were related positively to livestock density (LD) and negatively to feed self-sufficiency. As noted previously, N and P loading negatively impacts biodiversity of grasslands and whole farms (Kleijn et al. 2009; UNEP 2010). Mondelaers et al. (2009), from their comparative meta-analysis, concluded that there were generally lower per hectare nitrate and P leaching losses from organic systems per hectare, but that differences were less significant if expressed per unit of product. Korsaeth (2008) produced similar conclusions.

Davis et al. (2012) contrasted a simple, high-input 2-year corn–soybean rotation with a more complex, low-input 4-year rotation and manure. Although inputs were significantly reduced in the 4-year rotation, grain yields and profits were similar to or higher than those in the 2-year rotation. Pollution related to freshwater toxicity was two orders of magnitude lower under the 4-year rotation, and weeds were managed effectively in all cases. The 4-year rotation was not an organic system, but it was well within the range of organic management options and illustrates how reduced pollution with similar profits and yields of high-input systems can be realized.

A few studies have examined the comparative impact (organic vs. conventional) of nutrient and other agrochemical losses on water quality and aquatic ecosystems. In England, Hathaway-Jenkins et al. (2011) found agrochemicals (pesticides and total N, P, and K) in soil (0–20 cm) water were lower on organic farms. In New Zealand, Magbanua et al. (2010) compared the effects of organic, conventional, and integrated beef and sheep farming systems on water quality and stream macroinvertebrate communities. Conventional operations generated higher levels of fine sediment, nitrates, and glyphosate in streams, with adverse consequences for invertebrate densities and biological trait representation.

Other studies have looked more at particular potential contaminants. In organic systems, the challenge of managing N to sustain adequate productivity and minimize N losses is significant (Lynch 2009; Askegaard et al. 2011; Gomiero et al. 2011; Lynch et al. 2012b). While N is relatively easily supplied in mixed crop and dairy farms (Halberg 1999), synchronizing N availability with crop demand in organic cash crop systems is difficult as the supply of N from organic amendments, green manures, and crop residues, plus residual soil mineral N (RSMN), varies with climatic conditions (Lynch et al. 2012b). This reality means that N losses from organic systems can be substantial, but not usually as high as those found in conventional operations.

In the United States, Pimentel et al. (2005) found, over a 12-year period, that differences in annual nitrate leached from maize systems were not significant across organic and conventional farms, though there were sporadic increases in nitrate leached in organic operations when drought conditions reduced maize growth and N uptake. Similar conclusions were drawn in Sweden (Kirchmann et al. 2007). In processing tomato and maize production systems in California's Sacramento Valley, Poudel et al. (2002) found similar crop yields across organic, low-input, and conventional farming systems over 5 years, but a lower potential risk of N leaching in the low-input and organic systems. In Denmark, Askegaard et al. (2011) found, in a 12-year study conducted at three locations, that inclusion of a grass–clover green manure in organic cereal rotations did not increase N leaching. Fall catch crops (Askegaard et al. 2011) and modified timing of green manure incorporation (Lynch et al. 2012b) have also been shown to help minimize N leaching losses in organic systems.

Another way to measure nitrogen use efficiency (NUE) is units of food produced per unit of N applied. In general, farming is inefficient in this regard (Erisman et al. 2008). Studies comparing NUE in organic and conventional systems have so far produced results favoring organic systems, though not definitively. In Sweden, Kirchmann et al. (2007), from an 18-year trial, found the organic system had a lower NUE. N leaching was not reduced and crop N uptake was more compromised by competition from weeds. But two German studies (Finckh et al. 2006; Möller et al. 2007) found improved NUE under low-input or organic potato production, possibly because of better crop light use efficiency associated with a limited N environment. Halberg (1999) and Halberg et al. (1995) found significantly higher NUE on organic dairy farms and lower N surplus per hectare. There is significant room for improving organic system performance (Lynch et al. 2012a), in particular, greater understanding of how to improve soil quality without generating greater N leaching and GHG emissions (Griffiths et al. 2010).

Phosphorus loading in aquatic systems is now a widespread and serious problem in many parts of the world. P is very difficult to manage in organic systems, and some studies have reported soil and crop P deficiencies on organic farms (see Chapter 4; Lynch et al. 2012a), but somewhat paradoxically, this makes the possibility of reduced P release to water bodies more likely in organic systems. Chapter 4 reports on a limited number of comparative studies that monitored soil P pools. Most of them showed a decrease in total P, soil test P, and labile P in organically managed systems, due largely to lower P additions to the organic systems and lower solubility of added P. Microbial biomass P and residual P are usually increased under organic production, but such pools are less readily liberated. In addition, as outlined in Chapter 4, organic farms use more tillage that tends to mix P pools and prevent buildup of surface layer available P. The absence of certain P-containing pesticides also limits P in the soil profile. Mondalaers et al. (2009) also generally found lower P balances on organic farms in the studies they reviewed.

Beyond soil P balances, although comparative studies examining this in detail are limited, earlier work showed that P losses are almost always lower in organic systems (Lotter 2003). Mondelaers et al. (2009) found inconclusive results among the three studies they examined, but P leaching was so low in all systems as to be insignificant. Thomassen et al. (2008) in the Netherlands conducted a detailed *cradle-to-farm gate* life cycle assessment (LCA) of pollution impacts on water quality (i.e., eutrophication). Results indicated improved environmental performance with respect to eutrophication potential kg milk^{-1} for the organic compared to conventional farms.

13.4 GHG EMISSIONS*

Organic standards impose a specific set of realities on farms that affect their GHG emissions, realities that differ from those on most conventional farms. As organic farms typically have more diverse crop rotations, different input strategies, lower livestock stocking densities, and different land base requirements, GHG emissions are affected, both on a production area and production volume basis. This section focuses on the international evidence in support of farm-level GHG benefits of organic production, with a particular view to implications for Canada. The wider-angle land use change (LUC) implications of these comparisons (e.g., Flysjö et al. 2012) are not directly considered here.

A quick review of meta-analyses provides an integrated picture of the organic–conventional comparison; then, for each commodity area, we look at the three main GHGs (carbon, methane, and nitrous oxide) and also examine intensive vs. extensive production studies, with an eye to interpreting European results for the NA context. Gomiero et al. (2008) and Goh (2011) consistently found that organic systems had significantly lower carbon dioxide equivalent (CO_{2e}) emissions than comparable conventional systems, when measured on a per area basis, though in some systems, that benefit was lost when measured per tonne of production, depending on yield differences (see details in tables and also analysis of energy efficiency in Section 12.4). Most of their review focused on European studies where the intensity of conventional production produces greater spreads in yields than those found in North American ones (MacRae et al. 2007). Mondalaers et al. (2009), in their meta-analysis involving some studies not covered in Gomiero et al. (2008), reached similar conclusions. The *per area* improvements resulted from lower concentrate feeding and stocking rates and the absence of synthetic nitrogen fertilizer.

Kustermann and Hülsbergen (2008) compared 33 German organic and conventional commercial farms for direct and indirect energy inputs, GHG fluxes, and C sequestration. They found that the mean GHG emission/ha was significantly lower (72% of conventional), but with higher variability on organic farms meaning that the upper range of emissions on the organic operations was comparable to conventional ones, though the lower range was significantly lower. Nitrous oxide and carbon dioxide emissions were clearly lower on organic farms and with much higher C sequestration. Lindenthal et al. (2010) conducted a comparative LCA of dairy, poultry meat, eggs, bread, and vegetables in Austria and found organic production to be consistently superior to conventional production on GHG emissions per kilogram of product.

---

* Much of this section is adapted from Lynch et al. (2011).

13.4.1   FIELD CROPS

An LCA modeling of Canada-wide conversion to organic canola, wheat, soybean, and corn production concluded that organic crops would "generate 77% of the global warming emissions, 17% of the ozone-depleting emissions, and 96% of the acidifying emissions [sulfur dioxide] associated with current national production of these crops. Differences were greatest for canola and least for soy, which have the highest and lowest nitrogen requirements, respectively" (Pelletier et al. 2008). In general, the absence of synthetic nitrogen fertilizer and associated net reductions in field emissions was the most significant contributor to emission reductions. The benefits would only be obviated if organic yields were far below typical averages.

Robertson et al. (2000), in the Midwest United States, compared the net global warming potential (GWP) of conventional tillage, no-till (NT), low input, and organic management of a corn–soybean–wheat system over 8 years. None of the systems provided net mitigation, and nitrous oxide (N_2O) production was the single greatest source of GWP. The NT system had the lowest GWP (14 g CO_2 m^{-2} year^{-1}), followed by organic (41), low input (63), and conventional (114). However, the lower results in NT may have been an artifact of a shallow soil sampling strategy (MacRae et al. 2010). Cavigelli et al. (2009) calculated GWP and greenhouse gas intensity (GHGI = GWP per unit of grain yield) for NT, chisel till (CT), and organic (Org3) cropping systems in Maryland, United States, producing results more favorable to organic than the Robertson et al. study. GWP (kg CO_2e ha^{-1} year^{-1}) was positive in NT (1110) and CT (2348) and negative in Org3 (−1069), primarily due to differences in soil C and secondarily to differences in energy use among systems. Despite relatively low crop yields in Org3, GHGI (kg CO_2e Mg grain^{-1}) for Org3 was also negative (−207) and significantly lower than NT (330) and CT (153). Org3 was thus a net sink, primarily by increasing soil C with legume cover crops and animal manures.

Meisterling et al. (2009), also in the United States, used a hybrid LCA approach to compare the GWP and primary energy use involved in the production process (including agricultural inputs) plus transport processes for conventional and organic wheat production and delivery in the United States. The GWP of a 1 kg loaf of organic wheat bread was found to be about 30 g CO_2e less than that for a conventional loaf. However, when the organic wheat was shipped 420 km farther to market, the two systems had similar impacts. Organic grain yields were assumed at 75% of conventional average yields of 2.8 t grain ha^{-1}. Soil C storage potential was assumed the same for both systems and was omitted as a mitigation credit. N and P fertilizer production in the conventional system was a significant contributor to the GWP total. N_2O emissions from soil were assumed to be a large contributor in both systems and were rated as equivalent. As noted by these authors, there is the greatest uncertainty with respect to soil C storage and N_2O emissions (uncertainty ranges were greater than the calculated GWP difference between the two systems), and "uncertainty and variability related to these processes may make it difficult for producers and consumers to definitively determine comparative GHG emissions between organic and conventional production" (Meisterling et al. 2009).

Nemecek et al. (2005), reported by Niggli et al. (2008), found after analyses of data from two long-term comparative cropping systems studies in Europe that the

GWP of all organic crops was reduced by 18% per unit product compared to the conventional production systems. Gomiero et al. (2008), in their meta-analysis, drew upon three studies of winter wheat cropping systems in Europe, also reported in Stolze et al. (2000). CO_2 emissions per land unit (kg CO_2 ha^{-1}) were lower in the organic systems by an average of 50%, while emissions per unit of grain production (kg CO_2 ha^{-1}) were found to be lower in two of the studies (by 21%) and greater in one (by 21%).

Although other studies confirm (with Chirinda et al. 2010 an exception, finding no differences) that in most organic field crop systems, total N inputs to soil and the potential for N_2O emissions are reduced compared to conventional systems (Lynch et al. 2011), an increased risk for N_2O emissions occurs in organic farms following the flush of soil N mineralization after incorporation of legume green manure or crop residues. Scialabba and Müller-Lindenlauf (2010) cited one German study in which emissions, while peaking at 9 kg N_2O ha^{-1} following legume incorporation, averaged 4 kg N_2O ha^{-1} for the organic system compared with 5 kg N_2O ha^{-1} for a conventional system. Also in Europe, Petersen et al. (2006) tracked N_2O emissions from five field crop rotation sequences and found N_2O emissions were lower in the organic rotation, ranging from 4.0 to 8.0 kg N_2O-N ha^{-1} across all crops as total N inputs increased from 100 to 300 kg N ha^{-1} year^{-1}.

With qualifications regarding variations in study methodology, uncertainty with N_2O emissions, and different study assumptions with respect to soil C storage, the literature suggests that organic field crop management also improves GWP both per ha and per unit product when compared to conventional production (see Table 13.1).

13.4.2   LIVESTOCK (INCLUDING PASTURE/FORAGE AS APPROPRIATE)

Since many animal production systems are very dependent on field cropping, we include a quick overview here. For animal production, fewer studies have been conducted and the comparisons are more difficult because of the dramatic differences in operations, particularly for hogs and poultry. There is tremendous scope for expanded research on organic livestock systems and GHG emissions.

13.4.2.1   Beef

Because of their location in food supply chains, beef production systems are known to be much less energy efficient than crop production, requiring seven times as many inputs for the same calorie output (Smil 2001). Compared to other commodities, beef production has higher GHG emissions than poultry, egg, and hog production; milk; and crops. Niggli et al. (2008) suggested that, in general, net GHG emissions from beef production are in the range of 10 kg CO_2e kg^{-1} meat product compared with 2–3 kg CO_2e kg^{-1} for poultry, egg, and hog production; 1 kg CO_2e kg^{-1} for milk; and typically less than 0.5 kg CO_2 equivalents kg^{-1} for crop production systems.

Sonesson et al. (2009a) noted, however, that there is usually great variation in GHG studies of beef, because of methodological differences, system boundaries,

TABLE 13.1
Field Crops: Summary of Organic vs. Conventional Comparisons (%Org − Conv/Conv)

Authors	Region	Type of Study	Measure	Org < Conv (%)	Org > Conv (%)
Pelletier et al. (2008)	Canada	LCA (of conversion)	CO_2e ha^{-1}	61	
			CO_2e product^{-1}	23	
Robertson et al. (2000)	United States	Comparative field trial	GWP (g m^{-2})	64[a]	
Cavigelli et al. (2009)	United States	Comparative field trial	E use (CO_2e ha^{-1})	57	
			GWP (CO_2e ha^{-1})	69[b]	
			GWP (CO_2e unit grain^{-1})	42[c]	
Miesterling et al. (2009)	United States	LCA	GWP (CO_2e) kg bread^{-1}	16	
Nemecek (2005) in Niggli et al. (2008)	Europe	Comparative field trials	GWP (CO_2e) per unit product	18	
Kustermann and Hülsbergen (2008)	Germany	Meta-analyses	E use (CO_2e ha^{-1})	64	
Gomiero et al. (2008)	Europe	Meta-analyses (incl three wheat studies)	GWP (CO_2e ha^{-1})	50	
			GWP (CO_2e kg grain^{-1})	21 (two studies)	21 (one study)
Hirschfeld et al. (2008) in FAO (2011)	Germany	LCA	CO_2e product^{-1}	62	
Venkat (2012)	California	LCA models (to farm gate)	Alfalfa hay (CO_2e product^{-1})	34	
Chirinda et al. (2010)	Denmark	Field studies	Field crops (N_2O/area)	0	

[a] The NT system surpassed the organic, however, with GWP of only 14 compared to the organic at 41 and conventional at 114.

[b] When compared to an NT treatment, this gain is 51%.

[c] When compared to an NT treatment, this gain is 61%.

and differences in production systems. They reported, summarized from studies conducted in Europe, Brazil, and Canada, a higher range than Niggli et al. (2008) (14–32 kg CO_2e kg^{-1} meat product). All the cited studies found that methane emissions accounted for 50%–75% of total GHG emissions. As noted by Niggli et al. (2008) and others, however, while the methane emitted by ruminants is the major limitation of their use, by allowing efficient use of often marginal land and generating cost-effective forage benefits on productive land, they play a critical role in global food security. Furthermore, the methane emissions of ruminants consuming forages only are at least partially offset by the sequestration of CO_2 by those same perennial forages. Much depends, however, on how forages are managed as animal feed, so it is not automatically the case that beef fed primarily forages will produce lower emissions (Pelletier et al. 2010).

In Ireland, Casey and Holden (2006) undertook a *cradle-to-farm gate* LCA approach to estimate emissions kg^{-1} of live weight (LW) leaving the farm gate per annum (kg CO_2 kg LW^{-1} $year^{-1}$) and per hectare (kg CO_2 ha^{-1} $year^{-1}$). Fifteen units engaged in suckler-beef production (five conventional, five in an Irish agri-environmental scheme, and five organic units) were evaluated. The authors concluded that moving toward more extensive production, as found in organic systems, could reduce emissions per unit product and there would be a reduction in area and LW production per hectare. The average emissions from the conventional units were 13.0 kg CO_2 kg LW^{-1} $year^{-1}$, from the agri-environmental scheme units 12.2 kg CO_2 kg LW^{-1} $year^{-1}$, and from the organic units 11.1 kg CO_2 kg LW^{-1} $year^{-1}$. The average emissions per unit area from the conventional units were 5346 kg CO_2 ha^{-1} $year^{-1}$, from the agri-environmental scheme units 4372 kg CO_2 ha^{-1} $year^{-1}$, and from the organic units 2302 kg CO_2 ha^{-1} $year^{-1}$. GWP increased in a linear fashion, both per hectare and per unit animal LW shipped as there was an increase in either farm livestock stocking density, N fertilizer application rate, or concentrates fed.

Flessa et al. (2002) reported on a German research station comparison of two beef management systems: one a conventionally managed confinement-fed system and the other an organic pasture-based system. Combined GWP per unit land base was 3.2 Mg CO_2e ha^{-1} and 4.4 Mg CO_2e ha^{-1} for the organic and conventional systems, respectively, with soil N_2O emissions as the largest contributor in both systems. The higher percentage of emissions associated with N_2O, rather than methane, in other studies was likely a product of the relatively intensive application of fertilizers, synthetic N, and slurry in the conventional operation and manure, slurry, and N fixation in the organic operation. When compared per unit product (i.e., per beef LW of 500 kg), there were no differences in GWP between the two systems, primarily as productivity was approximately 20% greater for the confinement-based system.

Sonesson et al. (2009a) noted that few systematic studies are available providing data on the GWP impact of different beef production systems in Sweden. Data on GWP per unit product, however, were presented from three studies of organic, *ranch systems* and Swedish *average beef* systems, respectively, conducted by the same group of Swedish researchers. GWP impact averaged 22, 24, and 28 kg CO_2e kg^{-1} meat for organic, ranch, and average production systems, respectively.

Very limited analysis is available on which to base a conclusion for this sector (Table 13.2), particularly from North America. While organic beef production appears to reduce GWP per hectare, this is not consistently evident when calculated per unit of meat product. Drawing conclusions is hampered by numerous management and measurement differences across farms, regions, and studies, including how forages and pastures are managed, manure management, the lifespan of animals in different systems, and the sources of off-farm feedstuffs (Pelletier et al. 2010).

TABLE 13.2
Beef: Summary of Organic vs. Conventional Comparisons

Authors	Region	Type of Study	Measure	Org < Conv (%)	Org > Conv
Casey and Holden (2006)	Ireland	LCA	GWP (CO_2e ha^{-1})	57	
			GWP (CO_2e kg meat^{-1})	15	
Flessa et al. (2002)	Germany	Comparative systems study	GWP (CO_2e ha^{-1})	27	
			GWP (CO_2e kg meat^{-1})	0	
Sonneson et al. (2009a)	Sweden	2 LCAs	GWP (CO_2e kg meat^{-1})	21	

13.4.2.2 Dairy

A modeling study in Atlantic Canada examining 19 different dairy production scenarios found that a seasonal grazing organic system emitted 29% less GHGs compared with the average of all other analyzed systems (Main 2001; Main et al. 2002). A different study comparing nonorganic seasonal grazing with confined dairying did not find such significant differences between the two systems, suggesting that additional organic management requirements provide some significant efficiency opportunities (Arsenault et al. 2009). This study conducted an LCA of dairy systems in Nova Scotia to compare environmental impacts of typical pasture and confinement operations. Uses of concentrated feeds, N fertilizers, transport fuels, and electricity were dominant contributors to environmental impacts. Somewhat surprisingly, grazing cows for five months per year (typical of pasture systems in Nova Scotia) had little effect on overall environmental impact. Scenario modeling suggested, however, that prolonged grazing is potentially beneficial.

Olesen et al. (2006) used the whole farm model FarmGHG to analyze conventional and organic dairy farms, located in five European agroecological zones, on relative GHG emissions. Farms were assumed to have the same land base of 50 ha and, in each region, to achieve the same milk yield per cow. LD was 75% higher on the conventional farms compared to the 100% feed self-sufficient organic farms. Livestock contributed an average of 36% of total emissions, while fields contributed about 39%. Of the GHGs, N_2O and methane (CH_4) dominated, accounting for an average of 49% and 42% of total farm emissions. GHG emissions per hectare (Mg CO_2e ha^{-1}) increased with production intensity (i.e., LD) and thus farm N surplus, for both types of farms, and were usually higher for conventional dairy farms. GHG emissions per unit milk product (or metabolic energy, kg CO_2e kg milk^{-1}), however, were inversely related to farm N efficiency.

Bos et al. (2007) assessed GHGs on organic and conventional model dairy farms in the Netherlands. Model farms were designed on the basis of current organic and conventional farming practices. Total GHG emissions per Mg of milk in organic dairy farming were found to average approximately 80% (per ha) and 90% (per kg milk), respectively, of that in conventional dairy farming.

Thomassen et al. (2008) in the Netherlands conducted a detailed *cradle-to-farm gate* LCA, including farm environmental impact with respect to GHG and pollution impacts on water quality (i.e., eutrophication). As also reported previously by Olesen et al. (2006), N_2O and CH_4 accounted for the bulk of emissions. In the conventional system, CO_2, N_2O, and CH_4 accounted for 29%, 38%, and 34% of total GHGs, compared to 17%, 40%, and 43%, respectively, for the organic dairy farm system. Results indicated improved environmental performance with respect to energy use and eutrophication potential kg milk^{-1} for the organic compared to conventional farms (3.1 vs. 5.0 MJ kg^{-1} FPCM, respectively). On the other hand, farming systems failed to differ with respect to GWP per unit milk produced. Overall recommendations from this study included reducing use of concentrates with a high environmental impact and reducing whole farm nutrient surpluses.

It should be noted that the studies of Olesen et al. (2006), Bos et al. (2007), and possibly Thomassen et al. (2008) may have overestimated N_2O emissions associated with legume nitrogen fixation (a key component of organic farm systems) as older IPCC coefficients and methodology were used in these studies.

Flachowsky and Hachenberg (2009) conducted a review of nine European studies reporting GHG emissions from conventional and organic dairy farms and discussed at some length the gaps and uncertainties in the data. While one study reported equivalent GWP per unit milk product (kg CO_2e kg milk^{-1}), for five of the studies, organic systems resulted in greater GWP (ranging from 1% to 27% increase), while organic reduced GWP (ranging from 5% to 8%) in the remaining three studies.

Organic ruminant livestock farms differ also from conventional with respect to the cross-breeding and management goals, which, as less intensive systems, often result in improved animal longevity. As noted by Niggli et al. (2008), methane emissions per product can thus be reduced when calculated on the total lifespan of organic cows because they have a higher ratio of productive (milking) to nonproductive (calf/heifer) phases than found in conventional production where animals are commonly culled within a few lactations. As comparative data on relative longevity across dairy production systems are limited, this consideration has yet to be included in farm system GWP comparisons.

In an Austrian study, Hörtenhuber et al. (2010) conducted a *life cycle chain* analyses of eight different dairy production systems representing organic and conventional farms located in alpine, upland, and lowland regions. Notably, and rather innovatively, the authors included an estimate for GHG impacts of the estimated LUC required to produce concentrates (which ranged from 13% to 24% of total feed intake for various farms), such as soybean production replacing tropical forests. Nitrogen fertilizer was assumed not used on any farms and only partially during external-to-the-farm production of concentrates. About 8% of total GHGs for the conventional farms was attributed to LUC associated with concentrates. In general, the study found that the higher yields per cow and per farm for the conventional farms did not compensate for the greater GHGs produced by these more intensive systems, with organic farms on average emitting 11% less GHGs (0.81–1.02 kg CO_2e kg milk^{-1} compared to 0.90 to 1.17 kg CO_2e kg milk^{-1}). Such an analysis of LUC is not necessarily applicable to North American operations since concentrates are typically produced domestically.

Sonesson et al. (2009b) summarized LCA studies from 10 OECD countries that found emissions up to the farm gate ranged from 1.0 to 1.4 kg CO_2e kg milk^{-1}. While there were minor differences between conventional and organic farms, the contribution of each GHG differed. In general, organic systems had higher methane emissions kg milk^{-1} but lower emissions of N_2O and CO_2 per unit product. An identified complication of the analysis is the intersection between dairy and beef production, as calves and culled cows end up as beef products.

On balance, organic dairying may provide some benefits on a per ha basis compared to conventional production, but with respect to GWP per unit product, there is no consensus in the data available to suggest organic dairy management is significantly beneficial (Table 13.3). It must be noted, however, that Canadian and North American data are particularly scarce.

TABLE 13.3
Dairy: Summary of Organic vs. Conventional Comparisons

Authors	Region	Type of Study	Measure	Org < Conv (%)	Org > Conv (%)
Main (2001)	Atl. Canada	Modeling of farming systems	GWP (CO_2e kg milk^{-1})	29	
Olesen et al. (2006)	Denmark/EU	Comparison of model farms	Mg CO_2e ha^{-1} kg CO_2e kg milk^{-1}	40	11
Bos et al. (2007)	Netherlands	Comparison of model farms	GWP (CO_2e kg milk^{-1})	10	
Thomassen et al. (2008)	Netherlands	LCA	GWP (CO_2e kg FPCM^{-1})	0	
Flachowsky and Hachenberg (2009)	EU	Review of nine studies	GWP (CO_2e kg milk^{-1})	0 (one study) 5–8 (three studies)	1–27 (five studies)
Gomiero et al. (2008)	EU	Review of five studies	GWP (CO_2e kg milk^{-1})	0	
Hörtenhuber et al. (2010)	Austria	LCA	GWP (CO_2e kg milk^{-1})	11	
Sonneson et al. (2009b)	Sweden	Review of LCAs	GWP (CO_2e kg milk^{-1})	0	
Hirschfeld et al. (2008) in FAO (2011)	Germany	LCA	CO_2e product^{-1}	10	
Flysjö et al. (2012)	Sweden	LCA (no LUC)	Organic vs. high yield conventional CO_2e product^{-1}		6

13.4.2.3 Hogs

Organic hog production may be the least energy efficient of the major animal systems (Kumm 2002), possibly because of frequently lower than optimal levels of pasturing hogs, inappropriate breeds for organic systems, and the failure to find the most efficient roles for hogs in mixed farming operations. For example, hogs can play a useful role in weed control postharvest or field renovation (Honeyman 1991) and even compost aeration, with the potential to, therefore, reduce energy expenditures for weed control.

In a comparison of conventional, natural (Red Label), and organic hog production in France, van der Werf et al. (2007) found, using a detailed LCA, that organic systems produced the lowest emissions of methane and carbon dioxide on a per ha basis, but not a 1000 kg pig basis, for which they were significantly outperformed by conventional production on nitrous oxide and carbon dioxide emissions. Only in methane production did organic maintain a reduction over conventional, but the natural system performed even better. Two Swedish LCA studies, in contrast, found emissions in the organic operations to be 50% less than this French study and concluded that reduced growth rates, inefficient feed production, and composting of manure, with subsequent low NUE and higher ammonia and indirect nitrous oxide emissions, likely explain the different results (Sonesson et al. 2009c). However, emissions kg meat^{-1} were higher in the organic studies compared to most of the conventional operations. Similar results were found for energy expended kg meat^{-1}. Degre et al. (2007) also looked at three comparable Belgian systems (organic, free range, and conventional) and found GHG emissions (CO_2e) pig^{-1} were lowest for the organic system followed by free range and conventional, with nitrous oxide the dominant gas. Organic system emissions were 87% of conventional, with slurry from conventional operations having much higher emissions than straw litter in the organic system. However, organic performance was inferior in some of the other environmental criteria assessed.

Williams et al. (2006), modeling UK systems, found lower emissions on a per tonne basis for organic systems (11% lower GWP100* emissions), but with 1.73 times greater land requirements t^{-1} of production.

Halberg et al. (2010) modeled standard LCAs on three different Danish organic hog systems and compared the results with the literature on conventional operations (Halberg et al. 2008). They found higher levels of GHG emissions (CO_2e pig^{-1}) on all organic operations because of higher N_2O and lower feed conversion efficiencies but concluded that if C sequestration associated with the organic rotations was included in the calculations (11%–18% reductions in CO_2e pig^{-1}), two of the three organic operations would outperform the conventional one (Halberg et al. 2008, 2010). Comparing the different conclusions of their work with those of van der Werf et al. (2007), Halberg et al. (2010) concluded that "methodological differences make a direct comparison between the two studies problematic. The French study also found that organic pig production had a better environmental performance compared with conventional when calculated per ha but worse when

---

* GWP defined over a 100-year time frame.

TABLE 13.4

Hogs: Summary of Organic vs. Conventional Comparisons

Authors	Region	Type of Study	Measure	Org < Conv (%)	Org > Conv (%)
van der Werf et al. (2007)	France	LCA	CH_4 ha^{-1}	69	
			N_2O ha^{-1}		33
			CO_2 ha^{-1}	13	
van der Werf et al. (2007)	France	LCA	CH_4 pig^{-1}	46	
			N_2O pig^{-1}		242
			CO_2 pig^{-1}		58
Sonneson et al. (2009c)	Sweden	2 LCAs	CO_2e kg $meat^{-1}$	6 (one study)	2–35 (six studies)
Degre et al. (2007)	Belgium	Expert ranking	CO_2e pig^{-1}	13	
Williams et al. (2006)	United Kingdom	Modeling	GWP100 t^{-1}	11	
Halberg et al. (2010)	Denmark	Modeled LCA	GHG100 kg^{-1} and C sequestration	4%–33% for 2/3 org. farms	7% for 1/3 org. farms

calculated per kg pig product. But they did not include differences in the soil carbon sequestration as in our study."

Given the range of things hogs can eat, low meat yields of pork in organic systems may be more efficient if a high percentage of human inedible feedstuffs are consumed. It is reasonable to postulate that too much reliance on high production leads to excessive consumption of human edibles as hog feed.

Sonesson et al. (2009c) concluded that although there are only a limited number of high-quality studies on hogs, there was sufficient information to set out a workable protocol for the Swedish Climate Labeling for Food scheme, focusing on individual operations (whether conventional or organic) rather than the organic sector as a whole.

On balance, comparative results were mixed for hogs (Table 13.4). Including carbon sequestration in the analysis, given the role of forages and cover crops in well-managed organic rotations, appears to create more positive comparisons for organic.

13.4.2.4 Poultry

Williams et al. (2006) used standard LCA to model typical conventional and organic production scenarios in the United Kingdom.* GWP from organic poultry meat production was up to 45% higher than conventional production. Bokkers and de Boer (2009) reached similar conclusions when examining Dutch organic and conventional operations, not necessarily surprising, given that some of their modeling was based on the work of Williams et al. (2006). The key comparative factor is the high feed conversion rates obtainable in conventional production.

---

* This work was updated in Leinonen et al. (2012a,b).

Sonesson et al. (2009d), from their review of five European studies including Williams et al. (2006), found that N_2O emissions from conventional feed, associated with N fertilizer and soil losses, presented the greatest opportunities for savings in well-designed organic systems. The design of barn heating systems would be another significant area for efficiencies, especially in hatcheries. Comparative data on organic vs. conventional poultry production are particularly sparse, especially for eggs (Sonesson et al. 2009e).

13.5 ADDITIONAL CONSIDERATIONS, NOT CAPTURED IN COMPARATIVE STUDIES

13.5.1 Landscape-Level Land Use

On a global basis, hundreds of millions of ha of existing arable land are underutilized or compromised by conventional agricultural practices. Given organic agriculture's better performance related to protecting soil resources (see Chapters 3 through 6 of this volume) and greater adaptability to less suitable moisture conditions, using organic production for cropland remediation, that is, overcoming underutilization associated with erosion or salinization, is a feasible sectoral development strategy. It also would help counter the claims that organic is not a viable production strategy because of lower yields. On a total land use basis, organic production is more efficient because it does not reduce the quality of the land base and requires lower levels of exogenous energy and nutrients. Consequently, organic farming is better placed to adapt to shocks. However, the potentially additional land requirement on a per farm basis (e.g., for greater forage production) means again that a mechanism is required for landscape-level land use shifts, beyond the decision making of individual operators. It is not feasible for the market to encourage suitable re-allocations, such as taking land out of animal production, shifting to human crops, or resuscitating less productive lands compromised by conventional practices. Related to this, an undetermined, but significant, amount of high-quality land area is devoted to nonfood production, including tobacco, floriculture, landscaping plants, and crops for beverage production. Many of such lands may be better suited to food crops, with nonfood crop production shifted to less valued locations.

13.5.2 Plant Varieties and System Efficiencies

Regarding plant varieties, the focus on high optimal harvest in conventional plant breeding may be reducing overall system efficiencies associated with the plant, and it also increases off-farm export of nutrients, making it more difficult to close nutrient loops on farms. Because organic farmers may make better use of the nonhuman edible parts of the plant—either for organic matter, for animal feed, for bedding, or for weed management (taller, more competitive plants with lower nitrogen requirements)—lower harvest indices are desirable, and such efficiencies could be augmented with further inquiry.

13.5.3 NUTRIENT RECOVERY AND RECYCLING

NUE of cereals globally decreased from 80% in 1960 to ~30% in 2000 because of inefficiencies related to synthetic N utilization (Erisman et al. 2008). What levels of overall nitrogen recovery from biological sources are feasible, including gains from recycling residues, crop rotation, biological organisms, reduced tillage, and reduced soil loss? What level of applied nutrient can be absorbed by the crop in an efficient system? Can 70% absorption be achieved? How might root/microbe interactions be optimized to improve this efficiency? This needs further work since there is a tremendous diversity in the microbial world that has yet to be tapped. Equally, figuring out ways to return nutrients when organic crops are exported as animal feed (and for energy) is a priority area for investigation. It is a significant issue in certain Canadian organic commodities, for example, export of soybeans to Europe/Japan and hay and alfalfa anywhere off-farm.

13.5.4 PHOTOSYNTHETIC EFFICIENCY

C4 photosynthetic pathway crops, such as corn and sorghum, have not fit well in Canadian organic systems although some Canadian farms report shifting to sorghum because of increasingly droughty conditions. Other C4 crops may have a significant role to play in energy crops in organic operations (MacRae et al. 2010). Consequently, the sector should continue exploration of perennial C4 plants in organic systems to maximize possibilities for photosynthetic efficiency, while also balancing potentially increased nutrient requirements. Nutrient efficiencies may be gained by planting heavy feeding C4 crops after legume plow downs or manure and compost applications.

13.6 CONCLUSION

A growing literature supports the conclusion that organic systems have greater cropping, floral, and habitat diversity; reduced agrochemical intensity and pollution; and lower GHG emissions compared to conventional operations. A more limited set of studies also concludes that these differences subsequently generate measurable environmental benefits. But if organic agriculture is to build on this comparative picture, further improvements are in order that should reflect a more comprehensive understanding of the variability within organic production systems. As well, given the key role of field margins and nonproduction areas in promoting biodiversity, revised organic regulations should outline requirements for their establishment and maintenance. Research that enhances understanding of which nonfarmed elements are more important, the relationship to landscape factors for optimizing populations of various taxa, and the sustainability of the farming system to the regional carrying capacity (Noe et al. 2005; Gabriel et al. 2010; Gomiero et al. 2011) will all be important to make such regulations workable. As organic management improves, along these lines, and greater landscape-level management efficiencies are implemented, even greater improvements in organic performance may result.

REFERENCES

Altieri, M.A. 1987. *Agroecology: The Scientific Basis of Alternative Agriculture*, 2nd edn. Boulder, CO: Westview Press.

Arsenault, N., P. Tyedmers, and A. Fredeen. 2009. Comparing the environmental impacts of pasture-based and confinement-based dairy systems in Nova Scotia (Canada) using life cycle assessment. *International Journal of Agricultural Sustainability* 7:19–41.

Askegaard, M., J.E. Olesen, I.A. Rasmussen, and K. Kristensen. 2011. Nitrate leaching from organic arable crop rotations is mostly determined by autumn field management. *Agriculture, Ecosystems and Environment* 142:149–160.

Bartram, H. and A. Perkins. 2003. The biodiversity benefits of organic farming. In *Organic Agriculture: Sustainability, Markets and Policies*, ed. OECD. Wallingford, U.K.: CABI Publishing, pp. 77–96.

Bengtsson, J., J. Ahnstrom, and A.C. Weibull. 2005. The effects of organic agriculture on biodiversity and abundance: A meta-analysis. *Journal of Applied Ecology* 42(2):261–269.

Birkhofer, K., T.M. Bezemer, J. Bloem et al. 2008. Long-term organic farming fosters below and above ground biota: Implications for soil quality, biological control and productivity. *Soil Biology and Biochemistry* 40:297–308.

Bokkers, E.A.M. and L.J.M. de Boer. 2009. Economic, ecological, and social performance of conventional and organic broiler production in The Netherlands. *British Poultry Science* 50:546–557.

Bonti-Ankomah, S. and E.K. Yiridoe. 2006. *Organic and Conventional Food: A Literature Review of the Economics of Consumer Perceptions and Preferences*. Truro, Nova Scotia, Canada: Organic Agriculture Centre of Canada.

Bos, J.F.F.P., J.J. de Haan, W. Sukkel, and R.L.M. Schils. 2007. Comparing energy use and greenhouse gas emissions in organic and conventional farming systems in The Netherlands. In *Proceedings of the Third QLIF Congress*, Hohenheim, Germany, March 20–23, 2007. Available online: http://orgprints.org/view/projects/int_conf_qlif2007.html (accessed on October 29, 2010).

Boutin, C., A. Baril, and P.A. Martin. 2008. Plant diversity in crop fields and woody hedgerows of organic and conventional farms in contrasting landscapes. *Agriculture, Ecosystems and Environment* 123:185–193.

Boutin, C., A. Baril, S.K. McCabe, P.A. Martin, and M. Guy. 2011. The value of woody hedgerows for moth diversity on organic and conventional farms. *Environmental Entomology* 40(3):560–569.

Casey, J.W. and N.M. Holden. 2006. Greenhouse gas emissions from conventional, agri-environmental scheme, and organic Irish suckler-beef units. *Journal of Environmental Quality* 35:231–239.

Cavigelli, M.A., M. Djurickovic, C. Rasmann, J.T. Spargo, S.B. Mirsky, and J.E. Maul. 2009. Global warming potential of organic and conventional grain cropping systems in the Mid-Atlantic Region of the US. In *Proceedings of the Farming System Design Conference*, Monterey, CA, August 25, 2009.

Chirinda, N., M.S. Carter, K.R. Albert et al. 2010. Emissions of nitrous oxide from arable organic and conventional cropping systems on two soil types. *Agriculture Ecosystems and Environment* 136(3–4):199–208.

Crowder, D.W., T.D. Northfield, M.R. Strand, and W.E. Snyder. 2010. Organic agriculture promotes evenness and natural pest control. *Nature* 466(7302):109–123.

Davis, A.S., J.D. Hill, C.A. Chase, A.M. Johanns, and M. Liebman. 2012. Increasing cropping system diversity balances productivity, profitability and environmental health. *PLoS ONE* 7(10):e47149. DOI: 10.1371/journal.pone.0047149.

Degre, A., C. Debouche, and D. Verheve. 2007. Conventional versus alternative pig production assessed by multicriteria decision analysis. *Agronomy and Sustainable Development* 27:185–195.

Erisman, J.W., M.A. Sutton, J. Galloway, Z. Klimont, and W. Winiwarter. 2008. How a century of ammonia synthesis changed the world. *Nature Geoscience* 1(10):636–639.

Finckh, M.R., E. Schulte-Geldermann, and C. Bruns. 2006. Challenges to organic potato farming: Disease and nutrient management. *Potato Research* 49:27–42.

Flachowsky, G. and S. Hachenberg. 2009. CO_2-footprints for food of animal origin-present stage and open questions. *Journal für Verbraucherschutzund Lebensmittelsicherheit* 4:190–198.

Flessa, H., R. Ruser, P. Dörsch, T. Kamp, M.A. Jimenez, J.C. Munch, and F. Beese. 2002. Integrated evaluation of greenhouse gas emissions (CO_2, CH_4, N_2O) from two farming systems in southern Germany. *Agriculture, Ecosystems and Environment* 91:175–189.

Flohre, A., M. Rudnick, G. Traser, T. Tscharntke, and T. Eggers. 2011. Does soil biota benefit from organic farming in complex vs. simple landscapes? *Agriculture Ecosystems and Environment* 141(1–2):210–214.

Flysjö, A., C. Cederberg, M. Henriksson, and S. Ledgard. 2012. The interaction between milk and beef production and emissions from land use change—Critical considerations in life cycle assessment and carbon footprint studies of milk. *Journal of Cleaner Production* 28:134–142.

Food and Agriculture Organization (FAO). 2011. Organic agriculture and climate change mitigation. A report of the Round Table on Organic Agriculture and Climate Change. Rome, Italy, December 2011.

Freemark, K.E. and D.A. Kirk. 2001. Birds on organic and conventional farms in Ontario: Partitioning effects of habitat and practices on species composition and abundance. *Biological Conservation* 101:337–350.

Fuller, R.J., L.R. Norton, R.E. Feber, P.J. Johnson, D.E. Chamberlain, A.C. Joys, F. Mathews, et al. 2005. Benefits of organic farming to biodiversity vary among taxa. *Biology Letters* 1:431–434.

Gabriel, D., S.M. Sait, J.A. Hodgson, U. Schmutz, W.E. Kunin, and T.G. Benton. 2010. Scale matters: The impact of organic farming on biodiversity at different spatial scales. *Ecology Letters* 13(7):858–869.

Gabriel, D. and T. Tscharntke. 2007. Insect pollinated plants benefit from organic farming. *Agriculture Ecosystems and Environment* 118(1–4):43–48.

Garratt, M.P.D., D.J. Wright, and S.R. Leather. 2011. The effects of farming system and fertilisers on pests and natural enemies: A synthesis of current research. *Agriculture Ecosystems and Environment* 141(3–4):261–270.

Gliessman, S. 1997. *Agroecology: Ecological Processes in Sustainable Agriculture.* Boca Raton, FL: CRC Press.

Goh, K.M. 2011. Greater mitigation of climate change by organic than conventional agriculture: A review. *Biological Agriculture and Horticulture* 27(2):205–229.

Gomiero, T., M. Paoletti, and D. Pimentel. 2008. Energy and environmental issues in organic and conventional agriculture. *Critical Reviews in Plant Science* 27:239–254.

Gomiero, T., D. Pimental, and M.G. Paoletti. 2011. Environmental impact of differential agricultural management practices: Conventional vs. organic agriculture. *Critical Reviews in Plant Science* 30:95–124.

Griffiths, B.S., B.C. Ball, T.J. Daniell et al. 2010. Integrating soil quality changes to arable agricultural systems following organic matter addition, or adoption of a ley-arable rotation. *Applied Soil Ecology* 46(1):43–53.

Halberg, N. 1999. Indicators of resource use and environmental impact for use in a decision aid for Danish Livestock Farmers. *Agriculture, Ecosystems and Environment* 76:17–30.

Halberg, N., R. Dalgaard, J.E. Olesen, and T. Dalgaard. 2008. Energy self-reliance, net energy production and GHG emissions in Danish organic cash crop farms. *Renewable Agriculture and Food Systems* 23:30–37.

Halberg, N., J.E. Hermansen, I.S. Kristensen, J. Eriksen, N. Tvedegaard, and B.M. Petersen. 2010. Impact of organic pig production systems on CO_2 emission, C sequestration and nitrate pollution. *Agronomy and Sustainable Development* 30:721–731.

Halberg, N., E. Kristensen, and I.S. Kristensen. 1995. Nitrogen turnover on organic and conventional mixed farms. *Journal of Agricultural and Environmental Ethics* 8(1):30–51.

Hathaway-Jenkins, L.J., R. Sakrabani, B. Pearce, A.P. Whitmore, and R.J. Godwin. 2011. A comparison of soil and water properties in organic and conventional farming systems in England. *Soil Use and Management* 27(2):133–142.

Hole, D.G., A.J. Perkins, J.D. Wilson, I.H. Alexander, F. Grice, and A.D. Evans. 2005. Does organic farming benefit biodiversity? *Biological Conservation* 122:113–130.

Honeyman, M.S. 1991. Sustainable swine production in the U.S. corn belt. *American Journal of Alternative Agriculture* 6:63–70.

Hörtenhuber, S., T. Lindenthal, B. Amon, T. Markut, L. Kirner, and W. Zollitsch. 2010. Greenhouse gas emissions from selected Austrian dairy production systems—Model calculations considering the effects of land use change. *Renewable Agriculture and Food Systems* 25:330.

Kirchmann, H., L. Bergstrom, T. Katterer, L. Mattsson, and S. Gesslein. 2007. Comparison of long-term organic and conventional crop-livestock systems on a previously nutrient-depleted soil in Sweden. *Agronomy Journal* 99(4):960–972.

Kleijn, D., F. Kohler, A. Baldi et al. 2009. On the relationship between farmland biodiversity and land-use intensity in Europe. *Proceedings of the Royal Society B—Biological Sciences* 276(1658):903–909.

Korsaeth, A. 2008. Relations between nitrogen leaching and food productivity in organic and conventional cropping systems in a long-term field study. *Agriculture, Ecosystems and Environment* 127(3–4):177–188.

Krauss, J., I. Gallenberger, and I. Steffan-Dewenter. 2011. Decreased functional diversity and biological pest control in conventional compared to organic crop fields. *PLoS ONE* 6(5):e19502.

Kumm, K. 2002. Sustainability of organic meat production under Swedish conditions. *Agriculture, Ecosystems and Environment* 88:95–101.

Kustermann, B. and K.J. Hülsbergen. 2008. Emission of climate relevant gases in organic and conventional cropping systems. Presented at the *16th IFOAM Organic World Congress*, Modena, Italy, June 2008.

Leinonen, I., A.G. Williams, J. Wiseman et al. 2012a. Predicting the environmental impacts of chicken systems in the United Kingdom through a life cycle assessment: Broiler production systems. *Poultry Science* 91(1):8–25.

Leinonen, I., A.G. Williams, J. Wiseman et al. 2012b. Predicting the environmental impacts of chicken systems in the United Kingdom through a life cycle assessment: Egg production systems. *Poultry Science* 91(1):26–40.

Lindenthal, T., T. Markut, S. Hörtenhuber, and G. Rudolph. 2010. Greenhouse gas emissions of organic and conventional foodstuffs in Austria. In *Proceedings of the VII International Conference on Life Cycle Assessment in the Agri-Food Sector*, Volume 1, Bari, Italy, pp. 319–324.

Lotter, D.W. 2003. Organic agriculture. *Journal of Sustainable Agriculture* 21:59–128.

Lynch, D.H. 2009. Environmental impacts of organic agriculture: A Canadian perspective. *Canadian Journal of Plant Science* 89(4):621–628.

Lynch, D.H., N. Halberg, and G.D. Bhatta. 2012a. Environmental impacts of organic agriculture in temperate regions. *CAB Reviews* 7(No. 010):1–17.

Lynch, D.H., R.J. MacRae, and R.C. Martin. 2011. The carbon and global warming potential impacts of organic farming: Does it have a significant role in an energy constrained world? *Sustainability* 3:322–362.

Lynch, D.H., M. Sharifi, A. Hammermeister, and D. Burton. 2012b. Nitrogen management in organic potato production. In *Sustainable Potato Production: Global Case Studies*, ed. H. Zhongi, R.P. Larkin, and C.W. Honeycutt. New York: Springer Science+Business Media, pp. 209–231.

Lynch, D.H., J. Sumner, and R.C. Martin. In press. Transforming recognition of the social, ecological and economic goods and services derived from organic agriculture in the Canadian context. In *Organic Food and Farming: Prototype for Sustainable Agriculture, eds.* S. Bellon and S. Penvern. Paris, France: Springer.

Lynch, D.H., R.P. Voroney, and P.R. Warman. 2004. Nitrogen availability from composts for humid region perennial grass and legume-grass forage production. *Journal of Environmental Quality* 33(4):1509–1520.

MacRae, R. 2011. A joined-up food policy for Canada. *Journal of Hunger and Environmental Nutrition* 6:424–457.

MacRae, R., R. Martin, A. Macey, P. Doherty, J. Gibson, and R. Beauchemin. 2004. Does the adoption of organic food and farming systems solve multiple policy problems? A review of the existing literature. Report funded by the Canadian Agriculture and Rural Develop (CARD) Program of Agriculture and Agrifood Canada. Truro, Nova Scotia, Canada: Organic Agriculture Centre of Canada.

MacRae, R.J., B. Frick, and R.C. Martin. 2007. Economic and social impacts of organic production systems. *Canadian Journal of Plant Science* 87:1037–1044.

MacRae, R.J., D. Lynch, and R.C. Martin. 2010. Improving the energy efficiency and GHG mitigation potentials of organic farming and food systems in Canada. *Journal of Sustainable Agriculture* 34(5):549–580.

Mader, P., A. Fleisbach, D. Dubois, L. Gunst, P. Fried, and U. Niggli. 2002. Soil fertility and biodiversity in organic farming. *Science* 296:1694–1697.

Magbanua, F.S., C.R. Townsend, G.L. Blackwell, N. Phillips, and C.D. Matthaei. 2010. Responses of stream macroinvertebrates and ecosystem function to conventional, integrated and organic farming. *Journal of Applied Ecology* 47(5):1014–1025.

Main, M.H. 2001. Development and Application of the Atlantic Dairy Sustainability Model (ADSM) to evaluate effects of pasture utilization, crop input levels, and milk yields on sustainability of dairying in Maritime Canada, MSc thesis. Halifax, Nova Scotia, Canada: NSAC and Dalhousie University.

Main, M.H., D. Lynch, R.C. Martin, and A. Fredeen. 2002. Sustainability profiles of Canadian dairy farms. Presented at the *IFOAM Scientific Congress*. Victoria, British Columbia, Canada, August 2002.

Meisterling, K., C. Samaras, and V. Schweizer. 2009. Decisions to reduce greenhouse gases from agriculture and product transport: LCA case study of organic and conventional wheat. *Journal of Cleaner Production* 17:222–230.

Möller, K., J. Habermeyer, V. Zinkernagel, and H. Reents. 2007. Impact and interactions of nitrogen and *Phytophthora infestans* as yield-limiting factors and yield reducing factors in organic potato (*Solanum tuberosum* L.) crops. *Potato Research* 49:281–301.

Mondelaers, K., J. Aertsens, and G. Van Huylenbroeck. 2009. A meta-analysis of the differences in environmental impacts between organic and conventional farming. *British Food Journal* 111:1098–1119.

Morandin, L. and M. Winston. 2005. Wild bee abundance and seed production in conventional, organic, and genetically modified canola. *Ecological Applications* 15(3):871–881.

Nelson, K.L., D.H. Lynch, and G. Boiteau. 2009. Assessment of changes in soil health throughout organic potato rotation sequences. *Agriculture Ecosystems and Environment* 131(3–4):220–228.

Nemecek, T., O. Huguenin-Elie, D. Dubois, and G. Gaillard. 2005. *Ökobilanzierung von Anbausystemen im Schweizerischen Acker- und Futterbau*. Schriftenreihe der FAL, Vol. 58. Zürich, Switzerland: FAL Reckenholz.

Niggli, U., H. Schmid, and A. Fliessbach. 2008. Organic Farming and Climate Change. A report prepared for the International Trade Centre (ITC) of the United Nations Conference on Trade and Development (UNCTAD) and the World Trade Organization (WTO). Geneva, Switzerland: ITC, UNCTAD/WTO, p. 30.

Noe, E., N. Halberg, and J. Reddersen. 2005. Indicators of biodiversity and conservational wildlife quality on Danish organic farms for use in farm management: A multidisciplinary approach to indicator development and testing. *Journal of Agricultural and Environmental Ethics* 18(4):383–414.

OECD. 2008. Agrienvironmental performance in OECD countries since 2008. Canada Country Report. Paris, France: OECD.

Olesen, J.E., K. Schelde, A. Weiske, M.R. Weisbjerg, W.A.H. Asman, and J. Djurhuus. 2006. Modelling greenhouse gas emissions from European conventional and organic dairy farms. *Agriculture, Ecosystems and Environment* 112:207–220.

Pan-European Common Bird Monitoring Scheme. 2011. Available from URL: http://www.rspb.org.uk/ourwork/projects/details/266243-paneuropean-common-birdmonitoring-scheme-pecbms (accessed August 7, 2011).

Pelletier, N., N. Arsenault, and P. Tyedmers. 2008. Scenario modeling potential eco-efficiency gains from a transition to organic agriculture: Life cycle perspectives on Canadian canola, corn, soy, and wheat production. *Journal of Environmental Management* 42:989–1001.

Pelletier, N., R. Pirog, and R. Rasmussen. 2010. Comparative life cycle environmental impacts of three beef production strategies in the Upper Midwestern United States. *Agricultural Systems* 103:380–389.

Petersen, S.O., K. Regina, A. Pollinger, E. Rigler, L. Valli, S. Yamulki, M. Esala, C. Fabbri, E. Syvasalo, and F.P. Vinther. 2006. Nitrous oxide emissions from organic and conventional crop rotations in five European countries. *Agriculture, Ecosystems and Environment* 112:200–206.

Pimentel, D., P. Hepperly, J. Hanson, D. Douds, and R. Seidel. 2005. Environmental, energetic, and economic comparisons of organic and conventional farming systems. *Bioscience* 55:573–582.

Ponce, C., C. Bravo, D.G. Leon, M. Magana, and J.C. Alonso. 2011. Effects of organic farming on plant and arthropod communities: A case study in Mediterranean dryland cereal. *Agriculture Ecosystems and Environment* 141(1–2):193–201.

Poudel, D.D., W.R. Horwath, W.T. Lanini, S.R. Temple, and A.H.C. van Bruggen. 2002. Comparison of soil N availability and leaching potential, crop yields and weeds in organic, low input and conventional farming systems in northern California. *Agriculture Ecosystems and Environment* 90(2):125–137.

Power, E.F. and J.C. Stout. 2011. Organic dairy farming: Impacts on insect–flower interaction networks and pollination. *Journal of Applied Ecology* 48(3):561–569.

Purtauf, T., I. Roschewitz, J. Dauber, C. Thies, T. Tscharntke, and V. Wolters. 2005. Landscape context of organic and conventional farms: Influences on carabid beetle diversity. *Agriculture Ecosystems and Environment* 108(2):165–174.

Roberts, C.J., D.H. Lynch, R.P. Voroney, R.C. Martin, and S.D. Juurlink. 2008. Nutrient budgets of Ontario organic dairy farms. *Canadian Journal of Soil Science* 88:107–114.

Robertson, G.P., E.A. Paul, and R.R. Harwood. 2000. Greenhouse gases in intensive agriculture: Contributions of individual gases to the radiative forcing of the atmosphere. *Science* 289:1922–1925.

Scialabba, N.E. and M. Müller-Lindenlauf. 2010. Organic agriculture and climate change. *Renewable Agriculture and Food Systems* 25:158–169.

Smil, V. 2001. *Feeding the World: A Challenge for the Twenty-First Century*. Cambridge, MA: MIT Press.

Sonesson, U., C. Cederberg, and M. Berglund. 2009a. Greenhouse gas emissions in beef production: Decision support for climate certification. 2009-4 Climate change for food. Available online: http://www.klimatmarkningen.se/in-english/underline-reports/ (accessed on October 29, 2010).

Sonesson, U., C. Cederberg, and M. Berglund. 2009b. Greenhouse gas emissions in milk production: Decision support for climate certification. 2009-3 Climate change for food. Available online: http://www.klimatmarkningen.se/in-english/underline-reports/ (accessed on October 29, 2010).

Sonesson, U., C. Cederberg, and M. Berglund. 2009c. Greenhouse gas emissions in hog production: Decision support for climate certification. 2009-5 Climate change for food. Available online: http://www.klimatmarkningen.se/in-english/underline-reports/ (accessed on October 29, 2010).

Sonesson, U., C. Cederberg, and M. Berglund. 2009d. Greenhouse gas emissions in chicken production: Decision support for climate certification. 2009-6 Climate change for food. Available online: http://www.klimatmarkningen.se/in-english/underline-reports/ (accessed on October 29, 2010).

Sonesson, U., C. Cederberg, and M. Berglund. 2009e. Greenhouse gas emissions in egg production: Decision support for climate certification. 2009-7 Climate change for food. Available online: http://www.klimatmarkningen.se/in-english/underline-reports/ (accessed on October 29, 2010).

Stockdale, E.A., M.A. Shepherd, S. Fortune, and S.P. Cuttle. 2002. Soil fertility in organic farming systems—Fundamentally different? *Soil Use and Management* 18:301–308.

Stolze, M., A. Piorr, A. Häring, and S. Dabbert. 2000. *The Environmental Impacts of Organic Farming in Europe*. Volume 6 of the series *Organic Farming in Europe: Economics and Policy*. Stuttgart, Germany: University of Hohenheim.

Thomassen, M.A., K.J. van Calker, M.C.J. Smits, G.L. Iepema, and I.J.M. de Boer. 2008. Life cycle assessment of conventional and organic milk production in The Netherlands. *Agricultural Systems* 96:95–107.

UNEP (United Nations Environment Program). 2010. Assessing the environmental impacts of consumption and production: Priority products and materials. A report of the working group on the environmental impacts of products and materials to the International Panel for Sustainable Resource Management. Nairobi, Kenya: UNEP.

van der Werf, H.M.G., J. Tzilivakis, K. Lewis, and C. Basset-Mens. 2007. Environmental impacts of farm scenarios according to five assessment methods. *Agriculture, Ecosystems and Environment* 118:327–338.

Venkat, K. 2012. Comparison of twelve organic and conventional farming systems: A life cycle greenhouse gas emissions perspective. *Journal of Sustainable Agriculture* 36(6):620–649.

Wiebe, N. 2012. Crisis in the food system: The farm crisis. In *Critical Perspectives in Food Studies*, ed. M. Koc, J. Sumner, and A. Winson. Don Mills, Ontario, Canada: Oxford University Press, pp. 155–170.

Willer, H., J. Lernoud, and L. Kilcher (eds.). 2013. *The World of Organic Agriculture: Statistics and Emerging Trends 2013*. Bonn, Germany: IFOAM and FiBL.

Williams, A.G., E. Audsley, and D.L. Sandars. 2006. Determining the environmental burdens and resource use in the production of agricultural and horticultural commodities. Defra project report IS0205. London, U.K.: Department for Environment, Food & Rural Affairs (Defra).

Wiltshire, K., K. Delate, M. Wiedenhoeft, and J. Flora. 2010. Incorporating native plants into multifunctional prairie pastures for organic cow–calf operations. *Renewable Agriculture and Food Systems* 26(2):114–126.

Winqvist, C., J. Bengtsson, T. Aavik et al. 2011. Mixed effects of organic farming and landscape complexity on farmland biodiversity and biological control potential across Europe. *Journal of Applied Ecology* 48(3):570–579.

Wortman, S.E., J.L. Lindquist, M.J. Haar, and C.A. Francis. 2010. Increased weed diversity, density and above-ground biomass in long-term organic crop rotations. *Renewable Agriculture and Food Systems* 25(4):281–295.

14 Will More Organic Food and Farming Solve Food System Problems? Part II: Consumer, Economic, and Community Issues*

Rod MacRae, Derek H. Lynch, and Ralph C. Martin

CONTENTS

---

* Based on MacRae et al. (2004, 2007) and Lynch et al. (in press).

14.1 INTRODUCTION

Using an approach similar to Chapter 13, we examine consumer, economic, and community issues and the degree to which organic farming and food systems help solve current problems. In general, however, fewer studies have been conducted on these themes compared to environmental ones and often less definitive conclusions can be drawn.

14.2 CONSUMER ISSUES

14.2.1 CONSUMER CONFIDENCE

In the OECD world, consumer confidence in the food supply has been waning. Disease outbreaks, worries about new technologies, and concerns about the ability of regulatory systems to keep up with the changes in a globalized food system are all contributing factors. Whether real or perceived, the loss of consumer confidence is worrisome to governments and the food industry and is producing more onerous management and record-keeping requirements for many farmers and merchants.

Organic farming and food processing standards do not permit a number of inputs and practices perceived to be risky by many consumers and this, in part, explains consumer interest in organic foods (Magkos et al. 2006). For example, genetically modified seeds and production aids are not permitted. The use of synthetic preservatives and additives is severely restricted in organic processing, largely to materials derived from naturally occurring substances. Of the 500 or so additives in general use, only 30 or so are generally permitted in organic processing (Heaton 2002). Food irradiation is not permitted. In surveys, North American consumers have purchased organic foods primarily for reasons of personal health (Neilsen 2005; Lynch 2009), while European consumers have been motivated by a wider range of interests including fair prices to farmers and animal welfare (Zander and Hamm 2010).

Some of the key consumer anxieties about the food supply are reviewed here, including the safety of pesticides and horticultural practices, certain animal production safety and welfare issues, and the nutritional value of organic versus conventional foods. Our purpose is not to analyze the merits of consumer anxieties, but instead to review products and processes where significant anxiety exists and to summarize the state of evidence regarding the role of organic food and farming in reducing them.

14.2.1.1 Food Safety: Synthetically Compounded Pesticides

Almost all pesticides believed to have potentially significant negative health impacts on humans are not permitted in organic production. Some 50 million kg of pesticides is applied annually in Canada,* and of the more than 500 active ingredients registered, very few are permitted in organic production, and most of those permitted are essentially registered as low-risk products. Consequently, dietary residues of production pesticides are almost always lower in organic foods (FAO 2000;

---

* Estimates from World Wildlife Fund—Canada. http://www.wwfcanada.org/satellite/prip/factsheets/ PRIP_PesticidesManagement.pdf.

Bourn and Prescott 2002; Lu et al. 2006; Magkos et al. 2006; Tasiopoulou et al. 2007; Lairon 2010). Globally, fruit, vegetables, and cereals have been most commonly examined for residues in organic versus conventional food studies (Hansen et al. 2002). One study found that the organic foods examined had residues on 23% of the samples, while the conventional foods had residues on 75% of the samples. In both types of food, most detected residues lay below established safety limits (Baker et al. 2002). Using data over a 14-year period from the USDA pesticide residue testing program, Benbrook (2008) reported that nonorganic fruits and vegetables were 3.2 and 3.5 times more likely to contain a pesticide residue than organic fruits and vegetables. Baby foods may sometimes be an exception, where residues are often similar in both systems, in part because of higher-quality control measures applied by conventional baby food manufacturers. For example, Michigan-based Gerber has invested heavily in IPM and pesticide residue control among its contract growers. A Michigan study found no residue differences, consequently, between the production systems (Moore et al. 2000). Lairon (2010), in reviews conducted for the French Agency for Food Safety, found that 94%–100% of organic samples did not contain pesticide residues. At least two studies have found significantly lower levels of organophosphate pesticide metabolites in the urine of children fed a predominantly organic diet (Curl et al. 2003; Lu et al. 2006). Not all pesticides have been assessed, however, and studies examining fungicides and herbicides and pesticide cocktails are particularly lacking (Oates and Cohen 2011).

Organic certification is a process, not a product guarantee, and organic food is not residue-free since organic farmers are unable to control atmospheric deposition of airborne pollutants, and the 3-year conversion period may not always be sufficient to eliminate soil contaminants from conventional production (Woese et al. 1997; Magkos et al. 2006). However, in organic milk, there is some evidence of lower levels of synthetic chemicals associated with airborne transport, including chlorinated hydrocarbons, PCBs, DDT, and lindane (Stockdale et al. 2001). Other possible reasons for the presence of residues in organic foods include processing contamination, inadvertent or intentional mixing of organic and nonorganic ingredients in the distribution chain, and fraud (Lo and Matthews 2002).

Ultimately, the issue is whether these differences are biologically meaningful (Magkos et al. 2006), a complex subject beyond the scope of our review as it addresses the levels at which various contaminants have significant impacts on the human body. Certainly, many studies now suggest that dietary exposure can be significant (Oates and Cohen 2011) and some substances used in agriculture are active at lower doses than previously thought (cf. Colborn et al. 1996). Fortunately, in general, dietary pesticide risks are falling as pesticide registration systems focus more on low-risk products. Organic appears to represent the lowest-risk systems approach, as some studies have shown that dietary residue levels shift risks out of uncertainty ranges to negligible ones (e.g., Curl et al. 2003); however, no studies currently document that organic diets reduce pesticide-related diseases and conditions (Oates and Cohen 2011). Also important, organic certification provides for many of the traceability systems that are now emerging as food safety requirements for conventional products (Giovannucci 2003). Because of this capacity in Europe, inadvertent contamination of organic food has been found more quickly than in conventional foods (Helfter 2003).

14.2.1.2   Food Safety: Animal Production

In contrast to nonorganic farmers, organic farmers in North America are generally not permitted to use uncomposted or unaerated manure, except under very specific circumstances. The composting and aging process reduces pathogen levels and leaching of nutrients. During experiments, most, but not all, bacterial pathogens are killed following exposure to temperatures of 55°C–60°C for a few hours or less. Such temperatures are typically achieved and maintained in a compost pile for days to weeks (Patriquin 2000).

Growth hormones are not permitted, and animals must be a fed a diet for which their digestive system is adapted. Consequently, the digestive conditions associated with elevated *E. coli* 0157:H7 levels do not normally occur on organic farms (Couzin 1998; Diez-Gonzalez et al. 1998; FAO 2000).

Standards do not permit the use of antibiotics, unless the life of the animal is in jeopardy. Most standards then require that the animal be removed from the organic stream, although some permit its return following an extended withdrawal period. As a result, levels of antibiotic-resistant bacteria are likely to be lower in organic than conventional systems (Aarestrup 1995; Smith-Spangler et al. 2012). Sapkota et al. (2011) studied ten conventional and ten newly organic large-scale poultry houses in the United States, testing for *enterococci* bacteria in poultry litter, feed, and water. All farms tested positively, but the organic operations had significantly lower levels of antibiotic-resistant strains (among 17 common antimicrobials). Especially important were lower levels of multiresistant strains. Few differences have been reported in bacterial contamination of organic versus conventional animal foods (Magkos et al. 2006; Smith-Spangler et al. 2012).

Although BSE is not completely avoidable on organic farms, due to either random events or import of conventionally reared young stock into organic farms, if BSE is a product of feeding ruminants infected animal protein, then organic farms have provided some risk reduction possibilities since standards do not permit feeding of animal protein to ruminants.

Mycotoxin levels in cereals and animal feeds are similar in conventional and organic agriculture (Lairon 2010), and some European studies have found lower levels in organic than conventional milk (FAO 2000).

Infectious and zoonotic diseases are poorly studied in organic systems, although there is some evidence of no differences in hygienic quality between organic and conventional milk (Stockdale et al. 2001).

14.2.1.3   Food Safety: Vegetables

Vegetables are more studied than field crops at this point, so we include an overview here to provide a sampling of what might be pertinent to future field crop investigations.

Tissue nitrate levels are generally lower in organic foods of plants with significant nitrate accumulation capacity, such as leafy greens (Magkos et al. 2006; Lairon 2010), likely reflecting lower levels of available nitrogen in organic versus conventional fertilization strategies.

One study of organic vegetables in Northern Ireland found no confirmed pathogenic bacteria although the authors did find that *Aeromonas* bacteria, a possible pathogen, were present, at levels comparable to those found in conventional produce. They cautioned that ready-to-eat vegetables could contain more *Aeromonas* since conventional ready-to-eat products can be washed with bacteriacides like chlorine that are not permitted in organic production (McMahon and Wilson 2001). Although some believe that conventional farmers' dependence on synthetic fertilizers makes such operations less risky, there is limited empirical evidence to support this (Magkos et al. 2006), and the assertion also fails to account for the number of conventional operations that spread raw manure. Proximity of vegetable operations, whether conventional or organic, to cattle production can heighten risks, though there is some evidence that organic leafy greens are better able to withstand *E. coli* 0157:H7 and *Salmonella* infection and proliferation (Benbrook 2007). New protocols following the 2006 outbreak in California from leafy greens highlight the importance of applying well-composted manure (Benbrook 2007).

There are limited data on whether organic systems have an impact on plant uptake of heavy metals and the results are inconclusive (Magkos et al. 2006). One study examining cadmium, lead, chromium, and zinc suggests that factors other than production system may be more important, for example, inherent soil conditions and climatic conditions (Jorhem and Slanina 2000).

14.2.1.4 Improving Food Safety in Organic Systems

There are several areas of weakness that need to be addressed if organic systems are to provide significant public benefits:

- Although pesticides used in organic systems generally have lower environmental impact scores, some pesticides still permitted in some organic systems are hazards to either workers or some organisms (Edwards-Jones and Howells 2001; Magkos et al. 2006). Such products should progressively be removed from the permitted substances lists of the organic standards.
- Some disease organisms are reasonably well controlled in conventional systems, directly or indirectly with pesticides, so they or their associated disease hazards (e.g., fumonisins, aflatoxins) are generally under control. If organic systems are not properly managed to minimize their presence, there may be additional risks, although the theoretical and empirical support for such claims is controversial (Magkos et al. 2006). It is often argued by critics of organic farming that these hazards are inherently more elevated in organic systems, but most employ alternative control strategies to minimize risks. Unfortunately, there are limited data on the effectiveness of some of these alternative control measures, but increasingly, they are the subject of innovative ecological research (e.g., competitive fungi for *Fusarium*). This is a priority research area to improve organic food safety.
- Organic farmers and processors may need to exercise a higher degree of quality control than their conventional counterparts. Because it is an emerging sector, without the appropriate level of support, resources, information dissemination, and skills development, there is an argument that some in the organic sector do not have the skills or resources to practice

the necessarily level of quality control. However, in Denmark where organic is well supported relative to Canada, a study suggests that when comparing farmers practicing integrated production and organic, there are differences in the degree of compliance with the complex rules of both systems. In Denmark, both systems are inspected in similar ways by the same public agency. In the case of organic farming, between 0.0% and 0.2% of certified farms were deprived of certification every year between 1995 and 1999, while the percentage lay between 5.8% and 24.9% in the case of certified integrated farms, from 1996 to 1999 (Michelsen 2001). The author concluded that the higher entry and exit costs of organic farming inspired greater commitment to the rules (Michelsen 2003). The result suggests that greater quality control and management can be a product of both sector and state support. Such state support does not currently exist in Canada, though recent financing of the Organic Science Cluster suggests a possibility for the future.

- Because organic farmers use composted manure on a wider range of crops than conventional growers, there is a significant premium on good composting techniques and proper timing of application. There are insufficient data available to assess whether compost is always well managed on organic farms. A training and surveillance system would be a valuable addition to organic farming supports.

- Because animals spend more time outside in organic systems, they are sometimes subject, given the general state of environmental contamination, to additional exposure to pollutants, whether atmospheric or soil based. One Swedish comparative study found that cadmium levels were lower in organic than conventional feed, but higher in kidneys and manure, suggesting in part that rooting around outdoors might contribute to higher environmental exposure (Linden et al. 2001). For both systems, cadmium levels were lower than reported in other studies of conventional pig production. Outdoor conditions may also lead to higher levels of parasitic infections (Kouba 2003; Magkos et al. 2006). Clearly, keeping animals outdoors is a priority in organic systems, but additional inquiry on the implications of exterior pollution would round out the sector's understanding of the potential trade-offs, with potentially useful best practice recommendations.

14.2.2 Animal Welfare

Concerns about welfare of farm animals have been rising as confinement has increased and breeding programs and diets modified to increase productivity. The relationship between field cropping and animal diets is sufficiently significant that we include this issue here.

Agriculture is frequently pilloried by animal welfare organizations and some consumers have become more wary in their purchasing decisions. Animal welfare now appears in consumer surveys as a decision point for purchasing (AAFC 2009, 2011). Some governments in Europe have enacted animal welfare regulations forcing changes upon farmers. Canadian conventional farm organizations have been updating voluntary codes of practice on welfare issues in the hopes of getting ahead of consumer backlash and regulatory interventions (Bradley and MacRae 2010). In this environment, animal welfare organizations have worked with farm organizations, particularly organic farming groups, to further animal welfare objectives in organic production. Their interest is related to organic farming standards that require less confinement, usually lower stocking densities, different feeding regimes, and preventive approaches to health.

Evaluations of animal welfare on organic farms conclude that organic farms generally receive higher marks than conventional systems (Stockdale et al. 2001). Here are some examples why:

- By standard, organic farmers avoid many of the more egregious features of confined industrial livestock rearing, including broilers that cannot walk under their own power, animals that never leave their confined space, surgeries designed to reduce competitive behavior, and dramatically shortened life spans due to hyperproductivity. In Europe, organic standards exceed most of the standards for animal welfare set down by the EEC Council directives on the protection of animal welfare (Sundrum 2001).
- In European studies, organic dairy cows tend to have a longer average productive life than conventional dairy cows, and this may be associated with the intensity of management (Stolze et al. 2000). Similar conclusions have been drawn by Rozzi et al. (2007) in Ontario and Benbrook et al. (2010) in the United States. One contributing factor may be lower rates of metabolic disorders in organic dairy cows (Hovi et al. 2003).
- Later weaning in organically reared pigs likely has positive impacts on nutritional health and welfare, including lower incidence of weaning diarrhea (Hovi et al. 2003).

But this entire field remains understudied. A 2001 literature review, one of the few comprehensive surveys, found only 22 peer-reviewed papers on the subject, all dealing with health rather than broader welfare issues. The vast majority were on either dairy herd health or parasite management. None of the papers were from North America (Lund and Algers 2003). Consequently, the evidence of organic benefit remains tentative. In a number of areas, organic consistently demonstrates better results than conventional, but in other areas, results of many studies are mixed. Because there remains some significant debate among researchers regarding the best measures of animal welfare (Alroe et al. 2001), it is difficult to make definitive conclusions at this point.

Although research indicates that organic production standards are positive for animal welfare, and animals in organic production are generally found to be as healthy, or healthier, than those in conventional production (Lund and Algers 2003), there remains the potential for nutritional problems for ruminants solely grazing on forage plants, particularly a lack of selenium, sodium, cobalt, and iodine. Whether alternative remedies used by organic farmers are always adequate substitutes for chemical interventions is also an outstanding question (Jones 2003). Parasite control

TABLE 14.1

Five Freedoms of Animal Welfare

1. Freedom from hunger and thirst—by access to freshwater and a suitable diet
2. Freedom from discomfort—by providing an appropriate environment including shelter and a comfortable resting area
3. Freedom from pain, injury, or disease—by prevention or rapid diagnosis and treatment
4. Freedom to express normal behavior—by providing sufficient space, proper housing, and company of the animal's own species
5. Freedom from fear and distress—by ensuring living conditions and handling that avoid suffering

Source: The UK Farm Animal Welfare Council.

remains a challenge in many organic operations (Hovi et al. 2003). Feather pecking and cannibalism have been reported to be higher in organic layer operations compared to conventional, and it is believed that better stockmanship, nest design, and selection of breeds suitable for extensive operations (without debeaking) will reduce these problems (Hovi et al. 2003).

The Canadian organic sector is in the process of updating standards to bring more welfare dimensions into effect (see Animal Welfare Task Force 2011). Although no formal comparisons have been conducted of Canadian organic versus conventional approaches to animal welfare, even a cursory comparison of the work of the Animal Welfare Task Force and the National Farm Animal Care Council (Bradley and MacRae 2010) reveals a much greater commitment in organic standards to the five freedoms (see Table 14.1) as proposed by many animal welfare organizations than found in conventional farming (http://www.oacc.info/Docs/AnimalWelfare/BCSPCA/Five_Freedoms_Urton2008.pdf). The conceptual direction for integrating animal welfare and organic systems has been set out in "Animal Health and Welfare in Organic Agriculture" (Vaarst et al. 2004).

14.2.3 Nutrition

For at least 90 years, scientists have known that soil conditions affect some nutritional parameters of foods.* This knowledge led to refinements in such things as fertilization strategies to improve wheat milling quality or to lengthen the storage period of many foods. Because organic farmers employ fundamentally different soil management practices compared to conventional farmers, organic foods may have a more optimal nutritional profile than conventional foods, in particular those constituents that may have more subtle impact on health than deficiencies of protein and carbohydrates. Although some argue such questions are irrelevant, given the amount of food available to Canadians, data from historic nutrient files in Canada, the United States, and the United Kingdom suggest that levels of some micronutrients have fallen significantly in the past 50 years (Bergner 1997:46–75; Mayer 1997;

---

* See Commonwealth Agricultural Bureau annotated bibliographies on soil conditions and food quality.

Haliwell 2007).* Given that the Canadian population only scores between 54% and 65%, depending on age and sex, on Health Canada's Healthy Eating Index (Canadian Community Health Survey—Nutrition 2004), and some 10% of Canadians have previously reported being deprived sporadically of sufficient food due largely to poverty (Che and Chen 2001), it is possible that such nutrient losses could have an impact on the health of many.

Results of studies looking at plant foods, mostly vegetables, are highly variable, with some showing no difference, some showing organic to be superior in certain parameters, and some showing conventional to be superior. At this point, the official position of most health authorities is that evidence is insufficient to support a claim that organic is nutritionally superior. Review studies, where authors have examined a wide range of results, have also been divided in their assessments, some concluding that differences are minimal or nonexistent (Bourn and Prescott 2002; Dangour et al. 2010) and others determining that organic is superior, quite consistently, in a number of constituents (Worthington 2001; Benbrook et al. 2008). Even where there are nutritional differences favoring organic foods, authors have concluded that they are of minimal significance for human health (Williamson 2007; Dangour et al. 2010).

Most studies examine particular nutrients (e.g., vitamins, minerals) in organic versus conventional foods, and these have typically produced mixed results. The most consistent (but not definitive) results pertain to lower quantity but higher-quality grain protein and lower nitrate and higher vitamin C levels in many organic foods (Woese et al. 1997; Worthington 2001; Benbrook et al. 2008; Lester and Saftner 2011). The reason for the conflicting results in other components appears to be that nutrient levels are affected by a whole host of factors, including the type of soil, fertilization, tillage, the variety of plant, the particular microclimate, planting and harvesting dates, harvesting and handling techniques, and postharvest handling. Managing all these variables to identify the organic versus conventional comparison is very difficult, and many studies have not managed it well, making their conclusions suspect. Additionally, it may be the wrong comparison. The specific nutrient content may be less important for nutritional health than the ratios of nutrients. In other words, the body's ability to utilize nutrients may, within a specified range, be less related to absolute levels of a particular nutrient and more connected to the relationships between numerous nutrients. Consequently, studies that focus on comparing absolute levels of a small range of nutrients may not be particularly helpful.

A bit more consistency is found in studies where test animals are fed an organic versus conventional diet. In these studies, the researchers look at wider-angle health indicators, for example, the health and fertility status of the animal. In these studies, animals on an organic diet tend to perform better in fertility and infant morbidity parameters than those on a conventional diet (Plochberger 1989; Woese et al. 1997). However, the reasons for this result are not clear, although a recent study, using fruit flies as the test organism, fed extracts from organic versus conventional foods purchased at retail. It found higher longevity and fertility for most organic diets, and

---

* In Canada, CTV News and the *Globe* and *Mail* national newspaper reported extensively on the Canadian results in June 2002. There is some debate about whether these differences are a result in changes in measurement techniques over time.

the authors concluded that this result was due to higher levels of nutrients and more balanced nutrient profiles (Chhabra et al. 2013). Other studies, however, have not been able to determine whether their results were related to nutritional parameters or possibly lower levels of pesticide residues found on organic foods (see the discussion earlier). The limited number of studies on humans consuming organic versus conventional diets has produced mixed results, some showing fertility advantages for those on primarily an organic diet and others showing no significant differences (Kouba 2003; Smith-Spangler et al. 2012).

More recently, plant studies have focused on food components other than minerals and vitamins. In a number of comparative studies, higher levels of some nutritionally significant phytonutrients (including antioxidants) have been reported in organic foods (Baxter et al. 2001; Brandt and Mølgaard 2001; Asami et al. 2003; Grinder-Pedersen et al. 2003; Benbrook et al. 2008; Lairon 2010; Hallmann 2012; Hallmann and Rembialkowska 2012; Koh et al. 2013), and this may prove to be an interesting research area in the future as there is some indication that consumption of such antioxidants is beneficial for human health (Chhabra et al. 2013). An emerging approach is to examine the effects of consumption of organic foods on immune system function. One recent study concluded that organic carrots had a greater stimulating effect on immune function in rats, than conventional carrots (Roselli et al. 2012).

The nutritional quality of animal products is even less well studied than plants. Comparisons are complicated by additional factors—different growth rates, different feed, different breeds, limited micronutrient supplementation, and free range husbandry. Extensive versus intensive rearing practices (rather than organic vs. conventional) may be the biggest factor determining differences in quality (Stockdale et al. 2001). Lairon (2010) concluded, however, that organic animal products contain more polyunsaturated fatty acids, than conventional ones. A meta-analysis of 29 studies conducted by Palupi et al. (2012) concluded that organic dairy products contain significantly higher protein, total omega-3 fatty acid, and other essential fatty acids than conventional dairy products. Also observed was a significantly higher omega-3 to omega-6 ratio in organic dairy. Benbrook et al. (2010) arrived at similar conclusions from their literature review and modeling. Kuczyńska et al. (2012), however, reported more variable results, with some quality parameters favoring organic and many favoring conventional. Length of pasturing was a confounding factor that made the organic/conventional comparison more complicated to assess.

In summary, the evidence is encouraging regarding nutritional benefits of organic food, but would not yet qualify as definitive. Part of the difficulty is the different ways studies define significant health effects. Those employing a disease reduction framework are more likely to conclude there are no effects or no significance for human health than those with a broader and more dynamic interpretation of what creates health (Huber et al. 2012).

14.3 ECONOMIC ISSUES

Many farmers are in financial difficulty. Prices for many commodities are low, and the costs of inputs to maintain yields increase more rapidly than average price levels. In discussions about solving these problems, most of the attention has focused on the design of farm financial safety-net programs, the squeeze on prices associated with US and European Union (EU) subsidies, global market pressures, and the need for even greater productivity. In contrast, we present in this section comparative assessments of the economic merits of organic farming and the potential long-term benefits of cost internalization.

14.3.1 FARM-LEVEL CONSIDERATIONS

Relatively little attention has been devoted, by the dominant food sector, to input cost reductions or tapping into environmental markets with price premiums. Organic farming presents opportunities on both counts, though there is a long-standing debate in the economics literature about the best indicators to use for measuring the viability of organic farming systems (Wagstaff 1987; Nemes 2009). This debate lies beyond the scope of this chapter, but traditional measures of unit costs per unit of output are contested as suitable indicators of organic farming performance.

Despite this limitation, organic agriculture systems are usually more profitable than conventional farming systems and this has been recognized for some time (Lampkin and Padel 1994). They typically have higher gross revenues on a comparative basis, in part due to premium prices and different marketing strategies. That reality, combined with lower input costs, typically produces better net revenue. This situation exists despite generally lower yields. In Europe, most farm comparisons show profits for organic farms between ±20% of conventional (Nieberg and Oppermann 2003). In North America, profitability is typically even more favorable, particularly for field crops and some animal and mixed farming operations (see Table 14.2; also Delate et al. 2003; Delbridge et al. 2011). Profits and evenness of income were higher for organic treatments than for high-input and reduced-input conventional treatments on the Canadian prairies (Zentner et al. 2011). In the United States, the estimated net income of $20,249 per farm per year averaged over 14,540 organic farms was more profitable than the average of all US farms (Bowman 2010). We explore these phenomena in more detail in the succeeding text.

Regarding yields, the following is our assessment from worldwide higher-order studies attempting to integrate information from multiple sources (MacRae et al. 1990; Stanhill 1990; Lampkin and Padel 1994; Pretty 1995; Stockdale et al. 2001; Liebhardt 2003; Lotter 2003; Badgley et al. 2007; de Ponti et al. 2012; Seufert et al. 2012).*

Globally, plant yields in organic systems are, on average, 10% below conventional systems. Global averages do vary between extensive (large area) and intensive systems because the conventional comparator is different. In Europe, where conventional production is very intensive, organic system yields look comparatively weaker than in extensive systems like those found in North America. In these regions, organic crop yields generally range from 20% less to slightly more than that of conventional systems. In Europe, they can be 20%–40% less, respectively, except in forages where the range is more likely 0%–30% less (Stockdale et al. 2001; de Ponti et al. 2012).

---

* Note that these studies employ different methods that make comparisons difficult, and some have inappropriately characterized the organic system (e.g., Badgely 2007) or the conventional comparator (e.g., Seufert et al. 2012) which conflates the results.

TABLE 14.2

Comparing Yields in Organic versus Conventional Production

Product	Country/Region	Org/Conv	Reference
Tomato	California, through the 1990s	1.0	Liebhardt (2003)
Many crops	Developing countries, numerous projects	1.8–4.0	Pretty and Hines (2001)
Cereals	Europe	0.6–0.7	Nieberg and Oppermann (2003)
	United States and Canadian Prairies	0.7–1.0	Entz et al. (2001) Bromm (2002)
Corn	Major US corn regions 1990s; experiment station studies 34 studies	0.94 0.89 (0.6–1.4)	Liebhardt (2003) de Ponti et al. (2012)
Soybean	Five US states, past 10–15 years; experiment station studies Summary 16 US studies	0.94 0.92	Liebhardt (2003) de Ponti et al. (2012)
Flax	US and Canadian Prairies	0.8–0.9	Entz et al. (2001) Bromm (2002)
Wheat	Two research institutions, past 10–15 years 66 global studies, all wheats	0.97 0.73 (0.4–1.3)	Liebhardt (2003) de Ponti et al. (2012)
All plant foods	Developed world	0.91	Badgley et al. (2007)
Grass–clover	European studies	0.89 (0.77–1.1)	de Ponti et al. (2012)
Dairy	Europe	0.8–1.0/cow, 0.6–.8/farm[a]	Nieberg and Oppermann (2003)

[a] Limited data from Canada suggest yields are equivalent (Ogini et al. 1999).

Examples of yields of different crops in different regions are listed in Table 14.2. In general, leguminous and perennial field crops in extensive rain-fed systems perform better in organic farming than nonleguminous crops and annuals (Seufert et al. 2012). These favorable results presented from extensive farming systems have come about almost entirely without the support of institutions normally involved in agricultural development, suggesting the possibility of more promising comparisons with additional research, development, and extension.

Crop yields in North American prairie organic systems have not produced as favorable results as studies conducted in many other regions (Table 14.2). Weed management appears to be a significant challenge, and several factors may account for the results obtained to date: (1) still-emerging knowledge about how to manage weeds ecologically in a prairie environment; (2) limitations on management options imposed by the requirements of research protocols (i.e., the best organic weed management practices have not always been employed in trials); and (3) crop rotation design, with the associated challenge of incorporating perennial legumes in moisture-limited environments. Low soil phosphorus levels may also be an extensive problem contributing to reduced yields (Entz et al. 2001; Buhler 2005; Martin et al. 2007).

Generally, however, yields in organic systems continue to rise as understanding of them grows. For example, 2006 results from the Rodale Research Center long-term trials in Pennsylvania, United States, showed 40% greater corn yield in an innovative one-pass roll/plant organic no-till system compared to that in a conventional chisel-till system (Hepperley et al. 2007).

For animal products, yields are, on average, 20% below conventional, with the same caveats regarding comparisons. Badgley et al. (2007) and Badgley and Perfecto (2007) concluded that yields in all organic animal systems were 97% of conventional in the developed world, but this conclusion is drawn from a limited number of studies, and further examination of methodological challenges in animal studies would be required to improve confidence in this conclusion (see Seufert et al. 2012 supplemental material for a fuller critique of this study). For animal systems, comparisons are even more difficult than plant systems. For ruminants, yields per animal can be roughly equivalent. However, output per hectare is usually lower in organic systems because stocking rates are generally lower. The exception is comparison of dairy and beef systems where concentrate feeding is dominant from early in the animal's life, a practice considered only marginally acceptable in some organic production standards. There are limited data on chicken and swine systems, and comparisons are even more difficult because of the different ways these animals are integrated into organic versus conventional farm operations. Yields per animal in organic systems are generally significantly lower than conventional systems (Stockdale et al. 2001). Improving the profitability of organic livestock systems, particularly swine and chickens, is a significant priority.

Gross margins of organic enterprises are at least as good as, if not better than, those under conventional regimes, but input costs are typically lower. In more extensive systems like those practiced in North America, input cost reductions are often sufficient to maintain margins. In more intensive production systems such as are found in Europe, premiums are often required to offset yield declines (Stockdale et al. 2001). Four factors usually account for the positive end of these income results.

First, operating costs for organic farms may be up to one-third lower, particularly for energy, chemicals, and drugs. Variable input costs are 50%–60% lower for cereals and grain legumes, 10%–20% lower for potatoes and horticultural crops, and 20%–25% lower for dairy cows (Nieberg and Oppermann 2003). Second, where premium prices are available in organic markets, the likelihood of a superior net income situation is even greater. Third, many organic farmers achieve higher net income by making more direct links with consumers, which allows them to capture a greater percentage of the consumer dollar. Fourth, organic farms may be more resilient in the face of poor weather. For example, during 5 years of droughty conditions in the Rodale Research Center 22-year cropping system trials, organic corn yields surpassed conventional ones by an average of 28%–34% depending on the organic system. In years of "regular" conditions, organic yields have been comparable to that in conventional production (Pimentel et al. 2005).

In many European countries, government payments for environmental stewardship compensate for yield declines relative to conventional production. Organic farmer support payments have accounted for between 16% and 46% of farm profits depending on product and country (Jones 2003). Of course in Europe, almost all farmers rely on support payments of various kinds. In aggregate, government investments in organic farming produce savings in other farm subsidies because organic farmers, once supported by organic schemes, have not drawn on other support programs to the same degree as they might have when they were conventional producers (Lampkin et al. 1999).

Labor productivity, measured against yields produced, has generally been lower in organic farms than in conventional systems, and traditional economic analysis views this as problematic. Labor requirements have generally been higher in Europe and in more intensive production systems (Jansen 2000; Green and Maynard 2006). More extensive systems, however, often have not required additional labor (Wynen 2003). Average increases in jobs/ha have been in the 10%–64% range, with wide variability among cropping systems (Soil Association 2006). Field crop and mixed operations reported slightly higher labor requirements for organic production, and for organic horticulture, substantially higher. In dairy production, however, requirements have been comparable. Further processing and direct marketing may also be significant contributors to increased labor requirements. Labor demands on organic farms, however, have generally been falling relative to the early 1990s (Stockdale et al. 2001).

These increased labor requirements are sometimes absorbed within the farm family but may result in additional hirings (Lobley et al. 2005; Green and Maynard 2006). Interestingly, returns to total labor may be higher on organic farms, and wages may also be higher. There is also evidence that the quality of labor is more positive in organic farming because the work is more diversified and less repetitive (Jansen 2000). On the negative side, there is some evidence from the United Kingdom that a significant percentage of such additional labor demands are part-time and insecure (Lobley et al. 2005). There may also be issues of labor availability, given current difficulties finding farm workers in many regions, and commodities. Few studies discuss how these demands might affect labor supply and rural employment. Lobley et al. (2005) provide a summary of pertinent issues in the European context.

Research on the system-wide implications of these enterprise-level results is relatively immature. Earlier work was methodologically controversial (Youngberg and Buttel 1984; Madden and Dobbs 1990), but those studies concluded that significant benefits would result from expansion of organic farming, including improved food quality, enhanced environmental and human health, higher net farm income, and lower government subsidy payments and crop storage costs (Oelhaf 1978; USDA 1980; Langley et al. 1983; Vogtmann 1984; Cacek and Langner 1986; World Commission on Environment and Development 1987). The effect on consumer food prices was projected to be minimal (1% increase in total food expenditures [Oelhaf 1983]) or substantial (up to 99% increases in some commodities [Langley et al. 1983]). Farm employment and farmer numbers could increase (Cornucopia Project 1984; Enniss 1985) and small- to medium-size farms could become more viable (CAST 1980). There was concern about the availability of labor, however, as more conversions took place (USDA 1980; Langley et al. 1983). Premium prices would decline in the long term as more organic food entered the marketplace (Duffy 1987), yet this might not reduce net profits if input costs fell at the same time. Reliance on external inputs, and therefore operating costs, would decline as our understanding of agroecosystems increased.

Oelhaf (1978), for example, estimated the cost of the conversion period as 5%–20% of food prices, a cost that would decline with more information and support from agricultural institutions. Even with a price-depressing increase in the supply of organically produced food, consumer demand would grow for these products so as to moderate and even offset the supply effect. Food export potential would likely decline over time (Langley et al. 1983), which would cause economic dislocations because so much of the North American agricultural economy is geared to export.

More recent studies have provided a bit more clarity on these questions, but certainly, our understanding of these issues remains murky. A few studies have examined labor impacts of significant conversion in a region, finding increased employment in the 10%–100% range depending on the region, commodities involved, and scope of the food chain examined (Jansen 2000).

Halberg et al. (2006) modeled the impacts of organic and low-input system conversion on food availability and market prices at a global level for the year 2020. They compared the conversion to assumptions about conventional agricultural development from 1997 to that year. Organic yields were conservatively estimated at 50%–85% of conventional in the industrial world and low-input systems at 90%–150% of conventional in sub-Saharan Africa, lower than results in many studies. Conversion in the industrial world did not affect global food prices if productivity gains for organic were assumed, a feasible situation only if research investments increased leading to ecofunctional intensification, and better nutrient recycling and plant breeding. In sub-Saharan Africa, the model projected conventional yield increases of 2% annually for cereals and increased acres in production, but due to projected population increases, such growth was insufficient to reduce the need for imports and improve food security. The situation was more promising for conversion to organic production, with significant reductions in imports, especially in certain coarse grains and improvements in food security associated with domestic staple production.

Badgley et al. (2007) compared the present food production at a global level with a 100% organic scenario and concluded that the organic scenario would increase calories produced/capita by 75%. However, organic yields, especially for animal production, were overly optimistic for both industrial and global south situations (MacRae et al. 2007; Seufert et al. 2012). Interestingly, they also concluded that nitrogen supply would not be a limiting factor, even at these elevated yield levels, if the full potential of using legumes and livestock manure was realized.

Zanoli et al. (2000) took a different approach, using fuzzy scenario methods to set out future scenarios based on policy conditions in Europe in 2000. The scenarios were constructed from existing data sets and expert opinion. One scenario was labeled "organic paradise," an optimal set of political, policy, programmatic, economic, and consumer conditions to encourage rapid expansion of the organic sector in Europe. Relative to business as usual and gloomy liberalization scenarios, organic paradise set out the kinds of conditions that could resolve a significant conundrum, that being having reasonable returns for producers as consumer prices for organic fell.

In summary, farm-level results suggest that most farmers will do better in organic than conventional production, postconversion. But the implications of widespread adoption are less well understood.

14.3.2 Issues of Cost Internalization

Although organic farm enterprise performance is often superior, assessing the food system implications is complicated by the way conventional food prices do not reflect the real costs of producing, processing, and distributing food. This situation exists, in part, because the agrifood sector receives considerable direct and indirect subsidies. These include direct government payments, government subsidies to nonrenewable fuel exploration and development, and extensive environmental subsidies. Many costs of agricultural production and distribution are externalized to the environment. Most of these externalized costs remain unpaid (Tegtmeier and Duffy 2004). This situation distorts market signals, producing a dysfunctional food marketplace, where both producers and consumers do not behave as they would, were these externalized costs internalized. Although both conventional and organic systems generate externalized costs, studies to date indicate that those from conventional systems are higher (Pretty et al. 2005; MacRae et al. 2009; Lynch et al. 2012).

The theoretical arguments have been articulated by Bateman (1994). He identified how widespread adoption of organic farming and food distribution can better enlist consumers as allies in improving the environment and financial situation of growers.

Empirical work, however, is limited. There are a number of studies naming the billions of dollars in externalized costs of conventional food production and distribution (Pimentel et al. 1992, 1995; Pretty et al. 2000; Tegtmeier and Duffy 2004). Only a few studies have attempted to relate these costs to consumer food prices. Work by Pretty et al. (2005) in the United Kingdom, using a weekly market basket approach, conservatively suggested that the full environmental cost of an average British conventional shopping basket should be at least 12% higher. This estimate did not account for many difficult-to-quantify environmental costs or diet-related externalities.

If such accounting of real costs were applied (i.e., if costs were internalized), it would raise conventional costs and bring them more in line with current organic prices. However, current organic food prices can be problematic for some consumers and remain one of the limiting factors to more widespread (across product lines and consumer demographics) organic purchasing. A number of factors, in addition to market distortions caused by cost externalities, contribute to the higher prices of organic foods, and many of these are a function of the conventional food distribution chain, not organic production and distribution per se.

First, although price premiums for organic farmers do exist in some commodities and some regions, often, these are not the main contributors to higher consumer prices. Farmers have been price takers for a long time and have had little control over both input and output prices (Martinson and Campbell 1980). In Canada, according to Statistics Canada estimates, less than 20% of the average consumer dollar goes to the farmer (AAFC 2003), so much of the price premium goes to other players. Globally, distributors, shippers, and retailers for some time have been retaining two-thirds of the economic value of food, while the farm sector (9%) and input sector (24%) have shared the other third (Vorley et al. 1995). This explains, in part, why many organic farmers direct market to capture some of the value from the distribution system.

Second, corporate concentration, product research, development and promotion, and long distribution lines have traditionally contributed to higher food distribution costs and prices (Winson 1992; MacRae et al. 1993). This is particularly true for immature markets, like organic, where volumes are frequently too low to capture savings or to sufficiently spread costs.

Third, even if the transition to organic farming results in higher prices at the production end (a disputable conclusion), it need not result in higher consumer prices if appropriate changes are made to what is distributed and how. For example, German studies have shown that an organic diet need not be more expensive, and may even be cheaper, than a conventional one if consumers eat more at home, purchase from non-traditional sources, and are consuming lower levels of some animal products. Note that the motivation for reducing animal products in the diet is not related specifically to price, but more associated with health concerns. Even modest overall reductions in animal products produce this effect. Typically, the organic diet is higher in some dairy products and lower in meat and eggs (Brombacher and Hamm 1990; Meier-Ploeger 1992). A survey of packaged food in New York state stores, based on a 4-day menu plan, found that a mainly organic diet purchased in supermarkets and health food stores was no more expensive than brand-name shopping in a supermarket. Purchasing organic foods at a food co-op was by far the cheapest means of acquiring any of the menus of the study (Berthold-Bond 1995). Furthermore, household budgets could be balanced by reducing food waste, given high household waste factors for many foods (Peters et al. 2002, 2003).

Some of the price dilemmas can be resolved with supports from agricultural institutions. As more dollars are devoted to research, lower cost production, distribution, and processing strategies may emerge. As volumes increase, distribution costs per unit shipped will decrease. As more market players enter the field, increased competition will reduce prices in some cases. There is anecdotal evidence that production, distribution, processing, and retailing costs started to come down in the early 2000s (Jones 2003). It is likely, however, that organic prices will be higher than the artificially deflated prices that exist in some conventional products sold through the dominant distribution channels.

14.4   SOCIAL AND RURAL COMMUNITY ISSUES

Many rural communities are struggling, with depopulation, loss of community services and infrastructure, and declining incomes being the most visible signs of distress. As a response, numerous rural community development approaches have been proposed (Sumner 2005a). Organic farming has emerged out of philosophies and social movements that value rural community development. The conceptual terrain of this relationship has been set out by numerous authors, particularly from Europe (Marsden et al. 1999; Knickel and Renting 2000; Banks and Marsden 2001). Whether such philosophical commitments translate into significantly improved social and community potential, however, is an open question.

Marsden et al. (2002) have proposed that to qualify an activity as rural development, the following three conditions should be met: (1) the activity is a response to the cost–price squeeze, (2) the activity expresses new relationships between agriculture and the wider populations' needs and expectations, and (3) the activity combines rural resources in new ways. Rural revitalization was a pressing Canadian government priority and one of the pillars of the Agricultural Policy Framework (APF), subsequently, Growing Forward I and II. Although much of the attention is nonagricultural, the federal government's efforts to improve rural community viability might be complimented by more widespread adoption of organic farming.

Several earlier US studies suggested that sustainable (including organic) agriculture can contribute significantly to rural vitality (Lasley et al. 1993; Bird et al. 1995). A Nebraska study of an agriculture-dependent community concluded that if more farms were following sustainable (including organic) practices, total family income would be more than double, compared to a scenario where all the farms used conventional practices (Kleinschmidt et al. 1994). Note that Midmore and Lampkin (1994) also found positive income effects for pasture-based systems in the United Kingdom. Studying also the US Midwest, Kleinschmidt et al. (1994) found that the property tax base would be larger. More would be spent on supplies, utilities, feed, veterinary expenses, charity, food, and personal-care products. Less, however, would be spent on agrichemicals, fuel, hired labor, livestock purchased for resale, seed, taxes, and interest. Using data from farm-level studies, Lockeretz (1989) concluded that lower production levels in sustainable systems may reduce economic benefits for farming communities in the short term. However, because a greater percentage of the value of production remains in the community, greater long-term financial benefits might result from sustainable (including organic) systems, particularly as production methods improve.

A study of four communities in the US Midwest found that those with more sustainable (including organic) agriculture practitioners had a greater capacity to mobilize community resources for local development, including more active participation in local government and the creation of new community economic development structures and new businesses. This result was attributed, in part, to the problem-solving and self-reliance skills of sustainable (including organic) agriculture practitioners (Flora 1995). These results were not however confirmed in England, where there were few significant differences between organic and nonorganic farmer participation in local activities. However, UK results suggested that farmers involved in direct marketing were more likely to engage in on-farm processing if organic rather than nonorganic producers (Lobley et al. 2005).

A North Dakota study concluded that some economic sectors would be enhanced (transportation, utilities, business services, and nonmetal mining), but others would decline (construction, professional services, finance, retail trade, and certain kinds of conventional agricultural processing); a better infrastructure for new marketing, processing, and storage needs would ensure that the overall benefits were positive (Northwest Area Foundation 1994). Because many communities lack products and services required by sustainable farmers, significant local economic opportunities were lost (Goreham et al. 1995). This would appear to be less an issue in the United Kingdom where local purchases by organic and nonorganic farmers were comparable (Lobley et al. 2005).

An Ontario study (Sumner 2005b) provided some indication of why a concentration of organic producers in a region may contribute to additional local and regional economic activities: 56% of respondents were involved in direct sales to local businesses; 27% engaged in farm-gate, farm-store, or produce-stand sales; 26% in sales to family, friends, and local farmers; 21% ran a community-supported agriculture (CSA) project; and 14% were selling at the local farmers' market. In total, 88% sold their products locally via a range of marketing channels. A UK study also found that organic farmers were more likely than nonorganic producers to be involved in such local marketing schemes, though, because of scale and volume differences, this did not mean higher absolute levels of local economy participation (Lobley et al. 2005). In Ontario, 93% preferred to purchase inputs and household goods locally (Sumner 2005b). Seventy-six percent of organic farmers volunteered in their communities and 70% had joined a local club or organization and they were also active politically, with 76% engaging with their local government and 61% participating in local roundtables and panels.

Many of these findings were corroborated by a second study that found strong farm–community linkages among organic farmers (MacKinnon 2006). Organic farmers made economic contributions through local purchasing, job creation, and viable farms and also made social contributions through a number of channels. They demonstrated strong involvement in education, networking, and leadership, and they built social capital—the invisible social infrastructure thought to underlie a community's capacity for development. Similar evidence of enhanced networks and social embeddedness has been found in European studies, although the degree of embeddedness appears to be somewhat dependent on how many new entrants to farming exist in the community. Organic farming appears to attract more new entrants than nonorganic farming, a positive development, but new entrants also take some time to link to a new place (Lobley et al. 2005).

MacKinnon (2006) also suggested that (1) organic farmers are less dependent on off-farm income, (2) organic farmers appear to be more involved in direct marketing, and (3) direct marketing appears to be more connected to community involvement than selling through brokers and export. There is also evidence, particularly from intensive systems in Europe, that labor demands are generally higher on organic farms, although they vary considerably from enterprise to enterprise and activity to activity (see Section 14.2.1). There is some evidence that under conditions of good prices, wages are higher in organic systems as well, though this is far from universal (see again Section 14.2.1). In more extensive systems, however, organic farming may be laborsaving (Jansen 2000). One additional reported benefit is increased tourism in regions with significant numbers of organic farms, likely due to the touring public's more positive image of organic versus conventional agriculture (Jones 2003).

Regarding gender relations, in their study for the International Federation of Organic Agriculture Movements (IFOAM), Farnworth and Hutchings (2009) did not find that gender relations in organic agriculture differed substantially from conventional agriculture. Results in Canadian and US studies, however, suggest some differences. Maceachern (2008) found that the decision-making process on organic farms offered some opportunity for difference from conventional farms. Hall and Mogyorody (2007) found that a significant percentage of farmers working together in heterosexual couples (38%) reported that decisions were shared equally. In the United States, Trauger et al. (2010) concluded that "in sustainable agriculture systems, the construction of masculinity and femininity, and their relationships to work roles and decision making, are changing." The authors warn, however, that these changes are not total or transformative because "women still shoulder the burden of domestic work in addition to taking on more of the productive work of the farm."

Hall and Mogyorody (2007:313) and Sumner and Llewelyn (2011) characterized the situation in this way—that organic agriculture can offer an opportunity for addressing rural gender relations but that this opportunity has often been squandered. However, as Farnworth and Hutchings (2009) reported, women organic farmers continue to increase in numbers and make their presence felt in organic organizational processes and economic relations. The conventionalization of organic agriculture, however, may be problematic for these gradual shifts. As McMahon (2005:138) warns, "the organic movement does not recognize that the conventionalization of organic agriculture, like earlier developments in non-organic agriculture, is itself a gendered process."

Organic farmers also change social and knowledge relations because they rely more on their own and community resources, rather than purchasing knowledge from the marketplace. Their breadth and depth of knowledge and their commitment to social learning have been recognized by the FAO in its position paper on organic agriculture and food security:

> Inexperience and lack of adequate extension and training for knowledge-intensive management systems and location-specific science require long-term investments in capacity building. With the objective of creating a critical mass and the necessity to strive in settings with limited opportunities, many organic communities have responded by establishing collective learning mechanisms and have become innovators or ecological entrepreneurs. (El-Hage Scialabba 2007)

Although most existing studies conclude that organic farming has net positive community benefits, it is not a sufficiently widely studied topic to make more definitive conclusions. The shift to organic farming clearly causes disruptions in more traditional agricultural service sectors, so those disruptions must be managed to fully benefit from widespread adoption of organic systems.

14.5   CONCLUSIONS

Although not yet definitive, there is significant evidence from a wide range of studies that adoption of organic food and farming can contribute to reductions in consumer anxieties, improve farm economic performance, and contribute to more robust communities. On many themes considered here, organic performance is often, but not consistently, superior at the product or enterprise level, but the wider implications of this potentially improved performance are less well studied. For example, with organic foods demonstrating improved performance in several food safety and nutritional parameters, might there be a significant positive impact on human health? Does improved enterprise profitability always translate to wider positive regional economic benefits? Similarly, organic farmers often require more farm labor, but does that have a positive impact on rural community vitality? To answer these questions requires significant methodological sophistication but also an increase in scale and social and economic linkages if researchers are going to move from simply modeling possible scenarios to more empirical inquiries. In Canada, organic food and farming remain in the 1%–2% range of farmed acres and retail sales, respectively, and supply chain actors are sufficiently dispersed that scale effects cannot be readily documented. In Europe, however, such scale effects are more readily identifiable, with some regions exhibiting 10%–30% adoption rates by acres or commodity retail sales (Willer et al. 2013). At these levels, the wider implications may be more apparent, but the methodological challenges remain. But the preliminary evidence in favor of organic food and farming should be seen as sufficiently compelling as to encourage a more robust round of inquiry.

REFERENCES

AAFC (Agriculture and Agri-Food Canada). 2003. *An Overview of the Canadian Agriculture and Agri-Food System*. Ottawa, Ontario, Canada: AAFC.

AAFC (Agriculture and Agri-Food Canada). 2009. *An Overview of the Canadian Agriculture and Agri-Food System*. Ottawa, Ontario, Canada: AAFC.

AAFC (Agriculture and Agri-Food Canada). 2011. Socially conscious consumer trends: Animal welfare. Market Analysis Report. International Markets Bureau. Ottawa, Ontario, Canada: AAFC. AAFC No. 10749E, pp. 15.

Aarestrup, F.M. 1995. Studies of glycopeptide resistance among *Enterococcus faecium* isolated from conventional and ecological poultry farms. *Microbial Drug Resistance* 1: 255–257.

Alroe, H.F., M. Vaarst, and E.S. Kristensen. 2001. Does organic farming face distinctive animal welfare issues? A conceptual analysis. *Journal of Agricultural and Environmental Ethics* 14: 275–299.

Animal Welfare Task Force. 2011. Factsheets and guidance documents. http://oacc.info/Extension/ext_welcome.asp (accessed May 11, 2013).

Asami, D., Y.-J. Hong, D.M. Barrett, and A.E. Mitchel. 2003. Comparison of the total phenolic and ascorbic acid content of freeze-dried and air-dried marionberry, strawberry, and corn grown using conventional, organic, and sustainable agricultural practices. *Journal of Agriculture and Food Chemistry* 51: 1237–1241.

Badgley, C., J. Moghtader, E. Quintero, E. Zakem, M.J. Chappell, K. Aviles-Vazquez, A. Samulon, and I. Perfecto. 2007. Organic agriculture and the global food supply. *Renewable Agriculture and Food Systems* 22: 86–108.

Badgley, C. and I. Perfecto. 2007. Can organic agriculture feed the world? *Renewable Agriculture and Food Systems* 22(2): 80–85.

Baker, B., C.M. Benbrook, E. Groth, and K. Lutz Benbrook. 2002. Pesticide residues in conventional, IPM-grown and organic foods: Insights from three data sets. *Food Additives and Contaminants* 19: 427–446.

Banks, J. and T. Marsden. 2001. The nature of rural development: The organic potential. *Journal of Environmental Policy and Planning* 3(2): 103–121.

Bateman, D.I. 1994. Organic farming and society: An economic perspective. In *The Economics of Organic Farming*, eds. N. Lampkin and S. Padel. Oxon, U.K.: CABI Publishing, pp. 45–69.

Baxter, G., A.B. Graham, J.R. Lawrence, D. Wiles, and J.R. Paterson. 2001. Salicylic acid in soups prepared from organically and non-organically grown vegetables. *Journal of Nutrition* 40: 289–292.

Benbrook, C., X. Zhao, J. Yáñez, N. Davies, and P. Andrews. 2008. *State of Science Review: Nutritional Superiority of Organic Foods.* Boulder, CO: The Organic Centre. http://www.organic-center.org

Benbrook, C.M. 2007. *Critical Issue Report E. coli 0157: Preventing Outbreaks.* Boulder, CO: The Organic Center. http://www.organic-center.org

Benbrook, C.M. 2008. *Simplifying the Pesticide Risk Equation: The Organic Option.* Boulder, CO: The Organic Centre. http://www.organic-center.org

Benbrook, C.M. et al. 2010. *Critical Issues Report: A Dairy Farm's Footprint.* Boulder, CO: The Organic Centre. http://www.organic-center.org

Bergner, P. 1997. *The Healing Power of Minerals, Special Nutrients and Trace Elements.* Rocklin, CA: Prima Publishing Co.

Berthold-Bond, A. 1995. *Shattering a Myth. Organic Packaged Food Can Be the Cheapest Way. The Green Guide for Everyday Life*, Vol. 9. New York: Mothers and Others for a Livable Planet, pp. 1–3.

Bird, E., G.A. Bultena, and J.C. Gardner. 1995. *Planting the Future: Developing Agriculture That Sustains Land and Community.* Ames, IA: Iowa State University Press.

Bourn, D. and J. Prescott. 2002. A comparison of the nutritional value, sensory qualities, and food safety of organically and conventionally produced foods. *Critical Reviews in Food Science and Nutrition* 42: 1–34.

Bowman, G. 2010. *USDA Census Shows Profitability of Organic Farming.* Emmaus, PA: Rodale Institute. http://www.rodaleinstitute.org/20100319/nf_USDA-census-shows-profitability-of-organic-farming (accessed on December 18, 2011).

Bradley, A. and R. MacRae. 2010. Legitimacy and Canadian farm animal welfare standards development: The case of the National Farm Animal Care Council. *Journal of Agricultural and Environmental Ethics* 24: 19–47.

Brandt, K. and J.P. Mølgaard. 2001. Organic agriculture: Does it enhance or reduce the nutritional value of plant foods? *Journal of the Science of Food and Agriculture* 81: 924–931.

Brombacher, J. and U. Hamm. 1990. Expenses for nutrition with food from organic agriculture. *Ecology and Farming* 1: 13–15.

Bromm, J. 2002. An economic and productivity comparison of organic and conventional farming in Saskatchewan. Honours thesis. Thunder Bay, Ontario, Canada: Lakehead University.

Buhler, R.S. 2005. Influence of management practices on weed communities in organic cereal production systems in Saskatchewan. Master of Agriculture thesis. Saskatoon, Saskatchewan, Canada: University of Saskatchewan.

Cacek, T. and L. Langner. 1986. The economic implications of organic farming. *American Journal of Alternative Agriculture* 1: 26–29.

Canadian Community Health Survey—Nutrition. 2004. http://www.hc-sc.gc.ca/fn-an/surveill/nutrition/commun/cchs_guide_escc-eng.php (accessed May 11, 2013).

Che, J. and J. Chen. 2001. Food insecurity in Canadian households. *Health Reports* 12(4): August 15, 2001. Catalogue No: 82-003-XIE. Ottawa, Ontario, Canada: Statistics Canada.

Chhabra, R., S. Kolli, and J.H. Bauer. 2013. Organically grown food provides health benefits to *Drosophila melanogaster. PLoS ONE* 8(1): e52988.

Colborn, T., D. Dumanoski, and J.P. Myer. 1996. *Our Stolen Future: Are We Threatening Our Fertility, Intelligence, and Survival? A Scientific Detective Story.* New York: Penguin.

Cornucopia Project. 1984. *Jobs for Americans: The Untapped Potential for Employing More People in America's Largest Industry.* Emmaus, PA: Cornucopia Project.

Council on Agricultural Science and Technology (CAST). 1980. Organic and conventional farming compared, Report #84. Ames, IA: CAST.

Couzin, J. 1998. Cattle diet linked to bacterial growth. *Science* 281: 1578–1579.

Curl, C.L., R.A. Fenske, and K. Elgethun. 2003. Organophosphorus pesticide exposure of urban and suburban preschool children with organic and conventional diets. *Environmental Health Perspectives* 111: 377–382.

Dangour, A.D., K. Lock, A. Hayter, A. Aikenhead, E. Allen et al. 2010. Nutrition-related health effects of organic foods: A systematic review. *American Journal of Clinical Nutrition* 92: 203–210.

Delate, K., M. Duffy, C. Chase, A. Holste, H. Friedrich, and N. Wantate. 2003. An economic comparison of organic and conventional grain crops in a long-term agro-ecological research (LTAR) site in Iowa. *American Journal of Alternative Agriculture* 18: 59–69.

Delbridge, T.A., C. Fernholz, W. Lazarus, and R.P. King. 2011. Economic performance of long-term organic and conventional cropping systems in Minnesota. *Agronomy Journal* 103(5): 1372–1382.

de Ponti, T., B. Rijk, and M.K. van Ittersum. 2012. The crop yield gap between organic and conventional agriculture. *Agricultural Systems* 108: 1–9.

Diez-Gonzalez, F., T.R. Callaway, M.J. Kizoulis, and J.B. Russell. 1998. Grain feeding and the dissemination of acid-resistant *Escherichia coli* from cattle. *Science* 281: 1666–1668.

Duffy, M. 1987. The economics of conversion to biological farming. In *Proceedings of the Management Alternatives for Biological Farming Workshop*, Vol. III, ed. R.B. Dahlgren and E.E. Klaus. Ames, IA: Iowa State University, pp. 43–51.

Edwards-Jones, G. and Howells, O. 2001. The origin and hazard of inputs to crop protection in organic farming systems: Are they sustainable? *Agricultural Systems* 67: 31–47.

El-Hage Scialabba, N. 2007. *Organic Agriculture and Food Security*. Rome, Italy: Food and Agriculture Organization of the United Nations. Available at: www.fao.org/organicag

Enniss, J.L. 1985. The likely inter-industry effects of organic farming adoption in the United States. MS thesis. Ames, IA: Ohio State University.

Entz, M.H., R. Guilford, and R. Gulden. 2001. Crop yield and soil nutrient status on 14 organic farms in the eastern portion of the northern Great Plains. *Canadian Journal of Plant Science* 81: 351–354.

FAO. 2000. Food safety and quality as affected by organic farming. *Twenty Second FAO Regional Conference for Europe*. Porto, Portugal, July 2000.

Farnworth, C. and J. Hutchings. 2009. *Organic Farming and Women's Empowerment*. Bonn, Germany: International Federation of Organic Agriculture Movements (IFOAM).

Flora, C.B. 1995. Social capital and sustainability: Agriculture and communities in the Great Plains and Corn Belt. *Research in Rural Sociology and Development* 6: 227–246.

Giovannucci, D. 2003. Emerging issues in the marketing and trade of organic products. In *Organic Agriculture. Sustainability, Markets and Policies*, ed. OECD. Wallingford, U.K.: CABI Publishing, pp. 187–198.

Goreham, G.A., F.L. Leistritz, and R.W. Rathge. 1995. Community trade patterns of conventional and sustainable farmers. In *Planting the Future. Developing an Agriculture That Sustains Land and Community*, eds. E. Bird, G.A. Bultena, and J.C. Gardner. Ames, IA: Iowa State University Press, pp. 131–146.

Green, M. and R. Maynard. 2006. The employment benefits of organic farming. In *Aspects of Applied Biology 79. What will Organic Farming Deliver? COR 2006*, eds. C. Atkinson, B. Ball, D.H.K. Davies, R. Rees, G. Russell, E.A. Stockdale, C.A. Watson, R. Walker, and D. Younie. Wellesbourne, Warwick, UK: Association of Applied Biologists, pp. 51–55.

Grinder-Pedersen, L., S.E. Rasmussen, S. Bügel, L.V. Jørgensen, L.O. Dragsted, V. Gundersen, and B. Sandström. 2003. Effect of diets based on foods from conventional versus organic production on intake and excretion of flavonoids and markers of antioxidative defense in humans. *Journal of Agricultural and Food Chemistry* 51: 5671–5676.

Halberg, N., T.B. Sulser, H. Høgh-Jensen, M.W. Rosegrant, and M.T. Knudsen. 2006. The impact of organic farming on food security in a regional and global perspective. In *Global Development of Organic Agriculture: Challenges and Prospects*, eds. N. Halberg, H.F. Alrøe, M.T. Knudsen, and E.S. Kristensen. Wallingford, U.K.: CABI Publishing, pp. 277–322.

Haliwell, B. 2007. *Still No Free Lunch: Nutrient Levels in U.S. Food Supply Eroded by Pursuit of High Yields*. Boulder, CO: The Organic Centre. http://www.organic-center.org

Hall, A. and V. Mogyorody. 2007. Organic farming, gender, and the labor process. *Rural Sociology* 72: 289–316.

Hallmann, E. 2012. The influence of organic and conventional cultivation systems on the nutritional value and content of bioactive compounds in selected tomato types. *Journal of the Science of Food and Agriculture* 92(14): 2840–2848.

Hallmann, E. and E. Rembialkowska. 2012. Characterisation of antioxidant compounds in sweet bell pepper (*Capsicum annuum* L.) under organic and conventional growing systems. *Journal of the Science of Food and Agriculture* 92(12): 2409–2415.

Hansen, B., H.F. Alrøe, E.S. Kristensen, and M. Wier. 2002. Assessment of food safety in organic farming. DARCOF Working Papers No. 52. Copenhagen, Denmark: DARCOF.

Heaton, S. 2002. Assessing organic food quality: Is it better for you? In *UK Organic Research 2002: Proceedings of the COR Conference*, eds. Powell et al. Aberystwyth, Wales, March 26–28, 2002, pp. 55–60.

Helfter, M. 2003. Pollution threats to organic production and products. In *Organic Agriculture. Sustainability, Markets and Policies*, ed. OECD. Wallingford, U.K.: CABI Publishing, pp. 221–226.

Hepperley, P., R. Siedel, and J. Moyer. 2007. Year 2006 is breakthrough for organic no-till corn yield; tops standard organic for first time at Rodale Institute. New Farm [online]. Available at: http://www.newfarm.org/columns/research_paul/2007/0107/notill.shtml, May 25, 2007.

Hovi, M., A. Sundrum, and S.M. Thamsborg. 2003. Animal health and welfare in organic livestock production in Europe: Current state and future challenges. *Livestock Production Science* 80: 41–53.

Huber, M., M.H. Bakker, W. Dijk, H.A.B. Prins, and F.A.C. Wiegant. 2012. The challenge of evaluating health effects of organic food; operationalisation of a dynamic concept of health. *Journal of the Science of Food and Agriculture* 92(14): 2766–2773.

Jansen, K. 2000. Labour, livelihoods and the quality of life in organic agriculture in Europe. *Biological Agriculture and Horticulture* 17: 247–278.

Jones, D. 2003. Organic agriculture, sustainability and policy. In *Organic Agriculture. Sustainability, Markets and Policies*, ed. OECD. Wallingford, U.K.: CABI Publishing, pp. 17–30.

Jorhem, L. and P. Slanina. 2000. Does organic farming reduce the content of cadmium and certain other trace metals in plant food? A pilot study. *Journal of the Science of Food and Agriculture* 80: 43–48.

Kleinschmidt, L., D. Ralson, and N. Thompson. 1994. *Community Impacts of Sustainable Agriculture in Northern Cedar County, Nebraska*. Walthill, NE: Center for Rural Affairs.

Knickel, K. and H. Renting. 2000. Methodological and conceptual issues in the study of multifunctionality and rural development. *Sociologia Ruralis* 40(4): 512–528.

Koh, E., S. Kaffka, and A.E. Mitchell. 2013. A long-term comparison of the influence of organic and conventional crop management practices on the content of the glycoalkaloid α-tomatine in tomatoes. *Journal of the Science of Food and Agriculture* 93(7): 1537–1542.

Kouba, M. 2003. Quality of organic animal products. *Livestock Production Science* 80: 33–40.

Kuczyńska, B., K. Puppel, M. Gołebiewski, E. Metera, T. Sakowski, and K. Słoniewski. 2012. Differences in whey protein content between cow's milk collected in late pasture and early indoor feeding season from conventional and organic farms in Poland. *Journal of the Science of Food and Agriculture* 92(14): 2899–2904.

Lairon, D. 2010. Nutritional quality and safety of organic food: A review. *Agronomy for Sustainable Development* 30: 33–41.

Lampkin, N., C. Foster, and S. Padel. 1999. *The Policy and Regulatory Environment for Organic Farming in Europe: Organic Farming in Europe, Economics and Policy*, Vol. 1. Stuttgart, Germany: University of Hohenheim.

Lampkin, N.H. and S. Padel. (eds.). 1994. *The Economics of Organic Farming: An International Perspective*. Wallingford, U.K.: CABI Publishing.

Langley, J., K. Olsen, and E.O. Heady. 1983. The macro implications of a complete transformation of U.S. agriculture to organic farming practices. *Agriculture, Ecosystems and Environment* 10: 323–333.

Lasley, P., E. Holmberg, and G. Bultena. 1993. Is sustainable agriculture an elixir for rural communities? *American Journal of Alternative Agriculture* 8: 133–139.

Lester, G.E. and R.A. Saftner. 2011. Organically versus conventionally grown produce: Common production inputs, nutritional quality, and nitrogen delivery between the two systems. *Journal of Agriculture and Food Chemistry* 59: 10401–10406.

Liebhardt, B. 2003. What is organic agriculture? What I learned in my transition. In *Organic Agriculture. Sustainability, Markets and Policies*, ed. OECD. Wallingford, U.K.: CABI Publishing, pp. 31–49.

Linden, A., K. Andersson, and A. Oskarsson. 2001. Cadmium in organic and conventional pig production. *Archives of Environmental Contamination and Toxicology* 40: 425–431.

Lo, M. and D. Matthews. 2002. Results of routine testing of organic foods for agrochemical residues. In *UK Organic Research 2002: Proceedings of the COR Conference*, ed. J. Powell. Aberystwyth: Wales, March 26–28, 2002, pp. 61–64.

Lobley, M., M. Reed, and A. Butler. 2005. *The Impact of Organic Farming on the Rural Economy in England*. Exeter, U.K.: Centre for Rural Research, University of Exeter. http://socialsciences.exeter.ac.uk/media/universityofexeter/research/centreforruralpolicyresearch/pdfs/researchreports/Organic_Impacts_final.pdf (accessed May 12, 2013).

Lockeretz, W. 1989. Comparative local economic benefits of conventional and alternative cropping systems. *American Journal of Alternative Agriculture* 4: 75–83.

Lotter, D.W. 2003. Organic agriculture. *Journal of Sustainable Agriculture* 21: 59–128.

Lu, C., K. Toepel, R. Irish, R.A. Fenske, D.B. Barr, and R. Bravo. 2006. Organic diets significantly lower children's dietary exposure to organophosphorus pesticides. *Environmental Health Perspectives* 114: 260–263.

Lund, V. and B. Algers. 2003. Research on animal health and welfare in organic farming: A literature review. *Livestock Production Science* 80: 55–68.

Lynch, D.H. 2009. Environmental impacts of organic agriculture: A Canadian perspective. *Canadian Journal of Plant Science* 89(4): 621–628.

Lynch, D.H., N. Halberg, and G.D. Bhatta. 2012. Environmental impacts of organic agriculture in temperate regions. *CAB Reviews* 7(No. 010): 1–17.

Lynch, D.H., J. Sumner, and R.C. Martin. In press. Transforming recognition of the social, ecological and economic goods and services derived from organic agriculture in the Canadian context. In *Organic Food and Farming: Prototype for Sustainable Agriculture*, eds. S. Bellon and S. Penvern. Paris, France: Springer.

Maceachern, A.B. 2008. The work of women on organic farms: A gendered analysis. Unpublished Master's thesis. Guelph, Ontario, Canada: University of Guelph.

MacKinnon, S.L. 2006. Identifying and differentiating farm-community linkages in organic farming in Ontario. Unpublished Master's thesis. Guelph, Ontario, Canada: Department of Geography, University of Guelph.

MacRae, R., R. Martin, A. Macey, P. Doherty, J. Gibson, and R. Beauchemin. 2004. Does the adoption of organic food and farming systems solve multiple policy problems? A review of the existing literature. Report funded by the Canadian Agriculture and Rural Develop (CARD) Program of Agriculture and Agrifood Canada. Truro, Nova Scotia, Canada: Organic Agriculture Centre of Canada.

MacRae, R.J., B. Frick, and R.C. Martin. 2007. Economic and social impacts of organic production systems. *Canadian Journal of Plant Science* 87: 1037–1044.

MacRae, R.J., S.B. Hill, and J. Henning. 1993. Strategies to overcome barriers to the development of sustainable agriculture in Canada: The role of agribusiness. *Journal of Agricultural and Environmental Ethics* 6: 21–53.

MacRae, R.J., S.B. Hill, G.R. Mehuys, and J. Henning. 1990. Farm-scale agronomic and economic transition to sustainable agriculture. *Advances in Agronomy* 43: 155–198.

MacRae, R.J., R.C. Martin, J. Langer, and M. Juhasz. 2009. Ten percent organic within 15 years: Policy and programme initiatives to advance organic food and farming in Ontario, Canada. *Renewable Agriculture and Food Systems* 24(2): 120–136.

Madden, J.P. and T.L. Dobbs. 1990. The role of economics in achieving low input farming systems. In *Sustainable Agriculture Systems*, eds. C.A. Edwards, R. Lal, P. Madden, R.H. Miller, and G. House. Ankeny, IA: Soil and Water Conservation Society, pp. 459–477.

Magkos, F., F. Arvaniti, and A. Zampelas. 2006. Organic food: Buying more safety or just peace of mind? A critical review of the literature. *Critical Reviews in Food Science and Nutrition* 46(1): 23–56.

Marsden, T., J. Banks, and G. Bristow. 2002. The social management of rural nature: Understanding agrarian-based rural development. *Environment and Planning A* 34(5): 809–825.

Marsden, T., J. Murdoch, and K. Morgan. 1999. Sustainable agriculture, food supply chains and regional development: Editorial introduction. *International Planning Studies* 4(3): 295–301.

Martin, R.C., D.H. Lynch, B. Frick, and P. van Straaten. 2007. Phosphorous status on Canadian organic farms. *Journal of the Science of Food and Agriculture* 87: 2737–2740.

Martinson, O. and G. Campbell. 1980. Betwixt and between. Farmers and marketing of agricultural inputs and outputs. In *The Rural Sociology of the Advanced Societies*, eds. F.G. Buttel and H. Newby. Montclair, NJ: Allenheld, Osman, pp. 215–254.

Mayer, A.-M. 1997. Historical changes in the mineral content of fruits and vegetables: A cause for concern? *British Food Journal* 99: 10–31.

McMahon, M. 2005. Engendering organic farming. *Feminist Economics* 11: 134–140.

McMahon, M.A.S. and I.G. Wilson. 2001. The presence of enteric pathogens and *Aeromonas* species in organic vegetables. *International Journal of Food Microbiology* 70: 155–162.

Meier-Ploeger, A. 1992. Organic product quality. In *Trade in Organic Foods*, eds. B. Geier, C. Haest, and A. Pons. Tholey-Theley, Germany: IFOAM, pp. 65–68.

Michelsen, J. 2001. Organic farming in the regulatory perspective: The Danish case. *Sociologia Ruralis* 41: 62–84.

Michelsen, J. 2003. The role of research, information and communication. In *Organic Agriculture. Sustainability, Markets and Policies*, ed. OECD. Wallingford, U.K.: CABI Publishing, pp. 367–378.

Midmore, M. and N. Lampkin. 1994. Modelling the impact of widespread conversion to organic farming: An overview. In *The Economics of Organic Farming: An International Perspective*, eds. N. Lampkin and S. Padel. Wallingford, U.K.: CABI Publishing, pp. 371–380.

Moore, V., M.E. Zabik, and M.J. Zabick. 2000. Evaluation of conventional and "organic" baby food brands for 8 organochlorine and 5 botanical pesticides. *Food Chemistry* 71: 43–47.

Neilsen, A.C. 2005. Functional food and organics: A Global AC Neilsen Online Survey on consumer behaviour and attitudes. Available from URL: http://hk.acnielsen.com/news/20051219.shtml

Nemes, N. 2009. *Comparative Analysis of Organic and Non-organic Farming Systems: A Critical Assessment Of Farm Profitability.* Rome, Italy: FAO.

Nieberg, H. and N. Oppermann. 2003. The profitability of organic farming in Europe. In *Organic Agriculture. Sustainability, Markets and Policies*, ed. OECD. Wallingford, U.K.: CABI Publishing, pp. 141–152.

Northwest Area Foundation. 1994. *A Better Row to Hoe: The Economic, Environmental, and Social Impact of Sustainable Agriculture.* St. Paul, MN: Northwest Area Foundation.

Oates, L. and M. Cohen. 2011. Assessing diet as a modifiable risk factor for pesticide exposure. *International Journal of Environmental Research and Public Health* 8(6): 1792–1804.

Oelhaf, R. 1978. *Organic Agriculture: Economic and Ecological Comparisons with Conventional Methods.* Montclair, NJ: Allanheld, Osman.

Oelhaf, R. 1983. The economic feasibility of widespread adoption of organic farming. In *Sustainable Food Systems*, ed. D. Knorr. Westport, CT: AVI Publishing, pp. 156–172.

Ogini, Y.O., D.P. Stonehouse, and E.A. Clark. 1999. Comparison of organic and conventional dairy farms in Ontario. *American Journal of Alternative Agriculture* 14: 122–134.

Palupi, E., A. Jayanegara, A. Ploeger, and J. Kahl. 2012. Comparison of nutritional quality between conventional and organic dairy products: A meta-analysis. *Journal of the Science of Food and Agriculture* 92(14): 2774–2781.

Patriquin, D.G. 2000. Reducing risks from *E. coli* 0157 on the organic farm. Canadian *Organic Growers Reference Series #15*. Ottawa, Ontario, Canada: Canadian Organic Growers.

Peters, C., N. Bills, J. Wilkins, and R. Smith. 2002. *Vegetable Consumption, Dietary Guidelines, and Agricultural Production in New York State—Implications for Local Food Economies.* Ithaca, NY: College of Agriculture and Life Sciences, Cornell University.

Peters, C., N. Bills, J. Wilkins, and R. Smith. 2003. *Fruit Consumption, Dietary Guidelines, and Agricultural Production in New York State—Implications for Local Food Economies.* Ithaca, NY: College of Agriculture and Life Sciences, Cornell University.

Pimentel, D., H. Acquay, M. Biltonen, P. Rice, M. Silva, J. Nelson, V. Lipner, S. Giordano, A. Horowitz, and M. D'Amore. 1992. Environmental and economic cost of pesticide use. *Bioscience* 42: 750–760.

Pimentel, D., C. Harvey, P. Resosudarmo, K. Sinclair, D. Kurz, M. McNair, S. Crist, L. Shpritz, L. Fitton, R. Saffouri, and Blair, R. 1995. Environmental and economic costs of soil erosion and conservation benefits. *Science* 267: 1117–1123.

Pimentel, D., P. Hepperly, J. Hanson, D. Douds, and R. Seidel. 2005. Environmental, energetic, and economic comparisons of organic and conventional farming systems. *Bioscience* 55: 573–582.

Plochberger, K. 1989. Feeding experiments: A criterion for quality estimation of biologically and conventionally produced foods. *Agriculture, Ecosystems and Environment* 27: 419–428.

Pretty, J., C. Brett, D. Gee, R.E. Hine, C.F. Mason, J.I.L. Morison, H. Raven, M.D. Rayment, and G. van der Bijl. 2000. An assessment of the external costs of UK agriculture. *Agricultural Systems* 65: 113–136.

Pretty, J. and R. Hines. 2001. *Reducing Food Poverty through Sustainable Agriculture.* University of Essex, Colchester, U.K.: Centre for Environment and Society.

Pretty, J.M. 1995. *Regenerative Agriculture.* London, U.K.: IISD.

Pretty, J.N., A.S. Ball, T. Lang, and J.I.L. Morison. 2005. Farm costs and food miles: An assessment of the full cost of the UK weekly food basket. *Food Policy* 30: 1–15.

Roselli, M., A. Finamore, E. Brasili, G. Capuani, H.L. Kristensen, C. Micheloni, and E. Mengheri. 2012. Impact of organic and conventional carrots on intestinal and peripheral immunity. *Journal of the Science of Food and Agriculture* 92(14): 2913–2922.

Rozzi, P., F. Miglior, and K.J. Hand. 2007. A total merit selection index for Ontario organic dairy farmers. *Journal of Dairy Science* 90: 1584–1593.

Sapkota, A.R., R.M. Hulet, G. Zhang, P. McDermott, E.L. Kinney, K.J. Schwab, and S.W. Joseph. 2011. Lower prevalence of antibiotic-resistant enterococci on U.S. conventional poultry farms that transitioned to organic practices. *Environmental Health Perspectives* 119(11): 1622–1628.

Seufert, V., N. Ramankutty, and J.A. Foley. 2012. Comparing the yields of organic and conventional agriculture. *Nature* 485(7397): 229–232.

Smith-Spangler, C., M.L. Brandeau, G.E. Hunter, J.C. Bavinger, M. Pearson et al. 2012. Are organic foods safer or healthier than conventional alternatives? A systematic review. *Annals of Internal Medicine* 157: 348–366.

Soil Association. 2006. *Organic Works: Providing More Jobs through Organic Farming and Local Food Supply*. Bristol, U.K.: Soil Association.

Stanhill, G. 1990. The comparative productivity of organic agriculture. *Agriculture, Ecosystems and Environment* 30: 1–26.

Stockdale, E.A., N.H. Lampkin, M. Hovi, R. Keatinge, E.K.M. Lennartsson, D.W. Macdonald, S. Padel, F.H. Tattersal, M.S. Wolfe, and C.A. Watson. 2001. Agronomic and environmental implications of organic farming systems. *Advances in Agronomy* 70: 261–327.

Stolze, M., A. Piorr, A. Häring, and S. Dabbert. 2000. *The Environmental Impacts of Organic Farming in Europe*, Volume 6 of the series *Organic Farming in Europe: Economics and Policy*. Stuttgart, Germany: University of Hohenheim.

Sumner, J. 2005a. *Sustainability and the Civil Commons: Rural Communities in the Age of Globalization*. Toronto, Ontario, Canada: University of Toronto Press.

Sumner, J. 2005b. *Organic Farmers and Rural Development: A Research Report on the Links between Organic Farmers and Community Sustainability in Southwestern Ontario*. Toronto, Ontario, Canada: OISE, University of Toronto.

Sumner, J. and S. Llewelyn. 2011. Organic solutions? Gender and organic farming in the age of industrial agriculture. *Capitalism, Nature, Socialism* 22: 100–118.

Sundrum, A. 2001. Organic livestock farming: A critical review. *Livestock Production Science* 67: 207–215.

Tasiopoulou, S., A.M. Chiodini, F. Vellere, and S. Visentin. 2007. Results of the monitoring program of pesticide residues in organic food of plant origin in Lombardy (Italy). *Journal of Environmental Science and Health B* 42: 835–841.

Tegtmeier, E.M. and M.D. Duffy. 2004. External costs of agricultural production in the United States. *International Journal of Agricultural Sustainability* 2: 1–20.

Trauger, A., C. Sachs, M. Barbercheck, K. Braiser, and N.E. Kiernan. 2010. 'Our market is our community': Women farmers and civic agriculture in Pennsylvania, U.S.A. *Agriculture and Human Values* 27: 43–55.

USDA. 1980. *Report and Recommendations on Organic Farming*. Washington, DC: Superintendent of Documents.

Vaarst, M., S. Roderick, V. Lunde, and W. Lockeretz. (eds.). 2004. *Animal Health and Welfare in Organic Agriculture*. Wallingford, U.K.: CABI Publishing.

Vogtmann, H. 1984. Organic farming practices in Europe. In *Organic Farming: Current Technology and Its Role in a Sustainable Agriculture*, Vol. 46, ed. D.M. Kral. Madison, WI: American Society of Agronomy Special Publication, pp. 19–36.

Vorley, W.T., D.R. Keeney, D. Loechlin, M.M. Mayhew, E. Ozdemiroglu, D.W. Pearce, J.N. Pretty, and R. Tinch. 1995. Can the pesticide industry benefit from sustainable agriculture? Research and practice: Learning to build sustainable industries for sustainable societies. *Fourth International Conference of the Greening of Industry Network*, Toronto, Ontario, Canada.

Wagstaff, H. 1987. Husbandry methods and farm systems in industrialized countries which use lower levels of external inputs: A review. *Agriculture, Ecosystems and Environment* 19: 1–27.

Willer, H., J. Lernoud, and L. Kilcher. (eds.). 2013. *The World of Organic Agriculture: Statistics and Emerging Trends 2013*. Bonn, Germany: IFOAM and FiBL.

Williamson, C.S. 2007. Is organic food better for you? *Nutrition Bulletin* 32: 104–108.

Winson, A. 1992. *The Intimate Commodity*. Toronto, Ontario, Canada: Garamond Press.

Woese, K., D. Lang, C. Boess, and K.W. Bogl. 1997. A comparison of organically and conventionally grown foods—Results of a review of the relevant literature. *Journal of the Science of Food and Agriculture* 74: 281–293.

World Commission on Environment and Development. 1987. *Our Common Future*. Toronto, Ontario, Canada: Oxford University Press.

Worthington, V. 2001. Nutritional quality of organic versus conventional fruits, vegetables and grains. *Journal of Alternative and Complementary Medicine* 7: 161–173.

Wynen, E. 2003. What are the key issues faced by organic producers? In *Organic Agriculture: Sustainability, Markets and Policies*, ed. OECD. Wallingford, U.K.: CABI Publishing, pp. 207–220.

Youngberg, I.G. and F.H. Buttel. 1984. U.S. agricultural policy and alternative farming systems: Politics and prospects. In *Restructuring Policy for Agriculture*, eds. S.S. Batie and J.P. Marshall. Blacksburg, VA: Virginia Polytechnic Institute, pp. 45–66.

Zander, K. and U. Hamm. 2010. Consumer preferences for additional ethical attributes of organic food. *Food Quality and Preference* 21(5): 495–503.

Zanoli, R., D. Bambelli, and D. Vairo. 2000. *Organic Farming by 2010: Scenarios for the Future. Organic Farming in Europe: Economics and Policy*, Vol. 8. Hohenheim, Germany: University of Hohenheim.

Zentner, R.P., P. Basnyat, S.A. Brandt, A.G. Thomas, D. Ulrich, C.A. Campbell, C.N. Nagy, B. Frick, R. Lemke, S.S. Malhi, O.O. Olfert, and M.R. Fernandez. 2011. Effects of input management and crop diversity on economic returns and riskiness of cropping systems in the semi-arid Canadian Prairie. *Renewable Agriculture and Food Systems* 26: 208–223.

15 Ten Percent Organic within 15 Years

Policy and Program Initiatives to Advance Organic Food and Farming in Ontario, Canada*

Rod MacRae, Ralph C. Martin,
Mark Juhasz, and Julia Langer

CONTENTS

* Reprinted with permission from MacRae, R., Juhasz, M., Langer, J., and Martin, R.C., 2006. *Ontario Goes Organic: How to Access Canada's Growing Billion Dollar Market for Organic Food.* World Wildlife Fund Canada (WWF—Canada) and the Organic Agriculture Centre of Canada, Truro, NS. http://www.organicagcentre.ca/ResearchDatabase/res_oos_intro.asp

15.1 INTRODUCTION

The Canadian food and agriculture sector is facing some significant environmental, food safety, and financial difficulties. These difficulties are affecting the
market's perceptions of Canadian food, both domestically and internationally.
These realities explain, in part, the development of the 2002 agricultural policy
framework (APF), implemented in 2003–2008 by the federal, provincial, and
territorial governments, and the agreement to create a next generation of APF
programs (or APF2).

Despite these preoccupations, with the possible exception of Quebec, Canadian
governments have treated organic food and farming as a niche market, with very
limited attention given to it in the new suite of APF programs. In many parts of the
world, organic farming has been similarly promoted, but rapid growth rates in the
past decade suggest that providing policy supports on a niche market basis is misplaced. In several European countries, the organic sector has reached 10% of the
agri-food economy and/or production area.[1] At such levels, it is estimated that some
of the proposed benefits of organic farming[2,3] could be realized.

With growth in retail sales estimated at 15%–25% year[-1], organic food represents the only significant growth sector in Canada's food system.[4] This explains,
in part, the recent spate of organic firm acquisitions by conventional food companies.[5] However, only 15%–40% of the organic food consumed in Canada is produced
domestically.[4,6] The rest is imported, primarily from the United States (perhaps
70%–75%) and Europe.[6] In contrast, 70% of conventional foods consumed domestically are produced in Canada.[7] Consequently, Canadian farmers are missing out
on many of the market opportunities that organic demand presents, at a time when
net farm income in aggregate has been low.[7] Without domestic production to match
demand, the significant environmental, health, financial, and social benefits that can
be associated with organic food production, processing, and distribution are accru-
ing elsewhere. Global trade in organic food is also contributing to greenhouse gas
(GHG) emissions, causing many in the sector to question an export-/import-oriented
organic agricultural strategy and to call for a new postorganic or "beyond organics"
approach.[8,9] At the leading edge of this opposition to global organic is the UK Soil
Association, which has suggested in a discussion paper that air-freighted organic
may not in future be eligible for certification.[10]

Market demand, on its own, appears to be insufficient to rapidly attract new
Canadian organic producers and processors. It appears that even the presence of significant price premiums is insufficient to overcome the anxieties about and real challenges of the transition to organic production. European evidence suggests that only
with supportive government interventions will the supply of organic foods increase
relatively rapidly.[11]

Government involvement is justified for many reasons consistent with historical interventions in the food and agricultural economy. Organic agriculture is an immature industry and governments have supported infant agricultural industries in the past, in Canada notably the canola oil industry on the Prairies and the wine sector in Ontario. These government investments in the organic sector progressively correct market failures—the fact that current approaches to production, processing, and distribution do not reflect real costs, generating significant externalized costs the private landowners, the general public, and governments have to pay for later. In theory, as the social and environmental benefits grow with organic farming adoption, government liabilities for these unfunded externalized costs should decline.[12,13]

In earlier work (2002), the Canadian organic sector established a series of targets for growth, setting out some of the governmental and sectoral interventions considered necessary to expand organic food and farming.[14] This study was designed to further that analysis, focusing specifically on governmental and sectoral initiatives for the province of Ontario. Ontario was chosen for several reasons:

1. One of the most concentrated areas of market demand for organic products in Canada is the Greater Toronto Area (GTA),[4] and therefore, Ontario farmers are well placed to meet this local market.
2. Production remains very limited, despite significant demand for a wide range of organic foods. In 2004 (the most up-to-date information at the time of the study), Ontario only had about 489 certified organic producers covering about 24,000 ha of cropland. Farm gate receipts were estimated at over $25 million. Certified organic processors only numbered in the hundreds.[6] Data from 2005 showed very small increases from 2004.[15]
3. Although organic farming and processing data are generally weak in Canada, Ontario data are relatively better than many other provinces.
4. There is interest in organic agriculture at a political level. The Ontario government in the 2003–2007 period relied to some degree on external organizations to put forward detailed action plans that they might implement. Senior politicians expressed interest in seeing detailed ideas on advancing organic agriculture, as part of their consultations with the agricultural sector.

Since some countries in Europe have the most advanced organic sectors, the European experience developing the organic sector reveals key instruments that have been critical to success.[11] Most of the countries with significant development have used a mixture of supply-side and demand-side policies and programs, including the following:

1. Definitions of organic agriculture.
2. A uniform national (and for Europe, an EU level) standard, with political recognition of standards, certification, and accreditation. In the EU, there have been statistically positive impacts from introduction of the EU standard.
3. Financial support for transitional growers. Numerous studies show initial positive impacts from direct payments in the agri-environmental schemes;

however, modifications to the schemes in the more mature countries like Denmark and Austria appear to have accelerated existing organic farming growth, but not necessarily brought in significant numbers of new organic farmers.

4. Advisory services and training to support the adoption process.
5. Local institutional supports for organic farming.
6. Supports for the development of organic markets–supermarkets and institutional buyers are often drivers of demand in Europe.
7. Coordinating and advising institutions to advance organics with positive participation and interaction with the conventional farming sector.

The European experience has led many states to recognize the need for a more integrated and balanced mix of policy and program measures.[16] This has produced a number of national action plans (and an EU plan) with both supply (push) and demand (pull) instruments with proposals for coordinating and implementation bodies. Denmark was the first to develop a plan (1995), and now England, Finland, France, Germany, the Netherlands, Norway, Sweden, Wales, and Spain have plans. The plan rationales are to increase the size of the organic sector because of the public benefits that result. They normally include targets for adoption (typically 5%–10% by 2000/2005 or 10%–20% by 2010), direct financial support through the agri-environment/rural development programs, marketing and processing support, producer information initiatives, consumer education, and infrastructure support. A typical mix is 50% of expenditures for direct payments and 50% for a host of other infrastructure- and training-related supports. Some plans focus more on the demand-side interventions (e.g., the Netherlands), others on building information support systems for all players in the organic food chain (Germany), and others on increasing supply (England and Wales). Interestingly, almost all plans focus on the need for integrated farming systems and cooperation among all players in the food chain and for formal advisory bodies that guide government decision making on organic agriculture. Plans generally commit millions of dollars in public funds to implementation.

Among Canadian jurisdictions, only Quebec has a full strategic plan that rivals plans in Europe. A comparison of Quebec and European plans can be found at http://www.oacc.info/DOCs/Paper_Supports_Version2_rm.pdf.

The hypothesis of this chapter is that an Ontario plan can be designed and delivered that takes lessons from plans in other jurisdictions and adapts them to the Ontario environment, to accelerate the adoption of organic farming and food processing and to reap the associated benefits for Ontario farmers and consumers.

15.2 METHODS

Working from targets set out in the National Organic Strategic Plan (NOSP)[14] for adoption of organic farming and processing in different commodity areas, the study used an iterative process of analysis to identify specific and reasonable conversion targets in multiple commodity areas. Consistent with the NOSP, an overall target of 10% of cropped area was chosen but a wider range of targets was considered for individual commodities (see Table 15.4 for the list of commodities) in order to meet the overall target. Target setting allowed for estimates of the required number of new entrants to farming and conventional producers converting to organic. Programs and program initiatives were proposed to help meet those targets, based on successes elsewhere and expert opinion, and estimates of expenditures developed including the costs of a transition payments program that included payments for environmental services. Based on adoption targets, potential savings were identified in fertilizer, pesticide, and antibiotic use associated with those transition targets.

15.2.1 Setting Targets

The analysis determined

- The number of hectares to be converted to organic production
- Crop-by-crop contributions to the overall 10% organic production target
- The number of animal head to be converted
- An estimate of the overall number of farms to be converted and new entrants to farming

Crops were chosen based on data available on organic production (primarily production years 2003 and 2004). Specialty crops, such as ornamentals, herbs, bird seeds, and ginseng, were excluded from the analysis because of insufficient data. Data on organic wheat, corn, and hay/pasture had to be disaggregated, based on conventional ratios and expert information.[17] Organic vegetable production data were limited, so all vegetables had to be reported together, except potatoes. The fruit production analysis was based on availability of organic production data.

Animals were chosen based on data available on organic production. Specialty, smaller volume animal production and aquaculture were excluded from the analysis because of insufficient data. Conventional crop and animal production data were taken from 2004 Ontario Ministry of Agriculture and Food and Rural Affairs (OMAFRA) statistics,[18] unless otherwise noted in our earlier report.

To establish 5- and 15-year targets for Ontario, we started from the national targets set out in the NOSP.[14] These were modified to reflect Ontario conventional and organic production realities and to balance crop and animal production requirements. Target hectares and head were calculated by multiplying conventional hectares (or animal head) by the target percentage for organic conversion. Current data on organic production, organic head, and current organic farms were based on 2003 data provided by Macey[6] unless updated with 2004 data provided by Macey.[19] Area (and head) to be converted was calculated by subtracting current organic area (head) from 15-year target area (head). To estimate how many farms would be required to convert to a specific commodity production, we estimate the current size of an average organic operation. Where that information was not available, we used conventional averages. We also took account of the size dispersal of operations in conventional production, under the assumption that most converting operations would not be in the largest size classes. Estimated additional numbers of farms reporting conversion to that crop (animal) were calculated by dividing area (head) to convert by the average size of an operation.

The average number of new organic farming entrants was estimated from conventional entrants. In the 1996–2001 period, 50,000 new farms entered,[20] most existing farms under new management. In an average year, then, 10,000 new operations started up. Assuming an even distribution according to provincial farm ratios, then 24% of those entered in Ontario, meaning 2400 new farms annually. Assuming that 1% annually of those new entrants are organic, then 24 new farms year[-1] entered or 120 new farm entrants over a 5-year period. In Phase II, with a fuller suite of supports in place, we anticipated a doubling of new entrants to organic farming or 48 farms year[-1]. This would total 480 farms over 10 years, for a 15-year total of 600.

Since many operations are diversified, the farm totals for each commodity will not reflect accurately the total number of farms required for conversion. To estimate that, we found that in the 2001 Census of Agriculture, when adding up farms reporting crops in each commodity studied and comparing that to the total number of farms, the ratio was 2.5. In other words, each farm reported on average 2.5 of the studied crops. We assumed that the number would be higher in organic production, since these operations are usually more diversified, so we divided the total of all farms required by three to come up with our estimate. We did not add in animal production numbers since we assumed that all organic livestock operations would also report crops.

As a check on the merits of our preliminary targets, we also examined rotation patterns and feed requirements for cattle (assuming primarily forage-based diets) and adjusted our targets accordingly. Given the relatively low rates of nonruminant conversions, we did not anticipate any problems ensuring sufficient feed grain availability, although current shortages are acute in some regions and often a product of price differentials between human and animal feed markets.

One weakness of this analysis is that we were unable to account for dynamic changes in crop rotations, in part because Ontario does not collect sufficient crop rotation data. Organic farmers usually diversify and employ longer course rotations. This would cause shifts in the relationships between different crops for which we could not account in this study. For example, many producers converting from conventional to organic production would likely reduce hectares planted to corn, with more cropped area in small grains and forages. In this sense, our study assumed that organic farmers keep producing what they did as conventional producers. More dynamic modeling in future studies could correct for this inaccuracy.

15.2.2 Synthetic Fertilizer Savings

We calculated savings on nitrogen, phosphorus, and potassium fertilizer for farms converting as follows:

- Conventional fertilizer application rates were taken from OMAFRA recommendations,[21–23] focusing on midrange soil test results, loam soils, and midrange yield objectives, unless otherwise noted in our earlier report. We assumed that all hectares would have been fertilized at such rates prior to organic conversion.

- Fertilizer prices were taken from the Ontario Farm Input Monitoring Project.[21]
- The N price (Cdn$0.47 kg^{-1}) was an average of ammonium nitrate, anhydrous ammonia, urea, and nitrogen solution.
- The P price (Cdn$0.47 kg^{-1}) was an average of monoammonium phosphate (MAP), diammonium phosphate (DAP), and triple superphosphate.
- The K price (Cdn$0.35 kg^{-1}) was for muriate of potash 60%.

15.2.3 PESTICIDE SAVINGS

We undertook detailed calculations of pesticide applications avoided (in kg active ingredient) and the input savings for farms converted as follows:

- Pesticide use data came from the OMAFRA survey of pesticide use.[24] Adjustments were made to vegetable and fruit production use totals by subtracting Bt, copper hydroxide, and sulfur from the savings, as these actives are permitted in organic production. Use patterns of these materials would likely be different as organic farmers have limitations on their use of copper and sulfur, but we were unable to account for this in the estimates, so we assumed that the same levels would be used in organic production.
- Pesticide costs are provided on a use-weighted basis, using data from the Ontario Farm Input Monitoring Project Survey.[25]
- Since not all pesticides were listed in that survey, we used the ones available that generally accounted for 80% of the active ingredient (ai) applied, except for fruits and vegetables where they accounted for about 66%. We assumed that closely related products were the same price if they were not separately listed. The estimate of pesticide costs in vegetables is high due to the cost of rimsulfuron use in sweet corn. The estimates provided here likely underestimate pesticide savings.

15.2.4 AVOIDED MEDICATIONS IN FEED

This analysis was carried out to estimate the amount of subtherapeutic medication that would not be consumed in animal feed resulting from the transition to organic production of beef, swine, and chicken (broilers) only as these were the commodities for which sufficient data could be assembled.

Few Canadian data are available on consumption of medications in feed, so the analysis was adapted from a method used in a US study by Mellon et al.[26] The first part of the analysis required a comparison of subtherapeutic antibiotics approved in both Canada[27] and the United States.[26,28] The comparison was frequently straightforward, as the list of approved materials is slowly being harmonized. However, some medications are approved in the United States, but not in Canada, or they are approved as a slightly different formulation or on different animals or at different growth stages, or they are approved at different doses or for different lengths of time. For these, we made the following assumptions:

- We eliminated from our analysis any medication approved in the United States but not in Canada.
- We eliminated from the analysis any medication/growth stage combination that is not approved in Canada; the largest discrepancies occurred in the broiler analysis, so this is likely the most conservative estimate.
- When it appeared that a slightly different formulation was used in Canada, we considered the Canadian medication equivalent to the US formulation.
- No veal, pregnant animals, or breeding stock was included in our analysis, so no medications used exclusively on those animals were included.
- We substituted Canadian doses for US doses where they differed.
- When multiple dose options were provided in the Compendium of Medicated Ingredients Brochures (CMIB),[27] we used those most related to weight gain and efficiency, not options for treatment of acute conditions; if there was more than one option related to weight gain and efficiency, we chose the one closest to US use patterns.
- We did not include medications that appear to be approved in Canada but not in the United States, since we had no data on percentage of animals treated to support an analysis.

We used the Mellon et al.[26] estimates of percentage of animals treated, which assumes that treatment patterns between the two countries are consistent. Since there are no public Canadian data on treatment patterns, we do not know how accurate this assumption is. We used their estimates of average days on feed unless there was Canadian information[27] that indicated that a shorter period was required in Canada. We used their estimates of feed intake for swine and broilers. In a few cases for beef, the Canadian doses were reported in ways that required that we multiply them by average daily feed intake for a particular growth stage, so we used standard animal production guides of feed intake to determine those. We substituted the number of animals converted to organic production.

Other assumptions relative to the Mellon et al.[26] analysis include the following:

- No mortalities.
- No medication combinations in the broiler analysis for which the majority of medications in the combination are not approved in Canada.
- Because the size ranges used in different growth stages were sometimes different from those employed in the United States, no medications used in Canada were included that extended beyond US growth stage categories.

15.2.5 AVOIDED COST PAYMENTS

Canadian studies of the full costs of Canadian agriculture are lacking, but US and British studies attempting to account for a relatively full suite of costs have recently been completed.[12,13] Of these, the most pertinent is a US study that builds upon methodologies used in other research, and its extensive agriculture is closer to Canadian realities than those in Britain. The authors, US agricultural economists Tegtmeier and Duffy,[13] concluded that US externalized costs of conventional

TABLE 15.1
Multiple Benefits of Organic Agriculture

Regarding environmental degradation

1. Adopting organic farming helps governments address pollution problems and their costs.
2. Adopting organic farming can reduce Canada's GHG emissions and help farmers adapt to the negative effects of climate change.
3. Organic farming can improve biodiversity relative to conventional farming.

    Regarding the need to build consumer confidence in the food supply

4. Adopting organic farming builds consumer confidence by not using products, practices, and processes seen to be controversial by some consumers.
5. Organic farming can improve animal welfare.
6. Organic foods may be nutritionally superior to conventional foods.

    Regarding the farm financial crisis

7. Adopting organic farming can reduce financial pressures on farmers.
8. Adopting organic farming can decrease the need for government farm payments.
9. Organic food prices reflect internalization of historically externalized costs.
10. Adopting organic farming can help with rural community revitalization.

Source: MacRae, R. et al., *Does the Adoption of Organic Food and Farming Systems Solve Multiple Policy Problems? A Review of the Existing Literature,* Organic Agriculture Centre of Canada, Truro, Nova Scotia, Canada, 2004.

Note: See also Chapters 13 and 14 for an updated analysis.

agriculture ranged from Cdn$39.73 to 112.56 ha^{-1}, assuming an exchange rate of $1Cdn = $0.85US. We drew on previous work[2,3] and expert opinion to evaluate the degree to which organic production might reduce these costs (see Table 15.1 for a summary of organic farming benefits). We conservatively used the low-end range of the Tegtmeier and Duffy costs for two main reasons: the intensity of production in Canada is generally lower and government program expenditures are lower on a farm area basis. However, just because Canadian governments chose to allocate fewer resources to solving agricultural problems does not mean that they do not exist at a comparable level. For example, pesticide contamination of surface waters is largely viewed as a localized problem, but Canadian monitoring capacity remains limited, although a national indicator is now under development.[29] Given limited knowledge in this area, it is not clear whether water treatment facilities allocate sufficient resources to address what problems may exist.

To produce useful comparisons between organic and conventional production, it is important to focus on the entire farming system or larger food system dynamics as opposed to examining specific elements outside of their larger operating context. It is also important to compare systems that have common components, including comparable management capacities. Clearly, poorly managed organic and conventional systems generate problems. We were interested in structural comparisons, so we assumed good management in systems being compared. In doing so, we have used an agroecological framework[30] to analyze how the structure of organic farming offers benefits that are not necessarily associated with conventional farming.[2,3] We also took account of the strength of the current literature, which results, for example, in an assignment of zero reductions to human health costs associated with agricultural pathogens, since the literature in this area has produced divergent results.

15.2.6 TRANSITION PAYMENTS

The following assumptions guided calculations for our proposed Transition Risk Offset Program and Payments for Environmental Services:

- The payments are set at 10% of the gross revenue loss associated with average yield declines during the transition (see Table 15.2 for estimated average yield declines in organic commodities and Table 15.3 for sources providing justification for a range of production systems). This level was chosen to be slightly lower than Europe, where such payments range from 15% to 20% of foregone revenue,[31] but at a base minimum suggested for improving adoption of other low-input systems in US studies.[32]
- Payments to animal production are on a per animal basis, assuming the same conditions of yield loss and compensation.
- To receive payments, farmers would have to belong to a certification agency, to be actively committed to the transition process, and to be participating in mentoring and training programs. Since this element of the program starts in Phase II, farmers who convert in the first 5 years of this strategy would be eligible to receive payments retroactively, based on record keeping provided by the certification agencies.
- Transition year—we reported payments for each of the 3 years of required transition, except in animal production for which the period on farm is less than 3 years.
- Yield decline—estimates were derived from the literature and expert opinion (see Tables 15.2 and 15.3).
- Average 5-year yield and prices (2000–2004) were taken primarily from OMAFRA statistics,[33] with supplemental data provided by some commodity organizations.

15.2.7 PROGRAM EXPENDITURES

The elements of the program proposed in the succeeding text met the following criteria:

- They have been shown to work elsewhere and can be adapted to Ontario.
- They strike a balance between cost and positive effects.
- They are relatively straightforward to implement.
- Most have significant cost-sharing opportunities with other levels of government, industry, and NGOs.
- Most have the potential for third-party delivery, which reduces government overhead costs.

TABLE 15.2
Yield Reduction Averages Relative to Conventional Production during the 3-Year Transition to Organic in Ontario (Comparison with a Small- to Medium-Sized Conventional Operation)[a]

Commodity	Year 1	Year 2	Year 3	5–10 Years
Field crops				
Pasture[b]	0	0	0	Same
Hay and alfalfa[b]	0.10	0.05	Same as conv.	Same as conv.
Spring wheat	0.30	0.20	0.10	0.05
Winter wheat	0.30	0.20	0.10	0.05
Barley	0.30	0.20	0.10	0.05
Fall rye	0.20	0.10	0.05	Same as conv.
Oats	0.20	0.10	0.05	Same as conv.
Buckwheat	0.30	0.20	0.10	0.10
Corn for grain	0.30	0.20	0.15	0.10
Corn for silage	0.20	0.15	0.10	0.05
Canola	0.50	0.40	0.30	0.20
Soybeans	0.30	0.20	0.15	0.10
Flax	0.45	0.30	0.15	0.10
Edible beans	0.30	0.20	0.15	0.10
Other field crops	0.30	0.20	0.10	0.05
Vegetable				
Potatoes	0.40	0.30	0.25	0.20
Other roots	0.40	0.30	0.25	0.20
Tomatoes	0.40	0.30	0.25	0.20
Field & greenhouse				
Cucumbers	0.40	0.30	0.25	0.20
Field & greenhouse				
Leguminous	0.40	0.30	0.25	0.20
Sweet corn	0.40	0.30	0.25	0.20
Cole crops	0.40	0.30	0.25	0.20
Leafy vegetables	0.40	0.30	0.25	0.20
Tree fruits	0.50	0.40	0.30	0.25
Small fruits	0.50	0.40	0.30	0.25
Dairy[c]	0.20	0.15	0.10	0.10
Beef[d]	0.20	0.15	0.10	0.10
Chicken[e]				
Meat	0.35	0.30	0.25	0.20
Eggs	0.35	0.30	0.25	0.20

(*continued*)

TABLE 15.2 (continued)
Yield Reduction Averages Relative to Conventional Production during the 3-Year Transition to Organic in Ontario (Comparison with a Small- to Medium-Sized Conventional Operation)[a]

Commodity	Year 1	Year 2	Year 3	5–10 Years
Turkey	0.40	0.30	0.25	0.20
Pork[f]	0.50	0.45	0.40	0.30
Sheep[g]	0.10	0.05	0	0

[a] For most field crops, the assumption is that the transition does not start from a forage crop.

[b] Assumes that conventional farmers managed hay and pasture without excess fertilization.

[c] Assumes herd reduction and increased hectares to accommodate increased pasture and hay production.

[d] Assumes integrated operation, birth to slaughter, no feedlots, and major yield declines are associated with reduced weight gain. See Ag Ventures, February 2001. Agdex 420/830-3. Available at website: http://www1.agric.gov.ab.ca/$department/deptdocs.nsf/all/agdex3458/$file/420_830–3.pdf?OpenElement (accessed January 5, 2009).

[e] Assumes floor operations. Available at website http://www.acornorganic.org/pdf/poultryeggsprofile.pdf.

[f] Most difficult comparison: assumes small independent hog operation with most feed produced on the farm. Assumes that 100 sows/1000 market hogs versus 100 sows/2000 market hogs in conventional operation. Transition focuses on 6.5% of Ontario operators roughly in this size range. Farrow to finish.

[g] Assumes most conventional sheep operations are low input, so differences are largely due to stocking rates and yield reductions from changes in deworming agents. Meat only.

We made a number of assumptions:

1. Ontario would participate in the federal program of organic standards development and accreditation, that is, it would not set up its own system as has been done by Quebec.
2. Ontario would participate in a national organic logo and associated publicity campaign initiated at a federal level, that is, Ontario would not develop its own provincial organic logo.
3. For most food safety and quality programming, the organic sector would take advantage of existing federal funding programs[34] and develop a national food safety and quality improvement plan for the sector.
4. Additional elements would be added as existing organizations identified suitable grant programs to provide funding. For instance, consumer

TABLE 15.3
Comparing Yields in Organic versus Conventional Production: Higher-Order Studies

Product	Country/Region	Organic/Conventional
Tomato	California, past 10–15 years	1.0[a]
Many crops	Developing countries, numerous projects	1.8–4.0[b]
Cereals	Europe	0.6–0.7[c]
Corn	Major corn regions of the United States, past 10–15 years, experimental station studies	0.94[a]
Soybeans	Five US states, past 10–15 years, experimental station studies	0.94[a]
Wheat	Two research institutions, past 10–15 years	0.97[a]
Dairy	Europe	0.8–1.0 per cow, with stocking rates of 0.6–0.8 LU per farm.[c] Note that some limited data from Ontario suggest that yields can be equivalent[d] and stocking rates higher than Europe.[e]

[a] Liebhardt.[48]
[b] Pretty and Hines.[49]
[c] Nieberg and Oppermann.[50]
[d] Ogini et al.[51]
[e] Roberts et al.[52]

education is one area in which such opportunities could be explored, especially in Phase II.

5. Organic farmers would have participated in Environmental Farm Plan (EFP) programs consistent with the existing program.[35]
6. New rules for nutrient management[36] and source water protection[37] may require technical adjustments as it relates to organic producers, but we assumed these would not impose additional costs for government.
7. Although the programs were designed to accommodate all farmers, we assumed that it would tend to be small- to medium-sized farms that participated, because their transition challenges might be lower relative to large operations.
8. The distribution of costs by year is usually an average or a graduated increase with a fixed formula, since the rate of uptake of programs is difficult to predict at this point.

15.2.8 ESTIMATING ORGANIC MARKET SHARE

Limited data make pinpointing the current and possible future sizes of the organic food market difficult. We derived our estimates of Ontario's current organic market from multiple sources including national organic market estimates ($1.3 billion[4]), conventional ratios of food retail value/farm gate value, and industry estimates of the percentage of organic production that is exported and organic consumption that is imported. To estimate how much organic farmers contribute to the total Ontario organic retail market after 5 and 15 years, we conservatively assumed a 15% annual growth rate in organic retail sales over the first 10 years of the program, and 10% in the last 5 years. We assumed growth rates in the conventional food market of 1.5% annually. We also assumed modest increases in the portion of organic production that goes to domestic versus export markets.

15.3 RESULTS AND DISCUSSION

15.3.1 ORGANIC TARGETS

15.3.1.1 Production

Results of the analysis are provided in Table 15.4. In Phase I, supply increases would come from a combination of new organic farms and expansion of scale and/or enterprises among existing organic operations. For example, organic field crop producers might certify an existing conventional beef herd, or a producer whose home farm is certified organic might subsequently certify another farm within the total operation. Regarding new organic operations, many farms are already organic but not certified, or in the process of transition,[38] and many companies are working with conventional producers to gradually bring them into their organic supply chain. Additionally, some new entrants to farming would enter organic production directly.

By the end of Phase II (15 years), 5343 organic farmers would be producing organically in all major commodities. Organic production would occur on about 367,000 ha of land (about 10% of current crop area), and some 1.4 million animals would be reared organically. This figure includes existing and converted acres.

15.3.1.2 Processing

Processing targets were more difficult to establish because of the very limited current data on organic processing. Of the conservatively estimated 64 certified processors and handlers in 2003,[6] the majority fell into the following categories: 5 in dairy, 4 in bakery, 2 in flax, 3 in fruit, 2 in nuts, 5 providing meat processing, 3 doing beverages, 14 considered packers and handlers, and 4 animal feed. How many were exclusively organic, versus processing both conventional and organic foods, is unclear from the data. Data from the OMAFRA show that the following primary processing organic products are available from Ontario processors: flours, eggs, fruits and vegetables (including fresh-cut, roasted garlic and soybeans and seasoned beans), honey, maple syrup, alternative sweeteners, soup mixes (dry, containing beans), fluid milk, and meats. Available in the secondary processing category are breads, rolls and baked goods (including cookies and pitas and baking mixes), snacks and cereals (including seeds, chips and popped snacks, snack crackers, and nut/fruit/meal replacement bars), beverages (including alcohol, teas, coffees, fruit and grass juices and seltzers, and powder mixes), prepared soups, condiments (including ketchup, salad dressings, miso, sauces, nut and fruit butters, and jams and jellies), chocolates, processed dairy

TABLE 15.4
Five- and 15-Year Organic Conversion Targets

Production	5-Year Target, 2 × Current (ha)	15-Year Target (Fraction of Conventional)	Hectares to Be Converted (ha)	New Organic Farms Needed
Crop				
Pasture	10,210	0.08	62,467	Reported with hay
Hay	11,435	0.08	70,071	3411
Spring wheat	408	0.12	5,625	Reported with winter wheat
Winter wheat	2,550	0.12	35,130	2933
Barley	1,535	0.10	9,548	575
Fall rye	763	0.15	3,562	419
Oats	2,113	0.15	5,316	597
Buckwheat	989	1.00	1,910	152
Corn (grain)	2,747	0.05	33,013	Reported with silage
Corn (silage)	485	0.05	5,825	2341
Soybeans	14,465	0.10	88,849	2410
Flax	362	1.00	622	37
Edible beans	62	0.10	2,398	100
Mixed grains	2,107	0.15	8,348	665
Potatoes	202	0.10	1,476	281
Vegetables	993	0.10	6,389	1754
Apples	707	0.25	1,407	79
Grapes	34	0.05	307	108
Peaches	40	0.10	231	36
Strawberries	6	0.10	122	50
Sour cherries	8	0.05	42	9
Pears	1	0.05	37	23
Raspberries	2	0.10	37	46
Crop totals	52,217	0.10	342,704	5343
Animal	Head		Head to be converted	
Dairy	6,882	0.10	31,959	432
Beef	5,046	0.02	43,637	1148
Sheep	1,206	0.10	33,397	726
Pork	26,400	0.03	97,500	89
Broilers	11,504	0.003	606,239	909
Turkeys	100	0.01	83,590	7
Layers	25,918	0.05	468,041	904
Animal totals	77,056	0.006	1,364,363	4215[a]

[a] Farms reporting animals are not added to the crop total since it is assumed that all farms reporting animals would also report crops. The ratio of farms reporting animals to crops is slightly higher than the 2001 Census of Agriculture, but this is sensible given that organic producers tend to report livestock to higher degrees than conventional producers. (For details on calculations and assumptions, see http://www.oacc.info/Docs/OntarioOrgStrategy/TargetOOS_Statistics_sheet1.pdf.)

products, pasta, ethnic meals and prepared foods (curries, entrees, pates, and baby foods), and ingredients—including starches, gums, flavors, and extracts. What percentage, however, of the ingredients of these processed goods are purchased from Ontario producers is not known.[39]

In 2003, there were a total of 2300 Ontario food processors with registered employees and an additional approximately 800 operated by the owner and/or family members and/or contract employees. The total estimated value of agricultural shipments (2001) was $24.5 billion[40]; thus, organic processors would represent about 2% of the total by number of enterprises and even less by value of shipments. Doubling the number of firms processing organic food within 5 years would be a reasonable target. Subsector targets for year 15 were impossible to determine at this point.

A key question is where organic processors will come from. In most food subsectors, smaller firms represent the majority, with over half the firms having 20 registered employees or less.[40] Smaller processors tend to focus on local markets, which is the objective of the organic strategies set out in this chapter. However, to optimize production costs, many processors find that they must serve local markets but do so across the country. This often requires a shift to medium scale, or the firm may move to larger markets to accommodate a larger local market when product is perishable. This type of scaling up often has investment challenges.

Plants of smaller scale are often more flexible and accommodate a wider product range, with better capacity for new product introductions. But countering this reality, smaller firms often face cost-effectiveness challenges and often do not have the resources to invest in new products while running day-to-day operations.[41] Despite these challenges, it is likely that organic processing capacity will come from existing small operations, perhaps many without registered employees, and new small firms that start out as exclusively organic processors. There are five main areas of processing activity in the province—SW near Windsor, Grand River Region, Niagara, Toronto, and Quinte area—and organic firms will likely similarly concentrate.

15.3.2 AVOIDED COSTS

Based on the Tegtmeier and Duffy[13] analysis, we estimated that with widespread adoption, organic farming could avoid 56% of these externalized costs, or Cdn$22.25 ha^{-1}. As shown in Table 15.5, we estimate that organic farming can reduce from 0% to 100% of specific externalized costs, depending on the type of negative impact.

At $39.73 ha^{-1}, Ontario cropland (restricted to cropland to make it comparable with US estimates) in 2001 was generating $145.28 million in annual environmental and health costs, many of which are avoidable. These environmental and health costs are currently borne by three levels of government and private landowners.

15.3.3 INPUT SAVINGS

Over the 15 years of the program, converted organic farmers would reduce fertilizer applications by about 43 million kg, pesticide applications by about 296,000 kg active ingredient (8% of pesticides applied on studied crops in 2003), and 7,079 kg of antibiotics consumed in animal feed. Financial savings (2004–2005 prices) would

TABLE 15.5
Analysis of Conventional Agriculture Costs: How Much Does Organic Production Avoid?

Damage Category	Rationale for Level of Avoided Costs Associated with Organic Adoption	US Conventional Cost (Low Estimates Only) (Cdn$ ha⁻¹; Cdn$1 = US$0.85)	Organic Avoided Costs in Ontario (Cdn$ ha⁻¹)
1. Damage to water resources			
1a. Treatment of surface water for microbial pathogens	Dramatically lower pathogen loads in compost than slurry, reduce by 50%.	0.83	0.41
1b. Treatment for nitrate	Organic does not eliminate nitrate leaching but in most studies reduces it by 40%.	1.32	0.53
1c. Treatment for pesticides	Since organic standards do not permit most synthetically compounded pesticides, especially those with persistence, this need for treatment is eliminated.	0.78	0.78
2. Damage to soil resources	Organic farming reduces soil erosion by 40%.[a]	15.68	6.27
3. Damage to air resources			
3a. GHG emissions from crops	Erosion rates reduced by 40%; CO_2 emissions net 50% lower in organic systems due to no emissions from manufacture of synthetic N fertilizers; methane losses comparable; N_2O losses at 20% below conventional production. Net reduction of 50%.	1.98	0.99
3b. GHG emissions from livestock	CAFOs are effectively not permitted in organic production because such operations cannot meet organic requirements; composting significantly reduces total GHG emissions. Lower stocking rates and different diets also contribute. Reduction of 40%.	1.17	0.47

TABLE 15.5 (continued)

Analysis of Conventional Agriculture Costs: How Much Does Organic Production Avoid?

Damage Category	Rationale for Level of Avoided Costs Associated with Organic Adoption	US Conventional Cost (Low Estimates Only) (Cdn\$ ha^{-1}; Cdn\$1 = US\$0.85)	Organic Avoided Costs in Ontario (Cdn\$ ha^{-1})
4. Damage to wildlife and biodiversity			
4a. Honey and pollinator losses	Significantly higher populations in almost all comparative studies; since the US study focuses on pesticide-related losses, reduce costs by 90%.	2.87	2.58
4b. Loss of beneficial predators	Significantly higher populations in almost all comparative studies; since the US study focuses on pesticide-related losses, reduce costs by 90%.	4.66	4.12
4c. Fish kills from pesticides	Since no synthetically compounded pesticides are used, there would be limited fish kills, although a few permitted biologicals are toxic to fish, so we apply a 90% reduction.	0.15	0.14
4d. Fish kills from manure	Since liquid manure is rarely used in organic production, especially in large storage facilities, such kills would be dramatically reduced. However, there is some possibility of water contamination from organic operations, so this is reduced by 90%.	0.08	0.07
4e. Bird kills from pesticides	Since no synthetic pesticides are used, and biologicals are not associated with bird mortalities, this problem is eliminated.	0.24	0.24

5. Damage to human health; pathogens	Although there is some evidence that pathogen loads can be reduced in organic production, this is an insufficiently studied area to warrant a reduction in costs.	2.91	0
6. Damage to human health: pesticides	Because there can occasionally be occupational exposure problems associated with a limited number of biological pesticides, we reduce this cost by only 80%.	7.06	5.65
Summary of costs and avoided costs		$39.73; $145.28 million total in Ontario	$22.25

Source: Tegtmeier, E.M. and Duffy, M.D., *Int. J. Agric. Sustain.*, 2, 1, 2004.

Assumptions:

1. That organic adoption is sufficiently widespread to have an impact in an area.
2. That the comparison is between well-run organic and conventional operations.
3. That the averages are blended across a variety of production systems.
4. In the face of limited data, estimates are always on the conservative side.

[a] We are comparing here conventional systems with simple crop rotations and minimal soil cover and those under organic management with longer course rotations and significant soil cover. Earlier reports of long-term comparative studies[46,47] reported erosion reductions associated with organic production much higher than 40%, but we moderated those results on the assumption that soil erosion rates have generally come down under conventional management. We exclude arid land estimates of erosion reductions, which are usually lower than 40% because they are not pertinent to the Ontario situation.

amount to about \$18.3 million in saved fertilizer applications and \$9.1 million for pesticides.[42] Given current trends in input prices, these are conservative estimates.

15.3.4   PROGRAMS AND EXPENDITURES

The plan is organized into two phases. Phase I (a 5-year phase) involves programs to build information, research and development (R&D), market development, and technology transfer infrastructure. Phase II (years 6–15) is concerned with the provision of active supports for the process of converting from conventional to organic production. Some Phase I elements would continue into Phase II. In total, the plan comprises 30 elements. Brief program descriptions are provided in Table 15.6 and expenditures in Table 15.7.

15.3.5   TRANSITION RISK OFFSET AND ENVIRONMENTAL SERVICE PAYMENTS

As the largest proposed program expenditure, this element of the overall program warrants more specific discussion. The objective of this initiative would be to pay farmers some of the revenue lost during the transition period, typically the most difficult period for organic farmers. In addition, this initiative provides a one-time payment for environmental services, an amount that recognizes the farmers' contributions to internalizing some of the costs of conventional production.

With program expenditures of over \$39 million (Table 15.7, Phase II), this element will likely be the most challenging for the government to implement in the current policy climate. Despite recent pilot projects, such as the Alternate Land Use Services initiative in Norfolk County, Ontario, which is paying producers for environmental services,[43] many policy-makers are reluctant to adopt this EU-style approach to ensuring environmental improvements with an additional dose of farm financial security.

In this analysis, annual payment levels varied from 0 to \$883 ha^{-1}, depending on commodity and transition year. The payment for avoided environmental costs would be delivered 3 years after full organic certification (assuming no intervening loss of certification status posttransition). Following on the analysis provided in Table 15.5, this payment is set at \$22.25 ha^{-1} for all crops, except pasture. The benefits of transition, as defined in the Tegtmeier and Duffy[13] study, are much lower for pasture (and in fact, they did not include pasture lands in their analysis), so we have set the level at \$0.5 ha^{-1}, largely to recognize the potential for lower GHG emissions on organically managed pasture. No per head payments are provided for animals, as it is assumed that all converting animal producers have a cropping base for their farm.

We assumed a 30% reduction in financial safety net payouts based on historical payment patterns of net new program costs[44] and assumed that the province saves 12% on other costs once making payments for this program. We assumed that program delivery is carried out by existing agencies involved in farm financial safety nets, with additional administrative costs of \$200,000 annually. Additionally, to support record keeping, each certification agency would receive a one-time administrative payment per certified farm of \$100 × 4,854 new organic producers = \$485,400.

TABLE 15.6

Brief Program and Outcome Descriptions

Program Title	Brief Description	Specific Outcomes
5.1. Implementation structures		
5.1.1 Establish a roundtable implementation model	Multistakeholder roundtable, involving industry, NGOs, scientists, and government officials, to guide plan implementation, coordinated by the Organic Council of Ontario.	
5.1.2 Provincial interdepartmental committee	Formal structures to direct provincial implementation of strategic plan elements.	
5.2 Market data collection	Annual provincial contribution to market studies as part of a multipartite funding initiative.	A full data set on organic market, to assist with targeted sectoral strategies.
5.3 Organic research and development		Full research programs in areas where organic research is particularly weak at present.
5.3.1 Horticulture research program	5-year support for a research coordinator, one PhD student and one postdoctoral fellow.	
5.3.2 Animal production research program	Long-term support for primarily poultry, beef/sheep, and swine production researchers, currently the most underrepresented areas.	
5.3.3 Social studies	5-year support for researcher.	
5.3.4 Food processing	4-year support for research coordinator, cost shared with the federal government.	
5.3.5 Farm business management	Developing production budgets in major commodity areas, including transition budgets to support the transition planning services.	

(continued)

TABLE 15.6 (continued)
Brief Program and Outcome Descriptions

Program Title	Brief Description	Specific Outcomes
5.4 Training		
5.4.1 Internship programs for nonfarm youth	Funding program delivery for three apprenticeship networks that help nonfarm youth enter organic farming.	60 farm interns annually (300 total), with 25% successfully owning or managing organic farms within 5 years (75).
5.4.2 Incubator farming program	Funding of three third-party organizations that offer farmland on a lease basis to new farm start-ups, with infrastructure and mentoring.	Pilot supports 15 incubator farmers site^{-1} for 3 years: 45 farmers primarily in horticulture.
5.4.3 Universities and colleges degree/diploma programs	New academic positions (see 5.3) and course development at the University of Guelph's organic bachelor's program.	25–30 graduates year^{-1} after 10-year support period.
5.4.4 Short courses	Introductory and advanced courses for farmers, processors, and professionals, building on existing course design and delivery by NGOs and ecological farm organizations.	175–225 farmers annually taking intro organic courses; 60 farmers, 5 processors, and 30 professionals per year in specialized courses.
5.5 Certification assistance	A 2-year "quick start" certification subsidy for farms and processors never previously certified; government pays 50%.	50 farms and 5 processors certified.
5.6 Production safety nets		
5.6.1 Analysis of organic farmer participation in programs	Fund an analysis of organic farmer participation in production safety net (business risk management or BRM) programs.	
5.6.2 Expand organic production insurance program	Enhancing crop coverage in the emerging Ontario organic crop insurance scheme run by Agricorp, which does not add to government costs since a tripartite funded insurance scheme.	A full organic production insurance program, with equivalent coverage to that for conventional producers.

5.7 Supply management marketing board changes		50–75 new entrants to organic production in supply-managed commodities.
5.7.1 Temporary quota and loan programs	All supply-managed commodities develop temporary quota or small farmer licensing programs to encourage new organic entrants that do not already have quota. Costs absorbed by marketing boards.	
5.7.2 Small organic flock licensing programs	For chicken, egg, and turkey production, creating provisions for small organic flock direct to consumer sales. Costs paid by licensees.	
5.7.3 Check-off changes	To increase organic research, carry out a feasibility study of an organic commission collects dues rather than organic producers contributing to a conventional commodity group.	
5.8 Collaborations to advance food safety	Government food safety program staff work with organic sector, through the Organic Council, to help implement sector-wide food safety initiatives.	Sector-wide food safety plans.
5.9 Animating nonretail food distribution channels, especially for low-income markets	Using existing grant programs, a third party applies for 3-year funding to develop nonretail distribution infrastructure (e.g., box schemes, CSAs, and buying clubs).	Add 2500 low-income households to purchasing pool with an additional $1.5 million in annual demand.
5.10 Processor supports[a]		
5.10.1 Organic business development expertise	One full-time equivalent position for organic processors within the Food Industry Competitiveness Branch of OMAFRA.	
5.10.2 Resurrecting orphaned processing facilities	Feasibility study to determine whether abandoned facilities in horticulture and field crops can be adapted to organic requirements.	

(continued)

TABLE 15.6 (continued)
Brief Program and Outcome Descriptions

Program Title	Brief Description	Specific Outcomes
5.10.3 Incubator processing facility	Using an eco-industrial park model, conduct a feasibility study for small- and medium-scale organic processing.	
5.10.4 Capital fund for SME processing and handling	Study the feasibility of establishing a $20 million capital fund at market rates for organic processors in the $0.5–10 million sales range.	
5.11 Support for cooperative production, processing, distribution, and marketing	Technical assistance grants cost shared with the federal government 50/50.	
5.12 Institutional procurement	Start-up grants for groups of Ontario farmers (and processors) attempting to meet institutional food service requirements.	
5.13 Transition advisory service	Funding for a transition planning center with regional coordinators and a network of mostly peer transition planners to work with converting farms and processors.	Center supports about 450 new actively transitioning farmers year[1] (courses, plan development) and 10–20 processors.
5.14 Transition risk offset and environmental service payments	See main text for details.	See main text for details.
5.15 Consumer and public education campaigns		
5.15.1 Organic information hotline and website	Primarily for processors and retailers who are responsible for advertising it, with government cost sharing 50/50 with industry.	
5.15.2 Generic POS material for retail	Government pays for development of POS materials that industry buys at postdevelopment costs.	

[a] Additional details of processor supports can be found in a supplemental report by Christianson and Morgan.[39]

TABLE 15.7
Program Expenditures (Cdn$)

					Expenditure Summary (Net Costs)											
Years		Phase I									Phase II					
All Programs	1	2	3	4	5	6	7	8	9	10	11	12	13	14	15	**Total**
Program																
5.1 Coordination																
5.1.1 Implementation model	30,000	30,000	30,000	30,000	30,000											150,000
5.1.2 Interdepartmental team	10,000	10,000	10,000	10,000	10,000											50,000
5.2 Market data collection	30,000	30,000	30,000	30,000	30,000											150,000
5.3 Research and development																
5.3.1 Organic horticulture	195,000	195,000	195,000	195,000	195,000											975,000
5.3.2 Organic animal production		125,000	125,000	250,000	250,000	375,000	250,000	250,000	125,000	125,000						1,875,000
5.3.3 Social sciences	90,000	90,000	90,000	90,000	90,000											450,000
5.3.4 Organic food processing	69,125	69,125	69,125	69,125												276,500
5.3.5 Organic farm business management	21,000	21,000	21,000	21,000	21,000											105,000
5.4 Training																
5.4.1 Mentoring	195,000	195,000	195,000	195,000	195,000											975,000
5.4.2 Incubator farming			150,000	150,000	150,000											450,000
5.4.3 Universities and colleges	2,000	3,000	3,000	4,000	4,000	5,000	5,000	5,000	5,000	5,000						41,000
5.4.4 Short courses	23,000	50,500	57,400	57,400	57,400	28,200	8,200	8,200	8,200	8,200	8,200	8,200	8,200	4,000	4,000	339,300

(continued)

TABLE 15.7 (continued)
Program Expenditures (Cdn$)

Expenditure Summary (Net Costs)

Years	Phase I					Phase II										
All Programs	**1**	**2**	**3**	**4**	**5**	**6**	**7**	**8**	**9**	**10**	**11**	**12**	**13**	**14**	**15**	**Total**
5.5 Certification assistance	8,942.5	8,942.5														17,885
5.6 Production safety nets																
5.6.1 BRM analysis	30,000															30,000
5.6.2 Organic production insurance	0	0	0	0	0	0	0	0	0	0	0	0	0	0	0	0
5.7 Marketing boards																
5.7.1 Temporary quota	0	0	0	0	0	0	0	0	0	0	0	0	0	0	0	0
5.7.2 Small organic flock licenses	0	0	0	0	0	0	0	0	0	0	0	0	0	0	0	0
5.7.3 Check-off changes	25,000															25,000
5.8 Collaborating for food safety improvements	0	0	0	0	0											0
5.9 Facilitating nonretail distribution	0	0	0	0	0											
5.10 Processor supports																
5.10.1 Organic business development		70,000	70,000	70,000	70,000	70,000	70,000	70,000	70,000	70,000	70,000					700,000
5.10.2 Orphaned facilities	120,000	40,000														160,000
5.10.3 Processing incubator		225,000														225,000
5.10.4 Capital fund feasibility			40,000													40,000

	Phase I					Phase II										Total
5.11 Cooperative support	200,000	200,000	200,000	200,000												800,000
5.12 Institutional procurement	165,000	330,000	495,000													990,000
5.13 Transition advisory service						220,000	200,000	200,000	200,000	200,000	200,000	200,000	200,000	200,000	200,000	2,020,000
5.14 Transition risk offset and environmental service payments						5,475,966	2,346,843	3,520,264	4,302,545	5,475,966	5,475,966	4,302,545	3,520,264	2,737,983	1,955,702	39,114,045
5.15 Consumer education																
5.15.1 Hotline and website						200,000	200,000	200,000	200,000	200,000						1,000,000
5.15.2 Genetic POS							75,000									75,000
Total	1,214,068	1,692,568	1,780,525	1,371,525	1,102,400	6,449,166	3,080,043	4,253,464	4,910,745	6,084,166	5,754,166	4,510,745	3,728,464	2,941,983	2,159,702	51,033,730
Phase I/Phase II totals	7,161,085					43,872,645										

We also ran several scenarios for different farms to test total payments. This analysis is provided in Table 15.8. Of the four case studies presented, total payments to farmers would range from $13,000 to 25,000 spread out over 4 years, well within the typical range of average government payments to farmers under present conditions.[7]

TABLE 15.8

Case Studies of Transition Risk Offset and Environmental Payments: Estimates for a 3-Year Transition Period plus One-Time Environmental Service Payment 3 Years after Certification

Conventional Production	Organic Transition Payment	Environmental Payment	Total
Apple—10 ha	At 1297.99 ha⁻¹ = $12,979.99	At $22.25 ha⁻¹ = $222.50	$13,202.49
Mixed vegetable (except potatoes) 20 ha operation 12 ha vegetables 8 ha cover crops	12 at $2099.49 ha⁻¹ = $25,193.88; 8 ha at $6.75 ha⁻¹ = $54	At $22.25 = $44.50	$25,292.38
Cash cropping 300 ha operation 63 ha in winter wheat 114 ha in grain corn 123 ha in soybeans	75 ha winter wheat × $37.37 = $2802.75 90 ha grain corn × $63.98 = $5758.20 90 ha soybeans × $43.20 = $3888.00 45 ha alfalfa/grass hay[a] × $6.75 = $303.75 Total: $12,752.70	At $22.25 = $6675.00	$19,427.70
Dairy[b] 100 ha operation 9 ha in winter wheat 6 ha in barley 17 ha grain corn 10 ha silage 8 ha soybean 42 ha hay 8 ha pasture 63 dairy cows	After organic transition[c]: 100 ha and 57 dairy cows 9 ha winter wheat × $37.37 = $336.33 6 ha barley × $25.10 = $150.60 10 ha grain corn × $63.98 = $639.80 8 ha soybeans × $43.20 = $345.60 55 ha hay × $6.75 = 371.25 12 ha pasture × 0 = 0 57 dairy cows × $214.83 animal⁻¹ after 3 years = $12,245.31 Total: $14,088.89	100 ha × $22.25 = $2225.00	$16,313.89

[a] Sold to nearby livestock operators.
[b] Derived from Canadian Dairy Commission (CDC), Dairy Farmers of Ontario (DFO), University of Guelph.[53]
[c] Derived from Ogini et al.[51]

15.3.6   MARKET SHARE

Implementation of this plan would allow the Ontario organic sector to capture 51% of Ontario's organic consumption, up from the currently estimated 15%. Organic sales would represent 1.9% of the total retail market after 5 years and 5.3% of the total market after 15 years.

15.4   CONCLUSIONS

A two-phase, 30-point plan was developed to boost organic production to 10% of agricultural area within 15 years and to capture 51% of Ontario's organic consumption, up from the currently estimated 15%. As with other government efforts, the objective is to stimulate organic production and processing with government initiatives so that sufficient size and momentum allow it to evolve on its own after 15 years. Although the impacts of implementing this proposed strategic plan cannot directly be predicted, the mix of policy instruments is generally consistent with those traditionally used in Canadian agriculture,[45] with the exception of payments for environmental services and transition payments. These instruments, however, have been widely tested in other jurisdictions,[11] are adaptable to the Ontario scene, and are increasingly part of the policy discussion.[43]

This overall program would cost the provincial government about $51 million over 15 years. The net total program costs would be significantly lower than $51 million since farmers would have saved almost $28 million in synthetic chemical inputs and received premium organic prices for most of their goods sold. This will unavoidably reduce pressures on the farm financial safety net system and government costs.

Additionally, this program contributes significantly to eliminating the long-term externalized costs of current approaches to agriculture, conservatively estimated at $145 million annually or $2.18 billion over the 15-year life of the program. Not all those costs will be saved within 15 years, but this exceedingly modest investment in organic production, representing only 2.3% of these externalized costs, will generate savings in externalized costs far beyond this one-time investment.

ACKNOWLEDGMENTS

We are grateful to the Laidlaw Foundation and the World Wildlife Fund Canada (WWF–Canada) Conservation Fund for financing this research. Members of the Organic Council of Ontario, an organization working to implement many elements of this plan, played a significant role in refining earlier versions of this chapter. We very much appreciate the very helpful comments of two anonymous reviewers.

REFERENCES

1. Willer, H. and Yussefi, M. 2006. *The World of Organic Agriculture 2006—Statistics and Emerging Trends*, 8th revised edition. IFOAM, Bonn, Germany.
2. MacRae, R., Martin, R.C., Macey, A., Doherty, P., Gibson, J., and Beauchemin, R. 2004. *Does the Adoption of Organic Food and Farming Systems Solve Multiple Policy Problems? A Review of the Existing Literature*. Organic Agriculture Centre of Canada, Truro, Nova Scotia, Canada.

3. MacRae, R., Frick, B., and Martin, R.C. 2007. Economic and social impacts of organic production systems. *Canadian Journal of Plant Science* 87(5):1037–1044.

4. The Nielsen Company. 2006. Review of Certified Organic Food Products at Retail in Canada. Prepared for the Organic Agriculture Centre of Canada, November 20, 2006.

5. Howard, P. 2007. Organic industry structure. Michigan State University, East Lansing, MI. Available at website: http://www.msu.edu/~howardp/ (verified January 5, 2009).

6. Macey, A. 2004. 'Certified Organic': The State of the Canadian Organic Market in 2003. Report to Agriculture and Agri-Food Canada (AAFC), Ottawa, Ontario, Canada.

7. Agriculture and Agri-Food Canada (AAFC). 2006. *An Overview of the Canadian Agriculture and Agri-Food System*. AAFC, Ottawa, Ontario, Canada.

8. Howard, P.H. and Allen, P. 2006. Beyond organic: Consumer interest in new labelling schemes in the Central Coast of California. *International Journal of Consumer Studies* 30:439–451.

9. Moore, O. 2006. What farmers' markets say about the post-organic movement in Ireland. In G.C. Holt and M. Reed (eds.), *Sociological Perspectives of Organic Agriculture*. CABI Publishing, Oxfordshire, U.K., pp. 18–36.

10. Soil Association Press Release, January 26, 2007.

11. Michelsen, J., Lynggaard, K., Padel, S., and Foster, C. 2001. *Organic Farming Development and Agricultural Institutions in Europe: A Study of 6 Countries. Organic Farming in Europe: Economics and Policy*, Vol. 9. University of Hohenheim, Stuttgart, Germany.

12. Pretty, J.N., Brett, C., Gee, D., Hine, R., Mason, C.F., Morison, J.I.L., Raven, H., Rayment, M., and van der Bijl, G. 2000. An assessment of the total external costs of UK agriculture. *Agricultural Systems* 65:113–136.

13. Tegtmeier, E.M. and Duffy, M.D. 2004. External costs of agricultural production in the United States. *International Journal of Agricultural Sustainability* 2:1–20.

14. MacRae, R., Martin, R.C., Macey, A., Beauchemin, R., and Christianson, R. 2002. *A National Strategic Plan for the Canadian Organic Food and Farming Sector*. Organic Agriculture Centre of Canada, Truro, Nova Scotia, Canada.

15. Macey, A. 2006. *Certified Organic Production in Canada 2005*. Canadian Organic Growers, Ottawa, Ontario, Canada.

16. Lampkin, N.L. 2002. Develop of policies for organic agriculture. In C. Powell et al. (eds.), *UK Organic Research 2002: Proceedings of the COR Conference*, March 26–28, 2002. Aberystwyth, Wales, pp. 321–324.

17. Martin, H. Organic Crop Production Programme Lead at the Ontario Ministry of Agriculture, Food and Rural Affairs (OMAFRA), personal communication.

18. OMAFRA. Ontario Agriculture and Food Statistics. Available at website: http://www.omafra.gov.on.ca/english/stats/welcome.html (verified January 5, 2009).

19. Macey, A. 2005. *Certified Organic Production in Canada 2004*. Canadian Organic Growers, Ottawa, Ontario, Canada.

20. Statistics Canada. 2002. 2001 Census of Agriculture—Canadian farm operations in the 21st century. *The Daily*, May 15, 2002. Available at website: http://www.statcan.ca/Daily/English/020515/d020515a.htm (verified January 5, 2009).

21. OMAFRA. 2005. *Agronomy Guide for Field Crops*. Publication 811. OMAFRA, Guelph, Ontario, Canada.

22. OMAFRA. 2005. *Fruit Production Recommendations, 2005–2006*. Publication 360. OMAFRA, Guelph, Ontario, Canada.

23. OMAFRA. 2005. *Vegetable Production Recommendations, 2005–2006*. Publication 363. OMAFRA, Guelph, Ontario, Canada.

24. Gallivan, G.J., Berges, H., and McGee, B. 2004. *Survey of Pesticide Use in Ontario, 2003. Estimates of Pesticides Used on Field Crops, Fruit and Vegetable Crops, and Other Agricultural Crops*. OMAFRA, Guelph, Ontario, Canada.

25. McEwan, K. 2005. Ontario farm input monitoring project. Survey No. 4. Economics and Business Section, Ridgetown College, Ridgetown, Ontario, Canada. For details of fertilizer prices, see http://www.oacc.info/Docs/OntarioOrgStrategy/TargetOOS_Statistics_sheet2.pdf

26. Mellon, M., Benbrook, C., and Lutz Benbrook, K. 2001. *Hogging It: Estimates of Antimicrobial Abuse in Livestock.* Union of Concerned Scientists, Washington, DC. Tables B-1, B-2, and B-3.

27. Canadian Food Inspection Agency (CFIA). Compendium of Medicated Ingredients Brochures (CMIB). CFIA, Ottawa, Ontario, Canada. Available at website: http://www.inspection.gc.ca/english/anima/feebet/mib/mibtoce.shtml (accessed January 5, 2009).

28. Title 21—Food and Drugs, Chapter I—Food and Drug Administration, Department of Health and Human Services, Part 558—New Animal Drugs for Use in Animal Feeds. Available at website: http://www.access.gpo.gov/nara/cfr/waisidx_99/21cfr558_99.html (accessed January 5, 2009).

29. Cessna, A., Farenhorst, A., and MacQueen, D.A.R. 2005. Pesticides. In A. Lefebvre, W. Eilers, and B. Chunn (eds.), *Environmental Sustainability of Canadian Agriculture.* Agri-environmental Indicator Report Series—Report No. 2. Agriculture and Agri-Food Canada, Ottawa, Ontario, Canada, pp. 136–137.

30. Altieri, M. 1995. *Agroecology: The Scientific Basis of Alternative Agriculture.* Westview Press, Boulder, CO.

31. World Wildlife Fund Canada. 2000. *Making Pesticide Reduction a Reality in Canada: Funding Programs to Advance Biointensive IPM and Organic Farming.* World Wildlife Fund Canada, Toronto, Ontario, Canada.

32. Fernandez-Cornejo, J. and Castaldo, C. 1998. The diffusion of IPM techniques among fruit growers in the USA. *Journal Production Agriculture* 11:497–506.

33. OMAFRA. Field Crop Statistics. Available at website: http://www.omafra.gov.on.ca/english/stats/crops/index.html; Horticultural Statistics. Available at website: http://www.omafra.-gov.on.ca/english/stats/hort/index.html; Dairy Statistics. Available at website: http://www.omafra.gov.on.ca/english/stats/dairy/index.html; Livestock and Poultry Statistics. Available at website: http://www.omafra.gov.on.ca/english/stats/livestock/index.html (accessed January 5, 2009).

34. Agriculture and Agri-Food Canada (AAFC). Food Safety and Quality. Available at website: http://www4.agr.gc.ca/AAFCAAC/display-afficher.do?id=1178115503821&lang=eng (accessed January 5, 2009).

35. Canada–Ontario Environmental Farm Plan Programme. Available at website: http://www.omafra.gov.on.ca/english/environment/efp/efp.htm (accessed January 5, 2009).

36. OMAFRA. Nutrient Management Programme. Available at website: http://www.omafra.gov.on.ca/english/agops/index.html (accessed January 5, 2009).

37. OMAFRA. 2004. Water management legislation and guidelines protecting water resources. Available at website: http://www.omafra.gov.on.ca/english/environment/water/legislation.htm (accessed January 5, 2009).

38. Statistics Canada. 2007. Census of Agriculture, 2006. Available at website http://www.statcan.ca/english/agcensus2006/index.htm (accessed January 5, 2009).

39. Christianson, R. and Morgan, M.L. 2007. Grow local organic: Organic food strategy for Ontario: Value-added processing. Report to WWF Canada. Rhythm Communications, Campbellford, Ontario, Canada.

40. E-Conomics Consulting and Jayeff Partners. Work force ahead: A labour study of Ontario's Food Processing Industry. Report for the Alliance of Ontario Food Processors, Guelph, Ontario, Canada. Available at website: http://www.aofp.ca/Uploads/File/Workforce%20Ahead%20Full%20Report%202005–04%20.pdf (accessed January 5, 2009).

41. sWCM Consulting. 2004. Eastern Ontario food-processing sector competitiveness study. Report to the Ontario East Economic Development Commission, Kingston, Ontario, Canada. http://www.brockville.com/UploadedFiles/linktofile_377.pdf (accessed January 5, 2009).

42. For details, see Fertilizers. Available at website: http://www.oacc.info/Docs/Ontario OrgStrategy/TargetOOS_Statistics_sheet2.pdf; Pesticides. http://www.oacc.info/Docs/ OntarioOrgStrategy/TargetOOS_Statistics_sheet3.pdf; Antibiotics. http://www.oacc. info/Docs/OntarioOrgStrategy/Avoided MedicatedFeed.pdf

43. Tyrchniewicz Consulting. 2007. Alternative land use services: A preliminary overview of cost reductions and potential financial benefits to Canada. Tyrchniewicz Consulting, Winnipeg, Manitoba, Canada. Available at website: http://www.sdeltawaterfowl.org/ alus/TychniewiczConsultingALUSReportJan2007.pdf (accessed January 5, 2009).

44. Friesen, B. 2006. Presentation to the *Ecological Goods and Services Workshop*, Winnipeg, Manitoba, Canada, February 13–16, 2006.

45. Brinkman, G. and Grenon, E. 2005. *Income from the Market and Government Payments—A Canada/U.S. Aggregate Comparison.* Canadian Agri-Food Policy Institute, Ottawa, Ontario, Canada.

46. Reganold, J.P., Elliott, L.F., and Unger, Y.L. 1987. Long-term effects of organic and conventional farming on soil erosion. *Nature* 330:370–372.

47. Arden-Clarke, C. and Hodges, R.D. 1987. The environmental effects of conventional and organic/biological farming systems. 1. Soil erosion, with special reference to Britain. *Biological Agriculture and Horticulture* 4:309–357.

48. Liebhardt, B. 2003. What is organic agriculture? What I learned in my transition. In OECD (ed.), *Organic Agriculture: Sustainability, Markets and Policies.* CABI Publishing, Wallingford, U.K., pp. 31–49.

49. Pretty, J. and Hines, R. 2001. *Reducing Food Poverty through Sustainable Agriculture.* Centre for Environment and Society, University of Essex, Colchester, U.K.

50. Nieberg, H. and Oppermann, N. 2003. The profitability of organic farming in Europe. In OECD (ed.), *Organic Agriculture: Sustainability, Markets and Policies.* CABI Publishing, Wallingford, U.K., pp. 141–152.

51. Ogini, Y., Clark, E.A., and Stonehouse, P. 1999. Comparison of organic and conventional dairy farms in Ontario. *American Journal of Alternative Agriculture* 14:122–134.

52. Roberts, C.J., Lynch, D.H., Voroney, R.P., Martin, R.C., and Juurlink, S.D. 2008. Nutrient budgets of Ontario organic dairy farms. *Canadian Journal of Soil Science* 88:107–114.

53. Canadian Dairy Commission (CDC), Dairy Farmers of Ontario (DFO), University of Guelph. 2005. Ontario Farm Dairy Accounting Project (OFDAP): Annual Report 2004 and Statistics Canada 2001 Census of Agriculture.

16 General Discussion and Conclusions

Ralph C. Martin and Rod MacRae

CONTENTS

16.1 INTRODUCTION

We introduced this volume by naming the knock against organic field crop production, namely, that strategies to manage energy, nutrients, and weeds are evolving too slowly. We conclude from the work reported here that the tool box for organic producers continues to expand in an affordable way, that organic field cropping has clearly identifiable strengths, but certainly challenges remain. In this final chapter, we cut across the chapters to summarize those strengths and remaining challenges.

While organic field cropping is a substantial part of organic production, it does not encompass the entirety of organic systems. But there are certainly important links with organic horticulture and also to livestock. Some chapters include references to horticultural crops and others elucidate relationships between forage crops for livestock and soil health as well as the key role of livestock manure. In that sense, the book does open doors to production considerations across the organic sector.

All the chapters (first introduced in Chapter 2) framed organic agriculture as a model of food production guided by principles of agroecology, with a regulated and inspected production system driven by consumer demand domestically and internationally. And there is a robust appreciation among organic researchers for integrated and systems level research across multiple subject categories to bring practical application to these agroecological principles (discussed in more detail in Chapter 2).

In turn, this appreciation for systems level research is consistent with organic principles that emphasize the health of the soil, plants, and livestock as a means to create farming system resiliency. However, this type of research is not without significant challenges. It is very complex, and it can be difficult to parse specific causes and effects and thus can take years (typically beyond traditional scientific funding cycles) before positive results emerge for producers (the learnings reported in Chapter 9 are instructive in this regard).

Integrated practices that promote soil quality may have the most long-term impact, and much of this volume has dealt directly or indirectly with this theme. The health of the soil is typically regarded as critical within organic agriculture for promoting resilient, competitive, and productive crops without the aid of synthetic pest control products. Understandably then, the core lessons of the volume relate to soil health.

16.2   P AVAILABILITY

Section I directly addressed phosphorus (P) availability, sometimes described as Achilles' heel of organic agriculture, since the P status of organic field crop production systems may result in P deficits (see Chapter 4). P is notoriously difficult to extract from soil, and the arbuscular mycorrhizal (AM) symbiosis was explored as a tool to address the problem (Chapter 3).

AM fungi cannot import P into an agroecosystem, but they can improve the P efficiency of crop production. Organic crops demonstrated well-functioning AM symbioses, and inoculation with selected AM fungal strains had very little noticeable effect on efficiency of P uptake, whereas, in conventional crops, higher plant tissue P concentrations and substantial increases in plant productivity were observed after inoculation. It would appear that crop diversification in space and time supports diverse AM fungal communities, explaining why organic systems do not respond to AM inoculation in trials as strongly as conventional systems that are deficient in this regard. The diversified crops, including perennial crop plants, such as clover or alfalfa (often mixed with grasses), are common in organic rotations. Crop production affects AM fungal communities in cultivated soils; however, this effect is less pronounced in organic production systems, which typically function at low levels of available P, with higher plant diversity and greater soil cover. AM fungi, as obligate biotrophs, benefit from such perennial plant cover. Other chapters (Chapters 4 through 9 and 11) also review the advantages of perennial plant cover, a key consideration in sustaining organic and nonorganic systems. Consequently, minimizing soil disturbance (most weed control tillage in organic systems is shallow) and the occurrence of nonhost plants in crop rotation are general recommendations for the preservation of diverse AM fungal communities in production fields.

Chapter 4 discussed how whole farm nutrient budgeting can reveal the diversity of management strategies for field cropping that help improve farm P sustainability. Fundamental ongoing work on soil organic and inorganic P composition and their dynamics under long-term organic field cropping is also providing important insights on the degree to which biologically mediated organic P turnover, and mycorrhizal–plant relationships, can be relied on to compensate for the often low farm P flows and low soluble P status of organically managed soils. It will not be enough to depend on phosphate rock, but advances in inoculant technologies for mycorrhizae and phosphate-solubilizing microbial inoculants are expected to improve P fertility and productivity of both green manures and cash crops in organic field crop production systems.

Interestingly, in the long-term trials, reported in Chapter 9, only the organic forage–grain system was in any danger of a P deficiency, and a small decline in available P due to organic was observed in the grain-only rotation, while available P was actually higher in the organic compared with the conventional green manure system. The P mining effect of the organic forage–grain rotation extended beyond the 0–15 cm soil depth, to 60 cm. This is evidence that green manure crops solubilize P, thus making it more plant available.

Future research is expected to assess the potential for anaerobic digestates, which concentrate P more than manure and compost, for efficient transport, after energy generation, while also including additional sources of nutrients from municipal solid waste. Clearly, the organic standards do not permit sewage use in organic agriculture. However, as noted in Martin et al. (2007), large volumes of additional municipal and industrial wastes, such as source separated and composted municipal solid (i.e., food) wastes or composted forestry industry by-products (e.g., papermill biosolids), are generated in Canada. Most of these municipal and industry composts meet or exceed national composting industry standards. Emerging processes to extract P from sewage in the form of struvite, without pathogen contamination, may also warrant further research and evaluation for inclusion in the organic standards (Ward 2011). Phosphorus will eventually become limiting for all of agriculture. Research in organic systems to limit P exports, to improve P availability and recycling on farms and to recover P from off-farm organic materials will serve organic farmers in the near term, and all farmers in the long term.

16.3 N EFFICIENCY AND TIMING

Chapter 4 contextualized the intensification of mainstream agriculture over the past 40 years, including how widespread and intensive use of synthetically produced nitrogen (N) and P fertilizers is oversaturating many of the world's agricultural areas, affecting many nontarget ecosystems. Chapter 13 picked up this theme, reporting on studies showing that organic farming has reduced nutrient loading and pollution. But the ability of organic farmers to manage N efficiency in a deliberately constrained fertility environment remains a significant challenge.

Chapter 4 then reviewed efforts to synchronize N availability with crop demand in organic systems, since the supply of N to the crop from organic amendments, green manures, and crop residues varies with climatic conditions and across years. Further research is required to optimize production in organic systems with nitrogen-use efficiency (NUE), while minimizing N losses.

There appear to be benefits to soil labile N pools of green manures of varying type, and duration and more testing to gauge soil N supply with whole plant bioassays of N uptake are expected to be helpful. Innovative approaches to termination of green manures that reduce reliance on tillage (see Chapter 6 as well) are increasingly being explored in organic field crop production systems (sometimes with livestock), with promising multi-tactical strategies that also assist with weed management.

Biological N fixation differentiates the N problem from the P problem in organic agriculture. N from the atmosphere can be added to agroecosystems, and the problem becomes one of timing legume growth and incorporation, with applying amendments and possibly seeding catch crops or grazing livestock to reduce N losses and meet the subsequent N needs of following cash crops.

16.4   REDUCTION IN TILLAGE

It has been difficult to develop organic cropping systems with less tillage (Chapters 5 and 6). The economic benefits of less tillage include less machinery and fuel use (see also Chapter 12), while biological benefits result from soil and water conservation. Until recently, no-till systems were understood to have an obligate relationship to herbicides, which seemed to exclude adoption by organic farmers. Nevertheless, testing of no-till organic systems has taken place over several decades, including the wide-blade cultivator (Noble blade), the rod weeder, the Lister plow, and crop rotation with perennial and biennial crops.

The authors of Chapter 6 challenged producers to reduce reliance on tillage by using grazing animals for strategic vegetation management in organic grain production. Most field agronomists have been reluctant to include livestock directly in their field agronomy experiments, although organic crop farmers (see Chapter 11) are exploring the possibilities with livestock. Future research, and indeed future student training, should emphasize integrated systems that include a broader range of creatures in the production system. Combining grazing with no-till organic systems is now also being suggested as a solution to low N mineralization and to counter problems in cover crop systems (rye and soybean) where the physical and chemical attributes of the rye reduce organic no-till soybean yields. The blade roller or crimper roller has given organic farmers new options for reduced tillage, and future research to use it more effectively with appropriate plants may advance these options further (also see Chapters 5 and 6).

Unlike the more widely applicable solutions for nonorganic farms, the solutions on organic farms are site specific, relying on the ecology of the region, the farm, and the field. Therefore, organic researchers and farmers must be excellent observers and respond with flexible reduced tillage options. Note that most tillage to control weeds on organic farms is shallow, and its low impact on soil and soil C is distinct from primary tillage (e.g., moldboard plowing) or secondary tillage (e.g., disking). The complementary tools discussed in this volume—cover crops, wide-blade cultivators, mowers, blade rollers, and grazing animals—could be supplemented in the future with perennial grain crops and trees. These tools will need to be optimized for specific conditions, and additional research is needed to fully exploit their utility and effectiveness.

16.5   MANAGING DECOMPOSITION

Lynch, in Chapter 5, appropriately concluded that "organic producers are fundamentally managers of the processes of decomposition. Thus, perhaps the most fundamental future challenge is whether organic crop management can become refined enough in its understanding and practices to develop a suite of SOC fractions in soil, that both retain stored SOC, while at the same still "feeding the soil"; that is, providing inputs of energy and nutrients for soil microbial populations well synchronized to also provide both crop and broader ecosystem benefits."

Chapter 5 also addressed questions of soil quality and soil health, including total and labile soil organic matter, aggregate dynamics, water infiltration, water holding capacity, and soil biological activity and diversity. By advancing the health and biological characteristics of soil, decomposition and nutrient cycling can be improved. Surveys (see Chapter 2) of organic farmers also reflect these central interests of organic farm management.

Methods to more directly evaluate soil health will help farmers adjust practices in a timely manner. The Cornell Soil Health Test, a farmer-oriented soil quality assessment tool consisting of 15 soil indicators and the soil ecotoxicology test using the Collembola bioindicator *Folsomia candida*, will contribute to managing for soil health. Nevertheless, we are in the early stages of the inquiry to more directly measure soil health, and the heterogeneity of unmasked biological systems, while essential for resilience, can frustrate scientists seeking to characterize soil health.

Under drought conditions, which may become more common, organic crops can outyield conventional crops, attributed to decomposition and enhanced soil organic matter levels and thus improved water capture and water holding capacity of soils (see also Chapter 14). Much of this phenomenon is linked to extended and diverse rotations (see Chapter 9). Such rotations are not unique to organic production (though typically more common than in other systems), but clearly beneficial management practices are as, or more, influential on soil quality attributes than whether the system is organic or not.

16.6 INTEGRATED CROPPING DESIGN AND PERENNIAL PLANT COVER

Section III appropriately led with a review of Canada's longest organic rotation experiment, at Glenlea, Manitoba, in Chapter 9. The R.W. Emerson quote, "the years teach much which the days never know," succinctly expresses the value of long-term research.

Organic systems used less energy and produced fewer calories than conventional systems. Given that not all calories produced in the forage–grain rotation are human useable, the authors recommend that future research should focus on energy production of integrated vs. grain-based rotations where different proportions of the grain crops are used for livestock feed.

Microbial biomass C was higher in the organic forage–grain rotation than the conventional forage–grain rotation, whereas microbial biomass C in grain-only rotations was lower in the organic than in the conventional system. The authors remind readers that the plots were not regularly manured as many organic treatments are. The organic systems have slightly lower C stocks, although a more active C pool compared with conventional systems. Lower subsoil C in all organic systems is of concern, and more research and adaptations are warranted.

The organic forage–grain (with manure added) system produced about 85% as much human useable energy but performed better than the conventional grain-only system in almost every other category. Higher energy yield in the conventional system was based on the assumption that none of the grain crops, including soybean, were fed to livestock. This long-term research is revealing the limits of organic and conventional rotations and also how they can be optimized.

With the previous challenges in mind, Chapter 11 records the voices and experience of organic farmers, to contextualize these agronomic, economic, and knowledge-related issues discussed in earlier chapters. An organic crop farmer's evolution is never done. She/he must control weeds, manage soil fertility, and produce a marketable crop by dedicating time, commitment, and the willingness to experiment. One farmer summed it by saying "we didn't perfect the system by any means. But we learned the basic concepts of crop rotation and using green manures. Since then, we've been continually improving on the system."

16.7   ROLE OF GENETICS

Chapter 10 explained how organic cereal breeding involves complex considerations. Conventional cultivars are selected for high responsiveness to highly soluble fertilizers, and thus under organic conditions, they are usually low yielding. Organic cultivars, in contrast, are bred for enhanced efficiencies in N and P acquisition, sometimes with fine root hairs, and for competitiveness with weeds. Conventional cultivars may not respond well to AM fungi because they are selected to produce under high soil P and for resistance to fungal pathogens, genes important in supporting interactions with AM fungi. Organic breeders must therefore select simultaneously for both AM fungi associations for P uptake efficiency and resistance to pathogenic fungi.

Certain plant genotypes perform better in some environments than others, and there is usually a crop genotype by environment (GxE) interaction. Chapter 7 authors adroitly explained that the crop GxE interaction is further complicated when considering that weeds potentially have their own GxE interactions. Similarly, Chapter 3 authors discuss the selection of crop genotypes, which are highly compatible with effective AM fungi and how the soil environment may modify the performance of AM fungi. A well-functioning agroecosystem will include crop GxE interactions, weed GxE interactions, and AM fungi GxE interactions, as well as the interactions among genotypes of crops, weed, AM fungi, and other organisms. Not all of these genetic complexities can be managed, but awareness will help researchers and farmers observe and consider more options when evaluating the role of genotypes. Organic agriculturalists can recognize species adjust to one another in a dance of dynamic stability (Chapter 11) by managing with alacrity and not by masking complex resilience potential with systemic pesticides to achieve a singular purpose.

Inherent stresses characterize organic environments and there is reason for direct selection under these conditions. The availability of genetic variability for tolerance to these stresses found among genotypes makes breeding and selection possible. However, there are high costs for direct breeding programs for organic cultivars. Direct selection may be cost-effective on organic farms, with participating farmers, during the final stages in which adaptation to specific farm environments are screened for yield, a trait easily evaluated at the farm level. This suggests a value in participatory plant breeding, an approach used commonly in the global south (see, e.g., work by Resource Efficient Agricultural Production, REAP, and partners [www.reap-canada.com], and the newly initiated Bauta Family Initiative on Canadian Seed Security [www.seedsecurity.ca/en]).

In Chapters 3 and 10, there is a recommendation for both pure and cultivar mixtures in organic systems. Cultivar mixtures are flexible and thus can continually evolve and adapt to the ever-changing organic conditions. The use of cultivar mixtures has given rise to the need to loosen current legislation that defines cultivars in order to accommodate registration and marketing of cultivar mixtures. As mainstream agriculture also recognizes the benefits of cultivar mixtures in adapting to climate change, policy may accommodate a shift toward managing for dynamic stability (see Chapter 11).

16.8   WEED AND PEST MANAGEMENT

Section II began with the most frequent, specific concern voiced by organic farmers, weed control. Chapter 7 reviewed integrated weed management (IWM) as an effective approach. Although IWM may be seen as overly complex, in contrast to some relatively simple and highly effective herbicide options, it does offer options (emphasis on plural) to organic farmers who want to stay ahead of weeds. IWM is presented as a series of modifications to the agronomic practices that farmers already do. The authors argued that using multiple nonherbicidal weed control methods will result in a higher crop yield than can be obtained using only a single weed control method.

Nonorganic systems can also benefit from nonherbicidal weed control methods. As the authors point out, many cropping systems have weeds that are closely related to the crops and as a result are impossible to control with selective herbicides. This is becoming more problematic as weeds develop resistance to herbicides. Further research in IWM is worthwhile for all farmers.

The first principle of weed management in organic systems is to establish a vigorous crop stand that is able to preempt resources for weeds by occupying above- and belowground space. A higher than normal seeding rate combined with timely, selective mechanical weed control was assessed as the best tools available to most organic crop farmers. Although successful at controlling weeds, fall rye, a very competitive annual crop, is planted on very little farmland in Canada because its markets are limited. This is one part of a much larger issue, how to shift consumption patterns to support agroecological approaches at the farm level.

The authors of Chapter 7 evaluated combinations of methods and pulled out three take-home messages. Firstly, effective weed management in organic systems can be achieved only by integrating weed control methods. Secondly, different tactics in an IWM plan can be ranked based on their effect on weed control, allowing farmers to easily pick the important practices according to their feasibility. Thirdly, farmers can achieve substantial weed control simply by altering normal cultural practices.

In Chapter 8, integration was also put forward as a means to address pests, other than weeds. An integrated insect pest management (IPM) strategy tries to take advantage of the large number of similarities between an agroecosystem and a natural ecosystem by favoring natural mortality factors and eliminating, as much as possible, practices that could upset these factors. In another plug for diversity, the authors called for diversification of organic cropping systems by planting of diverse crops and varieties or the use of trap crops or companion plantings. These can contribute to greater long-term stability of the agroecosystem with the proviso that a fundamental understanding of the mechanisms is critical to diversify appropriately for each pest and each crop. The use of organic mulches and amendments to improve soil health reduces density-dependent competition for nutrients between individual crop plants and maximizes plant tolerance to insect pests.

The landscape of conventional farms influences the role that conservation biological control can play in reducing insect pest populations in neighboring organic farms. For example. *Trichogramma*, released inundatively in fields of conventional sweet corn, could easily be transferred to organic fields of corn. The benefits are not limited to the farm but extend beyond to the health of the environment. Such themes were also taken up in Chapter 13, with a growing literature supporting a conclusion that organic systems have greater cropping, floral, and habitat diversity than conventional ones, with reduced agrochemical intensity and pollution.

The risk of resorting to curative methods rather than prevention increases as demand for organic products grows. It is difficult to rescue a crop from insects that are established and affecting the crop's development. Remedial actions often result in a major disruption of the agroecosystem, particularly but not limited to their negative impact on beneficial arthropods, leading to outbreaks of secondary pests and often a rebound in numbers of pest species. These changes are having inevitable impacts on the insect communities of these crops and their agroecosystems and creating interesting challenges for maintaining insect pests at an equilibrium position. Because insect pest protection in an organic production system relies essentially on ecological/behavioral modification methods, there will be an increased need for a better understanding of insect pest behavior as it relates to dispersal and host plant selection as well as parasitoid and predator behavior. Detailed ecological studies of plant–arthropod and predator–prey interactions leading to specific applications for the management of key pests in particular crops are required.

16.9 ENERGY AND INPUT USE

Numerous chapters (Chapters 4, 9, 11, and 12) highlight the ways in which organic systems typically (though not exclusively; see Chapter 12) reduce energy use, much of it a result of fewer synthetic inputs, in particular, N fertilizer. The comparative analysis provided in Chapter 12 revealed that organic agriculture has greater potential for long-term energy efficiency and conservation than conventional production, largely because of this reduced requirement for inputs. Greater energy efficiency gains could emerge with more emphasis on vegetable crops and field crops and less on temperature-controlled greenhouse horticulture and production (and consumption) of livestock products. However, as Chapter 12 concluded, "differences in energy analysis methods, and animal species and crop variety differences, as well as variability in climatic and other growing conditions, can complicate economic comparisons of energy use in agriculture." Ultimately, system-wide energy efficiency impacts could scale up with sustained growth in organic agriculture, but this would require significantly greater support from governments.

Related to energy use, the comparative analysis of Chapter 13 also concluded that with the exception of poultry, hog, and some horticultural systems where data remain limited, there is substantial evidence that organic systems produce less GHGs, particularly on a per hectare basis. In general, the comparisons still favor organic when contrasted on a per product basis, but not as strongly. As with energy, reduced input use was also a key reason for better organic performance.

16.10   POLICY OPTIONS

From literature and analysis provided in Chapters 13 through 15, we conclude that organic agriculture, once scaled up and out, can solve many of the pressing environmental, economic, consumer, and community issues facing the agricultural sector and rural areas. In Canada, supports remain modest with the federal commitments to the Organic Science Cluster, starting in 2009 with 27 research activities in 9 subprojects and over 50 researchers (many of them contributors here), being one of the more substantial recent contributions.

In Chapter 13, the case is made that organic systems have greater cropping, floral, and habitat diversity; reduced agrochemical intensity and pollution; and lower GHG emissions compared to conventional operations. There is also a recommendation for revised organic regulations to outline requirements for the establishment and maintenance of field margins and nonproduction areas in promoting biodiversity. A policy aspiration is to facilitate the clustering of organic farms in organically managed watersheds, to improve landscape-level synergies and organic performance with management efficiencies (Lynch et al. 2013). There is movement toward creating Canada's first National Urban Park located conveniently on the eastern edge of Toronto (www.agrifoodhub.org). The Rouge Park will include over 100 farms (70% of the park's area) and could provide a policy opportunity to integrate several organic farms in one contiguous area.

Chapter 14 posits several unanswered questions, which warrant more research. With organic foods demonstrating improved performance in several food safety and nutritional parameters, might there be a significant positive impact on human health? Does improved enterprise profitability always translate to wider positive regional economic benefits? Similarly, organic farmers often require more farm labor, but does that have a positive impact on rural community vitality? To answer these questions requires significant methodological sophistication, but also an increase in scale and social and economic linkages if researchers are going to move from simply modeling possible scenarios to more empirical inquiries. In Canada, organic food and farming remain in the 1%–2% range of farmed acres and retail sales, respectively, and supply chain actors are sufficiently dispersed that scale effects cannot be readily documented. Perhaps clustering organic farms as opportunities arise (see previous) will improve management efficiency and branding options. Chapter 15 explores a more traditional growth approach for the increasingly successful sector organic agriculture is becoming.

Chapter 15 developed a scenario whereby Ontario could reach a target of 10% of Ontario's cropped acres in organic production within 15 years. At a projected cost of $51 million over 15 years (averaging 3.4 million year^{-1}), it was calculated that the following reductions could be realized: (1) fertilizer applications by about 43 million kg (saving \$18.4 million year^{-1}), (2) pesticide applications by about 296,000 kg active ingredient (saving \$9.1 million year^{-1}), and (3) 7079 kg of growth-promoting antibiotics/medications consumed in animal feed. Additionally, this program would contribute significantly to reducing the externalized costs of current approaches to agriculture, conservatively estimated at \$145 million annually or \$2.18 billion over the 15-year life of the program. Implementation of this plan would allow domestic producers to capture 51% of Ontario's organic consumption, up from the currently low-range estimate of 15%. The policy and program supports required to facilitate the scaling up of organic agriculture (noted previously) are multifaceted and not, ultimately, that expensive.

Organic agriculture not only benefits those engaged in the organic value chain but all agricultural value chain participants across Canada. Clearly, organic proponents also appreciate the learning opportunities from mainstream agriculture. For example, the benefits of no-till were too obvious for organic farmers to ignore. In turn, nonorganic agriculturalists are becoming increasingly aware of methods and systems analyses that organic agriculture and food systems can offer to help mainstream agriculture adapt as challenges to sustain food production become more apparent. Some methods that are obligate in organic crop management (e.g., cover cropping) have the potential to be more broadly adopted now, and other methods (e.g., roller crimper if herbicide resistance advances) may be more universally employed in the longer term.

16.11   LAST WORDS FOR PRACTITIONERS

In Chapter 11, the term "indigenous" was shown to include knowledge intrinsically linked to geography. Understanding the unique agroecological features of a specific geographic location (e.g., soil type and fertility, microclimate, topography, moisture patterns, landscape interactions) enables organic farmers to make effective use of outside knowledge. Authors of other chapters also discussed the specific and localized adaptations, which are necessary for organic tillage and organic fertility and pest management. Knowledge and networks for the transmission of knowledge emerged as important resources for participants in this research, enabling them to better approach both agronomic and marketing challenges.

Mainstream agriculture seeks to minimize and control variables within the farming system, and in contrast, organic agriculture acknowledges and accounts for unpredictable variables to achieve resilience. The authors of Chapter 11 described organic agriculture as a system that assumes change and explains stability, instead of assuming stability and explaining change. Adaptiveness and creativity were seen as essential components to successful, long-term organic farming. As farmers learn to anticipate challenges and uncertainties, their farming systems gain resilience, and they are better equipped to farm in a manner that is agronomically, financially, ecologically, and socially sound.

In Canada, pioneers, mostly practitioners, of the organic sector struggled in the mid to latter parts of the twentieth century to gain recognition for organic systems. Although all agriculture had been organic prior to 1900, there was a twentieth-century period when views of progress drew practitioners and researchers to emphasize production with all available technologies. Organic producers have shown and continue to show that organic agriculture is possible. The stories and research reviews of this volume affirm that the indigenous and scientific ways of knowing can together expand possibilities for organic agricultural systems to sustain healthy Canadian communities.

REFERENCES

Lynch, D.H., J. Sumner, and R.C. Martin. 2013. Transforming recognition of the social, ecological and economic goods and services derived from organic agriculture in the Canadian context. In *Organic Farming: Prototype for Sustainable Agriculture*, H. Guyomard, J.M. Meynard, and S. Bellon, eds. Springer, New York.

Martin, R.C., D.H. Lynch, B. Frick, and P. van Straaten. 2007. Phosphorus status on Canadian organic farms. *J. Sci. Food Agri.* 87:2737–2740.

Ward, A. 2011. Phosphorus limitation of soybean and alfalfa biological nitrogen fixation on organic dairy farms. MSc thesis, NSAC, Bible Hill, Nova Scotia, Canada.

Index